G000122861

LASERS AND ELECTRO-OPTICS RESEARCH AND TECHNOLOGY

OPTICAL FIBERS

NEW DEVELOPMENTS

LASERS AND ELECTRO-OPTICS RESEARCH AND TECHNOLOGY

Additional books in this series can be found on Nova's website under the Series tab.

Additional e-books in this series can be found on Nova's website under the e-book tab.

LASERS AND ELECTRO-OPTICS RESEARCH AND TECHNOLOGY

OPTICAL FIBERS

NEW DEVELOPMENTS

MARCO PISCO
EDITOR

New York

For permission to use material from this book please contact us:
Telephone 631-231-7269; Fax 631-231-8175
Web Site: http://www.novapublishers.com

NOTICE TO THE READER

Library of Congress Cataloging-in-Publication Data

Library of Congress Control Number: 2013944380

ISBN: 978-1-62808-425-2

Published by Nova Science Publishers, Inc. † New York

CONTENTS

FOREWORD

Since the ancient "smoke signal" that transmits signals in free space by color of the smoke over long distance, the principle and technique for communication has been one of the most concerned subjects of sceince and technology, among which optical methods have been playing important roles by encoding signals into visible light and redirecting the "optical signal" to certain receivers. The invention of highly coherent light source —"laser" —by Theodore Maiman in 1960 had no doubt led to the revolution of human's ability of manipulating light for numerous applications, among which the idea of guiding laser pulses in low-loss optical fibers proposed by Kao and Hockham in 1966, initiated the long-haul and high-bandwidth modern optical communication, as well as many other applications including fiber-optic sensing, imaging and power delivery.

As one of the most important invention in the 20th century, optical fibers have been very successful in handling light passively or actively for a variety of applications covering almost all aspects of optical, or more generally photonic technology from the generation, propagation, amplification, modulation, conversion and detection of light. While the conventional fiber-optic technology have been well-established, the growing demand on better fiber-optic systems and the rapid progress in related technology (e.g., nanotechnology), have spurred great efforts for pushing forward the fiber optics. It is clear that, with the increasing confinement of light from free-space propagation to guided modes in an optical fiber, the bandwidth of optical communication drastically increases. This is also true for enhancing the sensitivity of optical sensing, the resolution of optical imaging, and many other performances of optical technology. Usually, a better confinement of light yields a more powerful optical technology.

For fiber optics, a better confinement enables better manipultion of light on time, spatial and energy scales, e.g., faster response, smaller footprint, and lower power consumption. Technically, the better confinement in an optical fiber may come from either a new waveguiding mechnisms or a new structure (or material), as have been constantly demonstrated since the early years of fiber optics. The contents of this book continues the renovation of fiber optics in a graceful manner. By enclosing waveguiding mechanisms from linear optical waveguide theory, near-field interaction to soliton dynamics in optical fibers, the Book covers a wide range of recent progresses in fiber optics. Especially, the newly emerging microfiber optics, the fiber-based MEMS devices, new advances in fiber Bragg gratings, polymer fibers, soliton propagation in optical fibers, fiber optical sensors, new WDM access for the last-mile technology and fiber connection techniques, will attracting a broad readship — the relevance of fiber optics to a range of disciplines, including physics, materials, photonics, electronics and nanotechnology. The broad interests, will in turn, push farward the fiber optics and technology in multidisciplinary areas, and bring a brighter future for manipulating light with optical fibers.

Limin Tong

State Key Laboratory of Modern Optical Instrumentation, Department of Optical Engineering
Zhejiang University
Hangzhou, China

PREFACE

The optical fibers are the technological innovation that revolutionized the way we communicate. The development of the optical fiber technology started from the earliest studies and pioneering experiments on total internal reflection and passed through several technological advancements, concurring to the achievement of modern optical fiber devices and systems for communication and sensing applications.

In 1840s Colladon and Babinet demonstrated the propagation of light down through jets of water via total internal reflection. Immediately, their discovery had a low impact and the main exploitation was limited to the realization of decorative fountains. In the following years, Tyndall brought wide attention to this light guidance phenomenon. In his lectures at the British Royal Society (1854), Tyndall produced fascinating optics experiments elucidating the total internal reflection of the light through stream of falling water.

Tyndall's experiments, although featured by educational purposes, had the merit to stimulate research efforts to guide light with more control than could be achieved in a jet of water.

Attempts to exploit total internal reflection in the medical industry were also successively performed. Bent rods of glass were used to illuminate internal organs by Roth and Reuss at Vienna (1888). In the following years, the possibility to guide light in glass rods was variously proposed for lighting and imaging applications.

Nevertheless, the fiber optic technology made significant steps forward only in the second half the nineteen century. In the 1950s there was the first all-glass fibrescope for medical applications (H. H. Hopkins and N. S. Kapany (1954)).

In the 1966, C. K. Kao concluded its study on "*Dielectric-Fibre Surface Waveguides for optical frequencies*" envisaging that "*a fibre of glassy material (...) represents a possible practical optical waveguide with important potential as a new form of communication medium*". He was talking about the low losses optical fibers (or better about a "*dielectric-fibre waveguide (...) with a circular cross-section*") as presently used in optical fiber communication systems. The term "fiber optic" was not still widely used even if it has been introduced shortly before by N. S. Kapany in an article in Scientific American in 1960.

Successively, the fiber optic technology experienced a considerable grown pushed by the research efforts to improve the light transportation capability. Driven by the ever-increasing global communications demands, the manufacture of optical fibers led to the modern single mode optical fibers featured by high transparency, great flexibility, broad transmission window, high reliability and durability.

As optical fibers cemented their position in the telecommunications industry and its commercial markets matured, significant efforts were carried out by a number of different research groups around the world to exploit them also in sensing applications.

In parallel with optical fiber technological developments, novel optical fiber components were proposed and investigated (fiber Bragg gratings, optical fiber interferometers,..) to control the optical wave propagation and to produce new effects in the optical fiber (Raman and Brillouin scatter, solitons propagations, fast and slow light, …) with the aim to manage the interaction of the light, either by improving the optical fiber communication characteristics and by offering optical fiber sensing devices with new functionalities and superior performances.

Furthermore, a large set of special optical fibers were proposed and demonstrated, ranging from plastic optical fibers, Bragg fibers to specific fibers such as double core optical fibers and D shaped fibers and many others constructed for specific applications. Among them, the Photonic Crystal Fibers (PCFs) deserve particularly to be mentioned, because they attracted great attention and generated renewal research efforts on the optical fiber technology since the end of the nineteen century. Shortly after the first solid core PCFs, brought to the general attention by Knight et al. in 1996, the first example of Hollow-core Optical Fibers (1999), exploiting photonic bandgap guidance (instead of total internal reflection), was also demonstrated.

In the recent years, a new concept is emerging in the scientific community dealing with the possibility to use optical fibers as platform to develop all-in-fiber multimaterial and multifunctional optical devices and systems for both communication and sensing applications. The key feature of these new optoelectronic devices relies on the proper integration of specific materials into the same optical fiber in order to attain advanced functionalities within a single optical fiber.

All-in-fiber optical devices are also increasingly investigated as highly integrated solutions for the development of components and sub-systems suitable to be incorporated in advanced optical communication and sensing systems.

Nowadays, the research and the development on the optical fiber technology is very wide and diversified. A multitude of aspects are singularly investigated by high number of researches with different scientific background. Topics intensively addressed involve the design and development of novel configurations of optical fibers, the exploitation of novel light manipulation mechanisms with optical fiber platforms, the assessment of novel fabrication process as well as the realizations of optical fiber components and systems for communication and sensing applications. The integration of advanced functional materials with the optical fiber is also addressed for the realization of optical fiber sensors and actuators.

This book is a collection of contributions by renowned scientists in the optical fiber technology field, covering a wide range of recent progresses pertaining to various topics such as special optical fibers, non-linear effects in optical fiber, components and devices for communication systems as well as optical fiber sensors. Inevitably many aspects are omitted but each chapter is a representative example of the latest trends and results in a rapidly evolving research scenario.

This collection is enriched by a few expert commentaries on selected topics, highlighting recent innovative developments and a future outlook in the optical fiber technology.

I would like to thank all the contributors of this book and particularly to Limin Tong, Masaaki Hirano, Stefan Webnitz and Albert Ferrando for their support.

Marco Pisco

Optoelectronics Division - Engineering Department
University of Sannio
Benevento, Italy

EXPERT COMMENTARIES

In: Optical Fibers: New Developments
Editor: Marco Pisco

ISBN: 978-1-62808-425-2

EVOLUTIONS OF TRANSMISSION FIBER

Masaaki Hirano
Optical Communications R&D Laboratories,
Sumitomo Electric Industries, Ltd.

Transmission optical fibers are an essential infrastructure to support today's Internet Age. Optical fiber is composed of a 125-μm-diameter silica-based glass and a 250-μm-diameter resin coating, which has features including high transparency, great flexibility, broad transmission window expanding 40 THz at maximum, high reliability and durability. Along with the rapid spread of bandwidth-hungry applications including social networking, internet video broadcasting, smartphone and cloud storage, large volumes of data are required to be transmitted over a long distance. As a corollary, the demand for global telecommunication networks continues to increase radically, at about 2 dB/year these days. A straightforward way to keep up with the explosive traffic growth is to increase the transmission capacity per a transmission fiber with continuous evolutions on transmission systems and fibers. Today, 10 Tb/s optical systems based on 100 Gb/s symbols have been actively deployed and operated. Considering a lot of systems are still based on 10 Gb/s symbols, this upgrade increases a capacity by a factor of 10, which may be considered as huge upgrades. Even at 10 Tb/s systems, however, the "capacity crunch" will become a possible reality in this or the next decade [1] if capacity demands keep increasing on pace, and therefore, continuous evolutions on both fibers and systems are strongly desired.

In this commentary, following historical evolutions, state-of-the-art and future prospects of transmission optical fibers will be discussed.

Historical evolutions of optical fibers shall be reviewed before discussing the future. In short, fiber developments are the results for overcoming the dispersion, optical loss and nonlinearity along with the innovation of optical, electrical devices and transmission system. In the earliest days in the 1970's, *multi-mode fiber (MMF)* having a step-index profile was deployed. Then, *graded-index MMF (GI-MMF)* was developed to minimize the modal dispersion. In a MMF, many guided modes are propagating through the fiber with different velocity, which causes a difference of delay time between modes, modal dispersion that severely distorts signal pulses. A GI-MMF can reduce the modal dispersion but cannot eliminate perfectly. In the 1980s, *standard single-mode fiber (SSMF)* was developed. As the name suggests, only the fundamental mode can propagate over a SSMF, and of course there is no modal dispersion.

SSMF has a very simple structure with a single 10-μm-diameter core. A SSMF has the zero-dispersion wavelength around 1300 nm and the lowest attenuation around 1550 nm. In order to minimize both the chromatic dispersion and attenuation, *dispersion-shifted fiber (DSF)* that has the zero-dispersion wavelength around 1550 nm was developed in the late 1980s. For example, DSFs were deployed in Japanese backbone networks based on a time division multiplexing (TDM) technology. From the mid-1990s, with a realization of Er-doped fiber amplifier (EDFA), a wavelength division multiplexing (WDM) technology utilizing the 1.55-μm-wavelength band enabled the transmission capacity to explosively increase. In WDM systems, the four wave mixing (FWM), one of Kerr nonlinear phenomena, is easily generated through a DSF that has the zero-dispersion in the signal wavelength band, and the FWM severely distorts signal pulses. Therefore, various *non-zero dispersion shifted fibers (NZ-DSF)* were developed to mitigate the Kerr nonlinearities. At around the same time, *dispersion compensating fiber (DCF)* was developed for compensating the accumulated chromatic dispersion of a transmission fiber almost perfectly in the EDFA bandwidth. Today, the majority of transmission media is a SSMF, especially in the Metro/Access networks. In long-haul backbone networks today, SSMF + DCF and NZ-DSF are mainly used.

State-of-the-art fibers today are low-attenuation and low-nonlinearity fibers. As mentioned above, WDM-systems based on 100 Gb/s have been actively deployed, in which digital coherent technologies with quadrature phase shift keying (QPSK) are applied. In this technology, a digital signal processor (DSP) that can equalize practically any amount of linear transmission impairments is utilized [2]. Accumulated chromatic and polarization-mode dispersions are no longer obstacles, and therefore optical dispersion compensation in the transmission line came to be unneeded; in fact, the larger chromatic dispersion improves the transmission capacity and distance [3]. With such electrical signal processing, required performances for optics came to be simple, the increasing of optical signal to noise ratio (OSNR) [4]. As for fiber characteristics, the lowering of attenuation and nonlinear coefficient (γ) are preferable, because the lower fiber attenuation can increase the output power after propagating through the fiber, and the lower γ can allow the higher input power managing the signal impairment due to Kerr nonlinearities. *Pure-silica core fiber (PSCF)* has the attenuation of 0.15 to 0.17 dB/km at 1550 nm [5], which is significantly lower than one around 0.19 dB/km of a SSMF doped with GeO_2 in the core. The γ is inversely proportion to the effective area (A_{eff}), and therefore, the enlargement of the A_{eff} is the key issue. Indeed, various *low-attenuation and large-A_{eff} fibers* have been developed. The challenge in realization of A_{eff}-enlarged fibers is to cope with poor macro- and micro- bending loss performance. In order to improve bending loss performance, large-A_{eff} fibers employing low Young's modulus primary coatings and appropriate refractive index profiles including trench-assisted and W-cladding were proposed [6-10].

As a result of developments of both PSCFs and digital coherent systems for an ultra-long haul link, e.g. submarine network, an attenuation of "standard" submarine fiber today has come to be 0.16 dB/km [11]. The enlarged A_{eff} more than 110 μm^2 has been actually deployed in the depth of sea for more than 10 years, and we have never experienced troubles. For research submarine fibers, the lowest attenuation is 0.1484 dB/km [6] and a huge A_{eff} of 211 μm^2 [7] have been reported. As for the A_{eff}, practical values will be limited in the range of 130 - 150 μm^2 because of unacceptable bending losses [8-10] with today's manufacturing technologies. Here the question is, "which is the best fiber for high capacity and long-haul transmission?" In order to answer the question, a fiber figure-of-merit (FOM) that can predict the degree of improvement on system performance should be known. An analytical fiber FOM was developed [12] using the Gaussian noise model for nonlinear interaction [13]. In a system with the span length of L [km], the FOM for a fiber having the A_{eff} [μm^2], attenuation of α [dB/km], chromatic dispersion of D [ps/nm/km] and coupling loss to a repeater α_{sp} [dB; >0] is described as [12]

$$FOM[\text{dB}] = 10/3 \cdot log\{Aeff^2\alpha \mid D \mid\} - 2/3\alpha L + 10 log[L] - 2/3\alpha_{sp}, \quad (1)$$

which are well consistent with transmission experiments. In a submarine link, a practical EDFA output is generally limited from +16 to +18dBm because of a limitation of electric power supply and broad WDM-bandwidth. In a case for 100 channels, launched power per a channel in an actual operation would be limited to -2 dBm/ch. Considering this practical limit of launched signal power, it is strongly suggested that the most preferable fiber for a ultra-long haul transmission would be the fiber having the possible low attenuation and an appropriate A_{eff} in the range of 110 - 140 μm^2, from an analytical Q-factor calculation as a function of launched signal power [14]. This is because the lower α results in the lower span loss, and therefore, the OSNR is simply improved. On the other hand, FOM improvement with the enlarging of A_{eff} (or the decreasing of γ) is depending on the signal power enhancement. However, signal cannot be launched higher than its limit determined by the EDFA output power, and FOM is saturated at a certain value of an A_{eff}. As a corollary, mass-productions of fibers with the record-low attenuation of 0.15 dB/km in [6] and with an appropriately-enlarged A_{eff} are has been strongly desired, and will be possibly has successfully realized in the very near future, I believe very recently [15].

Future prospects of transmission fibers are discussed finally. A PSCF with the attenuation of 0.16 dB/km and the A_{eff} of 130 μm^2 has about 3 dB higher FOM than one of SSMF's [12]. If this PSCF is applied, a 16 QAM-based 30 Tb/s system through a single fiber over 3,000 km transmission distance would be realized. This huge upgrade will be very helpful, but may not be sufficient, if we consider the exponential growth of global internet traffic. Therefore, it is anxiously awaited that the advent of innovative fibers over the next decade in order to prevent the transmission capacity from crunch, in collaboration with innovative transmission schemes. At major conferences for optical communications, various innovative fibers have been actively reported along with novel transmission configurations. Hereafter, examples of innovative transmission fibers will be briefly reviewed.

Multi-core fiber (MCF) is considered as a strong candidate of a next-generation transmission fiber, and most actively studied today. A MCF has several cores ranging from 3 to 19 in a single glass cladding, and a space-division multiplexing (SDM) utilizing signals individually propagating through each core of a single MCF is expected to dramatically increase the transmission capacity. In order to apply a MCF-SDM system to a long-haul transmission, there are many technical challenges, such as the reduction of inter-core cross talks, the realization of a multi-core EDFA, the realization of a compact fan-in fan-out device, and the splicing of MCF to MCF, which has been actively studied to find ways to be solved.

One of benefits of MCFs is that MCFs can be employed with "matured" low-attenuation production technologies originally developed for PSCF. A MCF having very low attenuation of 0.17 dB/km, enlarged Aeff of 120 μm^2, and negligibly small inter-core crosstalks less than -42 dB over 80km-long has been reported [16]. The record capacity over a single cladding fiber today is 1.01 Pb/s (= 1,010 Tb/s) through a 52km-long 12-core MCF [1517]. As for a multi-span transmission, 28.9 Tb/s over 6,160 km through 7-core MCF and 7-core EDFA has been demonstrated [1718]. These transmission experiments indicate a great potential capacity of MCF-SDM systems. Furthermore, real-time DSPs developed for today's digital transmission technologies can be fully applied to SDM schemes– if a practical way of demultiplexing of SDM signals is found.

Few-mode fiber (FMF) is another strong candidate. A FMF has 2 to 7 propagation modes in a single core, and a mode division multiplexing (MDM) can enhance the total capacity. Propagation modes are complicatedly coupled with each other through transmission and splicing points, and therefore, a multiple-input and multiple-output (MIMO) algorithm is applied to uncouple the modes into the originally launched modes. In FMF-MIMO system, many challenges similar to ones of MCF have been yet to be solved. The biggest challenge will be an implementation of MIMO-algorithm having high complexity to a real-time processor. Skews due to a modal dispersion result in the higher complexity, and designs of FMF to minimize the modal dispersion have been actively studied. For example, a FMF supporting LP_{01} and LP_{11} modes having low loss (0.198 dB/km for LP_{01} and 0.191 dB/km for LP_{11}), low modal dispersion (-0.08 ps/m) and low mode coupling (-25 dB) was developed [1819]. 73.7 Tb/s (96 channels x 3 modes x 256 Gb/s) mode-division-multiplexed transmission over 119km-long through 3-modes FMFs with inline MM-EDFA was demonstrated [1920].

Hollow core photonic bandgap fiber (HC-PBGF) has been considered as a functional fiber previously, and has been applied to a transmission fiber recently. In HC-PBGF, a signal propagates through the air surrounded by honeycomb microstructured glass walls. Compared to a solid-core fiber in which a signal propagates through silica-based glass, HC-PBGF has great promises of the ultra-low nonlinearity (two orders of magnitude lower) and ultra-low propagation delay time or latency (50 % lower). The lowest attenuation bandwidth is around 2 μm, which is suitable for silicon photonics. The 2 μm-bandwidth is, however, far from today's telecom window from 1.3 to 1.6 μm, and therefore, all of devices including transmitter, receiver, amplifier, coupler, feed-through fiber and fusion splicer have to be developed. Overcoming some of above difficulties, the first data transmission at 2 μm through HC-PBGF with the attenuation of 4.5 dB/km at 1.98 μm was demonstrated [2021]. The biggest challenge would be the lowering of attenuation to a comparable value to that of silica transmission fibers, even though the expected attenuation is 0.13 dB/km [2122]. One would recall the failure of fluoride fibers theoretically predicted to have a low attenuation. However, the development of HC-PBGF for transmission is just at the starting line; it is promising that the performance improvement will be dramatically proceeded.

Even for innovative ones, transmission fibers shall meet basic requirements from fiber users including fiber-cable installers, equipment suppliers and telecom carriers. There are many requirements, and some examples are listed as below.

i) high volume manufacturing: it is said that the global production of transmission fibers is about 250,000,000 km per a year today. It is far longer than the distance between the Sun and the Earth of about 150,000,000 km.
ii) extremely low-cost production: roughly estimated, a price of SSMF may be around 10 $/km today. You can imagine how cheap it is comparing the price of fishing lines (~10 $/100m) or shoelaces (~10 $/10m).
iii) ultimately high reliability and durability: "ancient" optical fibers installed in '80s have been still operating for 30 years long. Telecom carriers would hesitate replacing such aged fibers because the installation of fiber-cable is very costly.
iv) compatibility with existing systems: if innovative fibers need completely new equipments that cannot be applied to existing fibers, telecom carriers will have to replace installed fibers. It would be unrealistic; "standard" transmission fibers have been increased the length by 250 milions km year by year!
v) mixability with existing fibers: because of IV), innovative fibers may be going to be deployed together with existing various types of SSMF, DSF, and NZDSFs using the same system equipment. Dissimilar splicing of an innovative fiber to an existing fiber can be one of the issues.

Finally, it should be noted that MCF, FMF and HC-PBGF have not yet satisfied all of user's requirements mentioned above. However, transmission fibers and systems have been improved in their performance every 6 months at the major conferences of OFC/NFOEC and ECOC, and therefore, the state-of-the-art and innovative fibers stated here will become old-fashioned in the very near future.

REFERENCES

[1] A. Chralyvy, "The Coming Capacity Crunch," *Proc. ECOC2009*, Vienna, Austria, 2009, Plenary Paper 1.0.2.

[2] R.-J. Essiambre, et al., "Capacity Limits of Optical Fiber Networks, "*J. Lightw. Technol.*, vol. 28, no. 4, pp. 662-702, 2010.

[3] V. Curri, et al., "Performance Evaluation of Long-Haul 111 Gb/s PM-QPSK Transmission Over Different Fiber Types," *IEEE. Photon. Technol. Lett.*, vol. 22, no.19, pp.1446-1448, 2010.

[4] D. van den Borne, et al., "POLMUX-QPSK Modulation and Coherent Detection: The Challenge of Long-haul 100G Transmission," *Proc. ECOC2009*, Paper 3.4.1 2009.

[5] Y. Chigusa, et al., "Low-Loss Pure-Silica-Core Fibers and Their Possible Impact on Transmission Systems," *J. Lightw. Technol.*, vol. 23, no. 11, pp. 3541-3550, 2005.

[6] K. Nagayama, et al., "Ultra-low-loss (0.1484 dB/km) pure silica core fiber and extension of transmission distance", *Electronics Letters*, Vol. 38, Issue 20, p 1168-1169, 2002.

[7] M. Tsukitani, et al., "Ultra Low Nonlinearity Pure-Silica-Core Fiber with an Effective Area of 211 um2 and Transmission Loss of 0.159 dB/km," *Proc. ECOC2002*, paper 3.2.2, 2002.

[8] M. Hirano, et al., "Aeff-enlarged Pure-Silica-Core Fiber having Ring-Core Profile," *Proc. OFC/NFOEC2012*, paper OTh4I.2, 2012.

[9] S. Bickham, "Ultimate Limits of Effective Area and Attenuation for High Data Rate Fibers," in proc. *OFC/NFOEC2011*, paper OWA5 2011.

[10] P. Sillard, et al., "Micro-Bend Losses of Trench-Assisted Single-Mode Fibers," *Proc. ECOC2010*, paper We.8.F.3 2010.

[11] "Z–PLUS fiber®," at http://global-sei.com/fttx/product_e/ofc/fiber.html#A06

[12] V. Curri, et al., "Fiber Figure of Merit Based on Maximum Reach," *Proc. OFC/NFOEC2013*, paper OTh3G.2., 2013.

[13] P. Poggiolini, "The GN Model of Non-Linear Propagation in Uncompensated Coherent Optical Systems," *J. Lightw. Technol.*, vol. 30, no. 24, pp.3857-3879, 2012.

[14] M. Hirano, et al., "Analytical OSNR Formulation Validated with 100G-WDM Experiments and Optimal Subsea Fiber Proposal," *Proc. OFC/NFOEC2013*, paper OTu2B.6, 2013.

[15] HM. TakaraHirano, et al., "Record Low Loss, Record High FOM Optical Fiber with Manufacturable Process1.01-Pb/s (12 SDM/222 WDM/456 Gb/s) Crosstalk-managed transmission with 91.4-b/s/Hz Aggregate Spectral Efficiency," *Proc. OFC/NFOEC2013*, paper PDP5A.7, 2013.

[16] *ECOC2012*, paper Th.3.C.1, 2012.

[17] T. Hayashi, et al., "Uncoupled multi-core fiber enhancing signal-to-noise ratio," *Opt. Express*, vol. 20 Issue 26, pp. B94-B103, 2012.

[18] H. Takara, et al., "1.01-Pb/s (12 SDM/222 WDM/456 Gb/s) Crosstalk-managed transmission with 91.4-b/s/Hz Aggregate Spectral Efficiency," *Proc. ECOC2012*, paper Th.3.C.1, 2012.

[19] H. Takahashi, et al., "First Demonstration of MC-EDFA-Repeatered SDM Transmission of 40 x 128-Gbit/s PDM-QPSK Signals per Core over 6,160-km 7-core MCF," *Proc. ECOC2012*, paper Th.3.C.3, 2012.

[20] L. Grüner-Nielsen, et al., "Few Mode Transmission Fiber with low DGD, low Mode Coupling and low Loss," *Proc. OFC/NFOEC2012*, paper PDP5A.1, 2012.

[21] V.A.J.M. Sleiffer, et al., "73.7 Tb/s (96X3x256-Gb/s) mode-division-multiplexed DP-16QAM transmission with inline MM-EDFA," *Proc. ECOC2012*, paper Th.3.C.4, 2012.

[22] M. N. Petrovich, et al., "First Demonstration of 2μm Data Transmission in a Low-Loss Hollow Core Photonic Bandgap Fiber," *Proc. ECOC2012*, paper Th.3.A.5, 2012.

[23] P. J. Roberts, et al., "Ultimate low loss of hollow-core photonic crystal fibres," *Opt. Express*, vol. 13, no. 1, pp. 236-244, 2005.

In: Optical Fibers: New Developments
Editor: Marco Pisco

ISBN: 978-1-62808-425-2

EMERGING TOPICS IN NONLINEAR FIBER OPTICS

Stefan Wabnitz[*]
Dipartimento di Ingegneria dell'Informazione,
Università di Brescia, Brescia, Italy

Optical fibers have provided an ideal test-bed for nonlinear optics experiments: in spite of the relatively weak cubic nonlinearity of silica, the very low level of linear fiber losses permits to increase the effective nonlinear interaction lengths up to tens of km. Therefore the Kerr nonlinearity of silica optical fibers exhibits a very high figure of merit, coupled with an ultrafast response time. Indeed, the intensity-dependent refractive index change of fibers due to electronic distortion has a response time of the order of a few fs, which permits for example optical pulse compression approaching the single cycle limit. As a matter of fact, one of the most successful applications of nonlinear fiber optics is the generation of optical solitons, which represent a balance between pulse temporal spreading owing to anomalous dispersion, and pulse compression owing to the nonlinear index induced self-phase modulation, as proposed by Hasegawa and Tappert as far as forty years ago [1]. The nonlinear compensation of dispersive pulse spreading led to their proposal of using solitons as ideal communication bits in optical communication systems. As a result, significant experimental efforts were dedicated both in the academia and the industry for the demonstration of the applicative potential of optical solitons in ultra-long haul communication links in the '90s [2]. The advent of dispersion compensation techniques, either in-line using dispersion compensating fibers, or via electronic pre or post-processing, has later on substantially reduced the practical interest in soliton-based communications. On the other hand, the optical pulse shaping leading to robust soliton waveforms has been very successful in the field of ultrashort pulse fiber lasers, and it represents until today one of the most relevant applications of nonlinear optics.

In spite of its maturity of age, the story of nonlinear fiber optics is not quite over yet: exciting new developments are currently been investigated in the scientific community. In this commentary we would like to briefly review recent progress in two emerging topics,

[*] E-mail: stefano.wabnitz@ing.unibs.it.

namely the all-optical control and self-organisation of the state of polarization (SOP) of light, and the generation of rogue or freak waves in optical fibers.

As well known, the cross-interaction among intense beams in a Kerr nonlinear medium leads to the mutual rotation of their SOPs. As a result, in a fiber of finite length one may observe bistability and multistability in the steady-state distribution of the light SOP [3-4]. At the same time, spatial [5] and temporal [6] polarization instabilities and chaos were also predicted. The general spatio-temporal stability properties of the nonlinear polarization evolution of interacting beams in nonlinear Kerr media were associated by Zakharov and Mikhailov [7] to the universal or topological properties of the polarization coupling process, in formal analogy with the generation of stable domains of spin waves in ferromagnetic materials. Indeed, the formation of stable domains of mutual SOP arrangements may lead to all-optical polarization switching phenomena in optical fibers, as experimentally confirmed by Pitois et al.[8-9].

A related intriguing phenomenon, known as *polarization attraction*, resulting again from the nonlinear SOP interaction of counterpropagating waves, has been explored over the latest ten years [10-13]. To explain polarization attraction in simple terms, let us consider a backward (say, pump) beam which is injected at one end of the fiber with a well-defined SOP. Under certain conditions for the relative powers of the two waves and fiber interaction length, it may occur that a forward (or signal) beam, when it is injected at the opposite end of the fiber with an *arbitrary* SOP, emerges with nearly the same SOP as the pump (see Figure1). Thus we may say that the nonlinear polarization cross-interaction leads to the effective *attraction* of the signal output SOP towards the input SOP of the pump. The experimental demonstration of polarization attraction using two CW beams in long telecommunication fiber spans [13] has paved the way for conceiving a new class of devices which enable the all-optical and ultrafast control of the light SOP in practical (i.e., at relatively low CW power levels) optical communication and laser systems.

Only recently the equations ruling the SOP interaction of counterpropagating waves in randomly birefringent telecom fibers have been derived [14-16]. This enabled a good fit of the experimental observations of polarization attraction in long fiber spans [17]. From the analytical side, the study of the stationary SOP distributions could reveal that polarization attraction is closely linked with the existence of singular tori or separatrix solutions for the underlying mathematical models [18]. Yet, a full analytical treatment which explains the relaxation of the wave SOPs towards the polarization attractors in a purely conservative (i.e., lossless) physical system remains largely elusive to date. Indeed, the main difficulty in the analysis of the associated hyperbolic nonlinear partial differential equations is due to the presence of appropriate boundary conditions in a medium of finite length.

Very recent experiments involving the back-reflection of a single intense beam in a span of nonlinear telecom fiber with a feedback mirror have also unexpectedly revealed the effect of self-polarization attraction or stabilization. In this configuration the beam interacting with the back-reflected wave is self-attracted towards circular polarization states at the fiber ouput, largely independently of the input SOP orientation [19].

The dynamics of extreme waves, known also as freak or rogue waves, is currently a hot subject of intensive study in different fields of application, notably including nonlinear optical fibers [20-21]. In oceanography, rogue waves are known as a sudden giant deep-water waves which lead to ship wreakages. A relatively less theoretically explored, but potentially even more devastating manifestation of rogue waves occurs in shallow waters, such as for example the run-up of a tsunami towards the beach. In addition, the crossing of two waves with opposite speeds may also lead to giant and steep humps of water [22].

The generally accepted model for describing the generation and propagation of deep-water rogue waves is the Nonlinear Schrödinger Equation (NLSE). In this case, the creation of a rogue wave is closely linked with the modulation instability (MI) of a CW [23], whose nonlinear development is represented by temporally periodic solutions known as Akhmediev breathers [24]. A different striking outcome of MI is provided byt the Peregrine soliton, an isolated pulse of finite duration in both the spatial and the temporal domains [25]: the first unambiguous experimental observation of a Peregrine soliton in any physical medium was carried out using a nonlinear optical fiber [26]. Moreover, the statistics of spectral broadening in optical supercontinuum generation has been associated with the presence of extreme solitary wave emissions [27].

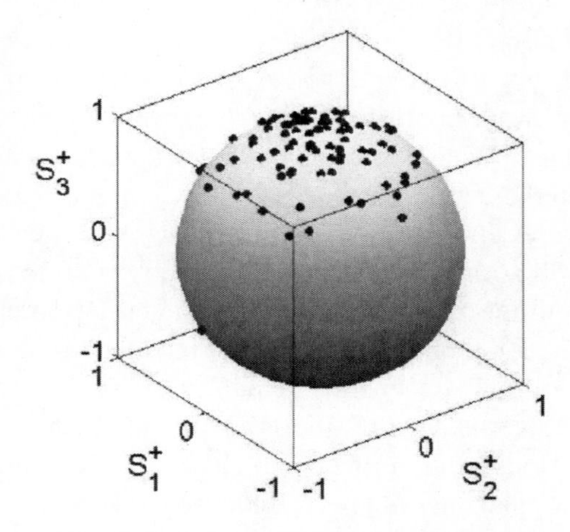

Figure 1. Distribution of a set of signal SOPs on the Poincaré sphere at the fiber output, showing polarization attraction to the pump SOP (north pole).

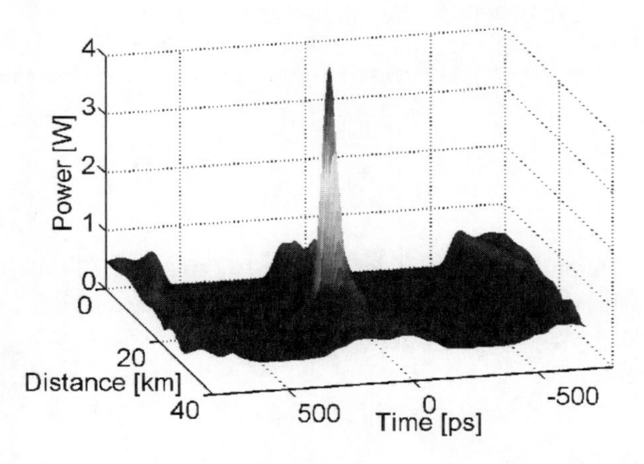

Figure 2. Rogue wave peak generated in a fiber with normal dispersion from an initial periodic frequency modulation of a CW beam.

Very recent work has revealed that rogue waves in optical fibers may also be generated in the normal group-velocity dispersion (GVD) regime of pulse propagation, in spite of the fact that MI is forbidden here [28]. Indeed, nonlinearity driven pulse shaping in this case may be described in terms of the semi-classical approximation to the NLSE, which leads to the so-called nonlinear shallow water equation (NSWE) [29-30]. Since the CW state of the field is stable, shallow water optical rogue waves may only be generated as a result of particular setting of the initial or boundary conditions [30]. Inded, one may exploit the initial temporal modulation of the optical frequency or prechirp, which is analogous to the collision between oppositely directed currents near the beach, or the merging of different avalanches falling from a mountain.

As discussed in refs. [28] and [31], a proper initial step-wise frequency modulation of a CW laser may lead in the normal dispersion regime of a fiber to the generation of intense, flat-top, self-similar and chirp-free pulses. The intriguing property of such pulses is their stable merging upon mutual collision into either a steady or an extreme, transient intensity peak (see Figure 2). Such type of extreme waves may also occur in optical communication systems, whenever different wavelength channels are transported on the same fiber.

In a related work it has also been recently predicted that, in analogy with the case of ocean waves that run-up to the beach, the shoaling of prechirped optical pulses may occur in the normal dispersion regime of optical fibers [32]. Namely, a vertical pulse edge or front develops in the power profile, prior to the occurrence of any significant wave breaking or fast temporal oscillations. In addition, it has been predicted that third-order dispersion in a tapered fiber with varying along the length normal dispersion may lead to the occurrence of extreme high-intensity pulse peaks or optical tsunamis, in analogy with the dramatic run-up and wave height amplification of a sea tsunamis as the coast is approached and the water depth is progressively reduced. The observation of optical shallow water rogue waves in the normal dispersion regime of optical fibers has not been achieved yet.

We foresee that the experimental characterization of these extreme wave phenomena, which so far has been hampered by the difficulty of carrying out real-time experiments in the context of water waves, could once again greatly benefit by exploiting the test-bed which is provided by optical fibers, thanks to the possibility of precisely setting the initial wave conditions as well as the optimizing and engineering the fiber dispersion profile during the fiber drawing process.

REFERENCES

[1] Hasegawa A, Tappert F. Transmission of stationary nonlinear optical pulses in dispersive dielectric fibers. I. Anomalous dispersion. *Appl. Phys. Lett.* 1973, 23: 142–144.

[2] Zakharov VE, Wabnitz S., eds. Optical solitons: theoretical challenges and industrial perspectives: Les Houches workshop, September 28-October 2, 1998. Vol. 12. *Springer*, 1999.

[3] Kaplan A.E., and Law C.T. Isolas in four-wave mixing optical bistability. *IEEE J. of Quantum Electronics* 1985, 21:1529-1537.

[4] Gauthier DJ, Malcuit MS, Gaeta AL, Boyd RW. Polarization bistability of counterpropagating laser beams. *Phys. Rev. Lett.* 1990, 64:1721-1724.

[5] Gregori G., Wabnitz S. New exact solutions and bifurcations in the spatial distribution of polarization in third order nonlinear optical interactions. *Phys. Rev. Lett.* 1986, 56:600-603.

[6] Gaeta AL, Boyd RW, Ackerhalt JR, and Milonni PW Instabilities and chaos in the polarizations of counterpropagating light fields. *Phys. Rev. Lett.* 1987, 58:2432-2435.

[7] Zakharov VE, Mikhailov AV. Polarization domains in nonlinear optics, *JETP Lett.* 1987 45:349-352.

[8] Pitois S, Millot G, Wabnitz S Polarization Domain Wall Solitons with Counterpropagating Laser Beams. *Phys. Rev. Lett.* 1998, 81:1409-1412.

[9] Pitois S, Millot G, and Wabnitz S Nonlinear polarization dynamics of counterpropagating waves in an isotropic optical fiber: Theory and experiments. *J. Opt. Soc. Am. B* 2001, 18:432-443.

[10] Pitois S, Sauter A, Millot G Simultaneous achievement of polarization attraction and Raman amplification in isotropic optical fibers. *Optics Letters* 2004, 29:599-601.

[11] Pitois S, Picozzi A, Millot G, Jauslin HR, Haelterman M Polarization and modal attractors in conservative counterpropagating four-wave interaction. *Europhysics Letters* 2005, 70:88-94.

[12] Pitois S, Fatome J, Millot G Polarization attraction using counter-propagating waves in optical fiber at telecommunication wavelengths. *Optics Express* 2008 16:6646-6651.

[13] Fatome J, Pitois S, Morin P, Millot G, Observation of light-by-light polarization control and stabilization in optical fibre for telecommunication applications. *Optics Express* 2010, 18:15311-15317.

[14] Kozlov VV, Wabnitz S Instability of Optical Solitons in the Boundary Value Problem for a Medium of Finite Extension. *Letters of Mathematical Physics* 2011, 96:405-413.

[15] Kozlov VV, Nuno J, Wabnitz Theory of lossless polarization attraction in telecommunication fibers. S, *J. Opt. Soc. Am. B* 2011, 28:100-108.

[16] Guasoni M, Kozlov VV, Wabnitz Theory of polarization attraction in parametric amplifiers based on telecommunication fibers. S, *J. Opt. Soc. Am. B* 2012, 29:2710-2720.

[17] Kozlov VV, Fatome J, Morin P, Pitois S, Millot G, Wabnitz S Nonlinear repolarization dynamics in optical fibers: transient polarization attraction. *J. Opt. Soc. Am. B* 2011, 28:1782-1791.

[18] Assemat E, Picozzi A, Jauslin HR, Sugny D Hamiltonian tools for the analysis of optical polarization control, *J. Opt. Soc. Am. B* 2012, 29:559-571.

[19] Fatome J, Pitois S, Morin P, Sugny D, Assemat E, Picozzi A, Jauslin HR, Millot G, Kozlov VV, Wabnitz S A universal optical all-fiber omnipolarizer *Scientific Reports* 2012, 2:938-1-8.

[20] Dysthe K, Krogstad HE, Muller P Oceanic Rogue Waves. *Annu. Rev. Fluid Mech.* 2008, 40:287-310.

[21] Akhmediev N, Pelinovsky E Editorial – Introductory remarks on "Discussion & Debate: Rogue Waves – Towards a Unifying Concept? *Eur. Phys. J. Special Topics* 2010, 185:1-4.

[22] Soomere T Rogue waves in shallow water *Eur. Phys. J. Special Topics* 2010, 185:81-96.

[23] Benjamin TB, J.E. Feir JE The disintegration of wave trains on deep water. Part 1. Theory *J. Fluid Mech.*1967 27:417-430.

[24] Akhmediev N, Korneev VI, Modulation instability and periodic solutions of the nonlinear Schrödinger equation *Theor. Math. Phys*. 1986, 69:1089-1093.

[25] Peregrine DH, Water waves, nonlinear Schrödinger equations and their solutions *J. Australian Math. Soc. Ser. B* 1983, 25:16-43.

[26] Kibler B, Fatome J, Finot C, Millot G, Dias F, Genty G, Akhmediev N, Dudley JM The Peregrine soliton in nonlinear fibre optics *Nat. Phys.* 2010, 6:790-795.

[27] Solli DR, Ropers C, Koonath P, Jalali B, Optical rogue waves *Nature* 2007, 450:1054-1057.

[28] Wabnitz S, Finot C, Fatome J, Millot G, Shallow water rogue wavetrains in nonlinear optical fibers. *Physics Letters* 2013, 377:932-939.

[29] Kodama Y, Wabnitz S Analytical theory of guiding-center nonreturn- to-zero and return-to-zero signal transmission in normally dispersive nonlinear optical fibers, *Opt. Lett.* 1995, 20:2291-2293.

[30] Didenkulova I, E. Pelinovsky, Rogue waves in nonlinear hyperbolic systems (shallow-water framework) *Nonlinearity* 2011, 24:R1:R18.

[31] Biondini G, Kodama Y On the Whitham equations for the defocusing nonlinear Schrodinger equation with step initial data *J. Nonlinear Sci.* 2006, 16:435-481.

[32] Wabnitz S. Optical tsunamis: shoaling of shallow water rogue waves in nonlinear fibers with normal dispersion, *J. Opt.* 2013, 15:064002.

In: Optical Fibers: New Developments
Editor: Marco Pisco

ISBN: 978-1-62808-425-2

DESIGNING SUPERCONTINUUM FIBER LIGHT SOURCES "À LA CARTE" FOR BIOMEDICAL APPLICATIONS

Albert Ferrando*
Departament d'Òptica, Universitat de València,
Burjassot, València, España

Supercontinuum (SC) generation in microstructure optical fibers is probably one of the most influential phenomena discovered in recent years in nonlinear optics. In descriptive and qualitative terms, SC generation is considered to occur when a narrow spectral input pulse experiences an extreme spectral broadening due to the intricate nonlinear dynamics involved in its evolution [1]. The gigantic frequency span, which can include the entire visible and infrared windows, that even a moderately narrow spectrum of an input pulse can experience throughout propagation in a relatively short optical fiber, is an intriguing phenomenon that has been the subject of intensive research for more than a decade [2]. However, the existence of the SC phenomenon in bulk media is known as early as in the late '60's, as reported in the classical paper by Alfano and Shapiro [3]. Since then, the phenomenon was extensively studied. It was quite clear from the beginning that this extraordinary conversion from a quasi-monochromatic light into a "white light source", sharing features of common laser light such as coherence, high degree of collimation and remarkable intensity, where also extraordinary for multiple applications [4]. However, it was not until the conjunction of ultrafast nonlinear temporal optics light sources with the development of a new fiber technology, leading to the invention of the photonic crystal fiber (PCF) or microstructure fiber that the SC topic could not see the splendorous rebirth we are still witnessing. SC in PCF's was first reported in an experiment using a microstructure fiber with small solid core pumped with 100 fs pulses at a wavelength of 790 nm, slightly higher than that of the zero dispersion wavelength (ZDW) of the fiber [5]. An important feature stressed by the authors here was that the fiber exhibited anomalous dispersion at visible wavelengths. This fact already pointed out the importance of the PCF dispersion properties for SC generation. PCF's were invented by Russell and co-workers only a few years earlier [6]. It was soon seen that the rich geometrical structure of

* E-mail: albert.ferrando@uv.es.

PCF's provided a lot of room for the design of their dispersion properties, as compared to standard fibers [7]. The physical mechanism behind SC generation in bulk media was provided mostly in terms of self-phase modulation (SPM) [4]. In the case of SC in PCF's, however, the interplay among SPM, dispersion properties of the fiber and Raman soliton self-frequency red-shift played an essential role [1]. Although these ingredients were also present in standard fibers, in which SC generation was also reported previously, SC in PCF's was especially simple to obtain and offered a wider variety of spectral features. Indeed, the subtle interaction between PCF dispersion and nonlinearities defines a panoply of potential output spectra by playing with the features of the input pulse and the fabrication parameters of the fiber. The discovery of PCF's provided then a great opportunity for the design and optimization of ultra-broadband spectral light sources. The fact that SC spectra could be also obtained by means of conventional tapered fibers added tapering as an extra design parameter [8].

Biomedical Applications of Supercontinuum Sources

The possibility of having ultra-broad spectral sources with similar properties to those of conventional lasers in terms of intensity, coherence and beam collimation was of nearly immediate interest for researchers in biomedical imaging. The existence of this new type of "white light" laser source could circumvent the traditional restrictions of conventional sources. On the one hand, because of the enormous bandwidth of the SC source there was no constraint, unlike in monochromatic lasers, in the available frequencies available for different applications. On the other hand, the SC source was highly competitive with other conventional "white light" sources, such as those based on LED technology, concerning the "quality" of the emerging light. The properties of some SC sources are comparable to that of monochromatic lasers regarding coherence, pulse duration or output power, while preserving an ultra-broadband spectrum. However, biomedical applications of SC sources can rely on diverse features of the SC spectrum, which are not necessarily similar for different applications. As mentioned before, SC spectra present an immense diversity of potential output profiles. It is for this reason that there exists a lot of room for the optimization of the spectral features of a SC source as a function of the specific biomedical application under consideration. We provide next some remarkable examples of the current state of the art of the utilization of SC sources for biomedical applications.

Until recently, flow cytometry relied on standard monochromatic lasers as a source of excitation for fluorescent probes. The coherence and power of the laser source is essential for this application but the restricted amount of available frequencies when using standard lasers limited its applicability. The appearance of SC sources was a clear step forward to develop this application [9]. In a similar way, fluorescence lifetime imaging microscopy (FLIM) was an earlier example of the use of SC sources for the excitation of fluorescent samples in cells in microscopy [10].

Another important imaging technique, multiple Coherent anti-Stokes Raman scattering (CARS) microscopy, can be also significantly benefitted from using SC sources [11]. The ability of CARS to identify an increasing number of chemical compounds in biological samples, by nonlinearly combining a pump beam with a second one (Stokes) to generate the recorded anti-Stokes signal beam, significantly depends on a wide spectral bandwidth for the

Stokes light. Thus a SC source is a perfect match for the Stokes beam. Another important application of SC sources is confocal microscopy. We can find applications of SC sources in many different implementations of confocal microscopy. It has been reported their use in interference-reflection microscopy (IRM) [12], confocal laser scanning fluorescence microscopy (CLSM) [13], confocal reflection microscopy [14] or confocal light absorption and scattering spectroscopic (CLASS) microscopy [15].

Special mention deserves the use of SC sources in Optical Coherence Tomography (OCT) (see [16] and references therein). OCT is essentially an interferometric technique for the acquisition of high resolution images in depth (i.e., in the axial direction) of biological tissues. There are two essential parameters characterizing optically the OCT: the axial resolution and the penetration depth [17]. The former depends on the inverse of the spectral bandwidth of the source as well as on the shape of its spectrum. Because the amount of scattered light within the biological sample depends on frequency (according to Rayleigh formula, it grows quickly with frequency), the penetration depth notably depends, roughly, on the average frequency of the spectrum, thus tending to be smaller for higher frequencies. Obviously, it is also dependant on the brightness of the source. Since SC sources simultaneously provide both wide spectral widths and high intensities, it is not surprising that OCT technologies have adopted this new laser "white-sources" as perfect partners.

Designing Optimized Supercontinuum Sources for Biomedical Applications

We have just seen the multiple and diverse aspects that SC sources can have in order to fulfill the technical requirements of so many different applications. It is not the same to provide a suitable source for, say, confocal reflection microscopy, in which a uniform spectrum is desirable, than for OCT, where best axial resolutions are obtained using perfect Gaussian spectral profiles [17]. Moreover, together to the specificity of the optical technique under consideration, it is also highly relevant to add in the design strategy the particular optical nature of the biological samples one wishes to analyze. Thus, every specific biomedical application is susceptible to be improved by optimizing distinguishing features of the output SC spectral of the light source. This is the procedure we define as *design "à la carte"*, which is implemented in three basic well-defined steps:

Defining the Target

As mentioned above, the first step prior to initiate any mathematical optimization strategy is to define "the suitable target spectrum" associated to the required biomedical application. This question, which in principle appears to be simple and naïve, is, in many cases, one of the most controversial parts of the optimization program. The whole optimization procedure can fail if the target spectrum is not well identified in both optical and biomedical terms. Defining a target spectrum to be optimized from biomedical information is not necessarily an easy task. Biomedical information relevant for diagnose must be provided by specialized experts. This raw information, many times of qualitative nature, must be then transformed into quantitative optically meaningful information in the form of a spectral function to be optimized. This requires, in many occasions, a considerable amount of translational work involving close interactions among biomedical and optical experts. The specificity of the spectral function, optimal for the required biomedical application, is the origin of the design “à la carte” procedure.

Defining the Right Strategy

Once the preliminary step is satisfactory fulfilled, the optimization process, properly speaking, can start. Up to this point, the process can be summarized in a single statement: one biomedical application, one target spectrum. But the optimization of a SC process, which has an inherent nonlinear nature that confers a high degree of complexity to him, is not necessarily straightforward even if the target function has been perfectly defined in both its biomedical and optical terms. As mentioned in the first section, the output spectrum profile of a PCF fiber is extremely sensitive to the fiber dispersion properties as well as to the nature of the input pulse. The subtle and complex interplay between the fiber dispersion and nonlinearities can make a “brute force” optimization process, based exclusively on standard optimization numerical techniques, unfeasible. The complexity of the sample space and the potentially high number of relevant parameters that can play a role in determining the output spectrum, makes a “blind” search not a good strategy to obtain reliable results. This fact determines that en essential part of the optimization procedure is the proper understanding of the physical scenario necessary to obtain the output spectrum that fits the desired target. One should be able to figure out, at least qualitatively, under which realistic physical conditions the target spectrum provided in the first step can be achieved. This goes beyond performing purely numerical simulations in a trial-and-error process and it demands a deep knowledge of the nonlinear dynamics involved in SC generation.

In practical terms, there are many possible scenarios for the output spectrum -which can be remarkably different- that are qualitative explained invoking a careful and nontrivial theoretical analysis. The theory involves the complex dynamics of the nonlinear interaction between soliton and dispersive waves, strongly dependant on both the dispersion properties of the fiber and the input pulse characteristics (see [2] and [18] for recent reviews). Consequently, given a target spectrum for a specific biomedical application, it is almost essential to visualize the physical scenario needed to reproduce qualitatively what kind of SC mechanism can give rise to the desired target spectrum.

The determination of the correct physical mechanism provides the right strategy to proceed. It will be immediate then to determine what are the physical parameters that need to be optimized. In terms of physical implementation, there are two types of parameters susceptible of being optimized. It is possible to optimize external and structural parameters. External parameters are those that can be changed in real time such as, for example, the temporal width, wavelength, or power peak of the input pulse. Structural parameters are, for example, the non-tunable fabrication properties of the fiber, such as its dispersion or nonlinear properties. The theoretical analysis will provide us with a reasonable set of parameters of the previous kind, which are relevant to achieve the desired optimization of the target spectrum.

Additionally, it will assist us to define another essential part of the optimization process. It will help us to establish the adequate range of values for these parameters, thus defining the right searching region in the sample space, on which the optimization algorithm will act. This strategy will ensure the success of the optimization procedure before performing the numerical, and often heavy, task.

Finding an Adequate Optimization Algorithm and a Suitable Computer Implementation

As a general rule, the better the theoretical analysis the lesser the computational optimization effort. However, even numerical simulations based on an excellent theoretical analysis cannot always provide *alone* optimal values to fulfill the requirements of a target spectrum. If the biomedical application requires an optimal spectrum fulfilling many conditions, the target spectrum can be difficult to achieve only using this approach, so a supplementary and efficient numerical exploration of the search parameter space will be needed. Here is the point where the optimization algorithm enters. A fitness function will be included to measure, in functional space, the difference between the desired target spectrum and that obtained by simulating the pulse propagation numerically for a set of fixed parameters (external or structural). The theoretical analysis provides this set of parameters as well as the range of values to perform the search. Therefore, it determines the sample space for optimization. At each point of this space the fitness function takes a value. The optimal value for the target spectrum will correspond to the absolute minimum (or the best approximation to it) of the fitness function. The values of the parameters that provide this minimum will be the optimal ones. At this point, it is clear that an efficient optimization algorithm is crucial.

Depending on the topography of the fitness function map in parameter space, the search for the minimum can be a delicate matter. A non-guided search of minima can be very inefficient. The subtleties of the nonlinear mechanisms of the SC, together with the biomedical requirements of the target spectrum, can make the behavior of the fitness function perverse and thus the topography of the sample space intricate. This can sometimes require dense discretizations of the parameter space, which implies an unmanageable amount of potential simulations. Besides, the dimensions of the sample space grow as a power law with the number of optimization parameters and the average time of SC simulations in standard CPU can become long (sometimes, a few hours) in some situations. A random, or non-guided, search in these circumstances is not feasible in terms of computing time. In order to set a general strategy to obtain the minimum, it is very convenient, on the one hand, to have a flexible optimization algorithm (probably, more than one) and, on the other hand, a scalable computational platform for this algorithm to run. The selection of a suitable optimization algorithm is very dependant on the form of the fitness function topography. Heuristic algorithms, such as genetic algorithms, are good at exploring different regions of the sample space in order to avoid sticking around local minima. Besides, they are well fit to its deployment into distributed computing platforms such as the Grid [9, 20]).

Grid technology permits to scale-up the optimization procedure to larger computational infrastructures under demand. Basically, if your algorithm has been developed with the adequate software to run in Grid platforms, you do not need to re-build it when your software is running in distributed computer facilities, no matter if they are small (as a local cluster) or big (as a set of remote large supercomputing facilities). In this way, if a particular application requires increasing the number of parameters to optimize, the deployment in a Grid platform gives you the option of scaling-up the computational infrastructure in a transparent way.

Conclusion and Perspectives

Finding the adequate light source spectrum for a specific biomedical application can be a challenge. On the other hand, the extraordinary properties of SC sources, even if they are not optimized, have shown to be very successful in many biomedical applications, in such a way their number keeps growing day after day. This fact raises the question if optimization is really necessary. The possibility of the design "à la carte" is an option based on the fact that specificity and diversity are inherent to biomedical samples. Even using the same optical imaging technique, the optical response of different biochemical compounds of a biological sample can be very different. Thus, although an *all-purpose* SC source can do a good job and improve the performance of the technique as compared to conventional sources, it is clear that there is some additional space for the optimization of the optical response by designing more specific spectra. The SC source "à la carte" can then complement the results obtained by a more general scheme based on standard SC sources.

From the optical point of view, the design "à la carte" process represents in some cases a real intellectual challenge. Although some optimizations are pretty straightforward to perform, so that a standard numerical optimization can provide reasonable answers, in other occasions the theoretical analysis acquires an essential relevance. On the other hand, the accuracy of the biomedical analysis to provide an adequate target spectral function requires a close conjunction of biomedical and optical knowledge. The combination of deep physical understanding of soliton physics together with translational work linking optical concepts to biomedical ones, makes this type of projects very enriching from the intellectual point of view. The formation of efficient, broad vision, interdisciplinary teams is then crucial for success.

As mentioned before, the field of applications of SC sources in biomedicine is quickly growing and there is no reason to believe this trend will change in the near future. On the contrary, as the possibilities of manufacturing and developing more complex microstructure and PCF fiber devices grow, the degrees of freedom for the design "à la carte" of SC sources will increase, thus offering us more options and flexibility to fulfill new requirements dictated by biomedical applications. In this direction, recent advances in taper technology, showing to what extent the tapering profile can be used to control the spectral output [21, 22], proofs how much room still exists to optimize the spectral response of SC sources.

References

[1] J. M. Dudley, G. Genty, S. Coen, *Rev. Mod. Phys.* 78, 1135 (2006).

[2] J. M. Dudley, J. R. Taylor, *Supercontinuum generation in optical fibers* (Cambridge University Press, Cambridge; New York, 2010), pp. xiii, 404.

[3] R. Alfano, S. Shapiro, *Phys. Rev. Lett.* 24, 584 (1970).

[4] R. R. Alfano, *The supercontinuum laser source: fundamentals with updated references* (Springer, New York, ed. 2nd ed., 2006), pp. xx, 537.

[5] J. K. Ranka, R. S. Windeler, A. J. Stentz, *Opt. Lett.* 25, 25 (2000).

[6] J. C. Knight, T. A. Birks, P. S. J. Russell, D. M. Atkin, *Opt. Lett.* 21, 1547 (1996).

[7] A. Ferrando, E. Silvestre, P. Andrés, J. J. Miret, M. V. Andrés, *Opt. Express* 9, 687 (2001).

[8] T. A. Birks, W. J. Wadsworth, P. S. J. Russell, *Opt. Lett.* 25, 1415 (2000).

[9] W. G. Telford, F. V. Subach, V. V. Verkhusha, *Cytometry A* 75, 450 (2009).

[10] C. Dunsby *et al.*, *J. Phys. D: Appl. Phys.* 37, 3296 (2004).

[11] T. W. Kee, M. T. Cicerone, *Opt. Lett.* 29, 2701 (2004).

[12] L. D. Chiu, L. Su, S. Reichelt, W. B. Amos, *J Microsc* 246, 153 (2012).

[13] G. McConnell, *J. Phys. D: Appl. Phys.* 38, 2620 (2005).

[14] M. Booth, R. Juskaitis, T. Wilson, *Journal of the European Optical Society - Rapid publications; Vol 3 (2008)* (2008).

[15] I. Itzkan *et al.*, *Proc Natl Acad Sci U S A* 104, 17255 (2007).

[16] L. Froehly, J. Meteau, *Optical Fiber Technology* 18, 411 (2012).

[17] A. F. Fercher, W. Drexler, C. K. Hitzenberger, T. Lasser, *Rep. Prog. Phys.* 66, 239 (2003).

[18] D. V. Skryabin, A. V. Gorbach, *Rev. Mod. Phys.* 82, 1287 (2010).

[19] A. Ferrando *et al.*, *Proc. SPIE 7839, 2nd Workshop on Specialty Optical Fibers and Their Applications (WSOF-2)* 7839, 78390W (2010).

[20] G. Moltó *et al.*, *Proceedings of the 3rd Iberian Grid Infrastructure Conference (IberGrid 2009)* 137 (2009).

[21] U. Moller *et al.*, *Optical Fiber Technology* 18, 304 (2012).

[22] J. M. Stone, J. C. Knight, *Optical Fiber Technology* 18, 315 (2012).

CHAPTERS

In: Optical Fibers: New Developments
Editor: Marco Pisco
ISBN: 978-1-62808-425-2

Chapter 1

MICROFIBER COIL RESONATORS: THEORY, MANUFACTURE AND APPLICATION

Fei Xu, Wei Guo, Wei Hu and Yan-qing Lu
College of Engineering and Applied Sciences and National Laboratory of Solid State Microstructures, Nanjing University, Nanjing, P.R. China

ABSTRACT

The manufacture of tapers from optical fibers provides the possibility to get long, uniform and robust micrometer or nanometer size wire. Optical microfibers (MFs) are fabricated by adiabatically stretching conventional optical fibers and thus preserve the original optical fiber dimensions at their input/output pigtails, allowing ready splicing to standard fibers. Since MFs have a size comparable to the wavelength of the light propagating in it, a considerable fraction of power can be located in the evanescent field, outside the MF physical boundary. When a MF is coiled, the mode propagating in it interferes with itself to give a resonator. In particular, a 3D MF coil resonator (MCR) can be obtained by wrapping an MF on a low index rod, which is difficult for standard planar light circuit (PLC) technology due to the stereoscopic geometry. With recent improvements in fabrication technology of low loss MFs, the Q-factor of MCRs could potentially compete with the highest Q-factors currently achieved only in whispering gallery resonators. In this chapter the latest results on the theory and experiment of optical MCRs are presented. The fabrication model and techniques are discussed and the dispersion characteristics and influence of strong coupling of an MCR are investigated. The colorful applications of refract metric sensing are also introduced.

Keywords: Microfiber, resonator, fiber

1. INTRODUCTION

In the past 40 years, optical fibers and fiber devices with diameters larger than the wavelength of transmitted light have been widely used in optical communications, sensors and other applications.

Recent years have seen an increasing interest in the research on optical nanowires. Optical fiber nanowires or microfibers (MFs) are fibers with submicrometer- and nanometer-diameter (tens to thousands of times thinner than the commonly used micrometer-diameter waveguides) which can be used as air-clad wire-waveguides with small cores [1, 2]. They have the potential to become building blocks in the future micro- and nano-photonic devices since they offer a number of unique optical and mechanical properties. In particular, they allow for large evanescent fields, high nonlinearity, extreme flexibility and configurability and low-loss interconnection to other optical fibers and fiberized components [3, 4]. MFs are fabricated by adiabatically stretching optical fibers while keeping the original dimensions of the optical fiber at their input and output allowing ready splicing to standard fibers (thus MFs are tapers and are also called nanotapers). This represents a significant advantage when compared to small-core micro-structured fibers that always present significant insertion/extraction losses. However the fabrication of low-loss MFs remains challenging because of high precision requirements.Since 1999, several types of dielectric submicrometer- and nanometer- diameter MFs of optical quality have been obtained. According to the methods presented in these papers, the length of MFs can vary from 1mm to tens of millimetres, with loss as small 0.02dB/mm and radii in the range of tens of nanometres [5]. Although the measured loss was orders of magnitude higher than that achieved later with flame-brushing techniques, it was low enough to open the way to a host of new devices for optical communications, sensing, lasers, biology and chemistry [3, 4, 6-29].

MFs are of interest for a range of emerging fiber optic applications since they offer a number of enabling optical and mechanical properties. In particular, MFs can easily be bent and manipulated and yet remain relatively strong mechanically. Bent radii of the order of a few microns can be readily achieved with relatively low induced bend loss allowing for highly compact devices with complex geometry e.g. 2D and 3D multi-ring resonators [48].

MF resonators can be classified according to their dimensionality: loop and knot resonators are example of 2D structures, while the microcoil resonator belongs to the group of a 3Dassembly [12, 15, 19, 20, 22, 26, 29-43]. The MF Loop/Knot Resonator (MLR/MKR) is a miniature version of a conventional fiber loop resonator, which was created for the first time back in 1982 from a conventional single-mode fiber and a directional coupler [44]. Due to the bending losses of a weakly guiding single mode fiber and dimensions of the fiber couple, the maximum value of the free spectrum range of fiber resonator was limited to the order of a gigahertz. Later, the authors of Ref [45] demonstrated a 2 mm-diameter self-coupling MLR fabricated from an 8.5 μm-diameter optical fiber taper. The diameter of the fiber was too large to ensure sufficient interfiber coupling. In order to enhance the coupling efficiency, the MLR was imbedded into a silicone-rubber having a refractive index close to the index of the fiber. Recently, Sumetsky demonstrated some MLRs in air [12, 13, 31, and 32]. Figure 1 (a) shows the general shape of a MLR with the input and output ends touching each other. The surface attraction forces (Van der Waals and electrostatic), keeps the ends together and overcomes the elastic forces that would otherwise straighten out the MF [43]. The fabrication of an MLR consists in drawing the optical MF and then bending it into a self-coupling loop. The characteristic radii of an MF, used for the fabrication of an MLR, are in the range 300-500 nm, which is approximately uniform along several millimeters of its length. The Q-factors were about 1500-120000. A major drawback of the self-touching MLR in air is its geometrical stability: the coupling is strongly affected by the microcoil geometry and a small change in shape results in a large change in its transmission properties.

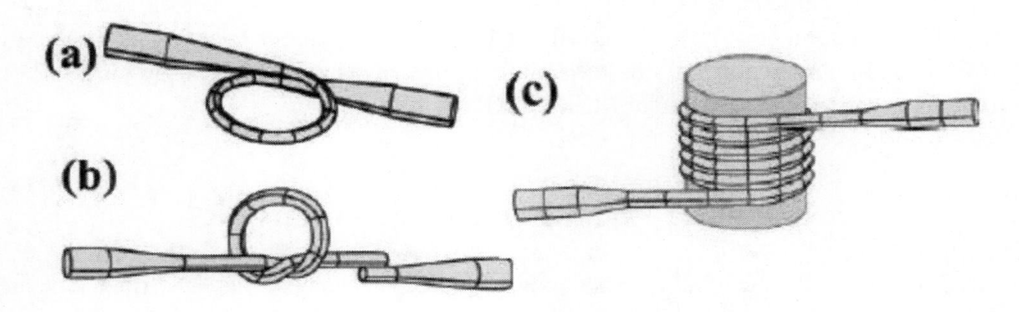

Figure 1. Illustration of the MCRs: (a) MLR, (b) MKR, and (c) MCR. The MLR/MKR can be taken as a one-turn MCR [43]. Reprinted with permission. Copyright 2010 Elsevier Limited.

Embedding the MLR in polymer has been the preferred solution to provide long term-stabilityeffect of a host polymer on an MF resonator, and fusing the loop contact points with a CO_2 laser [17] has also been proposed. An alternative approach to increase the MLR stability relies on the use of a copper support rod to preserve the MLR geometry [46]. Critical coupling has been demonstrated by tuning the resonator with thermal effects induced by current flowing in the conductor rod, achieving a Q-factor up to 4000 and an extinction ratio of 30 dB.A MKR has also been proposed and manufactured [15, 19, 41, 43], but its fabrication requires the MF to be broken in the minimum waist region as shown in Figure1(b). Because of their fabrication methodology, MKRs exhibit high input/output losses and relatively high difficulty in tuning the resonance to a specified wavelength.

In 2004, a 3D multiple-turn optical MF coil resonator (MCR) was proposed which consists of self-coupled adjacent loops in a helix arrangement [43, 47]. As illustrated in Figure 1(c), MCRs wrapped on a low index rod can overcome the problem of stability and have the potential of using this device as a basic functional element for the MF-based photonics. The resonance is formed by the interference of light going from one turn to another along the MF and returning back to the previous turn with the aid of weak coupling. It is interesting to note that MLR can be described as the simplest MCR with one single turn. MCR-based optical devices have two significant advantages over planar devices: smaller losses and greater compactness. The manufacture is still difficult because the MF is liable to be broken during wrapping. Prior to 2007, experimental demonstrations of MCRs were reported in air [34] and liquid [48]. Like all silica devices, bare MFs suffer from degradation when uncoated [49]: in order to obtain a practical device, the MCR was coated with low index polymer Teflon [31]. Yet, the determination of the coating thickness is a challenge because a thick coating layer limits the sensitive evanescent field, while a thin layer does not provide an appropriate protection to the device. In 2007, a solution was proposed to make the deployment of MCR as high sensitivity microfludic sensorspossible [26, 27]. After a review of resonators based on MF loops and microcoils, a summary of recent results on the use of resonators for sensing will be presented. Sensitivities as high as 700 nm/RIU were predicted (RIU is the refractive index unit). In 2008, the sensors were experimentally demonstrated [50]. Similar to the two-dimensional resonant microring structures, MCRs can perform complex filtering, time delay, and switching,too. In practical realizations, MCRs with more turns require longer MFs and a high degree of accuracy in positioning the MF coils adjacent to each other; thus they are much more difficult.

Because of the limitations imposed by transmission losses, in most cases only MCRs with a relatively small number of turns are considered. Finally, we want to mention again that the MLR/MKR can be treated as the one-turn MCR in theory. The relative theory on MCRs in the following sections are suitable for MLRs/MKRs.

2. Coupled Wave Equations for MCRs

The MCR properties can be analyzed using coupled wave equations. The fiber diameter is assumed to be smaller than or comparable to the wavelength of radiation. Introduce the local natural coordinate system (x, y, s) as shown in Figure 2, where s is the coordinate along the fiber axis and (x, y) are the coordinates along the normal, n, and binormal, b, of the axis, respectively. If the characteristic transversal dimension of the propagating mode is much smaller than the characteristic bend radius, then the adiabatic approximation of parallel transport can be applied. In this approximation and for a relatively small pitch of the helical microcoil, the transversal component is

$$E\ (x, y, s) = A(s)F_0(x, y)\exp\left\{i\int^{s}\beta(s)ds\right\}\exp(i\omega t) \tag{1.2.1}$$

Here the vector function $F_0(x, y)$ defines the local eigen mode corresponding to the propagation constant $\beta(s)$, $A(s)$ is the amplitude at the position s, t is the time, and is the frequency. It is convenient to define the amplitude of the field at a turn m as $\mathrm{A_m(s)}$ and to consider s as the common coordinate along turns, so that $0 < s < \mathrm{S_m}$, where is the length of the mth turn. If every turn is taken as a round with a radius R_m , s can be expressed as $R_m\theta$ ($0 \leq \theta \leq 2\pi$) in polar coordinates as shown in Figure 3(a). Figure 3(b) shows the cross-sections of pth and q^{th} turns, where n_f is the index of the MF and n_c is the index of the environment.

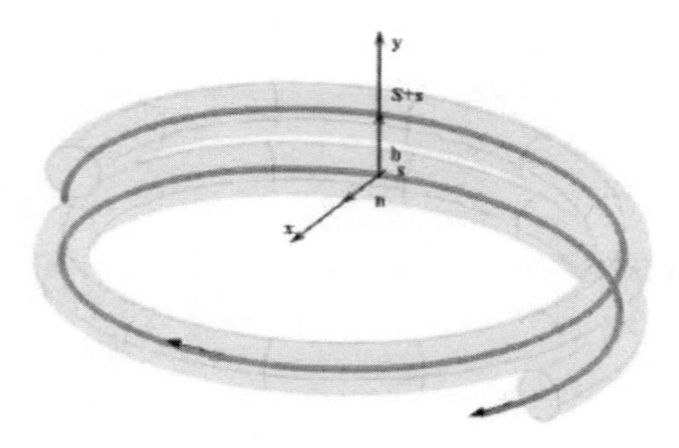

Figure 2. Illustration of local natural coordinate system of an MCR.

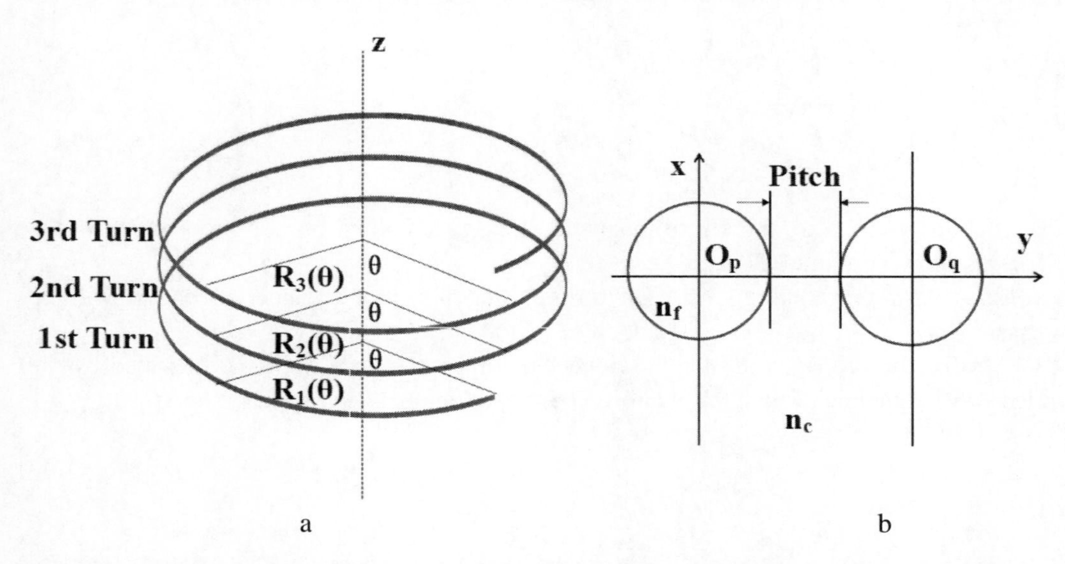

Figure 3. (a) Illustration of an MCR in cylindrical coordinates. (b) The cross-sections of two MFs.

The field in the mth can be expressed as:

$$E_m(x,y,\theta)=A_m(\theta)F_0(x,y)\exp\{\beta R\theta\} \tag{1.2.2}$$

Here we assume the MF is uniform and β is independent on s.

With the conventional coupled mode theory [51], the coupled equations for the coefficients $A_m(\theta)$ can be obtained on the condition of relatively slow variation of coefficients $A_m(\theta)$. As described in Ref [47], the propagation of light along the coil in a two-turn and three-turn MCR are described by the coupled wave equations.

In the two-turn MCR:

$$\frac{d}{d\theta}\begin{pmatrix}A_1\\A_2\end{pmatrix}=i\begin{pmatrix}0 & R_1(\theta)\chi_{12}(\theta)\\ R_2(\theta)\chi_{21}(\theta) & 0\end{pmatrix}\begin{pmatrix}A_1\\A_2\end{pmatrix} \tag{1.2.3}$$

In the three-turn MCR:

$$\frac{d}{d\theta}\begin{pmatrix}A_1\\A_2\\A_3\end{pmatrix}=i\begin{pmatrix}0 & R_1(\theta)\chi_{12}(\theta) & 0\\ R_2(\theta)\chi_{21}(\theta) & 0 & R_2(\theta)\chi_{23}(\theta)\\ 0 & R_3(\theta)\chi_{32}(\theta) & 0\end{pmatrix}\begin{pmatrix}A_1\\A_2\\A_3\end{pmatrix} \tag{1.2.4}$$

where $\chi_{pq}(\theta)=\kappa_{pq}(\theta)\exp\left\{i\int_0^{2\pi}\beta_p(\theta)rd\theta-i\int_0^{2\pi}\beta_q(\theta)rd\theta\right\}$ and $\kappa_{pq}(\theta)$ is the coupling coefficient between the turns p and q (p, q =1, 2, 3, …) [51]:

$$k_{pq} = \frac{\omega\varepsilon_0 \iint\limits_{O_p} (n_f^2 - n_c^2) E_p^* \cdot E_q dxdy}{\int_{-\infty}^{\infty}\int_{-\infty}^{\infty} u_z \cdot (E_p^* \cdot H_p + H_p^* \cdot E_p) dxdy} \tag{1.2.5}$$

and $E_p(E_q)$ and $H_p(H_q)$ are the electromagnetic mode fields when light is propagating in the MF$p(q)$, as shown in Figure 3(b).

In Eq.(1.2.4), the coupling between turns 1 and 3 is not considered, and it will be discussed later.

Generally, the propagation of light along the coil in a M-turn MCR is described by the coupled wave equations, which takes into account coupling between adjacent turns:

$$\frac{d}{d\theta}\begin{pmatrix} A_1 \\ A_2 \\ \cdots \\ A_m \\ \cdots \\ A_{M-1} \\ A_M \end{pmatrix} = i \begin{pmatrix} 0 & R_1\chi_{12}(\theta) & 0 & \cdots & 0 & 0 & 0 \\ R_2\chi_{21}(\theta) & 0 & R_2\chi_{23}(\theta) & \cdots & 0 & 0 & 0 \\ 0 & R_3\chi_{32}(\theta) & 0 & \cdots & 0 & 0 & 0 \\ \cdots & \cdots & \cdots & \cdots & \cdots & \cdots & \cdots \\ 0 & 0 & 0 & \cdots & 0 & R_{M-2}\chi_{M-1M-2}(\theta) & 0 \\ 0 & 0 & 0 & \cdots & R_{M-1}\chi_{M-2M-1}(\theta) & 0 & R_{M-1}\chi_{M-1M}(\theta) \\ 0 & 0 & 0 & \cdots & & R_M\chi_{MM-1}(\theta) & 0 \end{pmatrix} \begin{pmatrix} A_1 \\ A_2 \\ \cdots \\ A_m \\ \cdots \\ A_{M-1} \\ A_M \end{pmatrix}$$

(1.2.6.)

where for a uniform-waist optical MF, all β_p are independent of θ (p=1,2,…,M). In this chapter, Eqs. are solved with MATLAB.

The coefficients satisfy the continuity conditions

$$R_{m+1}(0) = R_m(2\pi) \tag{1.2.7}$$

$$A_{m+1}(0) = A_m(2\pi)\exp\left\{ i\int_0^{2\pi} \beta(\theta) R_m d\theta \right\}, m = 1,2,\ldots M-1. \tag{1.2.8}$$

The equation can be simplified by introducing the average radius R_0, coupling parameter K_{pq} and transmission amplitude T:

$$R_0 = \left(\sum_{m=1}^{M}\int_0^{2\pi} R_m(\theta) d\theta\right) \Big/ 2\pi M \tag{1.2.9}$$

$$K_{pq} = 2\pi R_0 \kappa_{pq}(\theta) \tag{1.2.10}$$

Finally the transmission coefficient was defined as:

$$T = \frac{A_M(2\pi)}{A_1(0)} \exp\left\{ i \int_0^{2\pi} \beta R_M(\theta) d\theta \right\} \tag{1.2.11}$$

When the loss or gain α is considered, β is replaced by a complex number $\beta+i\alpha$. If the propagation losses and gains are ignored, then the propagation constant, β is real α=0, and $|T|=1$. In this case, the coil performs as an all pass filter and the resonances of transmission coefficient appear in the group delay (t_d) only.

In order to simplify the formula, the MF diameter and the pitch between adjacent turns are assumed uniform, so the coefficient κ and β are independent on θ. R is taken independent on θ, which implies that R_i is constant in each ith turn. In this case, the coupling parameter can be expressed as

$$K = 2\pi R_0 \kappa \,, \tag{1.2.12}$$

where

$$R_0 = \frac{1}{M} \sum_{m=1}^{M} R_m \tag{1.2.13}$$

is the average radius.

By applying the transformations

$$B_m(\theta) = A_m(\theta) \exp\{i\beta R_m \theta\}, m = 1, 2, \ldots M-1. \tag{1.2.14}$$

the coupled wave equations can be written as

$$\frac{d}{d\theta}\begin{pmatrix} B_1 \\ B_2 \\ \cdots \\ B_m \\ \cdots \\ B_{M-1} \\ B_M \end{pmatrix} = i \begin{pmatrix} -R_1\beta & R_1\kappa & 0 & \cdots & 0 & 0 & 0 \\ R_2\kappa & -R_2\beta & R_2\kappa & \cdots & 0 & 0 & 0 \\ 0 & R_3\kappa & -R_3\beta & \cdots & 0 & 0 & 0 \\ \cdots & \cdots & \cdots & \cdots & \cdots & \cdots & \cdots \\ 0 & 0 & 0 & \cdots & -R_{M-2}\beta & R_{M-2}\kappa & 0 \\ 0 & 0 & 0 & \cdots & R_{M-1}\kappa & -R_{M-1}\beta & R_{M-1}\kappa \\ 0 & 0 & 0 & \cdots & 0 & R_M\kappa & -R_M\beta \end{pmatrix} \begin{pmatrix} B_1 \\ B_2 \\ \cdots \\ B_m \\ \cdots \\ B_{M-1} \\ B_M \end{pmatrix} \tag{1.2.15}$$

With the continuity conditions

$$B_{m+1}(0) = B_m(2\pi), m = 1, 2, \ldots M-1. \tag{1.2.16}$$

The transmission amplitude is defined as:

$$T = \frac{B_M(2\pi)}{B_1(0)} \tag{1.2.17}$$

The simplest case is the uniform MCR with the same radius and uniform pitches in all the M turns, $R_m = R_0$.

For a two-turn lossless uniform MCR which is similar to a ring resonator [47]:

$$T = \frac{e^{i\beta 2\pi R_0} - i\sin(K)}{e^{-i\beta 2\pi R_0} + i\sin(K)} \tag{1.2.18}$$

and the resonator conditions where the absolute of group delay $|t_d|$ is the maximum are:

$$K = K_u = (2u-1)\frac{\pi}{2} \tag{1.2.19}$$

$$\beta 2\pi R_0 = 2v\pi + \frac{\pi}{2} \tag{1.2.20}$$

where u, v are integers.

For a three-turn lossless uniform MCR [47]:

$$T = \frac{e^{-i\beta 2\pi R_0} - \sqrt{2}i\sin(\sqrt{2}K) - e^{i\beta 2\pi R_0}\sin^2(\sqrt{2}K)}{e^{i\beta 2\pi R_0} + \sqrt{2}i\sin(\sqrt{2}K) - e^{-i\beta 2\pi R_0}\sin^2(\sqrt{2}K)} \tag{1.2.21}$$

and the resonator conditions are [47]:

$$K = K_u^{(1)} = (2u-1)\frac{\pi}{\sqrt{2}}, \beta 2\pi R_0 = v\pi \tag{1.2.22}$$

or

$$K = K_{u\varepsilon}^{(2)} = \sqrt{2}[\varepsilon \arcsin(\frac{1}{\sqrt{3}}) + u\pi], \beta 2\pi R_0 = (2v + \frac{\varepsilon}{2})\pi, \varepsilon = \pm 1 \tag{1.2.23}$$

As an example, the group delay was calculated against the wavelength near 1550 nm when M=2, as shown in Figure 4, for different K.

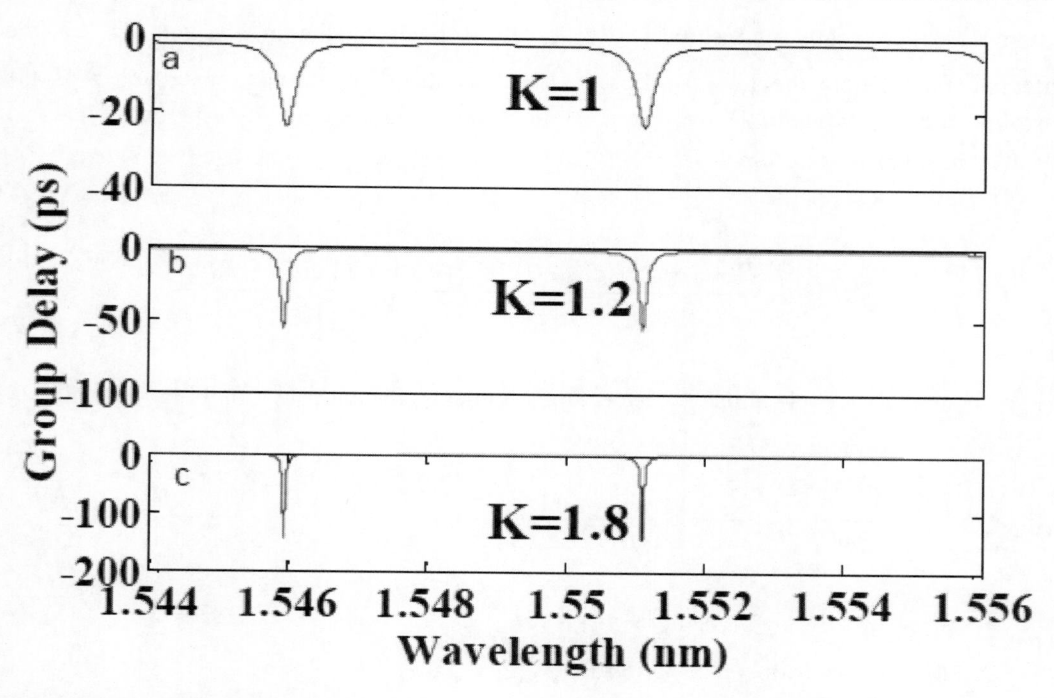

Figure 4. Group delay dependencies on the wavelength in a two-turn uniform MCR ;(a) K=1; (b) K=1.2; (c) K=1.8.

3. Strong Disperion Influence in MCRs

Similar to other kinds of micro-resonators, dispersion is one of the major limitations of MCRs in high-speed communication. Previous works have considered the waveguide dispersion of micro-resonators, but, to the best of our knowledge, the coupling dispersion is ignored in most cases. In an MCR, the waveguide dispersion is evident and possibly very large due to the very small MF diameter. However, the interaction between adjacent turns is another main source of dispersion because of the strong coupling strength and long coupling length. As a consequence, it will be very important and practical to investigate the dispersion characteristics of MCRs considering both waveguide and coupling contributions [52]. In this section, we calculate and investigate both the waveguide and coupling dispersions of a typical two-turn MCR. The possible influence and limitation of a high bit rate optical system are discussed, which is helpful for MCR based device design and applications [52]. Light propagating with loss around a uniform microcoil with two turns in the linear regime is described by the following equation on the transmission T:

$$T = \frac{[\cos\beta L + i(\sin\beta L - r^{-1}\sin K)]}{[\cos\beta L + i(-\sin\beta L + r\ (\sin K)]}, \gamma = \exp(-\alpha L) \qquad (1.3.1)$$

where L is the length of one turn, $\beta = 2\pi n_{eff}/\lambda$ is the propagation constant, n_{eff} is the effective index, α is the loss coefficient, $K = kL$ is the coupling parameter and κ is the coupling coefficient due to the overlap of the field modes between neighboring turns.

The resonance condition is $\beta L = 2m\pi + \pi/2$ and sin (Km) =1 where m is an integer.

The group delay is [52]

$$\tau = \frac{\partial\varphi}{\partial\omega} = imag\{\frac{2i\beta' L - \cos\beta L[i\xi\mu' + \beta' L\zeta\mu] + \sin\beta L[\zeta\mu' - i\beta' s\xi\mu]}{1+\mu^2 + i\mu[\zeta\cos\beta L + i\xi\sin\beta L]}\} \tag{1.3.2}$$

At resonating wavelengths, the dispersion parameter characteristic of an MCR is [52]

$$D_R = -\frac{2\pi c}{\lambda^2}\frac{\partial^2\varphi}{\partial\omega^2}|_{\omega_0} == \frac{D_f L(2-\xi\mu)}{1+\mu^2-\xi\mu} - \frac{\frac{d\beta}{d\omega}|_{\beta_0} L\frac{d\mu}{d\lambda}[2(2-\xi\mu)(2\mu-\xi) - i\zeta^2\mu]}{(1+\mu^2-\xi\mu)^2}\}$$

where $\xi = \gamma + \gamma^{-1}$, $\zeta = \gamma^{-1} - \gamma$, $\mu = \sin K$

For a lossless MCR, it can be simplified as [52]:

$$D_R = \frac{2D_f L}{1-\mu} + \frac{2\pi c}{\lambda_0^2}\frac{8\beta' L\mu'}{(1-\mu)^2} = D_1 + D_2 \tag{1.3.3}$$

where D_f is the group velocity dispersion (GVD) of MF. The dispersive properties of an MCR are very important. A higher dispersion may broaden the optical pulse width then deteriorate the system performance at a high bit rate B [52].

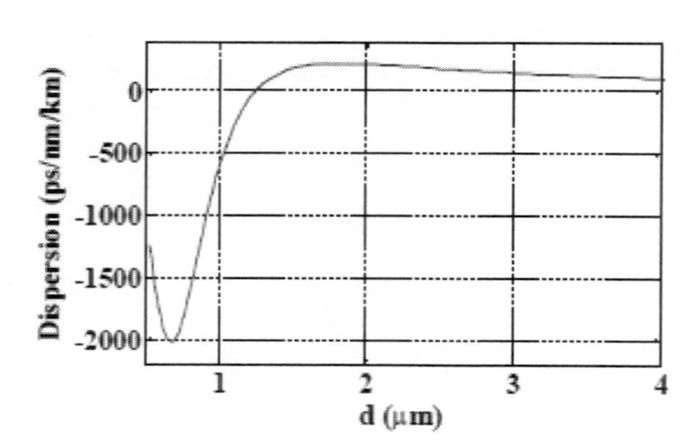

Figure 5. The GVD curve as a function of MF diameters.

There are two main contributors to the dispersion: the MF waveguide and material dispersions as shown in the first item of Eq. (1.3.3) and the coupling dispersion as shown in the second item. From Eq. (1.3.3), the coupling coefficient depends on the light frequency, which is easily understandable. Conventional work on micro-resonators ignore the dispersion of coupling, sometimes even waveguide and material dispersions. This approximation is reasonable in the cases of weakly coupling and low Q-factor. For very strong coupling and ultra high Q-factor, we believe these dispersions have to be considered totally [52].

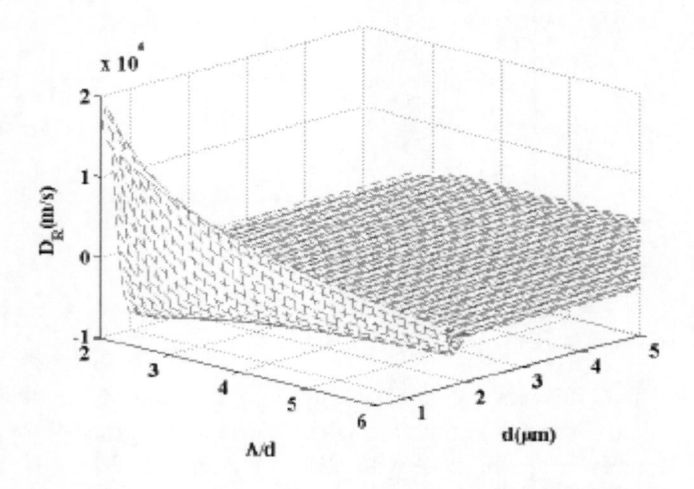

Figure 6. The total dispersion D_R as a function of d and Λ/d at nearly resonating.

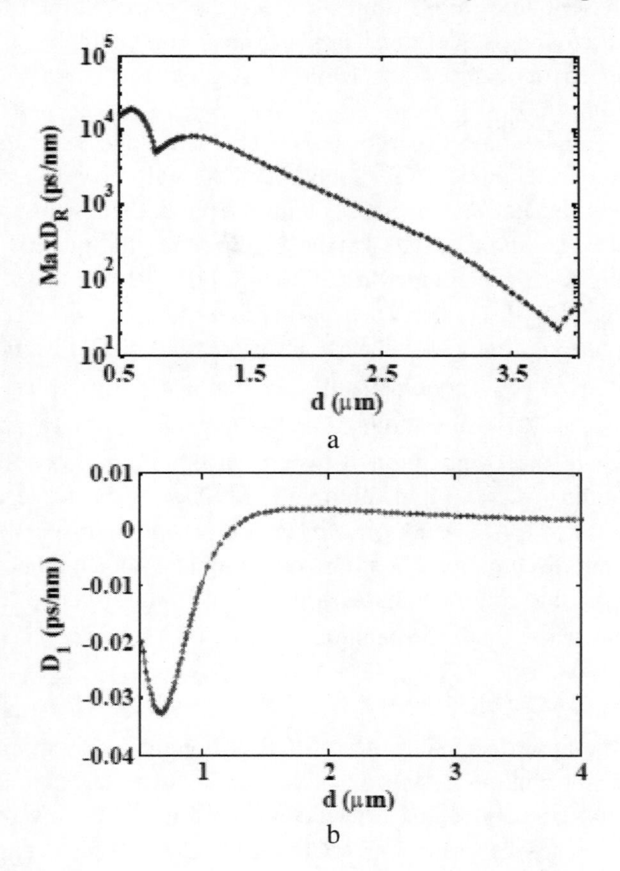

Figure 7. (Continued).

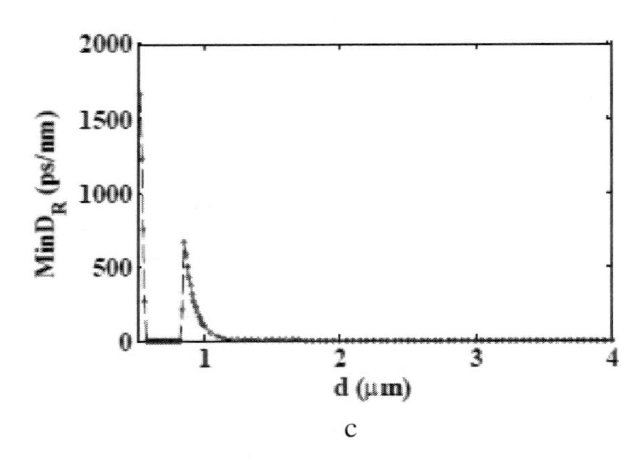

Figure 7. Comparing among Max D_R, D_1 and Min D_R as functions of d.

Figure 5 shows the GVD curve at $\lambda_0 = 1.55$ μm as a function of the diameter of an MF made out of fused silica. From the Figure, the GVD is sensitive to the MF diameter. It could be positive, negative or even around zero with a proper diameter. When the diameter is less than 1 μm, the GVD value changes remarkably while the flat GVD region corresponds to larger diameters. These interesting features reflect the competition of dispersions from different sources, which may have some applications. For example, a zero-dispersion is favorable for high-speed optical systems while a negative dispersion might be useful for dispersion compensating [52].

In most applications, we only need consider that d is around 1μm and Λ/d is around 2, which are the reasonable parameters for practical MCRs with current technique. Assuming $L = 1$ mm. For a given diameter d, K decreases with the pitch, the maximum is as large as 300 at d = 750 nm and $\Lambda = 2d$, and as small as nearly zero when the pitch is large enough. The dispersion of K, $dK/d\lambda$, has a wide range, from 0 to 400/μm [52].

Figure 6 shows the total dispersion D_R as a function of d and Λ/d. Here we only consider a high Q-factor resonator assuming K is nearly at resonating: sin(K) is about sin(1.1π/2) and the Q-factor $\sim \beta s/(K-K_m)^2$ is about 10^5. It can achieve as high as -1×10^4ps/nm and 2×10^4ps/nm and as low as 0. Because both β'' and β' depend on d only while k' depend on d and Λ, we may compare the competition between waveguide and coupling dispersions by comparing the maximum ($MaxD_R$) and minimum ($MinD_R$) at the same diameter. Figure 7 shows the curves of $MaxD_R$, D_1 (independent of the coupling dispersion) and $MinD_R$ at different d, the coupling dispersion D_2 at strong coupling is dominated because $D_1 \ll MaxD_R$. The advantage of a strong coupling dispersion is that with the coupling dispersion it is possible to cancel the wave guide dispersion: $MinD_R$=0, at $d\in[\sim0.6$ μm, ~0.8 μm] or [$\sim$1.5 μm, 4 μm]. When $d\in[\sim0.8$ μm, ~1.5 μm], $D_1 \ll MinD_R$, the coupling dispersion is totally dominated [52].

As we can see, the total dispersion of an MCR is dominated by its physical parameters such as MF diameter and ring pitch. Both positive and negative dispersion could be achieved by selecting suitable MCR parameters. There could be a possible way to engineer tailored dispersion curves by means of cascading MCRs. As the coupling strength between MCR turns is sensitive to the environmental index, even tunable dispersion might be obtained [52].

The strong dispersion causes serious effects on MCR performance, especially bit rate B (when D_R is zero, the third-order has to be considered) [52, 53]

$$B_{max} \approx \frac{1}{4}(\sqrt{D_R \frac{\lambda_0^2}{2\pi c}})^{-1} \tag{1.3.4}$$

The maximum bit rate is very poor (~1.6 Gb/s) when D_R~2x10^4ps/nm.

4. STRONG COUPLING INFLUENCE IN MCRS

For most MCRs, the light propagating in the MF generates a large evanescent field, which interacts with the surrounding medium (analytes).

It can work as a refractive index sensor by monitoring the resonation wavelength shifting with the analytes. The coupling strength between two turns of the microcoil depends on the MF diameter, the distance between the two coils (pitch), and on the choice of the surrounding material, which modifies the evanescent field decay length. The cross-section geometry and the effective index of MF would greatly influence the resonant wavelength and the sensitivity of the MCR. A smaller cross-section is preferred in the MCR to provide a large evanescent field. However, the coupling effect between the two segments in the coupling region can be very strong [52, 54] especially when the diameter is small and the coupling length is relatively long. In previous work, the resonant wavelength and sensitivity are always considered to only depend on the effective index of the solo MF and be unrelated to the coupling.

However, the coupling in the coupling region can be very strong, taking the benefit from the large evanescent fields, sufficient inter-waveguide coupling and relatively long coupling length. With the strong coupling effect it is possible to greatly influence the resonance condition and sensitivity; this should not be ignored. In this section, we investigate the resonance condition and sensitivity in the MCR by considering the strong coupling effect and our simulation shows that the coupling effect has a great influence on the resonant peak position and the sensitivity of the MFR [25].

In this section, for simplification, we mainly consider the MLR/MKR and two-turn MCR with simple two-wave coupling equations, which are very similar and coincide with the model as shown in Figure 8(a). Figure 8(a) shows a schematic of a typical MCR, which usually contains a coupling region where two pieces of MF are close to each other. The pitch between the two pieces of MF is fixed and it is not possible to change it into a loop or knot resonator. However, it can easily be tuned in an MCR.

We assume L to be the loop length and H to be the coupling length. The ratio $\eta = H/L$ (0 ~ 1) can be as large as 1 in a two-turn all-coupling MCR, which is shown in Figure 8(b). The cross-section of the MFR is also shown in Figure 8(c). We set P (= d in a loop/knot resonator, $\geqslant d$ in an MCR) to be the pitch between the two segments in the coupling region, n_f and n_c to be the refractive indices of the MF and the environment (or analyte), respectively. And if we denote the angular frequency as ω, the eigen modes in each segment before mode coupling as E_p and H_p (p = 1, 2), the refractive index distribution of the entire coupled segments and each segment as N and N_p, respectively, the electromagnetic field amplitudes A_1 and A_2 in the two MF segments of the coupling region are related by the coupled wave equations [25, 51]:

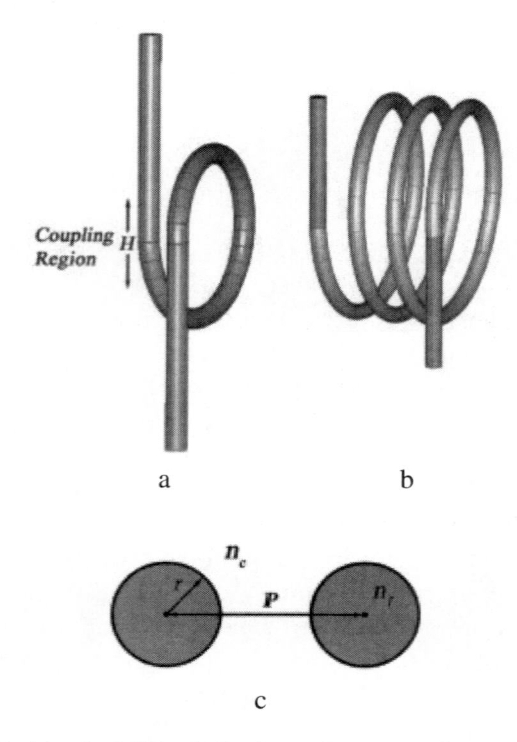

Figure 8. (a) Basic configuration of a MLR; (b) Basic configuration of a two-turn all-coupling 3D MCR; (c) the cross section of the coupling region.

$$\begin{cases} \dfrac{dA_1}{dz} + \sigma_{12}\dfrac{dA_2}{dz} - j\rho_1 A_1 - j\kappa_{12} A_2 = 0 \\ \dfrac{dA_2}{dz} + \sigma_{21}\dfrac{dA_1}{dz} - j\rho_2 A_2 - j\kappa_{21} A_1 = 0 \end{cases} \tag{1.4.1}$$

where

$$\kappa_{pq} = \frac{\omega\varepsilon_0 \int_{-\infty}^{+\infty}\int_{-\infty}^{+\infty} (N^2 - N_q^2) E_p^* \cdot E_q dxdy}{\int_{-\infty}^{+\infty}\int_{-\infty}^{+\infty} u_z \cdot (E_p^* \times H_p + E_p \times H_p^*) dxdy}$$

$$\sigma_{pq} = \frac{\int_{-\infty}^{+\infty}\int_{-\infty}^{+\infty} u_z \cdot (E_p^* \times H_q + E_q \times H_p^*) dxdy}{\int_{-\infty}^{+\infty}\int_{-\infty}^{+\infty} u_z \cdot (E_p^* \times H_p + E_p \times H_p^*) dxdy}$$

$$\rho_p = \frac{\omega\varepsilon_0 \int_{-\infty}^{+\infty}\int_{-\infty}^{+\infty}(N^2 - N_p^2)E_p^* \cdot E_p dxdy}{\int_{-\infty}^{+\infty}\int_{-\infty}^{+\infty} u_z \cdot (E_p^* \times H_p + E_p \times H_p^*)dxdy}$$

We assumed the two segments to be identical and we have$\kappa_{12} = \kappa_{21} = \kappa$, $\sigma_{12} = \sigma_{21} = \sigma$ and $\rho_1 = \rho_2 = \rho$.

And the pair of p and q are either $(p, q) = (1, 2)$ or $(2, 1)$. k is the coupling coefficient of the resonator, respectively.

In the previous work, the coupled mode equation is often simplified as

$$\begin{cases} \frac{dA_1}{dz} - jk_{12}A_2 = 0 \\ \frac{dA_2}{dz} - jk_{21}A_1 = 0 \end{cases} \tag{1.4.2}$$

because σ and ρ re assumed to be zero. By solving the equation (1.4.2) a traditional resonant condition of $\beta L=(2m+1/2)\pi$ can be obtained.

The resonant wavelength λ_R only depends on the effective index of the waveguide but is unrelated to the coupling effect.

However, if diameter d of the MF and pitch P between the two segments is small enough, there will be a large evanescent field and a strong coupling effect. In order to investigate the resonator strictly, we take σ and ρ into consideration and by solving Eq. (1.4.1) we find the modified resonant condition as:

$$\beta L + \Omega H = (2m + 1/2)\pi \quad m = 1, 2, 3...$$

where $\Omega = -\frac{\rho - k\sigma}{\sigma^2 - 1}$ and $N = -\frac{k - \sigma\rho}{\sigma^2 - 1}$.

Compared to the previous one, the coupling effect is included in the resonance condition with an added modified item ΩT. So next we calculate the values of Ω to see how important the coupling effect is.

In Figure 9 we assume n_c= 1.34 in water and a full vector finite element method is used to calculate the dependence of the k, σ and ρ profiles on the fiber diameter d and the pitch P, respectively. According to these simulation values of k, σ and ρ, we calculate the value of Ω in Figure 9(d).

In previous work σ and ρ are assumed to be zero and that will lead to a condition of $\Omega = 0$. However, according to our simulations results from Figure 9(d), the value of Ω can be as large as -15000 m^{-1} at d =1.28 μm and P = 1.2d and it can be comparable to the value of k at the same parameter.

Thus M cannot be ignored, especially when the pitch P is relatively small.

Because of the modified item *ΩT* between our theory and the tradition alone, the resonant wavelength is modified by $\Delta\lambda_R = \lambda_R(\Omega/\beta)\cdot\eta$. Here $\eta = H/L$ ($0 < \alpha \leqslant 1$) is the ratio of the coupling length to the ring length. In an all-coupling MCR, $\eta = 1$.

In Figure 10 we plot the values of $\Delta\lambda_R$ at different *P* and *d*. According to our simulation results, the resonant wavelength difference between our theory and the previous one can be up to -3.89 nm at $d = 1.28$ μm, $P = 1.2d$ and $\lambda \approx 1550$ nm.

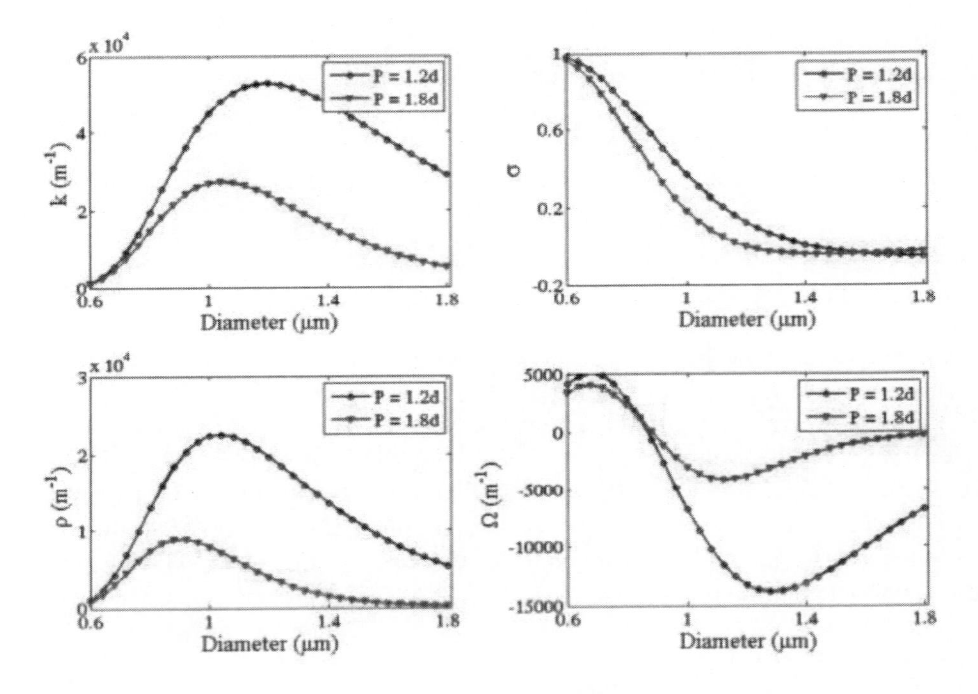

Figure 9. The calculated (a) *k*, (b) *σ*, (c) *ρ* and (d) *Ω* profile of a resonator with different diameter *d* and different pitch *P*.

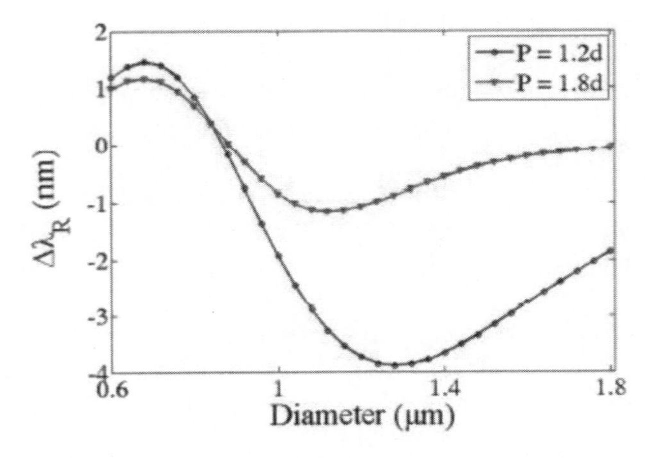

Figure 10. The calculated $\Delta\lambda_R$ profile of an MFR with different diameter *d* and different pitch *P*.

As the typical free spectra range is about several nanometers, the resonant wavelength shift caused by the coupling effect cannot be ignored.

Furthermore, it also shows a possible method to change the resonant wavelength by the control of pitch P and the diameter d.

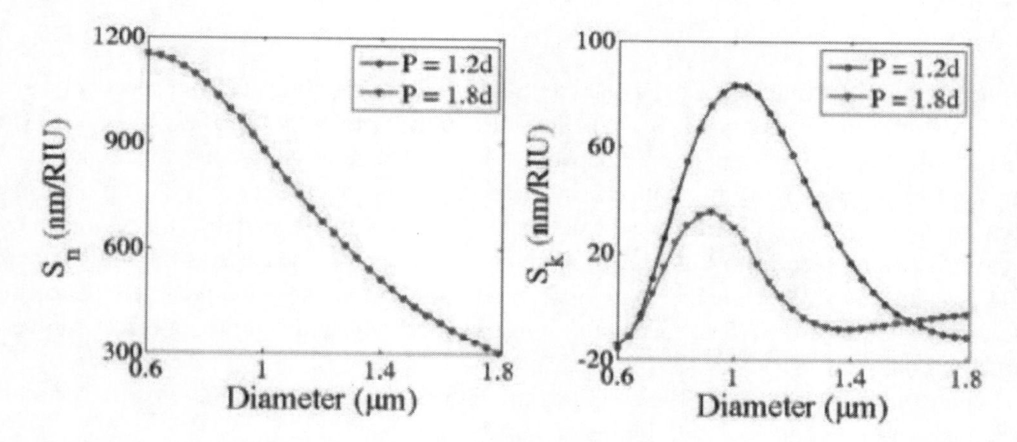

Figure 11. The calculated (a) S_n and (b) S_k profile of an MFR with different Diameter d and different pitch P.

In particularly, the pitch between adjacent turns in an MCR possibly changes with external vibration or pressure. It means that we can use the MCR as a vibration or pressure sensor and it will also give us an alternative external tuning technique by controlling the pitch and coupling strength.

MFRs have been widely used as refractometric sensors because of the large evanescent field. The sensitivity of an MFR is usually defined by the shift of the resonant wavelength as a function of the index change of the surrounding medium.

$$S_{RI} = \frac{\partial \lambda_R}{\partial n_c}$$

where λ_R is the resonant wavelength and the n_c is the refractive index of the surrounding medium. By taking the resonant condition $\beta L + \Omega H = (2m + 1/2)\pi$ into the equation (7) we obtain

$$S_{RI} = \frac{\partial \lambda_R}{\partial n_{sur}} = S_n + S_k \tag{1.4.3}$$

where

$$\begin{cases} S_n = \dfrac{2\pi}{\beta}\dfrac{\partial n_{eff}}{\partial n_c} \\ S_k = \eta\dfrac{\lambda}{\beta}\dfrac{\partial \Omega}{\partial n_c} \end{cases}$$

S_n and S_k is the contribution of the evanescent field and the coupling effect, respectively.

In the traditional theory of previous literature, only S_n is considered. It can be seen from the formula that the values of S_n only depended on the change of the propagating constant. We can see from Figure11 (a) that with the increase of the diameter of the MF, S_n drops. This is because the intensity of the evanescent field and the coupling between the two segments both decay. As can be seen from Figure 4(b), the values of S_k can be as large as 83 nm/RIU at d = 1.04 μm and P = 1.2d, while the value of S_n is 836 nm/RIU as the same parameter, according to our simulation results. That means the sensitivity caused by the coupling effect can be up to 10% of the total sensitivity [25].

Such a high S_k means that the coupling strength greatly influences the sensitivity and has to be taken into account when designing and analyzing an MFR [25].

Moreover, for simplification, we mainly consider the loop/knot MFR and two-turn MCR with a simple two-wave coupling equations, which are very similar and coincide with the model as shown in Figure 8(a). For a multi-turn MCR, it is more complicated because of the multi-resonant conditions and cross-coupling between different turns. However our results and conclusion are still applicable [25].

Finally, although we don't discuss the Q-factor, which is related to the detection limit, the coupling also has a great contribution on the Q-factor [25].

5. Fabrication and Application

The MF can be fabricated using a microheater. The 3D-MCR then can be manufactured with the set up shown in Figure 12. The MF had its pigtails connected to an EDFA and an OSA to check, in real time, the resonator properties during fabrication; then, with the aid of a microscope, the MF was wrapped on a low refractive index rod while one of its ends was fixed on a 3D stage; this process was carried out manually and the close positioning of the MF coils resulted from a combination of manually applied longitudinal tension (which kept the relative position of the coils and avoided considerable overlapping) and gravity (which translated vertically the new formed coil until it touched the coil beneath). Finally, the other microfiber end was fixed to another 3D stage and both microfiber ends were moved little by little to find the optimum resonator spectrum. This methodology is similar to that theoretically predicted for the design optimization of 3D microcoil resonators presented in Section 7.5. Because the coupling coefficient between the two adjacent microfibers is small, the microfibers need to be kept as close as possible.

In these experiments, the MCR was wrapped on a rod to maximize the MCR temporal stability and robustness. For microfibers with large evanescent fields or rods with high indices, it is possible that there are additional losses arising from the leakage loss because the fundamental mode becomes a leaky mode. If the loss is too high, it is difficult to optimize the microfiber position and the MCR geometry.

Generally, the loss can be minimized by increasing the microfiber thickness and the rod diameter, by using a low refractive index material for the rod and by improving the smoothness of the rod surface. In these experiments a rod was coated with Teflon @AF (DuPont, United States), to provide a low refractive index (n~1.3 at λ~1.55 μm) at the interface with the microfiber. The coating thickness is about tens of micrometers.

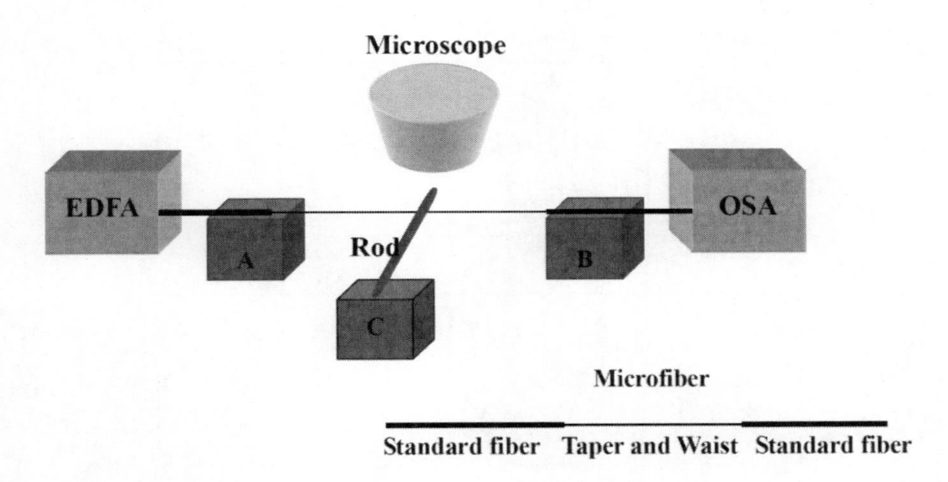

Figure 12. Set up used to manufacture 3D MCR. A, B and C are XYZ stages.

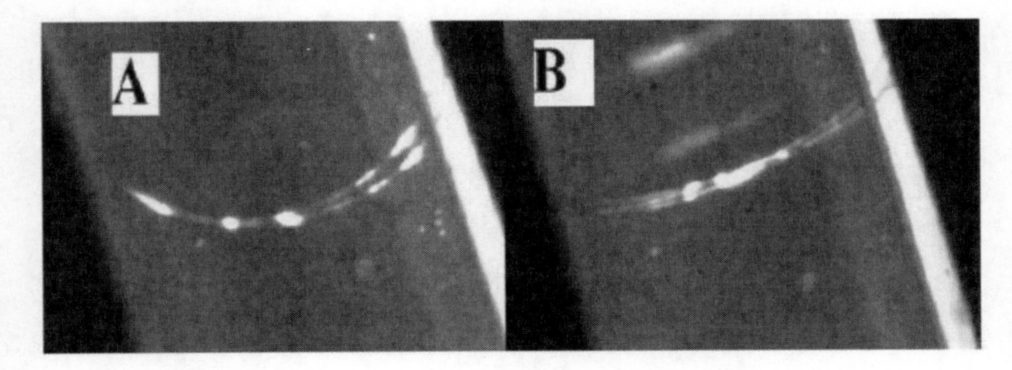

Figure 13. Pictures of MCRs. The number of turns in the MCR is two in (A) and three in (B).

Figure 13 shows the pictures of a two-turn and three-turn MCR made from the same MF. The MF radius and the length of the uniform waist region were ~1.5 μm and ~ 4 mm respectively. The diameter of the rod is ~ 560 μm. The manual fabrication of the MCR means that the pitches between adjacent turns are non-uniform and the MF coils present some degree of twist.

Figure 14 shows the resonator spectra of a straight MF not in contact with the rod, and of the three MCRs. The spectra in Figure 13A-B show a complicated profile because the coupling among the three or four turns is irregular and non-uniform. The maximum extinction ratios for the two-and three-turn MCR are 3 dB and 10 dB, respectively. While the spectrum of the two-turn MCR is simple, with a free spectral range (FSR) of 0.86 nm, the spectra of three-turn MCR show a complex profile. In particular, the spectrum of the three-turn MCR can be analyzed as a combination of two resonator modes, with the same FSR (about 0.94 nm).

The loss induced by wrapping the MF on the rod for the two-, three- and four-turn MCR was ~1 dB and 2.5 dB, respectively, which comes from the output difference of Figures (A)-(B) and Ref in Figure 14. The loss is possibly induced by surface roughness, mode discontinuities at the point of input/output, and leaky modes associated with the rod.

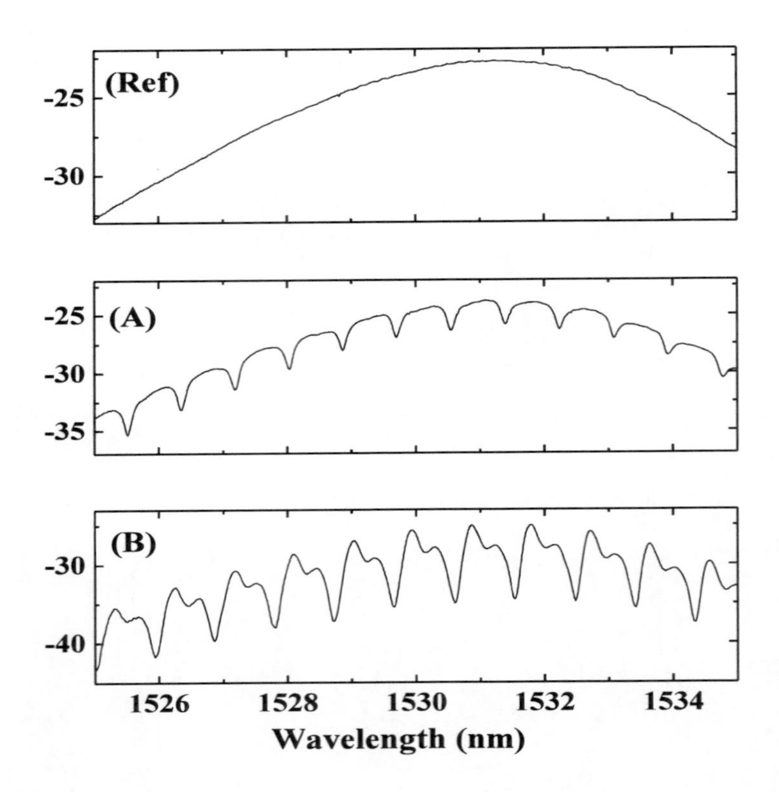

Figure 14. Spectra of MCR with (A) two turns (B) three turns. The top spectrum represents the reference obtained for a free-standing MF.

The largest Q-factor obtained is about 10,000, which is lower than that achieved with an MCR immersed in a liquid [48]; this can be possibly explained by the small coupling strength associated to the short coupling length (here we define the coupling length as the length where two adjacent turns are touching) and thick MF or by the additional loss at the MF/rod interface.

For practical reasons it is impossible to see the entire circumference of the MF coil, thus there are regions where the coils might overlap or their pitch increase. It is possible to tune the coupling by moving the pigtails position: slightly moving the two ends of the fiber up or down only changes the coupling area near the input and output ends without changing the coupling between inner turns.

Although the structure of an MCR wrapped on a rod is stable in air, sub-micrometric wires experience ageing when exposed to air for some days [49]; moreover, the MCR is not portable and can be easily damaged. For example, a demonstration of an MCR was recently reported for an MF immersed in a refractive index matching liquid [48]; still, the extinction ratio was smaller than 1.5dB, and the device was not portable. The embedding of an MCR in a low refractive index medium seems to be the best method to solve the reliability issues as it provides both protection from fast aging and geometrical and optical stability. In Ref [31], Xu demonstrated an embedded MCR with low-index Teflon. The MCR was manufactured by wrapping an MF on a rod having a low refractive index. The MF radius and the length of the uniform waist region were ~1.5 μm and 3.5 mm respectively.

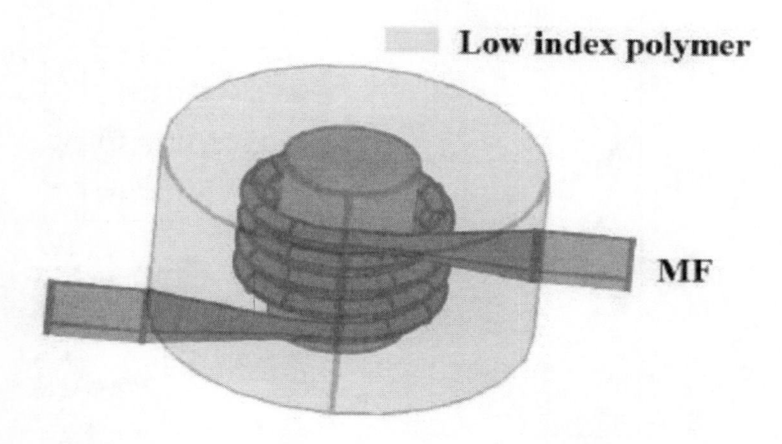

Figure 15. Schematic of the CMCRS.

The rod has a total diameter D~700 μm, with a silica core radius ~200μm. Then with the aid of a microscope, the MF was wrapped on the rod, finally the Teflon suspension was dropped onto the MF and the solvent was evaporated. Coating with Teflon provides a smooth surface and a homogeneous medium. Although some sporadic bubbles are observed in the polymer, their effect on the MCR is negligible because none of them is in proximity of the MF.

Coated MCR wrapped on a rod also makes the fabrication of high sensitivity biosensors possible. In 2007 and 2008, a coated MF coil resonator sensor (CMCRS) was proposed and demonstrated. It can be fabricated as follows: firstly, an expandable rod is initially coated with a layer of thickness t of a low-loss polymer such as Teflon, the rod is made from disposable materials such as PMMA (PolyMethylMethAcrylate), which is a polymer with an amorphous structure and is soluble in Acetone; then an optical MF is wrapped on the rod; next, the whole structure is coated with the same low-loss polymer, and finally the rod is removed. Figure 15 shows the final structure after the rod is removed: the MF is shown in blue, the analyte channel (in the space previously belonging to the support rod) in brown and Teflon in light green. It is a compact and robust device with an intrinsic fluidic channel to deliver samples to the sensor, unlike most ring or microsphere resonators, which require an additional channel. The embedded MF has a considerable fraction of its mode propagating in the fluidic channel, thus any change in the analyte properties is reflected in a change of the mode properties at the CMCRS output. Since the CMCRS is fabricated from a single tapered optical fiber, light can be coupled into the sensor with essentially no insertion loss: a huge advantage over other types of resonator sensors. Because of the interface with the analyte, the mode propagating in the coated MF experiences a refractive index surrounding similar to that of a conventional D-shaped fiber. The mode properties are particularly affected by two important parameters: the MF radius and the coating thickness t between the MF and the fluidic channel. We evaluate the effective index n_{eff} of the fundamental mode propagating in the MF by a finite element method.

Generally, n_{eff} increases with the index of analyte and increases more quickly with a smaller t since in this case a larger fraction of the mode is propagating in the analyte. The homogeneous sensitivity Scan be obtained by monitoring the shift of the resonant wavelength λ_0 corresponding to one of the propagation constants β.

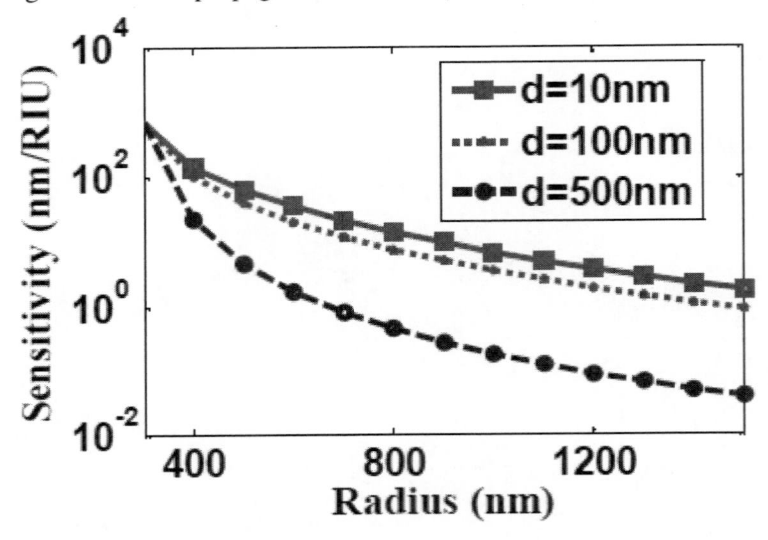

Figure 16. Sensitivity of the CMCRS versus MF radius for λ=970 nm and for different values of t.

Because water is the solvent for most analytes and the absorption of water at long wavelengths is high, we calculate the sensitivity near n=1.332 at short wavelengths (600 nm and 970 nm). Figure16 shows the dependence of S on the MF radius r for different t. S increases when d decreases or λ increases. Decreasing the MF radius r also increases sensitivity because this increases the fraction of the mode field inside the fluidic channel. S reaches 700 nm/RIU at r$\approx$300 nm for λ=970 nm.

CMCRS was first demonstrated in 2008 [55] from an MF with the length and diameter of the uniform waist region of 50 mm and ~2.5 μm respectively. The MF was wrapped on a 1mm-diameter PMMA rod and then repeatedly coated with the Teflon solution 601S1-100-6. The dried embedded microcoil resonator was then left in Acetone to dissolve the support rod. The whole PMMA rod was completely dissolved in 1-2 days at room temperature. At last the CMCRS sensor with a ~1mm-diameter microchannel and two input/output pigtails was obtained. Figure 17 shows a microscope picture of the manufactured sensor.

The sensor was connected to an Erbium-doped fiber amplifier and an optical spectrum analyzer and then immersed into the mixtures; its sensor sensitivity S was measured from the spectral shift observed inserting the sensor into a beaker containing different mixtures of Isopropyl and Methanol. In the seven mixtures, the Isopropyl fraction was gradually increased from 60% to 67.6% by adding small calibrated Isopropyl quantities to the Isopropyl/Methanol solution in a position far from the sensor. The refractive indexes of Isopropyl and Methanol at 1.5 μm were taken as 1.364 and 1.317 respectively [56] S was ~40 nm/RIU. This value is comparable with those reported previously for other resonating sensors like microsphere and microring [39, 57, 58].

The relatively low value for the sensitivity can be attributed to the small overlap between the mode propagating in the MF resonator and the analyte.

Another factor, which has probably affected the sensor sensitivity, is the lack of smoothness of the device surface in contact with the analyte; this is possibly caused by PMMA residues on the surface of the channel or it may originate from the original roughness of the PMMA support rod.

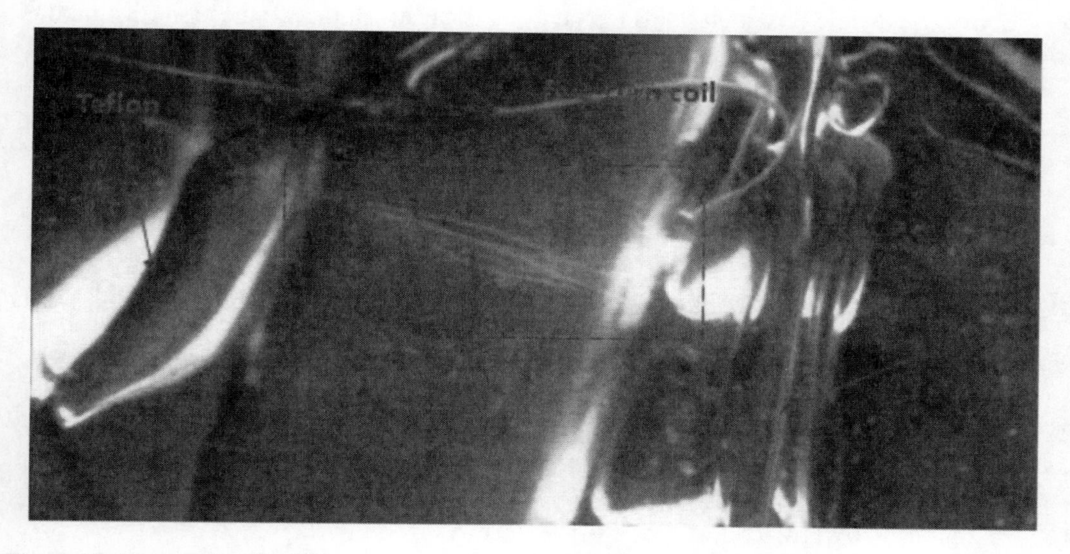

Figure 17. Picture of a CMCRS.

This roughness produced the moderately low Q-factor observed in the resonator ($Q{\sim}10^4$), which limited the interaction length between the mode and the analyte. It is believed that the overall sensitivity can be considerably improved to ~10^4 by using thinner MFs and by fabricating microcoil resonators with higher Q-factors. The minimum detectable refractive index change for this sensor is limited by the accuracy in the measurement of the peak wavelength (15pm). Assuming S=40 nm/RIU, the minimum detectable refractive index change results to be $\Delta n_{min}{\sim}0.015/40{\sim}4{\cdot}10^{-4}$. This value can be decreased with the use of better detection systems and more stable sources.

CONCLUSION

In this chapter, we presented the mathematical model, induced influence by strong dispersion and coupling, fabrication and applications of MCRs. Based on the full-coupled wave equations; we studied characteristics with strong coupling and waveguide dispersions. The dispersion-induced limitation on the performance of MCRs was discussed. We modified the previous theory on resonance condition and sensitivity of an MCR. According to our simulation results, the resonant wavelength and sensitivity strongly depend on the coupling effect, which has been disregarded previously. These results are very helpful for future device design and system applications. Finally, we introduced the manufacture method of MCRs and the application in refractive index sensing.

Acknowledgments

F. Xu and Y.-q Lu acknowledge the support from National 973 program under contract No. 2011CBA00200 and 2012CB921803, NSFC program No. 11074117 and 60977039, and the Priority Academic Program Development of Jiangsu Higher Education Institutions (PAPD).

References

[1] C. Caspar and E. J. Bachus, "Fibre-optic micro-ring-resonator with 2 mm diameter," *Electron. Lett.* 25, 1506-1508 (1989).

[2] G. Brambilla, V. Finazzi and D. J. Richardson, "Ultra-low-loss optical fiber nanotapers," *Opt. Express* 12, 2258-2263 (2004).

[3] G. Vienne, L. Yuhang and T. Limin, "Effect of Host Polymer on Microfiber Resonator," *IEEE Photonic. Technol. Lett.* 19, 1386-1388 (2007).

[4] G. Brambilla, F. Xu, P. Horak, Y. Jung, F. Koizumi, N. P. Sessions, E. Koukharenko, X. Feng, G. S. Murugan, J. S. Wilkinson, and D. J. Richardson, "Optical fiber nanowires and microwires: fabrication and applications," *Advances in Optics and Photonics* 1, 107-161 (2009).

[5] S. G. Leon-Saval, T. A. Birks, W. J. Wadsworth, P. S. J. Russell, and M. W. Mason, "Supercontinuum generation in submicron fibre waveguides," *Opt. Express* 12, 2864-2869 (2004).

[6] G. Brambilla and F. Xu, "Adiabatic submicrometric tapers for optical tweezers," *Electron. Lett.* 43, 204-206 (2007).

[7] T. Lee, N. G. R. Broderick and G. Brambilla, "Berry phase magnification in optical microcoil resonators," *Opt. Lett.* 36, 2839-2841 (2011).

[8] F. Xu, P. Horak and G. Brambilla, "Conical and biconical ultra-high-Q optical-fiber nanowire microcoil resonator," *Appl. Opt.* 46, 570-573 (2007).

[9] J. Yongmin, G. S. Murugan, G. Brambilla, and D. J. Richardson, "Embedded Optical Microfiber Coil Resonator With Enhanced High Q," *IEEE Photonic. Technol. Lett.* 22, 1638-1640 (2010).

[10] G. Brambilla, F. Xu and X. Feng, "Fabrication of optical fibre nanowires and their optical and mechanical characterisation," *Electron. Lett.* 42, 517-519 (2006).

[11] F. Xu, V. Pruneri, V. Finazzi, G. Brambilla, and IEEE, "High sensitivity refractometric sensor based on embedded optical microfiber loop resonator," *2008 Conference on Lasers and Electro-Optics and Quantum Electronics and Laser Science Conference*, Vols. 1-9, 564-565 (2008).

[12] F. Xu, G. Brambilla, J. Feng, and Y. Lu, "Mathematical model for manufacturing microfiber coil resonators," *Optical Engineering* 49(2010).
[13] F. Xu, P. Horak and G. Brambilla, "Optical microfiber coil resonator refractometric sensor: erratum," *Opt. Express* 15, 9385-9385 (2007).
[14] G. Brambilla and D. N. Payne, "The Ultimate Strength of Glass Silica Nanowires," *Nano Letters* 9, 831-835 (2009).
[15] X. S. Jiang, L. M. Tong, G. Vienne, X. Guo, A. Tsao, Q. Yang, and D. R. Yang, "Demonstration of optical microfiber knot resonators," *Appl. Phys. Lett.* 88, 223501 (2006).
[16] X. Guo and L. M. Tong, "Supported microfiber loops for optical sensing," *Opt. Express* 16, 14429-14434 (2008).
[17] P. Pal and W. H. Knox, "Fabrication and Characterization of Fused Microfiber Resonators," *IEEE Photonic. Technol. Lett.* 21, 766-768 (2009).
[18] M. Sumetsky, Y. Dulashko, J. M. Fini, A. Hale, and D. J. DiGiovanni, "The microfiber loop resonator: Theory, experiment, and application," *Journal of Lightwave Technology* 24, 242-250 (2006).
[19] T. Wang, X. H. Li, F. F. Liu, W. H. Long, Z. Y. Zhang, L. M. Tong, and Y. K. Su, "Enhanced fast light in microfiber ring resonator with a Sagnac loop reflector," *Opt. Express* 18, 16156-16161 (2010).
[20] J. Scheuer, "Fiber microcoil optical gyroscope," *Opt. Lett.* 34, 1630-1632 (2009).
[21] J.-l. Kou, F. Xu and Y.-q. Lu, "Highly Birefringent Slot-Microfiber," *IEEE Photon. Technol. Lett.* 23, 1034-1036 (2011).
[22] F. Xu, G. Brambilla, I. Chremmos, O. Schwelb, and N. Uzunoglu, "Microfiber and Microcoil Resonators and Resonant Sensors," *Photonic Microresonator Research and Applications* 156, 275-298 (2010).
[23] F. Xu, G. Brambilla, J. Feng, Y. Lu, and IEEE, "A Microfiber Bragg Grating based on a Microstructured Rod," *2009 Ieee Leos Annual Meeting Conference Proceedings*, Vols. 1and 2, 336-337 (2009).
[24] Jun-long Kou, Ming Ding, Jing Feng, Yan-qing Lu, Fei Xu, and G. Brambilla, "Microfiber-Based Bragg Gratings for Sensing Applications: A Review," *Sensors* (2012).
[25] W. Guo, Y. Chen, F. Xu, and Y.-q. Lu, "Modeling of the influence of coupling in optical microfiber resonators," *Opt. Express* 20, 14392-14399 (2012).
[26] F. Xu, P. Horak and G. Brambilla, "Optical microfiber coil resonator refractometric sensor," *Opt. Express* 15, 7888-7893 (2007).
[27] F. Xu, P. Horak and G. Brambilla, "Optical microfiber coil resonator refractometric sensor (vol 15, pg 7888, 2007)," *Opt. Express* 15, 9385-9385 (2007).
[28] M. Ding, M. Belal, G. Chen, R. Al-Azawi, T. Lee, Y. Jung, P. Wang, X. Zhang, Z. Song, F. Xu, R. Lorenzi, T. Newson, G. Brambilla, and B. Pal, "Optical microfiber devices and sensors," *Passive Components and Fiber-Based Devices* Viii 8307(2011).
[29] N. G. R. Broderick and T. T. Ng, "Theoretical Study of Noise Reduction of NRZ Signals Using Nonlinear Broken Microcoil Resonators," *IEEE Photon. Technol. Lett.* 21, 444-446 (2009).
[30] F. Xu, V. Pruneri, V. Finazzi, and G. Brambilla, "An embedded optical nanowire loop resonator refractometric sensor," *Opt. Express* 16, 1062-1067 (2008).

[31] F. Xu and G. Brambilla, "Embedding optical microfiber coil resonators in Teflon," *Opt. Lett.* 32, 2164-2166 (2007).

[32] S. S. Pal, S. K. Mondal, U. Tiwari, P. V. G. Swamy, M. Kumar, N. Singh, P. P. Bajpai, and P. Kapur, "Etched multimode microfiber knot-type loop interferometer refractive index sensor," *Rev. Sci. Instrum.* 82(2011).

[33] M. Sumetsky, Y. Dulashko and A. Hale, "Fabrication and study of bent and coiled free silica nanowires: Self-coupling microloop optical interferometer," *Opt. Express* 12, 3521-3531 (2004).

[34] F. Xu and G. Brambilla, "Manufacture of 3-D microfiber coil resonators," *IEEE Photon. Technol. Lett.* 19, 1481-1483 (2007).

[35] W. Guo, Y. Chen, F. Xu, and Y.-q. Lu, "Modeling of the influence of coupling in optical microfiber resonators," *Opt. Express* 20, 14392-14399 (2012).

[36] G. Brambilla, "Optical fibre nanowires and microwires: a review," *J. Opt.* 12, 043001 (2010).

[37] M. Sumetsky, Y. Dulashko, J. M. Fini, and A. Hale, "Optical microfiber loop resonator," In:*Lasers and Electro-Optics, 2005. (CLEO). Conference on*, 2005), 432-433 Vol. 431.

[38] F. Xu, P. Horak and G. Brambilla, "Optimized design of microcoil resonators," *Journal of Lightwave Technology* 25, 1561-1567 (2007).

[39] N. M. Hanumegowda, C. J. Stica, B. C. Patel, I. White, and X. D. Fan, "Refractometric sensors based on microsphere resonators," *Appl. Phys. Lett.* 87, 201107 (2005).

[40] Y. Chen, F. Xu and Y.-q. Lu, "Teflon-coated microfiber resonator with weak temperature dependence," *Opt. Express* 19, 22923-22928 (2011).

[41] S. W. Harun, K. S. Lim, S. S. A. Damanhuri, A. A. Jasim, C. K. Tio, and H. Ahmad, "Current sensor based on microfiber knot resonator," *Sensor Actuat a-Phys* 167, 60-62 (2011).

[42] X. Zeng, Y. Wu, C. L. Hou, J. Bai, and G. G. Yang, "A temperature sensor based on optical microfiber knot resonator," *Opt. Commun.* 282, 3817-3819 (2009).

[43] G. Brambilla, "Optical fibre nanotaper sensors," *Optical Fiber Technology* 16, 331-342 (2010).

[44] L. F. Stokes, M. Chodorow and H. J. Shaw, "All-Single-Mode Fiber Resonator," *Opt. Lett.* 7, 288-290 (1982).

[45] C. Caspar and E. J. Bachus, " Fibre-optic microring-resonator with 2mm diameter," *Electronnics Letters* 25, 1506–1508 (1989).

[46] X. Guo and L. Tong, "Supported microfiber loops for optical sensing," *Opt. Express* 16, 14429-14434 (2008).

[47] M. Sumetsky, "Optical fiber microcoil resonator," *Opt. Express* 12, 2303-2316 (2004).

[48] M. Sumetsky, "Demonstration of a multi-turn microfiber coil resonator," In: *Optical Fiber Communication Conference,*(San Diego, US, 2007).

[49] F. Xu and G. Brambilla, "Preservation of Micro-Optical Fibers by Embedding," *Japanese Journal of Applied Physics* 47, 6675-6677 (2008).

[50] F. Xu and G. Brambilla, "Demonstration of a refractometric sensor based on optical microfiber coil resonator," *Appl. Phys. Lett.* 92(2008).

[51] K. Okamoto, *Fundamentals of optical waveguides,* 2nd ed. (Elsevier, Amsterdam; Boston, 2006), pp. xvi, 561 p.

[52] F. Xu, Q. Wang, J.-F. Zhou, W. Hu, and Y.-Q. Lu, "Dispersion Study of Optical Nanowire Microcoil Resonators," *Ieee Journal of Selected Topics in Quantum Electronics* 17, 1102-1106 (2011).

[53] G. P. Agrawal, *Fiber-optic communication systems,* 3rd ed., Wiley series in microwave and optical engineering (Wiley-Interscience, New York, 2002), pp. xvii, 546 p.

[54] K. Huang, S. Yang and L. Tong, "Modeling of evanescent coupling between two parallel optical nanowires," *Appl. Opt.* 46, 1429-1434 (2007).

[55] F. Xu and G. Brambilla, "Demonstration of a refractometric sensor based on optical microfiber coil resonator," *Applied Physics Letters* 92, 101126 (2008).

[56] C. B. Kim and C. B. Su, "Measurement of the refractive index of liquids at 1.3 and 1.5 micron using a fibre optic Fresnel ratio meter," *Measurement Science and Technology* 15, 1683-1686 (2004).

[57] M. Adams, G. A. DeRose, M. Loncar, and A. Scherer, "Lithographically fabricated optical cavities for refractive index sensing," *Journal of Vacuum Science and Technology B* 23, 3168-3173 (2005).

[58] I. M. White, H. Oveys, X. Fan, T. L. Smith, and J. Y. Zhang, "Integrated multiplexed biosensors based on liquid core optical ring resonators and antiresonant reflecting optical waveguides," *Appl. Phys. Lett.* 89, 191106 (2006).

In: Optical Fibers: New Developments
Editor: Marco Pisco
ISBN: 978-1-62808-425-2

Chapter 2

GRADED-INDEX PLASTIC OPTICAL FIBERS: MATERIALS AND FABRICATION TECHNIQUES

***Kotaro Koike*[1,2*]*, Yasuhiro Koike*[1] *and Yoshiyuki Okamoto*[2]**
[1]Keio Photonics Research Institute, Keio University
[2]Polymer Research Institute, Polytechnic Institute of New York University

ABSTRACT

Graded-index plastic optical fibers (GI POFs) are a highly attractive transmission medium for short-range communications such as local area networks and interconnections. In addition to the extremely high bit-rates, which can be more than 100 Gbps over 100 m, these fibers offer many advantages derived from the nature of the polymeric materials: a greater flexibility, easier handling, and simpler installation. GI POFs are no longer just alternatives to other transmission media; they are becoming an indispensable option for many applications. This chapter reviews the development history and latest advances in the performance of GI POFs from the perspective of the materials and fabrication techniques that have supported this progress.

Keywords: Graded-index plastic optical fiber, preform-drawing, extrusion, poly(methyl methacrylate), perfluorinated polymer

1. INTRODUCTION

"Fiber-to-the-home" services that interconnect homes with a glass optical fiber (GOF) backbone are well established in advanced countries but internal building networks such as home networks have yet to be adequately developed. People have come to expect seamless, instantaneous, and continuous connection to communications networks using a myriad of devices such as smartphones, computers, tablet devices, high-definition TV, and game systems. Demand has increased for optical data-link connections not only to the end-users'

* Corresponding Author Email: kotarokoike@gmail.com.

buildings but also within them. For such short-distance applications, from optical local area networks for homes and offices to high speed backplanes, the extremely low attenuation of GOFs is unnecessary. Instead, a greater flexibility and higher reliability in the event of bending, shocks, and vibrations are considerably more valuable properties.

Another optical transmission medium that has been developed to meet the above requirements is the plastic optical fiber (POF). A particularly important advantage of POFs over GOFs is the large core size. Today, most GOFs are fabricated as single-mode structures. The core is designed to be 5–10 μm in diameter so that it carries only a single ray of light and enables an extremely high bandwidth. In addition, GOFs, by virtue of their usage, are required to not lose their flexibility, for which, they are restricted to be thinner than human hair. Thus, even with multi-mode GOFs, which are usually used for short-distance transmissions, the diameter is limited to a maximum of 62.5 μm. These GOFs are connected in a three-step process of cleaving the fibers using an expensive, specialized tool; epoxying them to the connector hardware; and polishing the face of the resultant assembled connector. In contrast, POFs can be enlarged to ~1 mm in diameter without loss in flexibility or ease of handling. They can be terminated using an inexpensive and simple tool, and connectors can just be crimped on. The operation takes a fraction of the time needed to connectorize GOFs. POFs are very easy to install and use, unlike conventional GOFs, especially for short-range communications such as local area networks and interconnections in data centers where many connections are required. Figure 1 shows schematically the difference in the cross-sectional areas of the fibers.

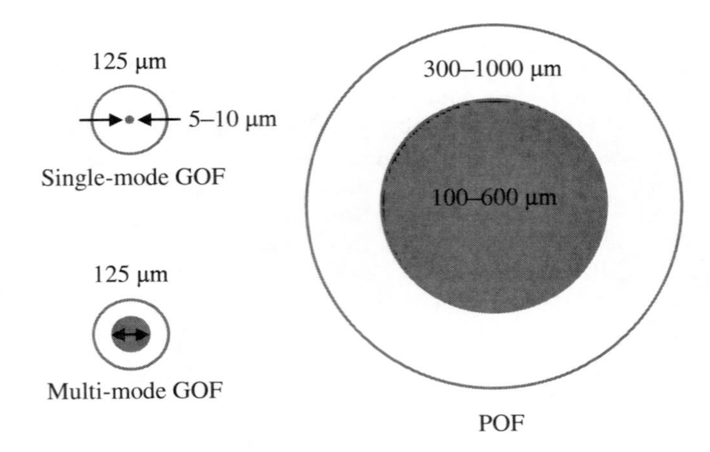

Figure 1. Cross-sectional views of the single-mode GOF, multi-mode GOF, and POF. Partially adapted from Koike, Y., Koike, K. (2011) Journal of Polymer Science Part B: Polymer Physics, 49 (1), 2–17, with permission of John Wiley & Sons.

However, POFs, which have been overshadowed by the success of GOFs in previous decades, underwent several advancements over a very long time before they gained their preset popularity in the communications field. For instance, any increase in the diameter would lead to the transmission bandwidth being severely limited owing to modal dispersion, which distorts the optical signals. Thus, for communication purposes, POFs were employed only in a limited number of applications, such as digital audio interfaces and factory automation systems, that do not require high bit-rates. What drastically changed the status of

POFs in the communications field was graded-index technology that improved the transmission speed to over 1 Gbps. The advent of the graded-index POF (GI POF) greatly expanded the potential applications of POFs and attracted many people into POF research and development. Today, after more than three decades, the POF capacity has exceeded 100 Gbps over 100 m. Behind this significant improvement, there has been tremendous progress made in terms of fabrication techniques and materials based on a fundamental understanding of photonics and polymer science.

This chapter traces the history of GI POFs from the perspective of developments in the fabrication techniques and materials. Section 2 briefly describes the earliest days of POFs and the significance of gradient refractive index profiles. GI POF fabrication methods can be classified into preform and extrusion methods, and in section 3, some representative principles for forming GI profiles in each process are presented. Section 4 discusses the basic ideas of how to choose or design the base polymer of a GI POF from a viewpoint of the fundamental properties of optical fibers: attenuation and bandwidth, and the characteristics and problems of typical base materials, poly(methyl methacrylate) (PMMA) and poly(perfluoro-4-vinyloxy-1-butene) (CYTOP®) are described. Finally, in section 5, recent material developments that have attempted to solve the difficulties of conventional polymers are summarized.

2. Background to Development

2.1. Reduction of Fiber Losses

The first POF, Crofon™, invented in the mid-1960s by DuPont, was a multi-mode fiber with a large core of PMMA and fluoroalkyl methacrylate as the cladding layer. Because of the abrupt refractive index change at the core-cladding interface, such fibers are called step-index (SI) POFs. The first commercialized SI POF was Eska™, which was introduced by Mitsubishi Rayon in 1975; subsequently, Asahi Chemical and Toray also launched Luminous and Raytela® in the market, respectively. However, the transparency of these early POFs was insufficient for communication, and their applications were limited to extremely short-range areas such as light guides, illumination, audio data-links, and sensors.

Fiber losses limit how far the signal can propagate in the fiber. As optical receivers need a certain minimum power for recovering the signal accurately, the transmission distance is inherently limited by the fiber losses. It is customary to express the attenuation α in units of dB/km using the relation

$$\alpha\ (\text{dB/km}) = -\frac{10}{L}\log_{10}\left(\frac{P_{out}}{P_{in}}\right), \tag{1}$$

where P_{in} is the power launched at the input end, and P_{out} is the output power of a fiber of length L. The attenuation of the first POF prototype exceeded 1000 dB/km; the lowest allowable limit in communications was less than a few hundred decibels per kilometer.

At the time, the mainstream production method for SI POFs was a continuous extrusion process [1, 2], open at the material input end. Kaino et al. [3] then developed a batch extrusion process that produced POFs in a completely closed system and achieved significant reductions in the fiber attenuation. Figure 2 shows a schematic diagram of the apparatus. The process starts by removing an inhibitor from a monomer by rinsing with alkali solution. The residual alkali is rinsed away with distilled water until the monomer is neutralized, and the monomer is then dried by adding Na_2SO_4 and CaH_2. Once dry, the monomer is distilled under reduced pressure, and the middle fraction is collected and placed in the monomer flask. After adding an initiator and chain transfer agent to another flask, both flasks are sealed and evacuated.

The monomer flask is cooled with liquid nitrogen and then melted by water under vacuum to eliminate the dissolved air. This process is repeated until air bubbles are no longer observed. This is the so-called freeze-pump-thaw cycle. The monomer is then distilled into the polymerization ampoule, and the ampoule is washed to remove dust and other contaminants attached to the inner wall. The monomer is transferred back into the flask by decantation and the process is repeated until no glitter from the dust at any point in the ampoule is detected by a He-Ne laser. Subsequently, the initiator and chain transfer agent are distilled into the polymerization ampoule. In the case of poly(styrene) (PSt), the reaction vessel is heated at 130 °C for 16 h to polymerize the monomer. The temperature is then raised slowly to 180 °C and kept at this temperature for another 16 h to fully polymerize the remaining monomer. Once the polymerization process is complete, the reactor vessel is heated further to 190 °C until the polymer becomes a fluid. Dry nitrogen gas is used to pressurize the reactor, which pushes the material through a nozzle to form the fiber. The fiber is pulled through another nozzle where it is coated with a cladding polymer. It was in this way that low-loss SI POFs were successfully fabricated. In 1981, Kaino et al. [3] reported a PSt-based SI POF with an attenuation of 114 dB/km at 670 nm, and later succeeded in obtaining a PMMA-based SI POF with an attenuation of 55 dB/km at 568 nm [4].

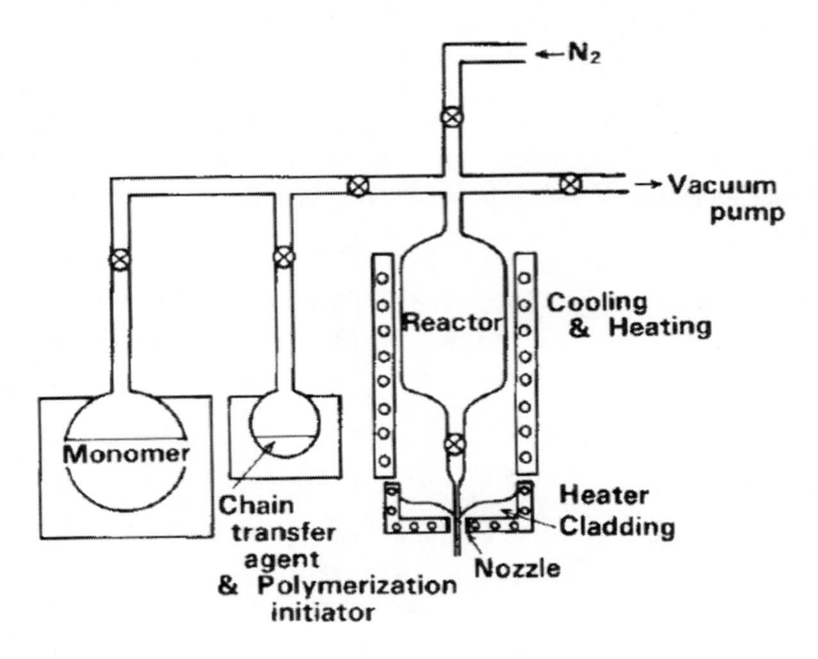

Figure 2. Batch extrusion process for SI POFs. Reprinted from Kaino, T., Fujiki, M., Nara, S. (1981) Journal of Applied Physics, 52 (12), 7061-7063, with permission of the American Institute of Physics.

2.2. Graded-Index Technology for Faster Transmission

After the studies of Kaino et al. [3, 4] on SI POFs based on several methacrylate and styrene derivatives, POFs became a more realistic solution for short-distance communications. However, despite these advances, SI POFs still exhibited limited bandwidth, which in turn indicated their limited capacity for transmission speed. The concept of a fiber bandwidth originates from the general theory of time-invariant linear systems. If the optical fiber can be treated as a linear system, its input and output powers should be related by the general relation

$$P_{out}(t) = \int_{-\infty}^{\infty} h\left(t - t'\right) P_{in}(t')dt' . \tag{2}$$

For an impulse $P_{in}(t) = \delta(t)$, where $\delta(t)$ is the Dirac delta function, $P_{out}(t) = h(t)$, and for this reason, $h(t)$ is referred to as the impulse response of the linear system. The Fourier transform,

$$H(f) = \int_{-\infty}^{\infty} h\left(t - t'\right) \exp(2\pi i f t)dt, \tag{3}$$

provides the frequency response and is known as the transfer function. In general, $|H(f)|$ falls off rapidly with increasing f, indicating that the high-frequency components of the input signal are attenuated by the fiber; the optical fiber acts as a bandpass filter. The fiber bandwidth $f_{-3\text{dB}}$ corresponds to the frequency $f = f_{-3\text{dB}}$ at which $|H(f)|$ is reduced by a factor of 2 or by −3dB:

$$|H(f_{-3\text{dB}})/H(0)| = \frac{1}{2}. \tag{4}$$

In the field of electrical communications, the bandwidth of a linear system is usually defined as the frequency at which electrical power drops by 3 dB. A narrower output pulse results in a larger $f_{-3\text{dB}}$.

The dominant factor degrading the bandwidth of SI POFs is modal dispersion. Modal dispersion can be understood from Figure 3 (a) and (b). In a single-mode fiber, a single ray travels along the center axis, whereas in an SI multi-mode fiber, different rays travel along paths of different lengths. Even if these rays are coincident at the input end and travel at the same speed inside the fiber, the rays are dispersed at the output end of the fiber. A narrow pulse would broaden considerably as a result of different path lengths, and the fiber bandwidth is a quantitative indicator of this broadening with respect to the original pulse. If the pulse broadens, then the number of pulses input to the fiber per unit time needs to be limited to avoid their overlap and interference. This is why SI POFs have a severe limitation in the possible transmission capacity. GI guides, however, have a lower modal dispersion and thus a greater transmission capacity. The refractive index of the core in GI fibers is not constant but decreases gradually from a maximum value n_1 at the core center to a minimum value n_2 at the core-cladding interface. Most GI fibers are designed to have a nearly quadratic decrease and are analyzed using a g-profile given by

$$n(r) = \begin{cases} n_1[1 - 2\Delta(r/R)^g]^{1/2}, 0 \le r \le R \\ n_1(1 - 2\Delta)^{1/2}. \quad R < r \end{cases} \tag{5}$$

Here, $n(r)$ is the refractive index n at radius r of the fiber, R is the core diameter, g is the refractive index profile coefficient, and Δ is the refractive index difference defined as

$$\Delta = \frac{{n_1}^2 - {n_2}^2}{2{n_1}^2} \cong \frac{n_1 - n_2}{n_1}. \tag{6}$$

From Figure 3 (c), it is easy to understand qualitatively why modal dispersion is reduced in GI fibers. As in the case of SI fibers, the path is longer for more oblique rays (the higher-order mode), but the ray velocity changes along the path because of variations in the refractive index. More specifically, ray propagation along the fiber axis takes the shortest path but has the slowest speed since the refractive index is a maximum along this path; oblique rays have a large part of their path in a lower refractive index medium in which they travel faster. The difference in the refractive index is small, e.g., 0.01, but sufficient to compensate for the time delay. This GI method was first proposed by Nishizawa for multi-mode GOFs and subsequently adapted for POFs.

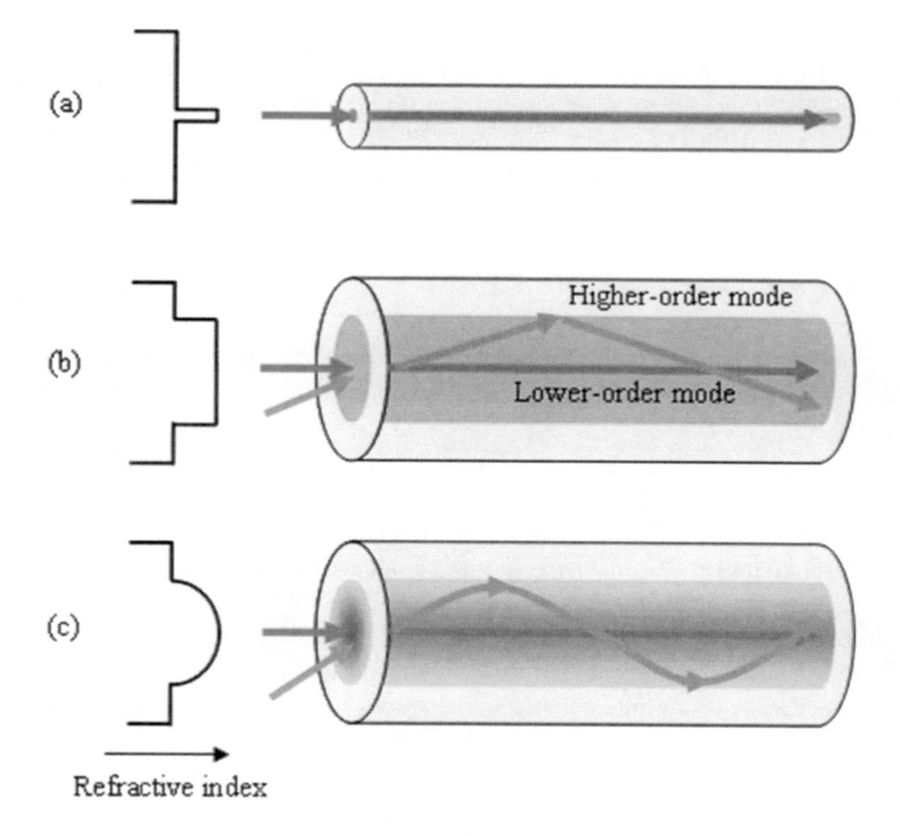

Figure 3. Ray trajectories through basic optical fibers: (a) single-mode, (b) step-index multi-mode, and (c) graded-index multi-mode. Reprinted from Koike, Y., Koike, K. (2011) Journal of Polymer Science Part B: Polymer Physics, 49 (1), 2–17, with permission of John Wiley & Sons.

3. PROGRESS IN FABRICATION TECHNIQUES

3.1. Preform-Drawing Method

The first important advance was announced by Ohtsuka in 1973 [5] who succeeded in forming a gradient refractive index profile in polymeric materials for the first time. The simple principle involves a cross-linked prepolymer rod that is immersed into a lower refractive index monomer. The monomers diffuse mutually and copolymerize. Since the composition of the lower index material increases with distance from the center of the rod, a quadratic refractive index profile is formed in the radial direction in accordance with the copolymer composition. One drawback of this method is that the prepolymer rod must be cross-linked to keep its shape during the immersion process. Thus, these preforms could not be heat-drawn into fibers.

The first GI POF was reported by Ohstuka and Hatanaka in 1976 [6]. Using a new method called photo-copolymerization [7, 8], they obtained a GI preform without any cross-linking agents. In copolymerization reactions, the monomer reactivity ratio is often referred to in order to discuss how the reaction proceeds. The monomer reactivity ratios r_{12} and r_{21} are defined as the ratios of the rate constant of reaction for a given radical added to its own monomer to the rate constant for its addition to the other monomer. Thus $r_{12} > 1$ means that the radical $M_1 \cdot$ prefers the addition of M_1, whereas $r_{12} < 1$ means that it prefers the addition of M_2. In the photo-copolymerization method, comonomers satisfying following requirements are used as the base material: $r_{12} > 1$, $r_{21} < 1$, and the refractive index of the M_1 homopolymer is lower than that of the M_2 homopolymer. A schematic diagram of the copolymerization process is shown in Figure 4. The monomer mixture, containing a specified amount of a photoinitiator and a thermal initiator, is sealed inside a glass tube. The tube is mounted on a rotating table that turns the tube around its axis. A UV lamp equipped with shadings is mounted on a vertical translation stage that moves upward at a constant velocity. Here, the vertical speed of the lamp is adjusted to ensure that the entire monomer in the tube is prepolymerized after the light source traverses its length. The reaction is carried out from the bottom to avoid forming a cavity caused by volume shrinkage during the copolymerization. Once the comonomer is exposed to UV light, the copolymerization reaction starts from the outside where the light intensity is the strongest, and a thin gel phase is formed along the inner wall; the reaction occurs preferentially in the gel phase than in the monomer. Consequently, the copolymer phase moves inward toward the center as the reaction proceeds. Here, because $M_1$is more reactive than M_2, the composition of M_1 at the periphery is greater than that in the middle. As a result, the refractive index of the rod decreases with distance from the center in accordance with the change in the copolymer composition (M_1 has a lower refractive index). After the photo-copolymerization process, the rod is heat treated to fully polymerize the remaining monomer. The GI preform is removed from the glass tube and heat-drawn into a fiber.

GI preforms can also be fabricated using an interfacial-gel polymerization technique [9], which was first developed in 1988. The principle is basically the same as that of the photo-copolymerization method discussed above except for the mechanism that forms the initial gel phase. In this method, the core solution is placed in a polymer tube rather than in a glass tube. The gel phase in the photo-copolymerization method referred to a prepolymer with a conversion of less than 100%, whereas in this method, the gel phase is the polymer layer on the inner wall of the tube swollen by the core monomer.

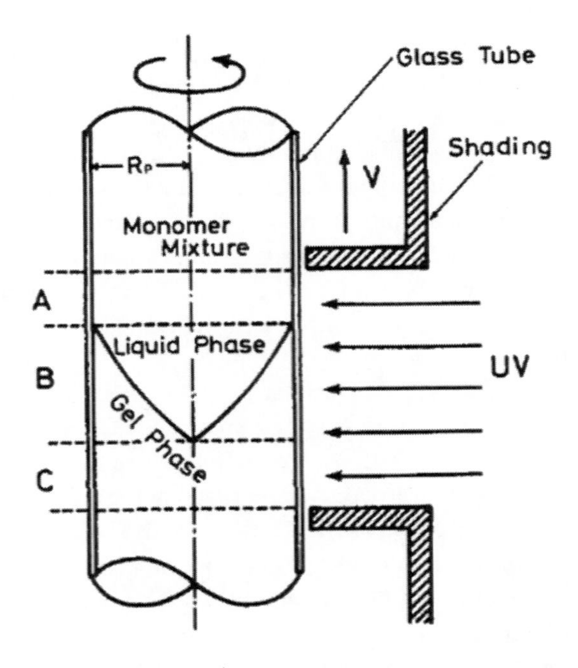

Figure 4. Photo-copolymerization process to obtain GI preforms: (A) liquid phase, (B) boundary between liquid and gel phases, and (C) gel phase. Reprinted from Ohtsuka, Y., Koike, Y., Yamazaki, H. (1981) Applied Optics, 20 (2), 280-285, with permission of the Optical Society of America.

After these developments, the photo-copolymerization and interfacial-gel polymerization methods were applied to every possible monomer combination, leading a variety of refractive index profiles and precise control. These studies also revealed the correlation between the monomer reactivity ratio and attenuation of the obtained GI POFs. In general, ideal random copolymers such as a copolymer of MMA and benzyl methacrylate (BzMA) are more transparent since they are more optically homogeneous. However, GI POFs based on copolymers whose monomer reactivity ratios are considerably different show less attenuation. This is because when either r_{12} or r_{21} is close to zero, the obtained polymer is close to being a homopolymer that simply contains the other monomer. For example, when MMA and vinyl benzoate (VB) are the monomers M_1 and M_2, respectively, $r_{12} = 8.52$ and $r_{21} = 0.07$. If the reaction is completed, the obtained polymer would be a blend polymer that strongly scatters light. However, most of the VB monomers do not actually copolymerize with MMA under the polymerization conditions employed in these two methods due to the significant difference in the reactivities. As a result, PMMA containing VB monomers is obtained by the reaction. Since the size of the VB monomer distributed in the PMMA is sufficiently smaller than the wavelength of the incident light, the light scattering loss of the polymer is as low as that of PMMA. In 1991, two low-loss GI POFs based on MMA-VB and MMA-vinyl phenyl acetate (VPAc) copolymers were successfully fabricated [10]. MMA (M_1) and VPAc (M_2) also have significantly different reactivity ratios of $r_{12} = 22.5$ and $r_{21} = 0.005$, respectively. The attenuations of the MMA-VB and MMA-VPAc systems at 652 nm were 134 dB/km and

143 dB/km, respectively, which at the time, were comparable with the attenuation of commercially available PMMA-based SI POFs.

Another noteworthy finding of these studies was that the driving force concentrating the second monomer into the middle of the core was not only a difference in the reactivities but also a difference in the molecular size. This was realized from a study of the MMA-BzMA copolymer system. Although the reactivity ratios for this system are almost unity and the copolymer is known to be an ideal random copolymer, a preform with a gradient refractive index change was successfully obtained. Because the BzMA molecule is larger than the MMA molecule, it is harder for the BzMA to diffuse into the gel phase and thus it concentrates into the middle of the core.

Based on these results, in 1993, nonreactive organic compounds started to be used as the high refractive index component instead of an M_2 monomer [11]. Bromobenzene (BB) was the first dopant used with MMA, and a GI preform was prepared with the interfacial-gel polymerization technique. Since the molecular size of BB is slightly larger than that of MMA, the BB molecules concentrated into the middle as the polymerization progressed. The procedure is illustrated schematically in Figure 5. By using a nonreactive dopant whose size is too small to scatter light, the attenuation was reduced to 90 dB/km at 572 nm [12].

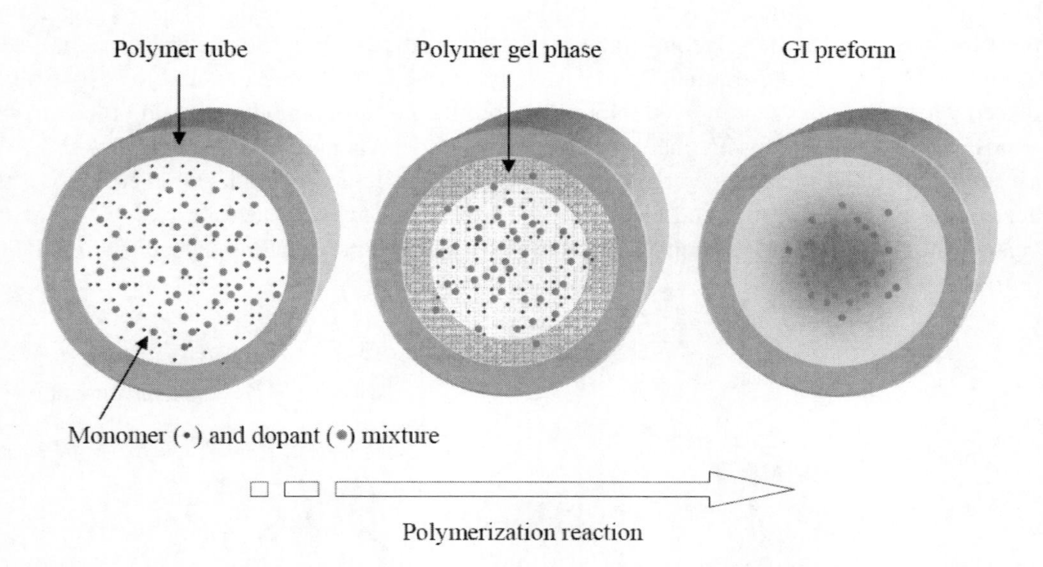

Figure 5. Interfacial-gel polymerization technique using a dopant. Modified from Koike, Y., Ishigure, T., Nihei, E. (1995) Journal of Lightwave Technology, 13 (7), 1475–1489.

Since this achievement, numerous dopant have been explored to improve the properties of GI POFs. Ideal dopants are relatively low molecular weight compounds that 1) are soluble in the base polymers and do not phase-separate or crystallize over time; 2) do not significantly increase the attenuation of the polymers; 3) do not reduce the glass transition temperature (T_g) of the polymers by an unacceptable degree; 4) provide large changes in the refractive indices at low concentrations, e.g., $n_1 - n_2 > 0.015$ for less than 15 wt% dopant; 5) are chemically stable in the polymers at the processing temperatures; 6) have a low volatility at the processing temperatures; and 7) are substantially immobilized in the glassy polymer in the operating environments. In particular, compared to previous copolymer systems, the biggest concern in the polymer-dopant system is that the dopant significantly decreases T_g due to the plasticization effect. Here, the T_g of a non-crystalline material is a critical temperature at which the material changes its behavior from being glassy to rubbery and vice versa. Since the operating temperature of a POF is basically dependent on T_g, this is considered to be one of the most important physical properties. Thus, it is preferable if the refractive index of the dopant is as high as possible so that a lower dopant concentration is capable of forming a GI profile with a sufficient numerical aperture (NA). For PMMA-based GI POFs, diphenyl sulfide (DPS) has been most suitable dopant identified so far.

The latest GI-preform method is the rod-in-tube method [13]. The idea, shown in Figure 6, is originally from a GOF fabrication method. In the rod-in-tube method, the core rod containing a high refractive index dopant and the cladding tube are prepared separately. The outer diameter of the rod and inner diameter of the tube are designed to be as close as possible, whereas the height of the rod is slightly higher than that of the tube so that any water and air in the interface can be easily removed. After washing with purified water, the rod is inserted into the tube, which is then covered with a heat-shrinkable tube. The original diameter of the heat-shrinkable tube is about 1.2 times larger than the outer diameter of the cladding tube. More importantly, when shrunk, the size of the heat-shrinkable tube must be smaller than the core rod diameter and thickness of the cladding tube combined. A Teflon® disc having almost the same diameter as the cladding tube is placed on the bottom and the end is completely closed by heating. Then, another Teflon® disc is placed on the other end and the assembly is placed vertically in a vacuum oven. Note that both the core rod and the cladding tube are still exposed to the air at this moment. After drying under vacuum at 40–50 °C for 2 h, the assembly is heated to a temperature above the softening points of each polymer. During the heat treatment, the core rod and cladding tube adhere because of the contractile force of the heat-shrinkable tube, and the dopant gradually diffuses into the cladding layer, forming the GI profile.

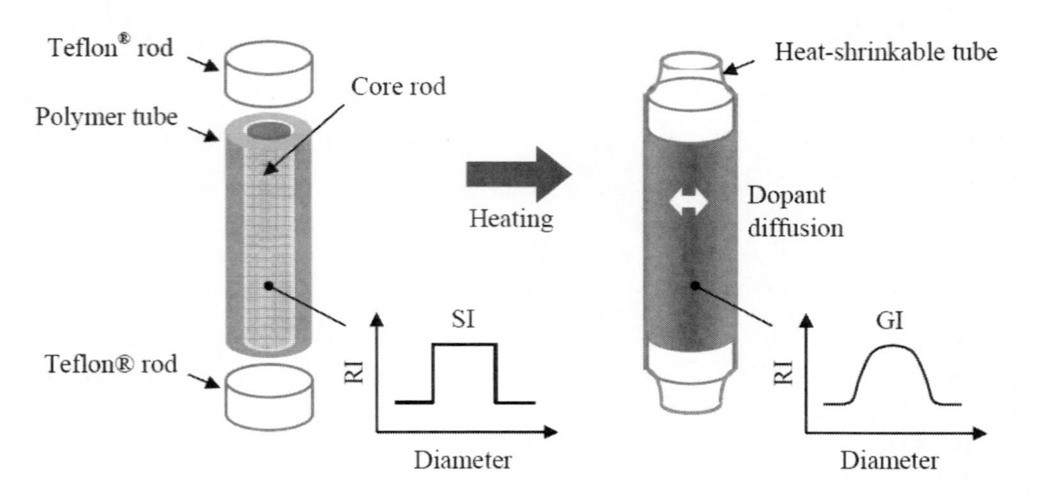

Figure 6. Rod-in-tube method. Reprinted from Koike, K., Kado, T., Satoh, Z., Okamoto, Y., Koike, Y. (2010) Polymer, 51 (6), 1377-1385, with permission of Elsevier.

The greatest advantage of this method is that GI preforms can be obtained from any core and cladding polymer combination as long as the two adhere well. In the case of the interfacial-gel polymerization technique, the core monomer is polymerized in the presence of a dissolved cladding polymer. Consequently, the core-cladding boundary becomes ambiguous unless the polymers are compatible and the refractive index difference is negligibly small. However, in the rod-in-tube method, the core and cladding layers are physically attached to each other, and the GI profile is formed by distributing the dopant in the radial direction. Thus there is no concern about an increment in the light scattering. Today, this method is most often utilized for laboratory-scale experiments.

3.2. Extrusion Method

The other common GI POF fabrication technique is the extrusion method. This was first proposed by Mitsubishi Rayon in 1989–90 [14–16]. Although the extruders described in these three patents are slightly different from each other, the principle for forming the GI profile is the same. Figure 7 shows the apparatus from [16]. In the system, a prepolymer containing a photoinitiator is first melted in a container and pressed into a die through a screw.

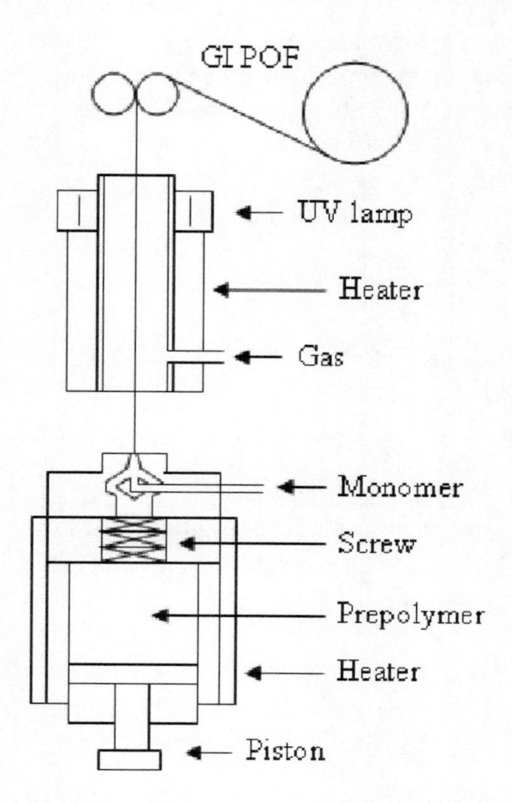

Figure 7. The original model of the GI POF extruder. Modified from Mitsubishi Rayon Co. Ltd. (1990) JP Patent 1990-33104.

Then, a different monomer that also contains the photoinitiator is added to the center of the extruded prepolymer through a different pipe. The polymer is pushed out of the nozzle and pulled by a capstan to draw the fiber.

As the fiber passes through the furnace between the nozzle and capstan, the added monomer diffuses into the surrounding prepolymer. Finally, the fiber is exposed to a UV lamp mounted on the end of the furnace, and the diffused monomer and remaining monomer in the prepolymer are photo-polymerized. Since the added monomer has a higher refractive index in its polymer state than the surrounding polymer, the refractive index gradually decreases from the center to the periphery with the change in the polymer composition. For instance, PMMA, with a conversion of 60%, and phenyl methacrylate were used as the prepolymer and diffusion monomer, respectively. A similar method was also reported by Ho et al. [17] and Chen et al. [18] in 1995–96 for fabricating an imaging lens.

Monomer diffusion was replaced by dopant diffusion in the extrusion method in a similar manner to the progress in preform techniques. While the requirements for the dopant are basically the same as in the preform methods, dopants used in extrusions also need to have a sufficiently high diffusivity in the polymers at the processing temperatures. The dopant extrusion technique was first patented in 1997 [19]. As shown in Figure 8, the first dopant diffusion extruder was quite simple. The polymer and dopant are placed in cylinders 1 and 2, respectively, and homogeneously heated in a furnace to a certain temperature. The molten polymer and dopant are pressurized by either nitrogen gas or a piston. The dopant is pushed out of nozzle 1 and diffuses into the polymer. The polymer with the concentration gradient of the dopant is pushed out of nozzle 2 and pulled by a capstan to draw the fiber. In this way, by diffusing a nonreactive dopant instead of monomer, the attenuation of extruded GI POFs was reduced to 90 dB/km at 570 nm.

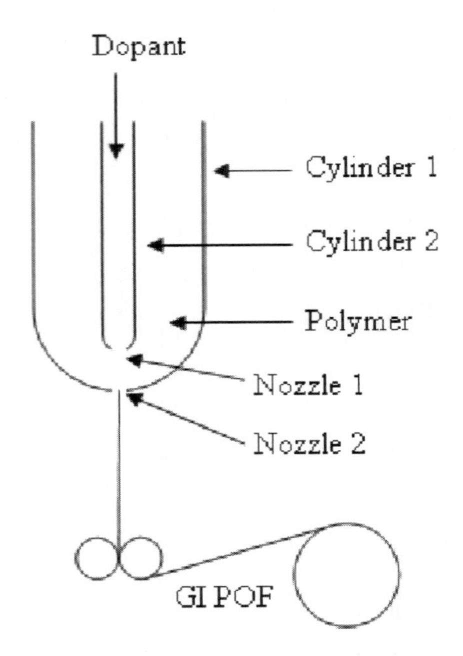

Figure 8. The first GI POF extruder using a dopant. Modified from Koike, Y., Nihei, R. (1997) US Patent 5,593,621.

Apart from using a dopant instead of a monomer, there is another important difference between this procedure and the previous exttrusion methods. In the previous methods, the GI profiles were formed after the polymers were drawn into fibers. Thus, large-sized heterogeneous structures generated during the monomer diffusion and the copolymerization processes remained in situ. In fact, there are no attenuation data in the relevant patents or papers. In contrast, the fiber drawing process is performed last in the dopant method. Since the profile is formed by dopant diffusion in this case, there is no concern about large-sized heterogeneous structures. However, as in the preform method, the stretching process is usually very efficient in terms of smoothing any heterogeneities in the fibers such as an irregularity in the core-cladding boundary or fluctuations in the diameter.

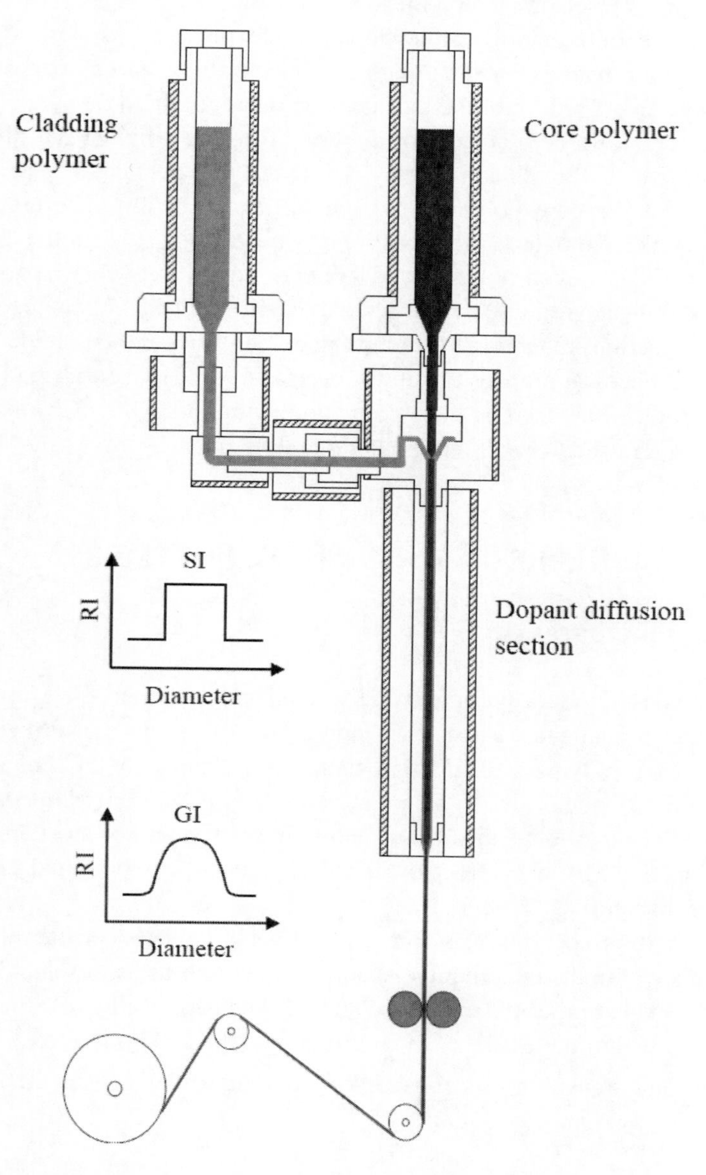

Figure 9. A general schematic of currently used GI POF extruders.

While the dopant extrusion enabled the fabrication of low-loss GI POFs, the method required either an intermittent stoppage or an extremely slow extrusion speed to allow the dopant sufficient time to diffuse due to the short distance between the nozzles. This was solved by introducing a dopant diffusion section into the process, as in the extruder patented by Lucent Technologies in 2001 [20]. Today, most GI POF suppliers more or less follow this method. The general setup is shown in Figure 9. All the tooling components are made from corrosion resistant materials, e.g., Hastelloy, and the surfaces that come into contact with the flowing melts are polished to promote a smooth flow with little material stagnation. All parts are machined to a high precision to ensure a good fit and thereby prevent leakage. The shaded areas in the figure represent heaters that are individually controlled by a programmable thermostat. The core and cladding polymers are fed into separate extruders, and the core polymer contains one or more diffusible dopants that provide the desired GI. The core polymer is fed from the core extruder through a connector hose into a coextrusion head, and the cladding polymer is fed into the coextrusion head through a connector pipe. The coextrusion head contains a core nozzle for directing the core polymer into the center of the diffusion section, while the cladding polymer from the connector pipe is distributed concentrically around the core polymer into the diffusion section. The core and cladding polymers flow together into the diffusion zone. The diffusion section is also heated to maintain the flow of the molten polymer and promote diffusion of the dopant; the length of the diffusion zone is designed to allow the desired extent of diffusion to occur. The polymers then flow from the section into an exit die, and the exiting fiber is then pulled by a capstan to draw the fiber at the necessary rate to obtain the desired final diameter. Using the new system, the coextrusion process has become capable of producing GI POFs at commercially useful speeds, e.g., at least 1 m/sec, for a 250 μm outer diameter fiber.

4. Representative Base Polymers

4.1. Poly(methyl methacrylate)

PMMA is a mass-produced commercially available polymer that provides excellent resistance to both chemical and weather corrosion. The transmittance of PMMA is the highest among general optical polymers; PMMA is known to transmit 93–94% of visible rays and reflect the remainder. Optical loss in fibers is caused by three factors: electronic transition absorption, molecular vibrational absorption, and light scattering. For most amorphous optical polymers including PMMA, the effect of the C–H molecular vibrational absorption is the dominant loss mechanism.

Electronic transition absorption is a result of transitions between the electronic energy levels of the bonds within the material; the absorption of photons causes an upward transition, which leads to the excitation of the electronic state. In the case of PMMA, the n–π* transition of the double bond within the ester group contributes to the absorption loss. The relationship between the electronic transition loss α_e (dB/km) and the wavelength of the incident light λ (nm) can be expressed by Urbach's rule [21]:

$$\alpha_e = A\exp(B/\lambda). \tag{7}$$

Here, A and B are substance-specific constants, which for PMMA are known to be 1.58 × 10^{-12} and 1.15 × 10^4, respectively [4]. The loss α_e decreases exponentially with an increase in the wavelength, and the value falls below 1 dB/km for wavelengths longer than 424 nm.

The scattering loss in polymers arises from microscopic variations in the material density. Using Einstein's fluctuation theory, the isotropic light scattering intensity V_V^{iso} from thermally induced density fluctuations in a structureless liquid is expressed as [22]

$$V_V^{iso} = \frac{\pi^2}{9\lambda_0^4}(n^2-1)^2(n^2+2)^2 kT\beta, \tag{8}$$

where λ_0 is the wavelength of light in vacuum; n, the refractive index; k, the Boltzmann constant; T, the absolute temperature; and β, the isothermal compressibility. The isotropic light scattering loss α_{iso} (dB/km) is related to V_V^{iso} (cm^{-1}) by

$$\alpha_{iso} = 4.342 \times 10^5 \pi \int_0^{\pi}(1+\cos^2\theta) V_V^{iso} \sin\theta \, d\theta. \tag{9}$$

The refractive index n and isothermal compressibility β of PMMA are 1.492 and 3.55×10^{-11} cm^2/dyn [23], respectively, at around the T_g point. Thus, assuming a freezing condition, the value of V_V^{iso} at room temperature is calculated to be 2.61×10^{-6} cm^{-1} at 633 nm. This gives the loss α_{iso} as 9.5 dB/km from Eq. (9), which is almost identical to the experimental value of 9.7 dB/km obtained by in a static light scattering measurement using a highly purified PMMA rod [24, 25].

The influence of molecular vibrational absorption on the fiber attenuation was extensively studied by Groh [26]. Figure 10 shows the spectral overtone positions and relative integral band strengths for the C–H, C–D, C–F, and C–Cl vibrations, which were calculated based on the Morse potential model. The scale on the vertical axis $E_\upsilon^{C-X}/E_1^{C-H}$ represents the vibrational energy ratio of each bond to the fundamental vibration of the C–H bond. The vibrational energy ratio is approximately correlated to the molecular vibrational absorption loss α_υ (dB/km) as follows:

$$\alpha_\upsilon = 3.2 \times 10^8 \frac{\rho N_{C-X}}{M}\left(\frac{E_\upsilon^{C-X}}{E_1^{C-H}}\right), \tag{10}$$

where ρ (g/cm^3) is the polymer density; M (g/mol), the molecular weight of the monomer unit; and N_{C-X}, the number of C–X bonds per monomer. In the case of PMMA, these values are ρ = 1.19 g/cm^3, M = 100 g/mol, and $N_{C-H} = 8$. Substituting these values into Eq. (10) with the vibrational energy ratios $E_\upsilon^{C-X}/E_1^{C-H}$ given in Figure 10, the absorption losses of 5th and 6th C–H overtones are equal to 2772 and 426 dB/km, respectively.

From these calculations and experiments, it is obvious that the optical loss in PMMA is almost all due to the C–H vibrational absorption. Figure 11 is a typical attenuation spectrum of a PMMA-based GI POF measured by a cut-back method. The fiber was prepared by heat-drawing a GI preform obtained by the interfacial-gel polymerization. The two peaks at around

730 and 620 nm correspond to the 5th and 6th C–H overtones, respectively, and their peak intensities agree very well with the approximated values.

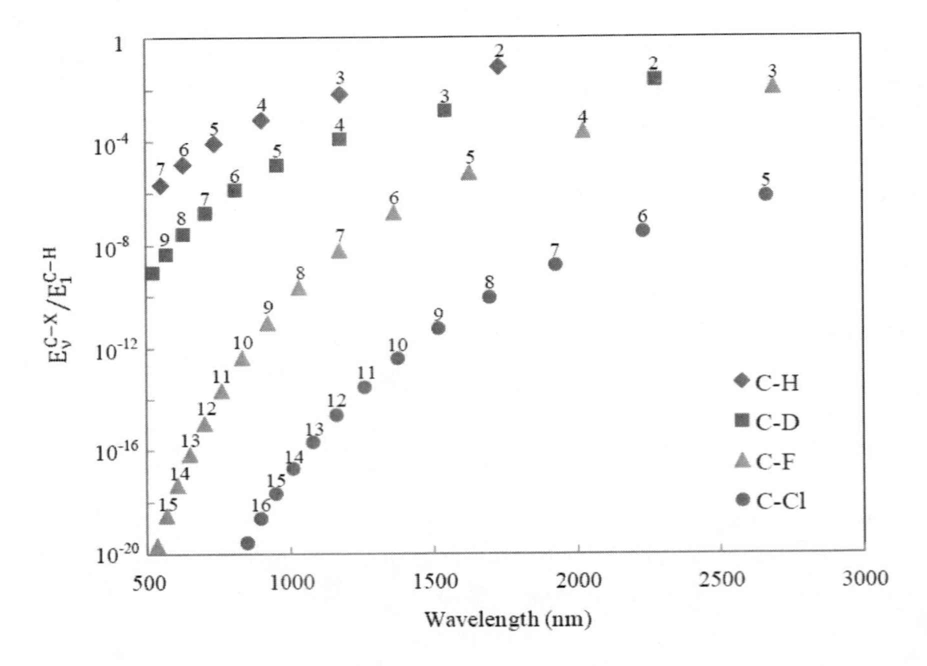

Figure 10. Calculated spectral overtone positions and normalized integral band strengths for the C–H, C–D, C–F, and C–Cl vibrations. Modified from Groh, W. (1988) Die Makromolekulare Chemie, 189 (12), 2861–2874.

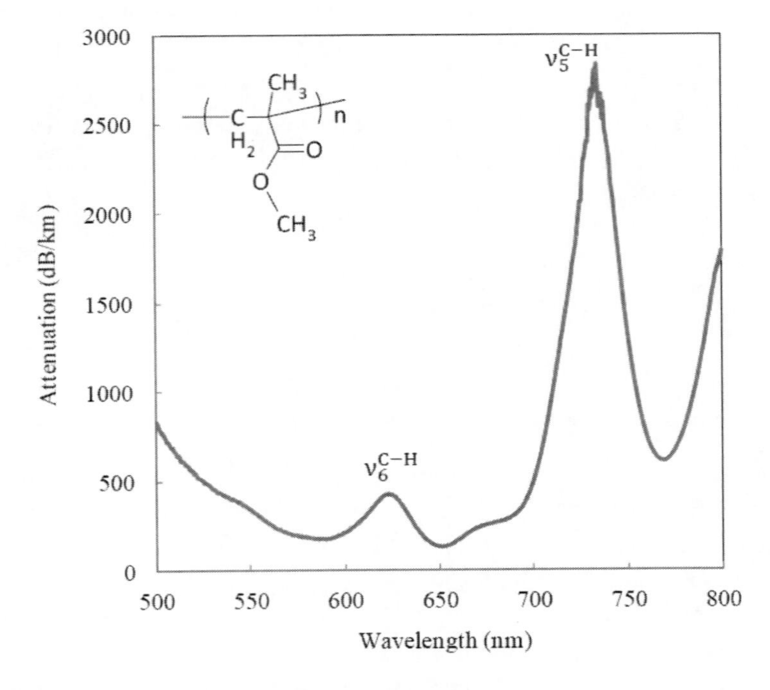

Figure 11. Attenuation spectrum of the PMMA-based GI POF.

Since the first POF commercialized by Mitsubishi Rayon in 1975, most SI POFs have been manufactured using PMMA. They have been used extensively in industrial field buses for controlling process equipment in rugged manufacturing environments and in automobiles to connect an increasing array of multi-media equipment. The increasing complexity of in-vehicle electronic systems in particular has led to PMMA-based SI POFs becoming indispensable to the automobile industry. Today, it is not uncommon to find 10 or 20 consumer electronic devices installed in cars such as e.g., DVD (blu-ray) players, navigation systems, telephones, Bluetooth interfaces, voice recognition systems, high-end amplifiers, and TV tuners. To meet all the necessary requirements for data transfer between such devices, PMMA-based SI POFs which offer up to several hundred megabits per second, have been an ideal solution.

However, if faster bit-rates of data transmission of more than 1 Gbps are required, attenuation becomes a problem in the PMMA fibers. Short-distance application systems have employed PMMA-based SI POFs as the transmission medium and red light-emitting diodes (LEDs) of 650 nm as the light source. As shown in Figure 11, 650 nm is in a low-loss window of the PMMA. On the other hand, to realize gigabit communications, it is not only the fiber has to be replaced with a PMMA-based GI POF but an appropriate light source with a higher modulation bandwidth is also necessary. At the present, vertical-cavity surface-emitting lasers (VCSELs), which offer bit-rates of up to 10 Gbps, are believed to be the most reasonable solution. VCSELs are a relatively new class of semiconductor laser that are expected to become key devices in gigabit Ethernet, high-speed local area networks, computer links, and optical interconnects. One issue, however, is that the emission wavelengths of VCSELs are longer than 670 nm where the attenuation of PMMA-based GI POFs exceeds 200 dB/km due to the C–H stretching vibrational absorption loss. Thus, the transmission distance is severely limited.

4.2. Perfluorinated Polymer, CYTOP®

Since the high attenuation of PMMA is caused by overtones of the C–H vibrational absorption, the most effective method for obtaining a low-loss POF is to substitute all the hydrogen atoms with heavier atoms such as fluorine. However, compared to the large number of radically polymerizable hydrocarbon monomers, only a few classes of perfluoromonomers can homopolymerize under normal conditions via a free-radical mechanism. The most typical example is tetrafluoroethylene (TFE), developed by DuPont in 1938. As is well known, poly(TFE) is opaque despite the lack of C–H bonds. In general, perfluorinated resins are rigid and easily form partially crystalline structures. Hence, light is scattered at the boundaries between the amorphous and crystalline phases, causing the haziness. To avoid the formation of the crystalline phase, introducing aliphatic rings into the main chain, which causes it to become twisted, is an effective method. The most well-known amorphous perfluorinated polymers are Teflon® AF, CYTOP®, and Hyflon® AD, which were developed by DuPont, AGC, and Solvay, respectively. The chemical structures of these three polymers are shown in Figure 12. They have excellent clarity, solubility in some specific fluorinated solvents, thermal and chemical durability, low water absorption, and low dielectric properties. In particular, the high transparencies arise from the cyclic structures existing in the polymer main chains. Teflon® AF is a copolymer of perfluoro-2,2-dimethyl-4,5-difluoro-1,3-dioxole, which posses a cyclic structure in its monomer unit, and TFE, while Hyflon® AD is a copolymer of perfluoro-2,2,4-trifluoro-5-trifluoromethoxy-1,3-dioxole and TFE. CYTOP®, however, is a homopolymer of perfluoro(4-vinyloxy-1-butene) and cyclopolymerization yields cyclic structures (penta- and hexa-membered rings) on the polymer backbone. Currently, only CYTOP® is utilized as a base material for GI POFs.

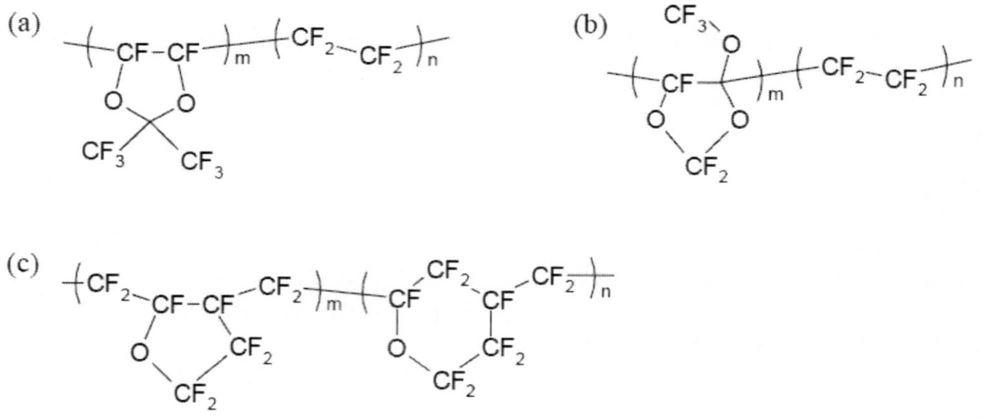

Figure 12. Chemical structures of (a) Teflon® AF, (b) Hyflon® AD, and (c) CYTOP®.

The first CYTOP®-based GI POF (Lucina™), produced by the preform-drawing method, was commercialized by AGC in 2000. The attenuation spectrum is shown in Figure 13 with that of the PMMA-based GI POF of Figure 11 reproduced for comparison. The theoretical attenuation shown in the figure was calculated using Morse potential energy theory and thermally induced fluctuation theory to consider the inherent absorption and scattering losses, respectively [27]. As can be seen in Figure 12, CYTOP® molecules consist solely of C–C, C–F, and C–O bonds. The wavelengths of the fundamental stretching vibrations of these bonds are considerably longer than that of the C–H bond; therefore, the vibrational absorption losses of CYTOP® in the visible to near-infrared region is negligibly small. Based on Eq. (10), the absorption loss due to, e.g., the seventh overtone of the C–F bond, which appears at around 1170 nm, is estimated to be 0.15 dB/km. For CYTOP®, $\rho = 2.03$ g/cm^3, $M = 278$ g/mol, and $N_{C-F} = 10$. In addition to the low molecular vibrational absorption loss, CYTOP® has a fairly low light scattering property because of its low refractive index (n_D = 1.34). The isotropic light scattering loss calculated from Eqs. (8) and (9) is 4.2 dB/km at 650 nm. This is less than half of the 9.5 dB/km at 633 nm of PMMA. The theoretical limit obtained from these calculations indicates that the attenuation can be lowered further by preventing extrinsic loss factors such as contamination during the fabrication process.

The excellent low-loss characteristics of CYTOP®-based GI POFs are sometimes far beyond the requirements for short-range networks. However, the real uniqueness of these fibers are the low material dispersion derived from the low dielectric constant. The bandwidth of multi-mode optical fibers is predominantly influenced by the modal dispersion. However, once the modal dispersion is minimized by forming a GI profile, the influence of the material dispersion on the bandwidth can no longer be ignored. Material dispersion is induced by the wavelength dependence of the refractive index and the finite spectral width of the light source.

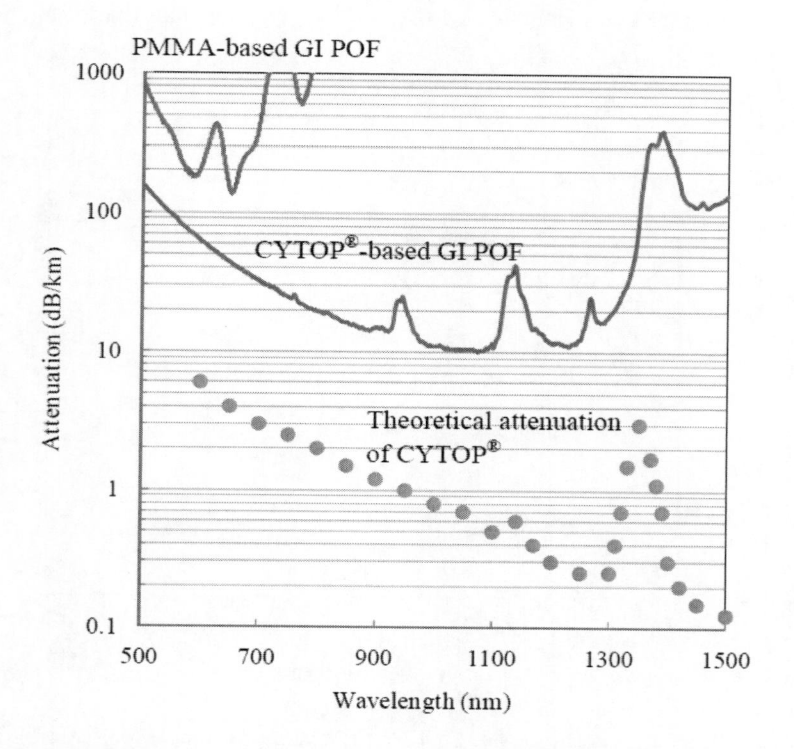

Figure 13. Attenuation spectra of the PMMA-based and CYTOP®-based GI POFs. Modified from Koike, Y., Ishigure, T. (2006) Journal of Lightwave Technology, 24 (12), 4541–4553.

When a pulse of light with a finite spectral width travels through a material, the pulse broadens as a result of the individual components traveling at different velocities. The width of the temporal spreading σ_{τ} after the pulse has travelled a length L in a fiber is given by

$$\sigma_{\tau} = \left| D_{\lambda} \right| \delta_{\lambda} L, \tag{11}$$

$$D_{\lambda} = -\frac{\lambda}{c}\frac{d^2 n}{d\lambda^2}, \tag{12}$$

where D_{λ} is the material dispersion coefficient; δ_{λ}, the initial source width; and c, the velocity of light. The wavelength-dependent refractive index is obtained by approximating several refractive indices measured at different wavelengths using the Sellmeier equation:

$$n^2 - 1 = \sum_i \frac{A_i \lambda^2}{\lambda^2 - \lambda_i^2}, \tag{13}$$

where A_i and λ_i are Sellmeier coefficients. In general, a very good fit is obtained for i = 3. Figure 14 compares the material dispersions of PMMA, CYTOP®, and silica glass. The absolute value of the material dispersion of CYTOP® is smaller than that of silica glass in a wavelength region of ~1 μm.

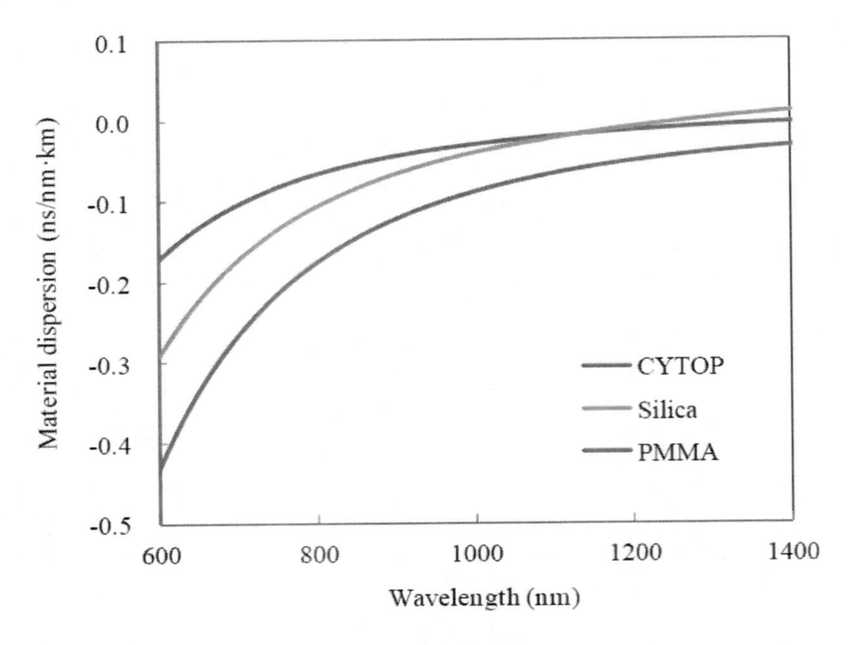

Figure 14. Material dispersion coefficients of CYTOP®, PMMA, and silica. Modified from Koike, Y., Ishigure, T. (2006) Journal of Lightwave Technology, 24 (12), 4541–4553.

The crucial factor defining the refractive index profile in GI POFs is the coefficient g in Eq. (5). Based on the Wentzel-Kramers-Brillouin (WKB) method, the pulse broadenings after a distance L in a GI POF due to the modal and material dispersions are correlated by the coefficient g as follows [28]:

$$\sigma_{modal} = \frac{Ln\Delta}{2c}\frac{g}{g+1}\left(\frac{g+2}{3g+2}\right)^{1/2}\left[S_1{}^2 + \frac{4S_1S_2\Delta(g+1)}{2g+1} + \frac{4S_2{}^2\Delta^2(2g+2)^2}{(5g+2)(3g+2)}\right]^{1/2}, \quad (14)$$

$$\sigma_{material} = \frac{L\sigma_s}{\lambda}\left[\left(-\lambda^2\frac{d^2n_1}{d\lambda^2}\right)^2 - 2\lambda^2\frac{d^2n_1}{d\lambda^2}(n\Delta)S_1\left(\frac{2g}{2g+2}\right) + (n\Delta)^2\left(\frac{g-2-\varepsilon}{g+2}\right)^2\frac{2g}{3g+2}\right]^{1/2}, \quad (15)$$

$$S_1 = \frac{g-2-\varepsilon}{g+2}, S_2 = \frac{3g-2-2\varepsilon}{2(g+2)}, \quad \varepsilon = \frac{-2n_1}{n}\frac{\lambda}{\Delta}\frac{d\Delta}{d\lambda}, n = n_1 - \lambda\frac{dn_1}{d\lambda}.$$

$$\sigma_{total} = \sqrt{\sigma^2_{modal} + \sigma^2_{material}}. \quad (16)$$

Here, the spectral width σ_s is the wavelength dispersion of the input pulse and Δ is defined in Eq. (6). Assuming that the output pulse waveform can be approximated as a Gaussian, the theoretical dependence of the −3 dB bandwidth on the wavelength is given by

$$f_{-3\,dB} = \sqrt{\frac{\ln 2}{2\pi^2}\frac{\ln 2}{2\pi^2}}\frac{1}{\sigma_{total}} = \frac{0.188}{\sigma_{total}}. \tag{17}$$

Figure 15 shows a comparison of the theoretical −3 dB bandwidths of a CYTOP®-based GI POF and a multi-mode GOF. The bandwidths were calculated for an optimal g at 850 nm; the optimum value g_{opt} for maximizing the bandwidth is given by

$$g_{opt} = 2 - \frac{2n_1}{N_1}P - \Delta\frac{\left(4-\frac{2n_1}{N_1}P\right)\left(3-\frac{2n_1}{N_1}P\right)}{5-\frac{4n_1}{N_1}P}, \tag{18}$$

$$N_1 = n_1 - \lambda\frac{dn_1}{d\lambda}, \tag{19}$$

$$P = \frac{\lambda}{\Delta}\frac{d\Delta}{d\lambda}. \tag{20}$$

The calculated result shows clearly that CYTOP®-based GI POFs can have a higher bandwidth than multi-mode GOFs and this was demonstrated in 1999 when AGC and Bell Laboratories reported an experimental transmission of 11 Gbps over 100 m using a CYTOP®-based GI POF [29]. In 2008, a group at the Georgia Institute of Technology and a collaboration between the University of South California and Keio University separately set a new record of 40 Gbps [30, 31]. Subsequently, a group at the Technical University of Eindhoven established a transmission of 47.4 Gbps in 2010 [32]. Furthermore, in 2012, a collaboration between NEC Laboratories and the University of Florida established a 112 Gbps transmission over 100 m [33]. These highly significant results demonstrate that GI POFs can achieve higher bandwidths than multi-mode GOFs.

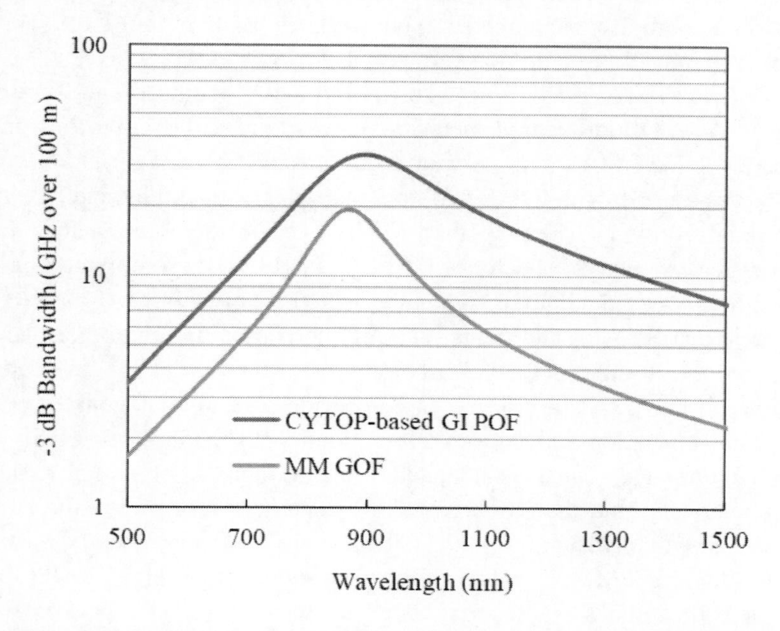

Figure 15. Dependence of the theoretical –3 dB bandwidth on the wavelength in the CYTOP®-based GI POF compared with that in the multi-mode GOF (σ_s = 1.0 nm). Modified from Koike, Y., Ishigure, T. (2006) Journal of Lightwave Technology, 24 (12), 4541–4553.

In 2010, AGC released another CYTOP®-based GI POF called Fontex™. By employing a double cladding structure (a thin layer with a considerably lower refractive index placed around the first cladding), the bending loss was further reduced while the high-speed capacity was maintained, thereby enabling various wiring designs. In addition, continuous fabrication of GI POFs was established by introducing the co-extrusion process.

5. Recent Material Studies

5.1. Partially Halogenated Polymethacrylate Derivatives

There is no doubt that CYTOP® is by far the best material for GI POFs currently available. CYTOP®-based GI POFs have been continuing to break records for the lowest attenuation and highest data transmission speed. However, despite the impressive performance, CYTOP®-based GI POFs have not been widely accepted since they were commercialized mainly due to the prohibitive costs. CYTOP® is a homopolymer made by the cyclopolymerization of perfluoro(butenyl vinyl ether) (PFBVE), prepared by a complex process outlined in [34]. The manufacturing costs are quite high, and CYTOP® resin typically sells for several dollars per gram, which makes it an expensive man-made polymer. However, many of the actual intended purposes and potential applications of POFs do not require the high specifications of CYTOP®. A low-cost fiber capable of a 1 Gbps transmission over ~30 m at the light source wavelength, e.g., 670–680 nm, is currently most in demand. This cannot be achieved with PMMA-based GI POFs due to the high attenuation, but it is not necessary to use CYTOP®. Thus, the partially fluorinated or chlorinated polymethacrylate derivatives discussed in this section have been studied as a possible alternative.

Recently, three partially halogenated methacrylates, shown in Figure 16, have been intensively investigated as a core base material for GI POFs: poly(2,2,2-trifluoroethyl methacrylate) (poly(TFEMA)) [35], poly(2,2,2-trichloroethyl methacrylate) (poly(TClEMA)) [36], and MMA-*co*-pentafluorophenyl methacrylate (PFPhMA) [13]. To form the refractive index profiles, poly(TFEMA) was doped with benzyl benzoate and the other two methacrylates were doped with DPS. All the materials are commercially available; the monomers are slightly more expensive than MMA but considerably cheaper than PFBVE. Figure 17 shows the attenuation spectra of GI POFs based on these polymers in comparison to that of the PMMA-based GI POF in Figure 11. The poly(TFEMA)-based GI preform-drawn fiber was prepared using interfacial-gel polymerization and the other two fibers were prepared using the rod-in-tube method. Further details of the fabrication procedure are given in the literatures [35, 36, 13]. The purpose of employing partially halogenated polymers is to decrease the fiber attenuation and the most important factor here is how many C–H bonds exist per unit volume. This can be calculated from the density of polymer, the molecular weight of monomer unit, and the number of C–H bonds per monomer unit. The C–H bond concentrations of poly(TFEMA) and poly(TClEMA) are 64% and 51%, respectively, of the corresponding value for PMMA. For a copolymer composition of MMA/PFPhMA=65/35 mol%, this value is 68%. These values roughly reflect the change in the peak intensities of 5th and 6th C–H overtones from those of the PMMA. Because of the considerable reduction in the C–H vibrational absorption losses, lower attenuations of less than 200 dB/km have been achieved for source wavelengths of 670–680 nm.

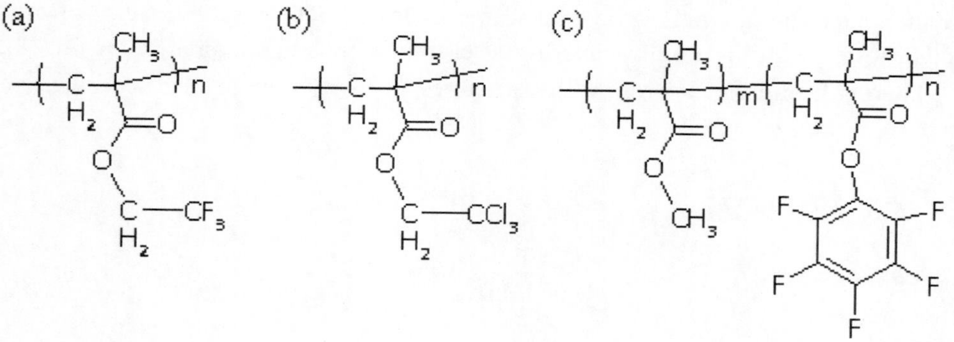

Figure 16. Chemical structures of (a) poly(TFEMA), (b) poly(TClEMA), and (c) MMA-*co*-PFPhMA.

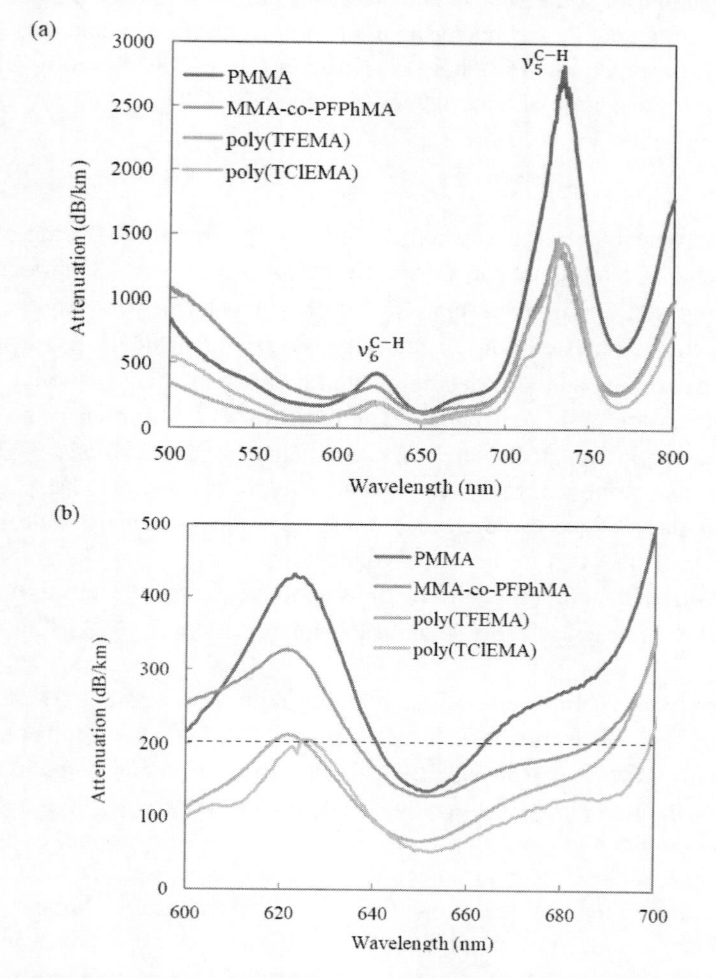

Figure 17. Attenuation spectra of the GI POFs based on PMMA, poly(TFEMA) [35], poly(TClEMA) [36], and MMA-*co*-PFPhMA (65/35 mol%) [13]: (a) 500–800 nm and (b) 600–700 nm.

The low attenuation of the copolymer-based GI POF is particularly noteworthy. As discussed in section 3.1, copolymers generally tend to have a high attenuation because dielectric fluctuations resulting from the copolymerization of monomers of different reactivities can cause non-negligible light scattering losses. This usually far exceeds the value expected from the intrinsic light scattering of each homopolymer expressed by Eq. (8). The light scattering from large heterogeneous structures V_{V2}^{iso} shows an angular dependence and is defined as [37]:

$$V_{V2}^{iso} = \frac{8\pi^3\langle\eta^2\rangle a^3}{\lambda_0^4(1+\nu^2 s^2 a^2)^2}, \tag{21}$$

$$\nu = 2\pi/\lambda, \tag{22}$$

$$s = 2\sin(\theta/2). \tag{23}$$

Here, $\langle\eta^2\rangle$ is the mean square average of the fluctuations of all the dielectric constants; a, the correlation length; and θ, the scattering angle relative to the incident direction. When a two-phase model is assumed such as that described in Figure 18, the correlation length a, which is a measure of the degree of heterogeneity, is given by [38]

$$a = \frac{4V}{S} V_A V_B, \tag{24}$$

where S is the total contact area of the A and B phases with different dielectric constants, V is the total volume, and V_A and V_B are the volume fractions of the A and B phases ($V_A + V_B = 1$), respectively. Eqs. (21-24) indicate that V_{V2}^{iso} can be reduced in two ways. The first is to increase the contact area S between the A and B phases (Figure 18 (b)). By increasing S, the correlation length a shortens and V_{V2}^{iso} decreases, and this is why ideal random copolymers or alternative copolymers are more transparent. The other way is to minimize the mean square average of the fluctuations of all the dielectric constants $\langle\eta^2\rangle$. If $\langle\eta^2\rangle$ is zero, V_{V2}^{iso} is also zero regardless of the monomer reactivity ratios. The copolymer of MMA and PFPhMA corresponds to this case. Since the dielectric constant is almost equal to the square of the refractive index, the value of $\langle\eta^2\rangle$ can be roughly estimated from the difference in the refractive indices of each homopolymer. In the copolymer case, the refractive indices are almost identical (PMMA: $n_D = 1.4914$, poly(PFPhMA): $n_D = 1.4873$), and thus, V_{V2}^{iso} is negligible.

This is an extremely useful method for obtaining highly transparent copolymers. Under actual use conditions, the transmission properties and the environmental resistance characteristics such as the heat resistance, weatherability, and chemical resistance must be considered; the mechanical properties are important for installation requirements. However, identifying homopolymers that meet all the physical properties while maintaining the ultimate transparency required for optical communications is a difficult task. Copolymerization is a general way of modifying the physical properties to meet specific needs but can result in high light scattering losses. Indeed, the number of ideal random copolymer and alternative copolymer combinations is severely limited. Furthermore, an accurate prediction of the monomer reactivity ratio is also quite difficult without undertaking the reaction. However, this method, which inhibits the excess scattering loss by adjusting refractive index of each component, is more realistic and relatively easy. The refractive index decreases by introducing fluorine, whereas it increases with the introduction of chlorine, sulfone, phenyl, and so on. Recently, several copolymer systems have been proposed as a novel GI POF base material in this way [39-41].

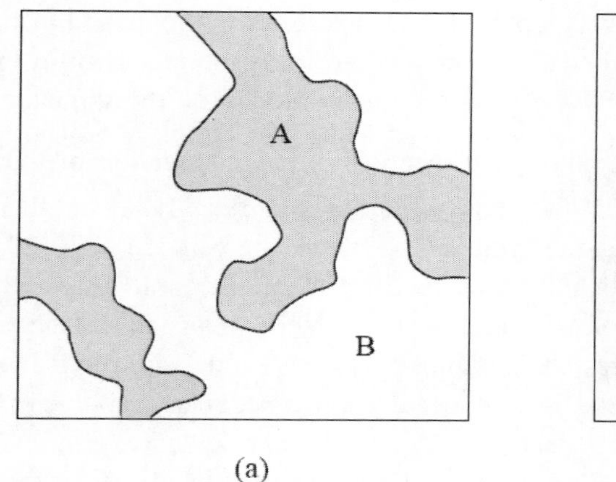

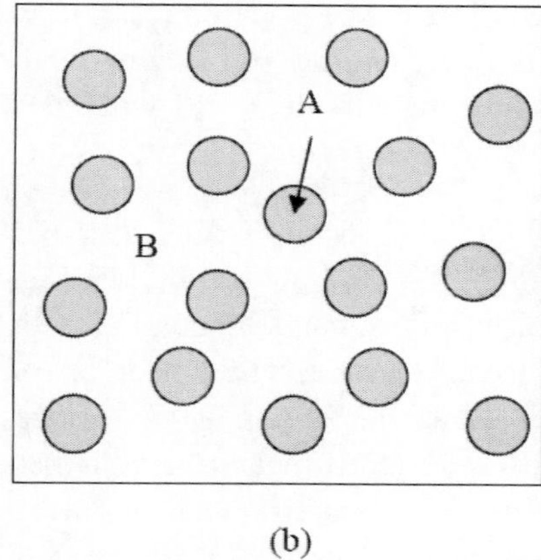

(a) (b)

Figure 18. Two-phase models of heterogeneous structures: the correlation length a is (a) long and (b) short.

In 2010, Sekisui Chemical commercialized a GI POF called GINOVER based on studies of poly(TClEMA). While the polymer had some disadvantages in that it depolymerized during the fabrication process and was considerably brittle, these problems were solved by copolymerizing TClEMA with a small amount of N-cyclohexyl maleimide (cHMI) [42, 43]. Continuous fabrication of this GI POF was also recently established by the co-extrusion process. The most distinctive feature of this fiber is the high temperature resistance. The attenuation of the copolymer-based GI POF, which is 132 dB/km at 660 nm, does not change after heating the fiber at 100 °C for over 2000 h. To the best of our knowledge, this is the first GI POF capable of maintaining a low attenuation at 100 °C for such a long period.

5.2. Partially Fluorinated Polystyrene Derivatives

PSt, another universal optical polymer, is one of the most widely used plastics today. In terms of its mechanical properties and chemical resistance, PSt is slightly inferior to PMMA, but is very attractive as a core base material and is also relatively cheap. As mentioned in section 2.1, PSt has been continuously and extensively studied for SI POFs since the earliest days of POF development. The first PSt-based low-loss SI POF had an attenuation of 114 dB/km at 670 nm [3]. This is the exact wavelength needed for gigabit data transmissions over GI POFs. PSt is a long chain hydrocarbon, and the attenuation in the visible to near-infrared region is also dominated by overtones of the C–H vibrational absorption, as is the case with PMMA. So, why is such a low attenuation possible? The styrene unit is composed of aliphatic and aromatic C–H bonds that resonate at different wavelengths. The absorption wavelengths of the aliphatic C–H bonds corresponding to the 5th, 6th, and 7th overtones are 758, 646, and 562 nm, respectively, whereas the same overtones of the aromatic C–H bonds appear at 714,

608, and 532 nm, respectively. In addition, the positions of the aliphatic C–H overtones from the main chain are shifted slightly to longer wavelengths compared to those for PMMA because of an induced effect from the benzene ring. As a result, the emission wavelength of a VCSEL (670–680 nm) is located at the lowest attenuation window, which is in the center of the 5th aromatic and 6th aliphatic C–H overtones.

In 2012, the first PSt-based GI POF was reported [44]. The fiber was obtained using the preform-drawing and the rod-in-tube methods. The attenuation spectrum is shown in Figure 19 alongside that of the PMMA-based GI POF. The nth overtones of the aliphatic and aromatic C–H bonds are labeled as ν_n and ν'_n, respectively. The attenuation of the PSt-based GI POF is 166–193 dB/km at 670–680 nm, which is significantly lower than the attenuation of the PMMA-based GI POF in the same region (approximately 240–270 dB/km). In spite of this advantage, the possibility of PSt-based GI POF had not been investigated until recently because the refractive index of PSt is as high as 1.59. While this is advantageous for SI POFs because there are a variety of possible cladding polymers, in the case of GI POFs, a higher refractive index means that a higher dopant concentration is necessary. In other words, the T_g of the doped core polymer is further decreased due to the plasticization effect. While the decrement was inhibited to some extent by employing a new dopant, the T_g at the core center where the dopant concentration is the highest is 80 °C and not sufficient for practical use.

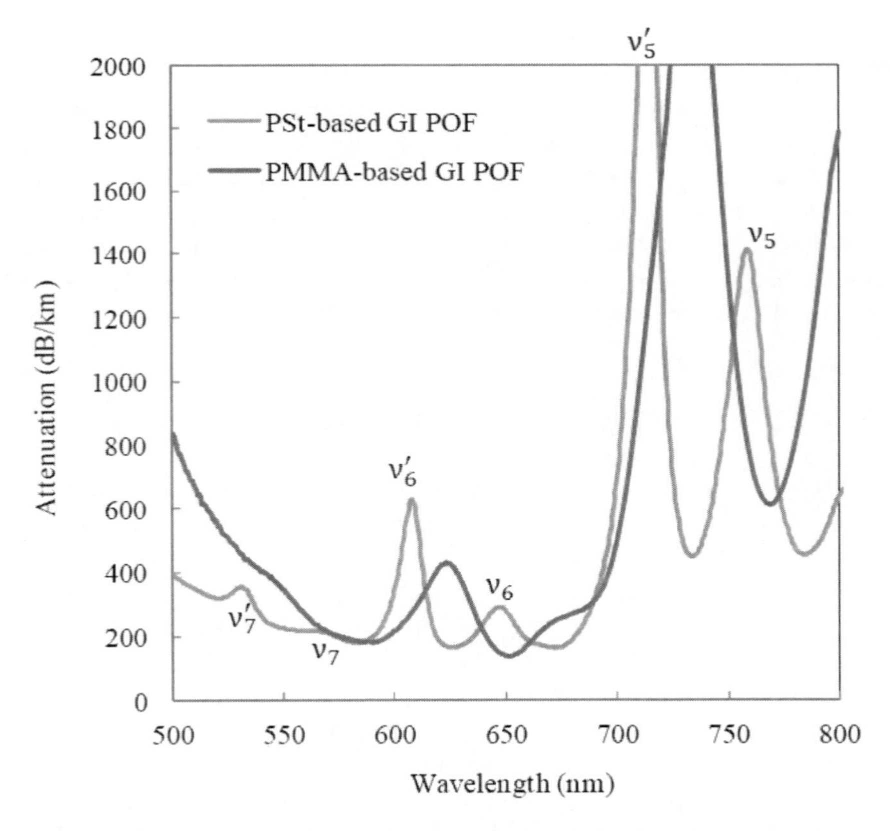

Figure 19. Attenuation spectra of the PSt- and PMMA-based GI POFs. Modified from Akimoto, Y., Asai, M., Koike, K., Makino, K., Koike, Y. (2012) Optics Letters, 37 (11), 1853–1855.

Introducing fluorine substituents is effective for lowering the refractive index, and recently, various partially fluorinated derivatives of PSt have been studied [45-47]. Figure 20 shows the refractive indices and T_g values of these polymers. The refractive index is dependent on the molar refraction [R] and a molecular volume V, as expressed by the Lorentz-Lorenz equation:

$$n = \sqrt{\left(2\frac{[R]}{V}+1\right)/\left(1-\frac{[R]}{V}\right)}. \tag{25}$$

Since the fluorine substituents result in a lower [R] and larger V than those for hydrocarbons, the refractive index is lowered roughly in accordance with the proportion of fluorine atoms contained in each monomer unit. Figure 20 also shows that the refractive index is not significantly affected by the substitution site. This indicates that the density of the polymer, or more specifically, the ratio of fluorine substituents per unit volume, does not significantly vary according to position.

The polymer T_g, however, is strongly affected by both the size and position of the substituent. When all the hydrogen atoms in the benzene ring are substituted with fluorine, i.e., poly(2,3,4,5,6-pentafluorostyrene), the T_g is almost the same as that of PSt. However, if some hydrogen atoms are substituted with CF_3 groups, then the polymer tends to show a higher T_g since such bulky substituents reduce the segmental mobility due to the steric hindrance. In particular, a substituent at the ortho position, which is the nearest to the main chain, is the most efficient in enhancing T_g. Currently, a novel GI POF based on poly(2-trifluoromethyl styrene), which has an extremely high T_g of 178 °C, is under investigation.

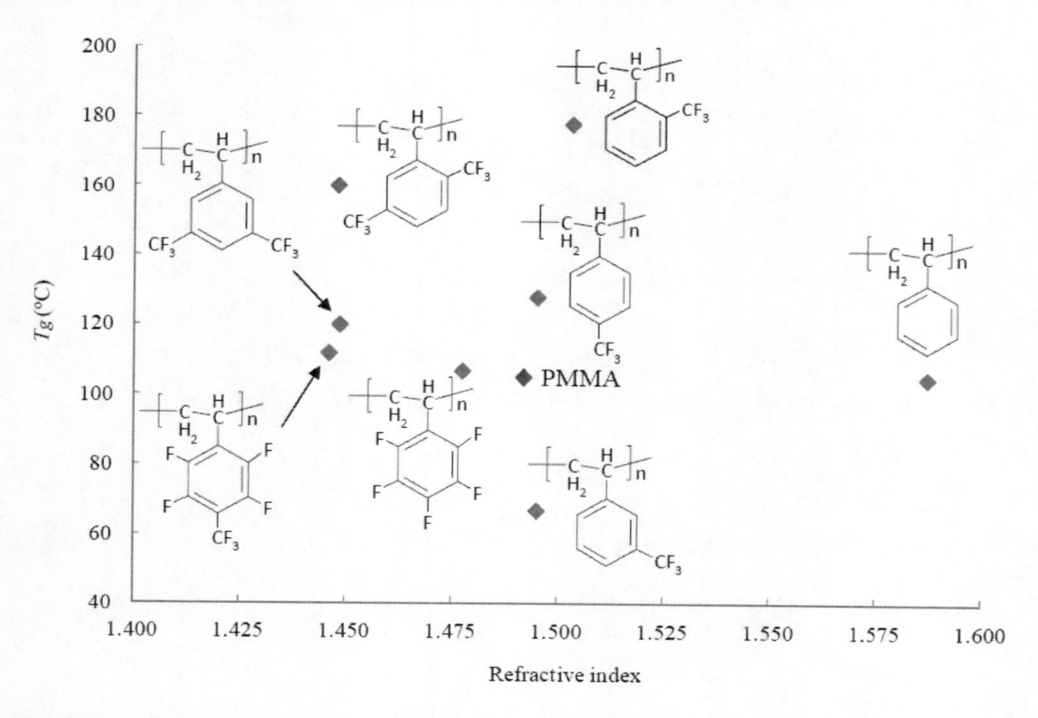

Figure 20. Refractive indices and glass transition temperatures T_g of the partially fluorinated derivatives of PSt.

5.3. Perfluorinated Polydioxolane Derivatives

The recent representative studies of partially halogenated polymers described in sections 5.1 and 5.2 were undertaken to obtain GI POFs with sufficiently low attenuations at specific light source wavelengths at low cost. These polymers can be categorized as an intermediate material between PMMA and CYTOP® from the viewpoints of both performance and material cost, which meet current demand.

Considerable effort has also been devoted to developing new perfluoropolymers as a substitute for CYTOP®. To realize less expensive amorphous perfluoropolymers, various structures and synthetic routes for each possible substitute have been studied. One of the most efficient and simplest routes is for poly(perfluoro-2-methylene-1,3-dioxolane) with various substituents at the 4 and 5 positions [48-50]. The synthesis route is shown in Scheme 1. The monomers can be obtained via direct fluorination of hydrocarbon precursors prepared from methyl pyruvate and diols. These perfluorodioxolane monomers are readily polymerized in the bulk using a perfluoroperoxide as a free radical initiator. The monomer structures and basic properties are summarized in Table 1. The most remarkable point is that some derivatives with relatively bulky substituents show excellent T_g values of up to 185 °C. Considering that the operating temperature of CYTOP®-based GI POFs is limited to 70 °C, there is no doubt that improving the thermal resistance is an important topic for future development. However, such high T_g amorphous polymers tend to be brittle, and these perfluorodioxolane polymers are no exception. Currently, various ways of improving the mechanical properties and preparing the fibers are under investigation.

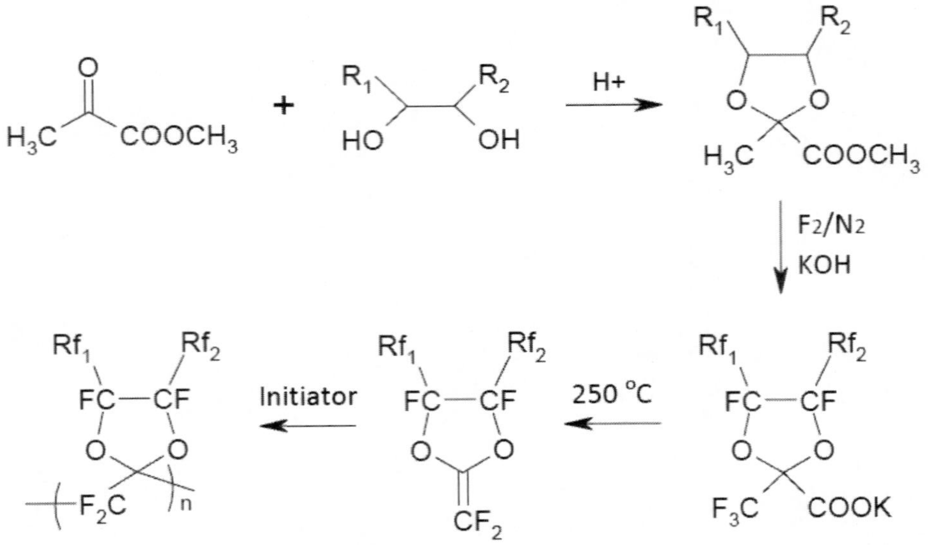

Scheme 1. Synthesis procedure of poly(perfluoro-2-methylene-1,3-dioxolane). Modified from Okamoto, Y., Teng, H. (2009) Chemistry Today, 27 (4), 46–48.

Table 1. Structures and basic properties of perfluoro-2-methylene-1,3-dioxolanes. Modified from Okamoto, Y., Teng, H. (2009) Chemistry Today, 27 (4), 46–48

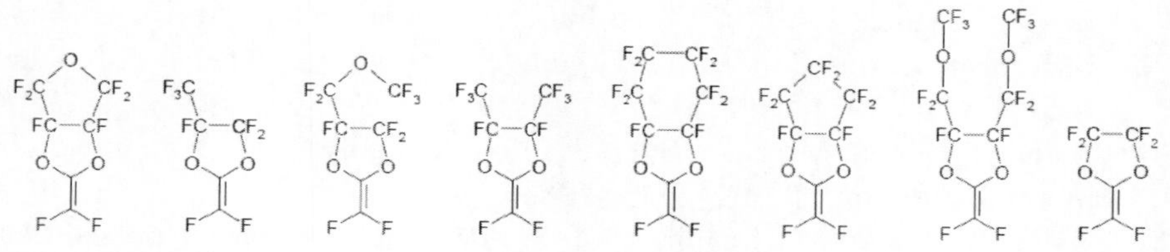

Monomer	A	B	C	D	E	F	G	H
Polymerization rate ($\times 10^4$ mol/Ls)[a]	1.66	1.56	1.40	0.15	0.25	0.18	1.02	1.63
Polymer T_g (°C)	168	135	101	165	165	185	93	110
Polymer $RI_{632\,nm}$	1.3570	1.3310	1.3328	1.3280	1.3420	1.3460	1.3520	1.3443

a) In FC113, [Monomer] =1.6 $molL^{-1}$, [perfluoro dibenzoyl peroxide] = 0.05 $molL^{-1}$, 41 °C.

CONCLUSION

This chapter has reviewed the status of GI POF developments in the last three decades from the perspectives of materials and fabrication techniques. Fundamental studies based on photonics and polymer science have slowly but steadily improved the performance of GI POFs. In addition to the many advantages derived from the nature of polymeric materials, GI POFs have acquired an extremely large capacity that exceeds those of multi-mode GOFs. GOFs, especially single-mode fibers, are highly effective for long-haul and metropolitan optical networks, but are too fragile for short-range communications such as local area networks and interconnections within data centers.

Compared with conventional copper cables, GI POFs provide a number of benefits such as superior performance, extended reach, physical flexibility for better space utilization, easier installation, and a reduced power consumption. GI POFs are no longer just alternatives to other transmission media; they have created a new transmission category in the communications field. However, there are still outstanding problems and areas for improvement. To be widely accepted in the market, development of less costly materials is indispensable. In addition, increasing and continuous effort must be devoted to developing high T_g materials that will expand the range of future GI POF applications.

ACKNOWLEDGMENT

This work was partly supported by the Japan Society for the Promotion of Science (JSPS) through its "Funding Program for World-Leading Innovative R&D on Science and Technology (FIRST Program)."

REFERENCES

[1] Mitsubishi Rayon Co. Ltd. (1974) *Light transmitting fibers and method for making same*. UK Patent 1,431,157, filed Jun. 20, 1974 and issued Apr. 7, 1976.

[2] Mitsubishi Rayon Co. Ltd. (1974) *Light transmitting filament*. UK Patent 1,449,950, filed Nov. 15, 1974 and issued Sep. 15, 1976.

[3] Kaino, T., Fujiki, M., Nara, S. (1981) Low-loss polystyrene core-optical fibers. *Journal of Applied Physics*, 52 (12), 7061-7063.

[4] Kaino, T., Fujiki, M., Jinguji, K. (1984) Preparation of plastic optical fibers. *Review of the Electrical Communication Laboratories*, 32 (3), 478-488.

[5] Ohtsuka, Y. (1973) Light-focusing plastic rod prepared from diallyl isophtalate-methyl methacrylate copolymerization. *Applied Physics Letters*, 23 (5), 247-248.

[6] Ohtsuka, Y., Hatanaka, Y. (1976) Preparation of light-focusing plastic fiber by heat-drawing process. *Applied Physics Letters*, 29 (11), 735-737.

[7] Ohtsuka, Y., Nakamoto, I. (1976) Light-focusing plastic rod prepared by photocopolymerization of methacrylic esters with vinyl benzoates. *Applied Physics Letters,* 29 (9), 559-561.

[8] Ohtsuka, Y., Koike, Y., Yamazaki, H. (1981) Studies on the light-focusing plastic rod. 6: The photocopolymer rod of methyl methacrylate with vinyl benzoate. *Applied Optics*, 20 (2), 280-285.

[9] Koike, Y., Takezawa, Y., Ohtsuka, Y. (1988) New interfacial-gel copolymerization technique for steric GRIN polymer optical waveguides and lens arrays. *Applied Optics,* 27 (3), 486-491.

[10] Koike, Y. (1991) High-bandwidth graded-index polymer optical fibre. *Polymer*, 32 (10), 1737-1745.

[11] Koike, Y. (1996) Optical resin materials with distributed refractive index, process for producing the materials, and optical conductors using the materials. US Patent 5,541,247, filed Jun. 17, 1993 and issued Jul. 30, 1996.

[12] Ishigure, T., Nihei, E., Koike, Y. (1994) Graded-index polymer optical fiber for high-speed data communication. *Applied Optics*, 33 (19), 4261-4266.

[13] Koike, K., Kado, T., Satoh, Z., Okamoto, Y., Koike, Y. (2010) Optical and thermal properties of methyl methacrylate and pentafluorophenyl methacrylate copolymer: Design of copolymers for low-loss optical fibers for gigabit in-home communications. *Polymer,* 51 (6), 1377-1385.

[14] Mitsubishi Rayon Co. Ltd. (1989) JP Patent 1989-189602, filed Jan. 25, 1988 and issued Jul. 28, 1989.

[15] Mitsubishi Rayon Co. Ltd. (1990) JP Patent 1990-16505, filed Jul. 5, 1988 and issued Jan. 19, 1990.

[16] Mitsubishi Rayon Co. Ltd. (1990) JP Patent 1990-33104, filed Jul. 23, 1988 and issued Feb. 2, 1990.

[17] Ho, B. C., Chen, J. H., Chen, W. C., Chang, Y. H., Yang, S. Y., Chen, J. J., Tseng, T. W. (1995) Gradient-index polymer fibers prepared by extrusion. *Polymer Journal*, 27 (3), 310-313.

[18] Chen, W. C., Chen J. H., Yang, S. Y., Cherng, J. Y., Chang, Y. H., Ho, B. C. (1996) Preparation of gradient-index (GRIN) polymer fibers for imaging applications. *Journal of Applied Polymer Science*, 60 (9), 1379-1383.

[19] Koike, Y., Nihei, R. (1997) Method of manufacturing plastic optical transmission medium. US Patent 5,593,621, filed Apr. 14, 1994 and issued Jan. 14, 1997.

[20] Lucent Technologies Inc. (2001) *Process for fabricating plastic optical fiber*. US Patent 6,254,808 B1, filed May 27, 1999 and issued Jul. 3, 2001.

[21] Urbach, F. (1953) The long-wavelength edge of photographic sensitivity and of the electronic absorption of solids. *Physical Review*, 92 (5), 1324.

[22] Einstein, A. (1910) Theorie der Opaleszenz von homogenen Flüssigkeiten und Flüssigkeitsgemischen in der Nähe des kritischen Zustandes. *Annalen der Physik*, 338 (16), 1275-1298.

[23] Hellwege, K. –H., Knappe, W., Lehmann, P. (1962) Die isotherme Kompressibilität einiger amorpher und teilkristalliner Hochpolymerer im Temperaturbereich von 20–250°C und bei Drucken bis zu 2000 kp/cm^2. Kolloid-Zeitschrift und Zeitschrift für Polymere, 183 (2), 110-120.

[24] Koike, Y., Tanio, N., Ohtsuka, Y. (1989) Light scattering and heterogeneities in low-loss poly(methyl methacrylate) glasses. *Macromolecules*, 22 (3), 1367-1373.

[25] Koike, Y., Matsuoka, S., Bair, H. E. (1992) Origin of excess light scattering in poly(methyl methacrylate) glasses. *Macromolecules*, 25 (18), 4807-4815.

[26] Groh, W. (1988) Overtone absorption in macromolecules for polymer optical fibers. *Die Makromolekulare Chemie*, 189 (12), 2861-2874.

[27] Tanio, N., Koike, Y. (2000) What is the most transparent polymer? *Polymer Journal*, 32 (1), 43-50.

[28] Olshansky, R., Keck, D. B. (1976) Pulse broadening in graded-index optical fibers. *Applied Optics*, 15 (2), 483-491.

[29] Giaretta, G., White, W., Wegmueller, M., Yelamarty, R. V., Onishi, T. (1999) 11 Gb/sec data transmission through 100 of perfluorinated graded-index polymer optical fiber. *Technical Digest of Optical Fiber Communication Conference and Exhibit*, February 21-26, 1999, San Diego, US, PD14/1-PD14/3.

[30] Polley, A., Ralph, S. E. (2008) 100 m, 40 Gb/s plastic optical fiber link. Proceeding of Conference on Optical Fiber Communication/National Fiber Optic Engineers Conference, February 24-28, 2008, San Diego, US, OWB2.

[31] Nuccio, S. R., Christen, L., Wu, X., Khaleghi, S., Yilmaz, O., Willner, A. E., Koike, Y. (2008) Transmission of 40 Gb/s DPSK and OOK at 1.55 μm through 100 m of plastic optical fiber. *Proceeding of 34th European Conference on Optical Communication*, September 21-25, 2008, Brussels, Belgium, We.2.A.4.

[32] Yang, H., Lee, S. C. J., Tangdiongga, E., Okonkwo, C., van den Boom, H. P. A, Breyer, F., Randel, S., Koonen, A. M. J. (2010) 47.4 Gb/s transmission over 100 m graded-index plastic optical fiber based on rate-adaptive discrete multimode modulation. *Journal of Lightwave Technology*, 28 (4), 352-359.

[33] Shao, Y., Cao, R., Huang, Y.-K., Ji, P. N., Zhang, S. (2012) 112-Gb/s transmission over 100 m of graded-index plastic optical fiber for optical data center applications. *Proceeding of Conference on Optical Fiber Communication/National Fiber Optic Engineers Conference*, March 4-8, 2012, Los Angeles, US, OW3J.

[34] Hung, M.-H., Resnik, P. R., Smart, B. E., Buck, W. H. (1996) Fluorinated plastics, amorphous. In Salamone, J. C. Editor, *Polymeric Materials Encyclopedia*: Vol. 4, 2466-2476, New York, US, CRC Press, Inc.

[35] Koike, K., Koike, Y. (2009) Design of low-loss graded-index plastic optical fiber based on partially fluorinated methacrylate polymer. *Journal of Lightwave Technology*, 27 (1), 41-46.

[36] Asai, M., Inuzuka, Y., Koike, K., Takahashi, S., Koike, Y. (2011) High-bandwidth graded-index plastic optical fiber with low-attenuation, high-bending ability, and high-thermal stability for home-networks. *Journal of Lightwave Technology*, 29 (11), 1620-1626.

[37] Debye, P., Bueche, A. M. (1949) Scattering by an inhomogeneous solid. *Journal of Applied Physics*, 20 (6), 518-525.

[38] Debye, P., Anderson, H. R., Brumberger, H. (1957) Scattering by an inhomogeneous solid. II. The correlation function and its application. *Journal of Applied Physics*, 28 (6), 679-683.

[39] Koike, K., Mikes, F., Koike, Y., Okamoto, Y. (2008) Design and synthesis of graded index plastic optical fibers by copolymeric. *Polymers for Advanced Technology*, 19 (6), 516-520.

[40] Koike, K., Mikes, F., Okamoto, Y., Koike, Y. (2009) Design, synthesis, and characterization of a partially chlorinated acrylic copolymer for low-loss and thermally stable graded index plastic optical fibers. *Journal of Polymer Science Part A: Polymer Chemistry,* 47 (13), 3352-3361.

[41] Lou, L., Koike, Y., Okamoto, Y. (2012) A novel copolymer of methyl methacrylate with N-pentafluorophenyl maleimide: High glass transition temperature and highly transparent polymer. *Polymer*, 52 (16), 3560-3564.

[42] Nakao, R., Kondo, A., Koike, Y. (2012) Fabrication of high glass transition temperature graded-index plastic optical fiber: Part 1-Material preparation and characterizations. *Journal of Lightwave Technology*, 30 (2), 247-251.

[43] Nakao, R., Kondo, A., Koike, Y. (2012) Fabrication of high glass transition temperature graded-index plastic optical fiber: Part 2-Fiber fabrication and characterizations. *Journal of Lightwave Technology*, 30 (7), 969-973.

[44] Akimoto, Y., Asai, M., Koike, K., Makino, K., Koike, Y. (2012) Poly(styrene)-based graded-index plastic optical fiber for home networks. *Optics Letters*, 37 (11), 1853-1855.

[45] Lou, L., Koike, Y., Okamoto, Y. (2010) Synthesis and properties of copolymers of methyl methacrylate with 2,3,4,5,6-pentafluoro and 4-trifluoromethyl 2,3,5,6-tetrafluoro styrenes: An intrachain interaction between methyl ester and fluoro aromatic. *Journal of Polymer Science Part A: Polymer Chemistry*, 48 (22), 4938-4942.

[46] Teng, H., Lou, L., Koike, K., Koike, Y., Okamoto, Y. (2011) Synthesis and characterization of trifluoromethyl substituted styrene polymers and copolymers with methacrylates: Effects of trifluoromethyl substituent on styrene. *Polymer,* 52 (4), 949-953.

[47] Koike, K., Teng, H., Koike, Y., Okamoto, Y. (2011) Trifluoromethyl-substituted styrene-based polymer optical fibers. *Proceeding of The 20th International Conference on Plastic Optical Fibers,* September 14-16, 2011, Bilbao, Spain, 031.

[48] Mikes, F., Yang, Y., Teraoka, I., Ishigure, T., Koike, Y., Okamoto, Y. (2005) Synthesis and Characterization of an Amorphous Perfluoropolymer: Poly(perfluoro-2-methylene-4-methyl-1,3-dioxolane). *Macromolecules*, 38 (10), 4237-4245.

[49] Mikes, F., Teng, H., Kostov, G., Ameduri, B., Koike, Y., Okamoto, Y. (2009) Synthesis and characterization of perfluoro-3-methylene-2,4-dioxabicyclo[3,3,0] octane: Homo- and copolymerization with fluorovinyl. *Journal of Polymer Science Part A: Polymer Chemistry,* 47 (23), 6571-6578.

[50] Okamoto, Y., Teng, H. (2009) Synthesis and properties of amorphous perfluorinated polymers. *Chemistry Today*, 27 (4), 46-48.

In: Optical Fibers: New Developments
Editor: Marco Pisco

ISBN: 978-1-62808-425-2

Chapter 3

BDK-DOPED POLYMER OPTICAL FIBER (BPOF) GRATINGS: DESIGN, FABRICATION AND SENSING APPLICATION

***Yanhua Luo*[1], *Qijin Zhang*[2] *and Gang-Ding Peng*[1]**
[1]School of Electrical Engineering and Telecommunications,
University of New South Wales, Sydney, NSW, Australia
[2]CAS Key Laboratory of Soft Matter Chemistry,
Department of Polymer Science and Engineering,
Anhui Key Laboratory of Optoelectronic Science and Technology,
University of Science and Technology of China, Hefei, Anhui, China

ABSTRACT

Benzil Dimethyl Ketal (BDK) doped Polymer Optical Fiber (POF) gratings have many advantages, such as lower Young's modulus, higher thermo-expansion and thermo-optic coefficients, better flexibility, over their silica fiber counterparts. Therefore it is important and necessary to study their development, properties and applications. In this chapter, the photosensitive BDK-doped POF (BPOF) and its gratings are reviewed with regard to design, fabrication, characterization to application. In particular, the applications based on both single mode (SM) and multimode (MM) BPOF gratings are presented to show the great potential of BPOF gratings for the temperature, stress and strain sensing.

1. INTRODUCTION

POF gratings have been attracted much attention for their unique properties such as lower Young's module, higher thermo-expansion and thermo-optic coefficient, better flexibility as well as their clinic favorability, compared with their counterparts in silica fibers [1, 2]. These properties allow the simpler and cheaper use with lower resolution, large yield limit and enable the small temperature change and stress sensing [1, 3-9]. And optical devices based on

POF gratings will allow device properties to be readily controlled thermally or mechanically with wide tuning range [5, 9]. Since the first demonstration in 1999 [10], POF gratings have been fabricated in POF with different fiber structures and materials employing different writing methods and mechanisms [1, 8, 11-18], and used in a series sensing fields, such as stress, strain, temperature, humidity, bend, etc. [4, 6, 7, 9, 19-28].

In our previous work [3, 18, 29-34], much attention has been paid on the design, fabrication of BPOF gratings and their sensing applications (temperature, stress and strain sensing), which starts from basic molecules to photonic devices. This short review presents results based on these work in an order as following:

1. Design of BPOF gratings [18, 29];
2. Fabrication of BPOF and its gratings [18, 29, 30, 35];
3. Sensing applications with BPOF gratings [3, 30-34, 36].

As the contents shown, this chapter mainly concentrates on the published work, although there is still much work performed in our group at present. However, by systematic analysis, it is realized that the development of high photosensitive POF, and novel POF gratings with higher stability and sensitivity from molecular level to optical device, is important for sensing applications of photonic polymer materials, especially POF. The principle will guide our future work in developing the new POF gratings and their photonic devices.

2. Design of BPOF Gratings

2.1. Photosensitivity of BPOF

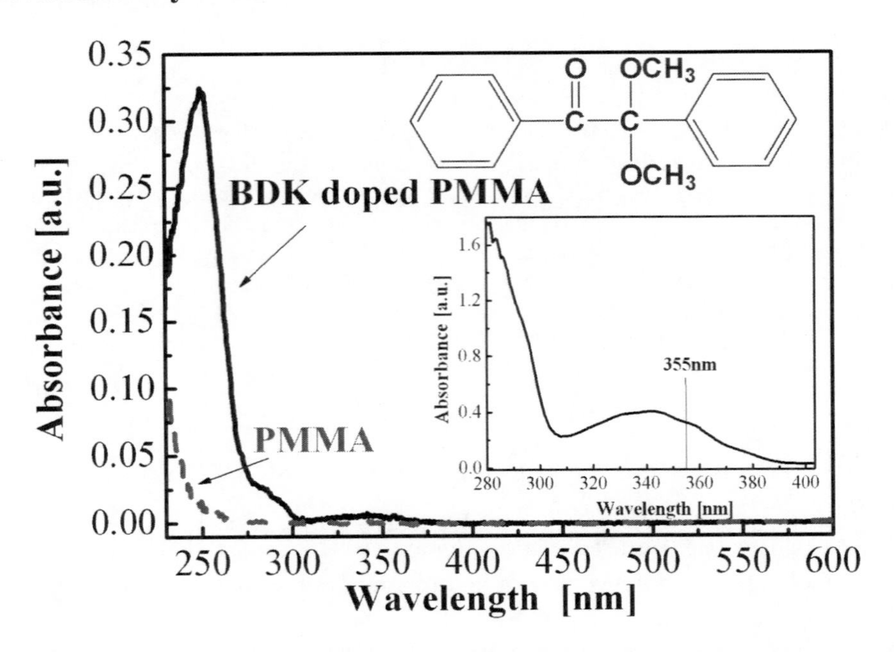

Figure 1. Absorption of PMMA and PMMA doped with BDK [18]. The inset is the absorption of BDK in tetrahydrofuran.

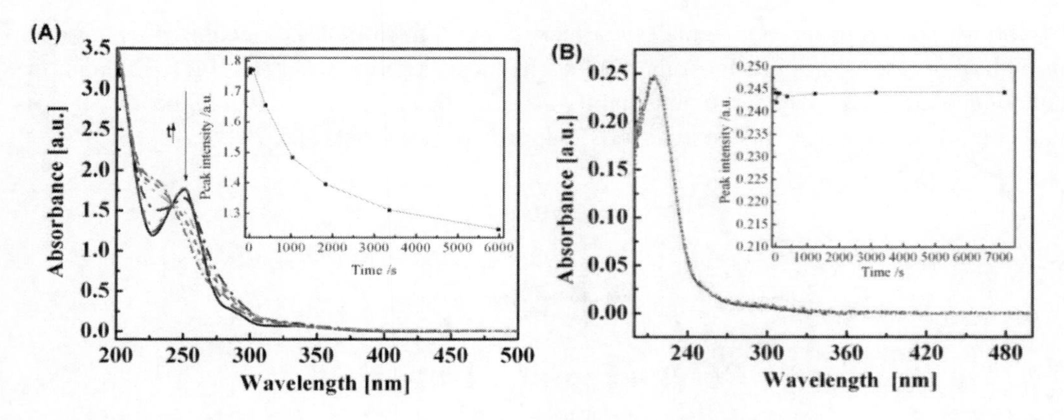

Figure 2. The absorption spectra of the PMMA film doped (A) with 10wt% BDK and (B) without BDK after varying 365 nm UV exposure. The inset is its corresponding absorption peak intensity with different exposure time [18].

In the past three decades, the innate photosensitivity of poly(methyl methacrylate) (PMMA) after irradiation with UV light was discovered by Tomlinson et al. [37], but the photosensitivity was too weak to make POF gratings with good performance and high strength because of the high power and long UV irradiation time [8, 38]. Generally, to enhance the photosensitivity of PMMA based POF, different photosensitive materials, such as trans-4-stilbenemethanol, methyl vinyl ketone (MVK), fluorescein, etc have been introduced into the core of POFs [1, 8, 14, 16, 18, 39, 40]. The mechanism of the photosensitivity can be classified into five kinds, including photobleaching [1], photopolymerization [10], photoisomerization [8], photodgradation [16, 18] and photoalignment [39, 40].

In this chapter, to fabricate the photosensitive POF gratings, *the first step* is to choose the photosensitive material. Here, an economical, classic and high photosensitive dye-BDK are introduced into the fiber core, which exhibits photodegradation into two radicals under a very wide band of UV illumination [18, 41, 42]. For better knowing the photosensitivity of BDK-doped PMMA based POF system, the UV absorption properties of PMMA and PMMA doped with BDK are studied as shown in Figure 1 and Figure 2 [18].

Two new absorption bands at about 250 nm and 344 nm are respectively attributed to the strong π-π^* transition and the weak n-π^* transition of BDK, demonstrated by the absorption of BDK in tetrahydrofuran shown in the inset of Figure 1 [18, 43]. Especially, an increase in absorption at 355 nm, linked with the n-π^* transition of BDK in comparison with PMMA without doping showing the absorption difference between the PMMA and PMMA doped with BDK [18]. In addition, the PMMA material has negligible absorption beyond 250 nm, which means that the light longer than 300 nm will transmit through the cladding material with low loss. And the absorption at ~355 nm is not that high, so using 355 nm UV light as writing source would be a choice to obtain bulk gratings in core materials.

Seen from Figure 2(a), after 365±5 nm exposure with 7.5 mW/cm^2 the absorption of PMMA doped with 10wt% BDK shows significant decrease at 250 nm. In comparison, the absorption of PMMA without doping shows no change as shown in Figure 2(b), showing evident photosensitivity of PMMA doped with BDK at 365 nm [18]. When BDK is irradiated by 365 nm UV light, it will decompose and produce two free radicals by an α-splitting [41, 42]. Then, a series of reactions will happen in the following as shown in Figure 3, including the polymerization of the rudimental monomers coming from POF fabrication [1], photolock of high index ketal fragments, other photochemical reaction of ketal [41, 42] and the photodegradation of polymer main chain [44]. All these reactions would be the factors inducing refractive index change in the BDK-doped core materials [18, 29].

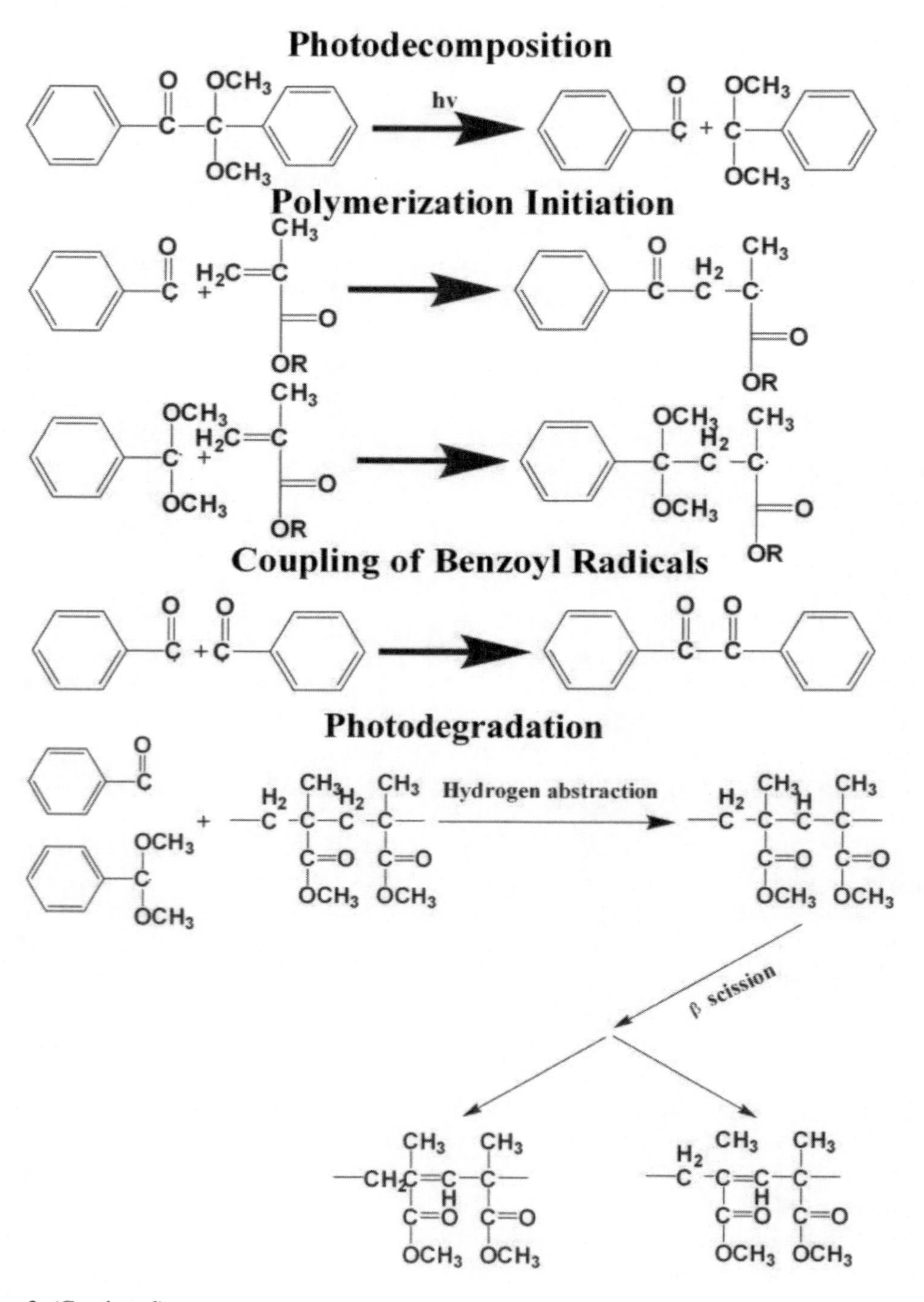

Figure 3. (Continued).

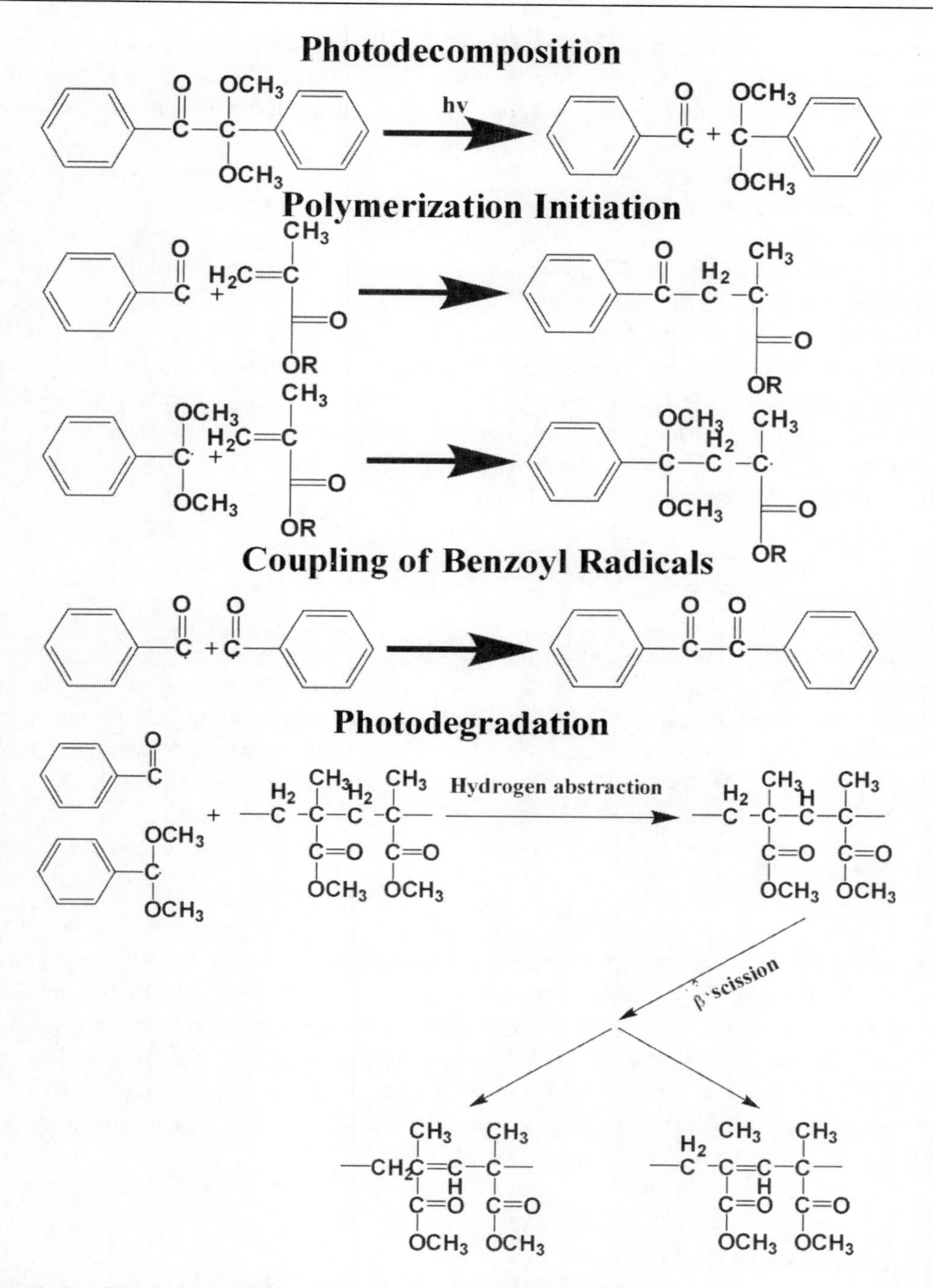

Figure 3. Possible photochemical reactions in the core of BPOF [29].

2.2. Composition of BPOF

When the photosensitive material has been decided, *the second step* is to confirm the main composition of the photosensitive POF, which is the most important factor to influence the refractive index and optical properties of POF. The refractive indices of the core and cladding materials are usually controlled by adding benzyl methacrylate (BzMA) and ethyl methacrylate (EMA) or butyl acrylate (BA) to the core and cladding, which vary with their compositions.

For the core of photosensitive BPOF, MMA, EMA, BzMA and BDK or MMA, BA and BDK are mixed to be polymerized as the core materials, where EMA and BzMA, or BA are used to control the refractive index, and BDK is used as the photosensitive agent [30]. For the cladding of photosensitive BPOF, two types of materials are used, which are the copolymers of EMA and MMA or BA and MMA.

According to the Lorentz-Lorenz equation, the Bragg condition, the weakly guiding condition and the experimental condition [29], two kinds of compositions of the core and cladding materials are selected as listed in Table 1, meanwhile, the refractive indices have also been confirmed. Besides, small quantity of initiators and chain transfer agents are introduced into the BPOF for the polymerization of the core and cladding, and the control of the mechanical and thermal properties of BPOF [29, 45].

2.3. Structure of BPOF

When the compositions have been decided, *the third step* is to confirm the structure parameters of BPOF. Generally, to obtain a high reflectivity and be compatible with the available photonic devices, a SM BPOF is more favorable for fabricating POF gratings [29]. According to the SM condition, the normalized frequency should obey the following equation [46]:

$$V = \frac{2\pi}{\lambda} \cdot a\sqrt{n_{co}^2 - n_{cl}^2} < 2.4 \tag{1}$$

where 2a and λ are the diameter of fiber core and the operating wavelength of BPOF gratings, respectively. From Eq. (1), it can be seen that the diameter of BPOF core should be controlled in terms of the difference of n_{co} and n_{cl}, which are mainly dependent upon their compositions. Eq. (1) also shows that the smaller the index difference between the core and cladding, the larger the BPOF core and the easier the BPOF fabrication. However, too low difference will bring the difficulty to constitute a waveguide [29], due to the diffusion of core materials during the interfacial-gel polymerization [30, 32].

2.4. Simulated BPOF Gratings

According to the compositions, the refractive indices and structure parameters of BPOF have also been confirmed, the BPOF grating reflection spectrum can be simulated in *the fourth step* to study the possible properties of BPOF gratings.

Table 1. Fiber parameters and grating fabrication conditions

Specification of BPOF for Grating A	Grating Fabrication for Grating A
Cladding(MMA:BA=68:32v%) Core(MMA:BA:BDK=62.6:29.4:8v%) $n_{co} - n_{cl}$: ~0.011 n_{co}: 1.479 D_{fiber}: ~290 μm D_{core}: ~21 μm	355nm frequency-tripled Nd:YAG pulse laser Frequency: 10 Hz Pulse width: 6 ns Average power intensity 673 mW/cm^2 Grating length: 6 mm Phase mask period: 1.0614 μm
Specification of BPOF for Grating B	Grating Fabrication for Grating B
Cladding(MMA:EMA=60:40v%) Core(MMA:EMA:BzMA:BDK=50:45:3:2 v%) $n_{co} - n_{cl}$: ~ 0.001 n_{co}: 1.482 D_{fiber}: ~230.4 μm D_{core}: ~11.8 μm	355nm frequency-tripled Nd:YAG pulse laser Frequency: 10 Hz Pulse width: 6 ns Average power intensity: 200 mW/cm^2 Grating length: 6mm Phase mask period: 1.0614 μm
Specification of BPOF for Grating C	Grating Fabrication for Grating C
Cladding(MMA:BA=68:32v%) Core(MMA:BA:BDK=62.6:29.4:8v%)) $n_{co} - n_{cl}$: ~0.011 n_{co}: 1.479 D_{fiber}: ~290 μm D_{co}: ~21 μm	355nm frequency-tripled Nd:YAG pulse laser Frequency: 10Hz, Pulse width: 6 ns One beam average intensity: 153 mW/cm^2 Grating length: 6mm Phase mask period: 1.0614 μm
Specification of BPOF for Grating D	Grating Fabrication for Grating D
Cladding(MMA: EMA = 60:40 v%) Core(MMA:EMA:BzMA:BDK = 50:45:3:2 v%) $n_{co} - n_{cl}$: ~ 0.001 n_{co}: 1.482 D_{fiber}: ~ 230 μm D_{core}: ~ 11 μm	355nm frequency-tripled Nd:YAG pulse laser Frequency: 10 Hz Pulse width: 6 ns Average power intensity:57 mW/cm^2 Grating length: 6 mm Phase mask period: 1.0614 μm
Specification of BPOF for Grating E	Grating Fabrication for Grating E
Cladding (MMA:BA=68:32v%) Core (MMA:BA:BDK=62.6:29.4:8v%)) $n_{co} - n_{cl}$: ~0.011 n_{co}: 1.479 D_{fiber}:~ 300 μm D_{core}:~23 μm	355nm frequency-tripled Nd:YAG pulse laser Frequency: 10 Hz Pulse width: 6 ns Average power intensity: 69 mW/cm^2 Grating length: 6 mm Phase mask period: 1.0614 μm

The power reflection coefficients R from a fiber Bragg grating (FBG) in BPOF can be given by [29, 46, 47]:

$$R = \frac{\kappa^2 \sinh^2(qL)}{\delta^2 \sinh^2(qL) + q^2 \cosh^2(qL)} \tag{2}$$

where κ is the coupling coefficient ($\kappa = \frac{\pi \delta n}{\lambda}$, where λ is the light wavelength and δn is the index variation in the fiber core), L is the length of grating region, δ is the difference between the actual propagation constant β ($\beta = \frac{2\pi}{\lambda} n_{eff}$) and the propagation constant at Bragg wavelength β_B ($\beta_B = \frac{2\pi}{\lambda_B} n_{eff}$) and $q^2 = \delta^2 - \kappa^2 > 0$ [29]. According to Eq. (2), the reflection spectrum of FBGs in BPOF can be simulated as shown in Figure 4. It can be seen from this figure that it has a maximum reflectivity of 69.5 % at 1571.4 nm with a full width at half maximum (FWHM) of 0.18 nm.

In addition, the maximum reflectivity R_{max} can be given by [29, 47]:

$$R_{max} = \tanh^2(\kappa L) = \tanh^2(\frac{\pi \delta n}{\lambda} L) \tag{3}$$

Seen from Eq. (3), the maximum reflectivity will increase with the increase of both δn and grating length L.

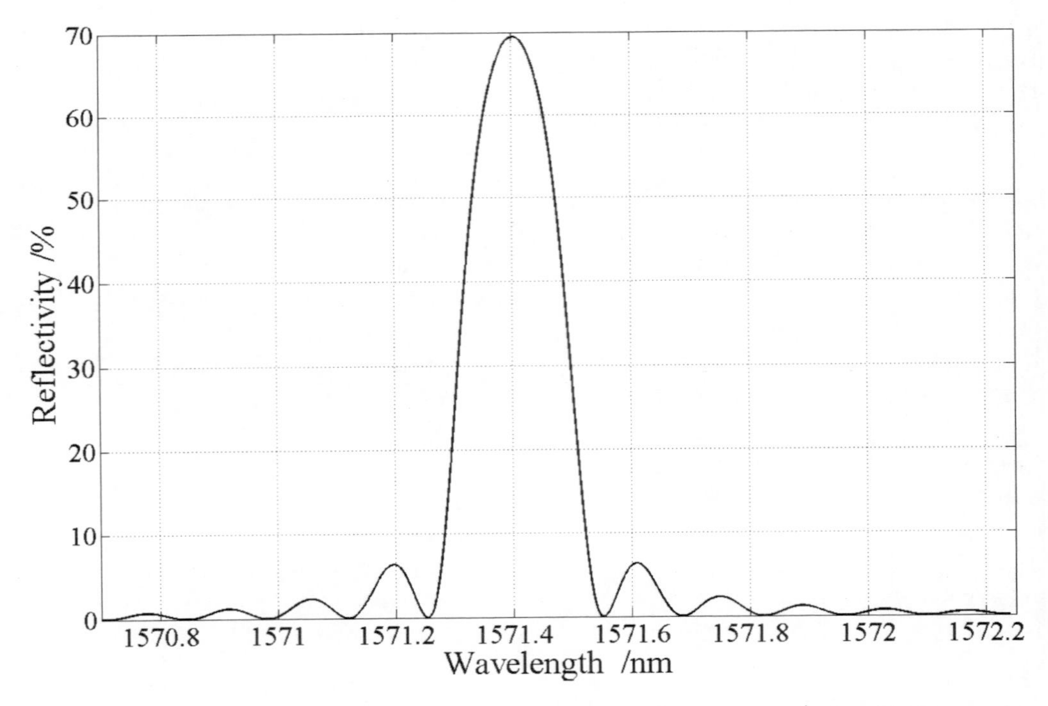

Figure 4. The simulated reflection spectrum of BPOF gratings with $\delta n = 1.0 \times 10^{-4}$, $\Lambda = 530.7$nm and L=6 mm and $n_{eff} = 1.4804$ [29].

Generally, δn varies with UV exposure time, especially for BPOF gratings, according to Ref. [41], δn of BDK doping PMMA under UV exposure with certain light intensity can be described with the following equation [29, 47]:

$$\delta\Delta = a\sqrt{t} - b \tag{4}$$

where a and b are the constants ($a=1.67\times10^{-4}$) [41]. Combining Eqs. (3) and (4), the theoretical maximum reflectivity after t mins UV exposure can be obtained as:

$$R_{max} = \tanh^2\left[\frac{\pi L(a\sqrt{t} - b)}{\lambda_B}\right] \tag{5}$$

Especially when constant b=0, R_{max} will increases fast and reach a saturated value (Figure 5), showing that R_{max} can't increase unrestrictedly with long time exposure.

3. Fabrication of BPOF Gratings

3.1. BPOF Fabrication

BPOF is fabricated by Teflon technique [1, 39, 45], introducing BDK into the fiber core as the photosensitive materials.

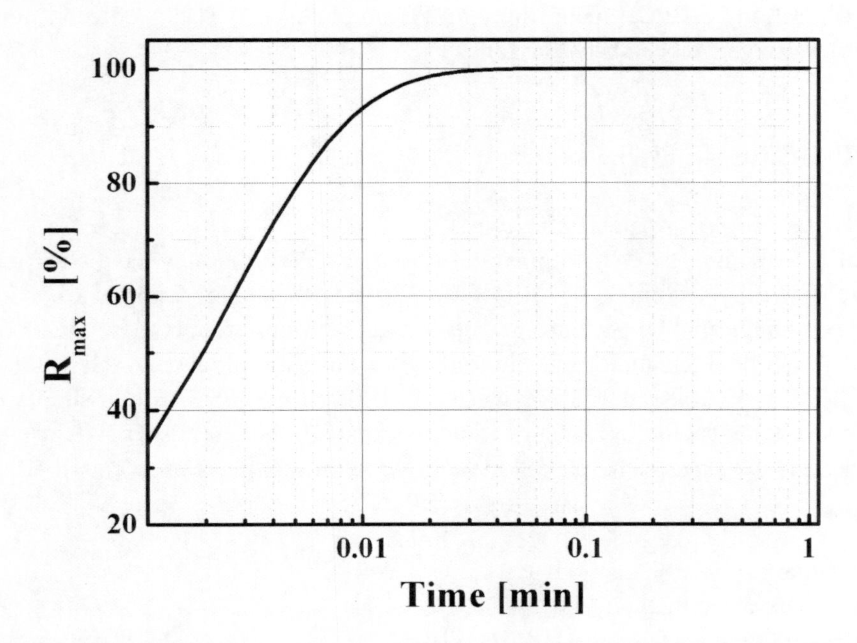

Figure 5. The maximum reflectivity vs illumination time when b=0, L=6 mm and $\lambda_B = 1571.4$ nm [29].

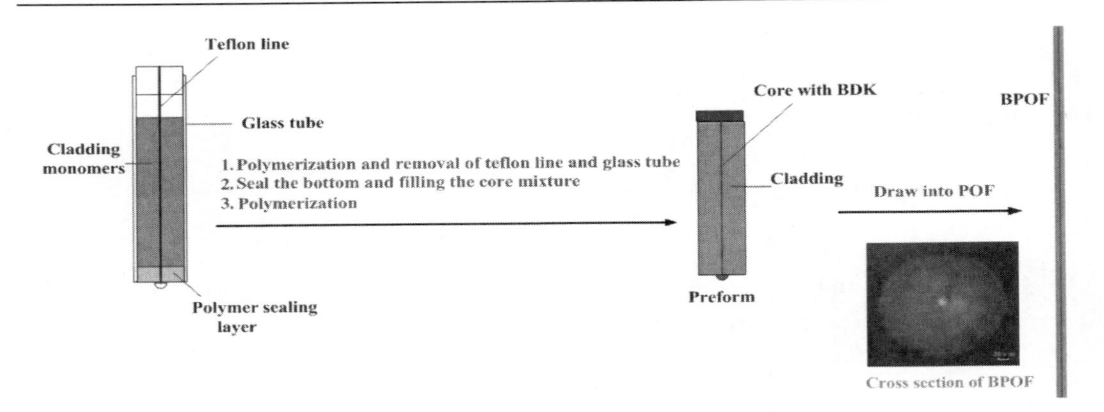

Figure 6. The procedure of BPOF fabrication with Teflon technique [29].

The details of the fabrication process are illustrated in Figure 6. First, using Teflon technique a hollow stick (cladding preform) is made by the copolymerization of the cladding monomers [39, 45], where the inner diameter of the glass tube is about 18mm and the diameter of the Teflon line is ~0.8 mm. Then, core monomers doped with BDK, initiator and chain-transfer agent are mixed in a vessel and the mixed solution is filled into the hollow stick as a core material under the minus pressure. Finally, the stick is put into an oven and the thermal polymerization is preformed, where the temperature is increased gradually from 36 ℃ to 88 ℃ in 4.5 days until solidification is fulfilled.

When the BPOF has been made, the refractive index profile of the POF perform is measured as reported before [48]. BPOF is heat-drawn from fiber preform at 225 ℃ by a taking up spool. According to Eq. (1), BPOF with different diameter has been made by tuning the diameter of tube/Teflon line and the drawing velocity. The inset of Figure 6 shows a typical cross section of BPOF made through Teflon technique. The parameters for all the BPOF samples used have been listed in Table 1.

3.2. BPOF Gratings Fabrication

3.2.1. BPOF Gratings Writing Technique

Since the invention of POF gratings in 1999, many POF gratings have been fabricated with different writing technique [1, 8, 12-18]. So far, there are mainly two methods to fabricate the gratings in POF, which are mechanical and optical methods. In 2006, M. P. Hiscoks et al. adopted a simple heat imprinting method for producing stable long-period gratings (LPGs) in microstructured polymer optical fiber (mPOF), which is simple and low cost, but only suitable for the LPGs fabrication instead of Bragg gratings [17]. However, most of the researchers use the optical method to writing gratings, which can be classified into four kinds:

1. The interferometeric method [1, 14];
2. The phase mask method [15];
3. The amplitude mask method [8, 16, 39];
4. The point to point method [11].

To overcome the effects of the zero-order diffraction on the grating writing in optical fibers, the modified sagnac interferometeric method is used to fabricate the first POF gratings in 1999 by Z. Xiong et al., which is suitable for the condition when you haven't the designated phase mask for the writing light source, but it requires good coherence of the light source [1]. Then in 2005, H. Dobb et al. use the phase mask method to write the gratings in mPOF. However, due to the low photosensitivity of the mPOF and non-designated phase mask used, the reflectivity is still not high [15]. In the same year, Z. Li et al. use the amplitude mask method to write the LPGs in MVK doped POF with high pressure mercury lamp, which has lowest requirement for the coherence of the writing source but only feasible for writing LPGs. But it has high requirement for the structure design of the POF to confirm the resonance of the core and cladding modes [16]. In 2009, David Sáez-Rodríguez et al. use the point to point method to write the LPGs in POF, where the coherence requirement of the writing source is not high but it requires the high focus technique. And it is also suitable to write the LPGs [11].

According to the present experimental condition (A phase mask with a period of 1.0614 μm designed for silica FBGs writing at 248 nm), to overcome the zero-order diffraction effects on the grating writing in optical fibers, the modified sagnac interferometeric method is adopted to inscribe BPOF gratings with a 355 nm frequency-tripled Nd:YAG pulse laser as shown in Figure 7 [29].

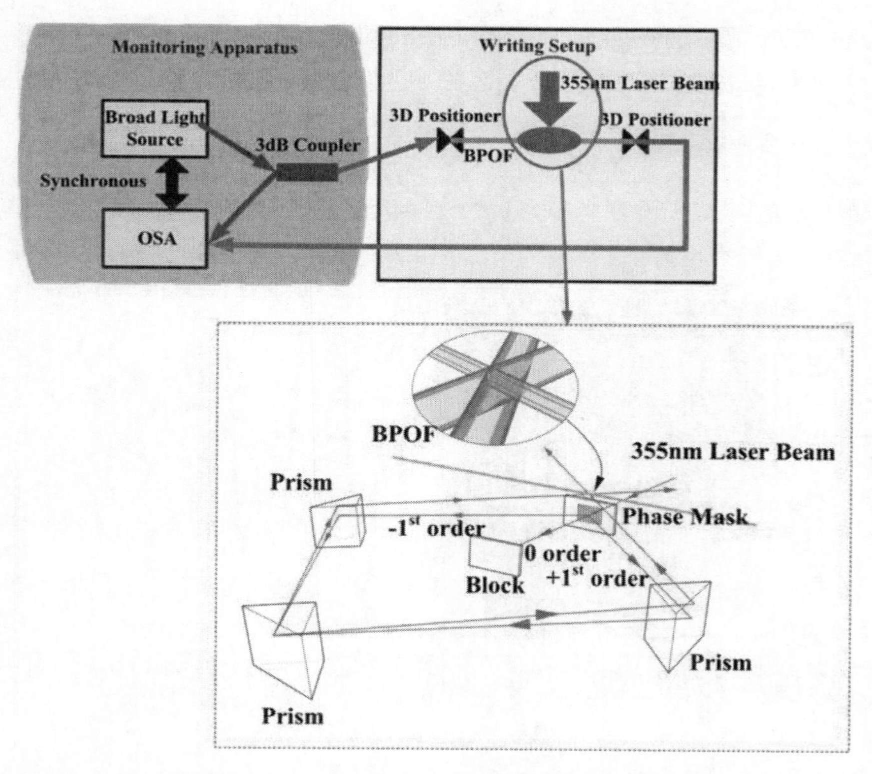

Figure 7. The diagram of BPOF gratings inscription with the modified sagnac interferometeric system [29].

It can be seen from Figure 7 that BPOF is mounted on the top of the phase mask. Three UV reflective prisms are aligned so that the counter-propagating coherence beams will be directed to the fiber core, interfere there, and finally write the gratings in the core of BPOF. Meanwhile, the zero order diffraction will be isolated and blocked.

To monitor the formation of BPOF grating, monitoring light coming from amplified spontaneous emission (ASE) (Thorlabs ASE-FL7002) or tunable laser (AQ 4321D of ANDO) is launched into BPOF through a 3 dB silica fiber coupler. On reaching the grating, portion of the light will be reflected while the rest will propagate through the grating and down the fiber. An optical spectrum analyzer (OSA) (Agilent 86140B OSA or AQ 4317C OSA) is used to detect the reflection and determine the reflection spectrum of the grating. Similarly, the OSA can detect the transmission of the grating and display the transmission spectrum [18, 29, 30]. Grating fabrication conditions for all BPOF gratings are listed in Table 1.

In addition, there is one issue need to mention that most of the present BPOF gratings operate at 1550 nm band, where BPOF has high loss. One key reason for this situation is that most of the related components, system and test techniques are readily available in the 1550 nm band [4, 5, 9, 27]. The visible or near IR wavelengths would be more desirable for POF applications, especially longer length applications such as distributed sensing, because of much lower material attenuation than the 1550 nm band. That is why a few research groups have started to develop POF gratings at visible or near IR regions [12, 29, 49, 50]. If the fabrication and measurement system exists, BPOF gratings at the optical window of POF, i.e. at 600–700 nm can also be designed and fabricated in future [29].

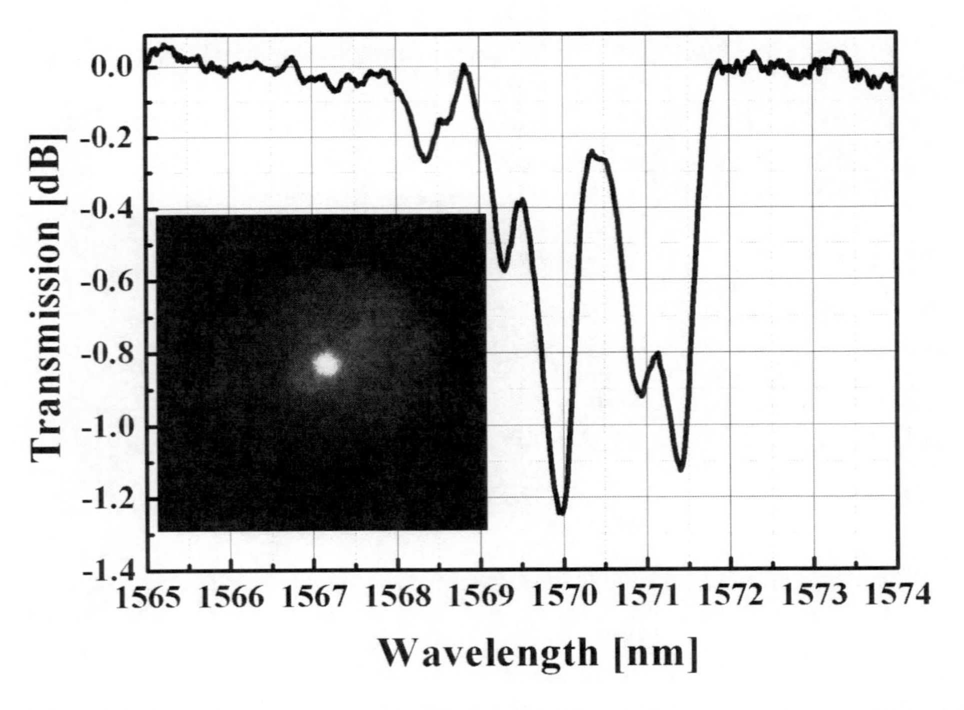

Figure 8. Typical transmission spectrum of **Grating A** with a grating length of 6mm and FWHM of 0.5 nm after 1.63 J 355 nm UV exposure. The inset is the near-field pattern of BPOF for **Grating A** at 633 nm [18].

3.2.2. High Power Writing

Although it is classified as the high power writing, the average power intensity for writing BPOF gratings is far lower than that for writing silica fiber gratings (10^4 mW/cm^2) [51]. With the modified sagnac interferometeric method, Grating A is inscribed in the BPOF, whose normalized frequency calculated at 1570 nm is ~5.4, larger than the maximum limit (2.405). Hence, multiple modes are expected as shown in Figure 8. The maximum reflectivity of Grating A is approximately 25% (1.25 dB) with FWHM of about 0.5 nm at 1570 nm. The index change in the grating region due to 16 mins 673 mW/cm^2 (1.63 J) 355 nm UV exposure is estimated to be 4.5×10^{-5}, far lower than the maximum value (2.4×10^{-3}) reported by H. Franke [41].

The lower value is attributed to several reasons: fewer residual monomers compared with H. Franke's work; the shorter actual grating length due to the non-uniformity of the laser beam; the instability of the interference pattern on the fiber core due to the temperature and mechanical drift and the low end coupling efficiency [18, 52].

In addition, the dynamic formation process of Grating A under varying 355 nm UV exposure is shown in Figure 9. Seen from Figure 9, the reflection peak position and its intensity change with the exposure dose. The former change might be attributed to the temperature and mechanical drift, resulting in the drift of the Bragg wavelength (the standard deviation ~0.24 nm). The latter change shows the growth of Grating A with UV exposure. A noticeable reflection peak can be observed after 0.10 J UV exposure; then the grating continues to grow stronger with further exposure. After 1.02 J UV exposure, the maximum reflection spectrum is obtained. Seen from Figure 9A, there are more than 5 modes reflected. The maximum reflection peaks at ~1570 nm with its intensity and FWHM of 12 dB and 0.4 nm, respectively. Compared with the transmission result in Figure 8, the maximum reflectivity should be larger than 25% after 10 mins 673 mW/cm^2 355 nm exposure (1.02 J). With further exposure, the reflection starts to decrease. The low reflection demonstrates rather small index changes for long time exposure. The growth of reflection spectrum at the initial stage is attributed to increase of refractive index change under UV exposure, while the decay at the later stage is due to the damage of core for long time UV exposure as well as the drift of the writing system shown in Figure 9B [13, 18, 53].

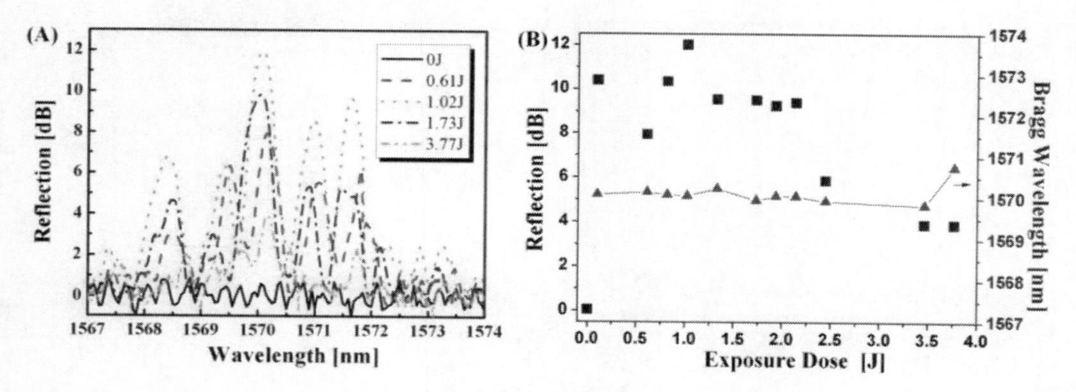

Figure 9. (A) The reflection spectra of **Grating A** after different dose of 355 nm UV exposure. (B) The maximum peak growth and its corresponding Bragg wavelength of **Grating A** versus the exposure dose.

3.2.3. Low Power Writing

3.2.3.1. SM BPOF Gratings Fabrication

To avoid the damage of the high power, the intensity of the writing beam is reduced to 200 mW/cm^2 and **Grating B** is fabricated. The normalized frequency of BPOF for **Grating B** at 1570 nm is ~0.9 [46]. Hence, single mode is expected as shown in Figure 10. After 0.29 J 355 nm UV exposure with lower writing power, the reflection spectrum of **Grating B** is shown in Figure 10. It can be seen that the Bragg wavelength is ~1573.2 nm with the reflection intensity of ~6.7 dB and FWHM of ~0.2 nm [18]. Compared with Figure 9A and 10, the reflection intensity of **Grating B** is smaller than that of **Grating A**, which may be due to both the low doping BDK and the low writing power.

3.2.3.2. MM BPOF Gratings Fabrication

With lower writing power of 153 mW/cm^2, **Grating C** is written in MM BPOF again. The reflection spectra of **Grating C** under different UV exposure are shown in Figure 11A [30]. It can be seen that there are mainly four modes reflected in **Grating C**. The reflection position varies a little with exposure dose as shown in Figure 11B, attributed to the temperature and mechanical drift. The intensity and position of the maximum reflection peak (around 1574.3 nm) vs exposure dose is plotted in Figure 11B. It can be seen that a noticeable reflection peak is observed after 0.03 J UV exposure; then the grating continues to grow stronger with further exposure. After 0.15 J UV exposure, the reflection spectrum is formed.

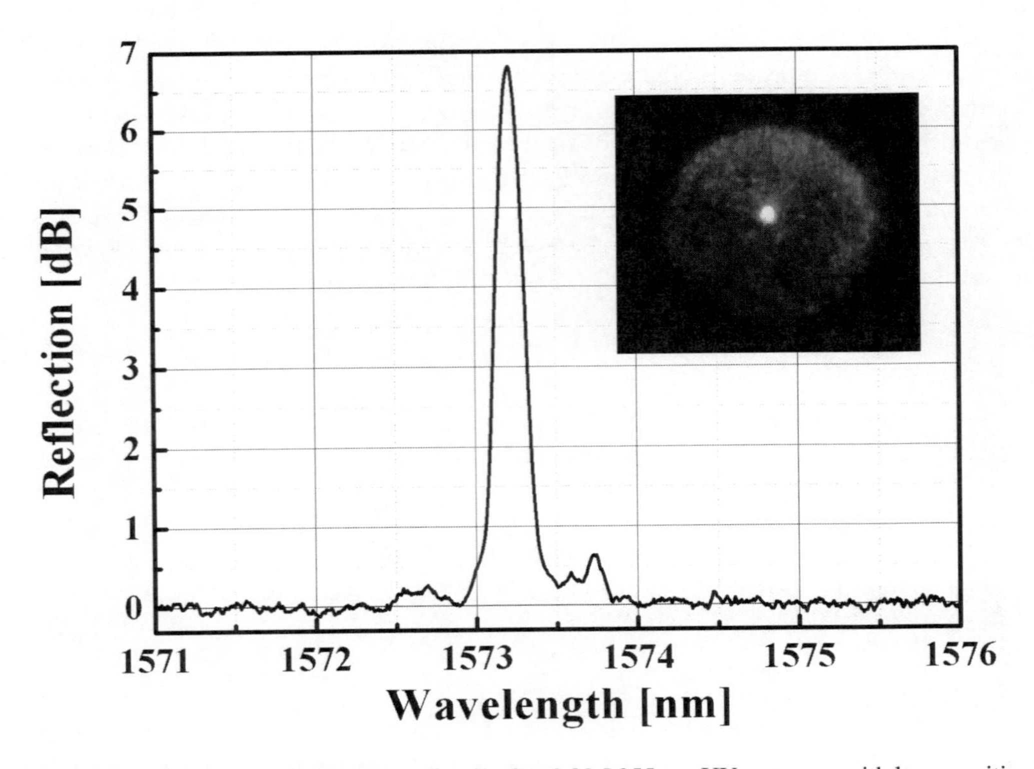

Figure 10. The reflection spectrum of **Grating B** after 0.29 J 355 nm UV exposure with lower writing power. The inset is the near field pattern of BPOF for **Grating B** at 633 nm [18].

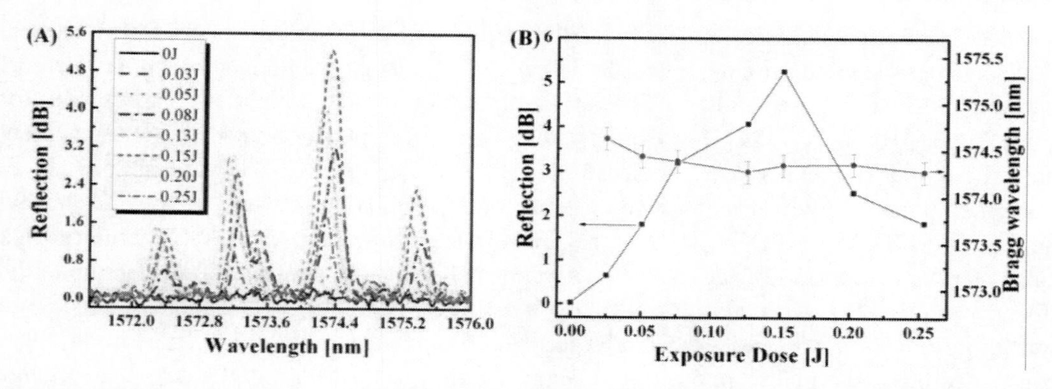

Figure 11. (A) The reflection spectra of **Grating D** after different 355 nm UV exposure at an average power intensity of 153 mW/cm^2. (B) The peak intensity and wavelength around 1574.3 nm versus the UV exposure dose.

The maximum reflection peaks at ~1574.3 nm with its intensity and FWHM of 5.1 dB and 0.3 nm, respectively. Then the grating reflection peak starts to decrease after 0.20 J UV exposure, indicating an onset of the grating decaying process. After 0.25 J UV exposure, the irradiation is OFF. The reflection of Grating C is not that high due to the multimode effect and drifting of writing system. Under the UV irradiation exposure, the growth and decay of Grating C under UV exposure displays again, due to the increase of refractive index change at the initial stage, and the damage of core after long time exposure and lower quality of the interference field at the later stage [30].

Compared with Figure 9B and 11B, the maximum reflection peaks should reach with very lower exposure dose (0.15 J), which suggests that optimization of the writing intensity and exposure dose play an important role to achieve high reflection in the BPOF gratings. In addition, the end coupling is very important for the detection of BPOF gratings due to the existence of the mechanical drift [30].

4. BPOF GRATINGS BASED SENSING APPLICATION

4.1. Temperature Sensing

POF gratings have higher temperature sensitivity than their silica counterparts due to large thermo-optic coefficient and large thermal expansion coefficient of polymers [4, 8, 9, 54]. Therefore, they are much more sensitive to temperature than those of silica, more than 10 times of the silica fiber gratings [7, 24]. For a given change of temperature, POF gratings temperature-sensors show a much greater wavelength shift than their silica counterparts, which either allows the use of lower resolution, simpler and therefore cheaper interrogation systems or alternatively enables the sensors to be used for the sensing of small temperature change in vivo. In addition, the higher temperature sensitivity of POF gratings based devices also allows device properties to be readily controlled thermally [5]. Moreover, advantage could be taken of the fact by making use of the huge range of organic chemistry techniques to modify the fiber materials [5] as well as lots of method or design to modify the fiber structure [6, 22, 33, 54, 55].

So far, the temperature sensitivity of the gratings in the step index SM PMMA based POFs has been studied previously [6, 33, 54], revealing a negative temperature coefficient of between -146 pm/°C and -360 pm/°C significantly larger in magnitude than that observed for silica FBGs(~10 pm/°C) [24]. For Bragg gratings created in perfluorinated cyclic transparent optical polymer (CYTOP) POF, the sensitivity is about -167 pm/°C [56]. And in 2007, K. E. Carroll et al. report a lower magnitude of temperature sensitivity between -52 pm/°C and -95 pm/°C in SM PMMA mPOFs [24]. For the FBGs in few mode TOPAS mPOF, there is even a calculated largest positive Bragg wavelength shift (810 pm/°C), which is dependent upon the fiber structure [22]. Recently, for SM TOPAS mPOF gratings the negative temperature sensitivity of -60 pm/°C and -36.5 pm/°C has also been measured [49, 57]. In addition, there exists a complex thermal response of the Bragg gratings in PMMA mPOFs due to its organic nature [5]. Recently, Chen et al. find that temperature response of SM PMMA POF with eccentric core also shows a negative sign with the thermal sensitivity of -50.1 pm/°C [55]. In addition, H. Dobb et al. find that for PMMA based POF the temperature sensitivity is also dependent upon the temperature range [58]. All these work demonstrates that the thermal response of POF gratings has great dependence upon their materials, fiber structure, as well as fabrication process [5, 6, 22, 24, 31, 49, 54-56, 58]. In this section, temperature sensing with SM & MM BPOF gratings will be reviewed, which will provide an effective guidance for the BPOF gratings used as a temperature sensor.

4.1.1. SM BPOF Gratings for Temperature Sensing

4.1.1.1. Thermal Property of SM BPOF Gratings

4.1.1.1.1. Differential Thermal Analysis

As one kind of the polymer materials, SM BPOF gratings will be required to operate under the glass transition temperature to assure good repeatability and stability. According to the differential scanning calorimetry result of BPOF [33], there is a small phase transition at 48.2 °C, originated to the disturbing orientation of the main chain of POF materials resulted from the BPOF drawing [59, 60], for which the heat content is ~0.0488 J/(g·°C). So it is necessary to anneal the BPOF at a little higher than 48.2 °C prior to grating inscription for the good stability and repeatability of BPOF gratings temperature sensor [49]. The glass transition temperature of BPOF is ~112 °C(similar to the previous value [61]), which is dependent on the thermal history of the fiber preform fabrication, the fiber materials, the fiber draw parameters and UV exposure transition of BPOF [59, 62]. In addition, from 81.7 °C BPOF starts to change from brittle to ductile of BPOF [33]. Therefore, the maximum operating temperature of BPOF gratings should be lower than 81.7 °C. Furthermore, the upper limit operating temperature of BPOF gratings is ~282 °C, which is the degradation temperature of POF [33, 61].

4.1.1.1.2. Thermal Mechanical Analysis

Thermal mechanical analysis (TMA) curve shows that BPOF expands when temperature increases from 25.43 to 79.95 °C, which can give the thermal expansion coefficient of ~ 78.51×10^{-6} /°C [33]. Especially, when temperature is higher than 79.95 °C, BPOF starts shrinking [33, 55], which clearly explains why the BPOF gratings can hardly survive up or higher than 100 °C [24].

4.1.1.1.3. Thermal Optical Analysis

According to the Ref. [33, 63], the refractive index of the BPOF core materials decreases with the temperature increasing. As PMMA and PEMA are the main materials of BPOF, thermal-optic coefficients of them are -1.3×10^{-4} /°C and -1.1×10^{-4} /°C, respectively [63]. So although there are other doping materials like BDK and BzMA, the thermal-optic coefficient only vary a little due to the small quantity of them [64]. Regardless the difference between the core and the cladding, dn/dT values of BPOF should be ~-1.36×10^{-4} /°C close to the value of the base material PMMA [33, 63] and ζ ~-0.92×10^{-4}/°C.

4.1.1.2. Theory of SM BPOF Gratings for Temperature Sensing

For SM BPOF gratings when the temperature changes only, thermal-optic effects will cause the change of n_{eff} (close to the core index) and the thermal expansion effects will cause the change of grating period Λ. According to the Bragg condition [29]:

$$\lambda_B = 2n_{eff}\Lambda \tag{6}$$

where λ_B is the Bragg wavelength. The Bragg wavelength λ_B will change due to the temperature change as given by [33, 65]:

$$\lambda_B(T) = \lambda_B(T_0) + 2\left(n_{eff}\frac{\partial\Lambda}{\partial T} + \Lambda\frac{\partial n_{eff}}{\partial T}\right)\Delta T \tag{7}$$

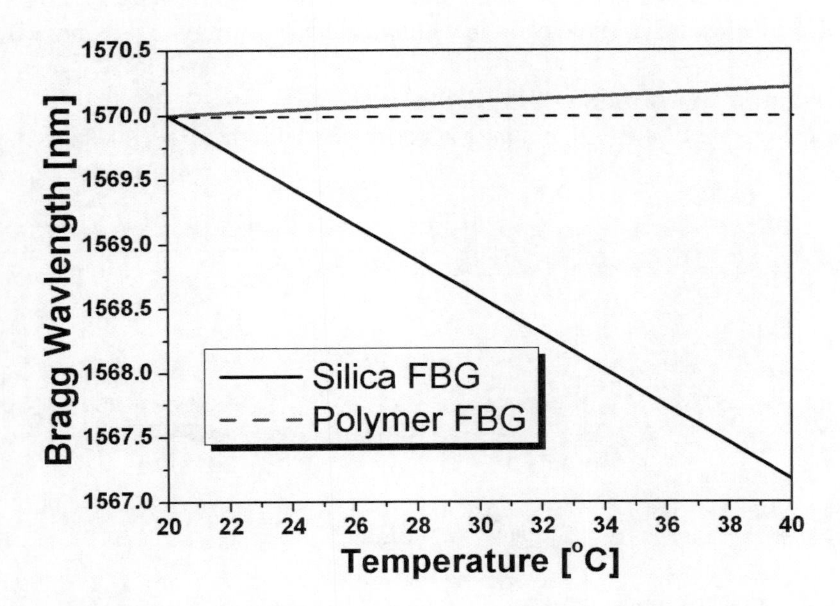

Figure 12. Temperature response of SM polymer FBGs and SM silica FBGs in theory.

In addition, it is often defined that $\alpha = \frac{1}{\Lambda} \cdot \frac{\partial \Lambda}{\partial T}$ and $\zeta = \frac{1}{n_{eff}} \cdot \frac{\partial n_{eff}}{\partial T}$ are core's linear thermal expansion coefficient and thermal-optic coefficient, respectively [33, 65].

So the relative Bragg wavelength shift can be given by [33, 65]:

$$\frac{\Delta \lambda_B}{\lambda_B} = (\alpha + \zeta)\Delta T \tag{8}$$

Generally, for polymer FBGs the thermal expansion coefficient α is ~10^{-5} order, while the thermo-optic coefficient ζ is typically -10^{-4} order [31, 66]. The accurate parameters depend on the POF materials. Assuming that polymer FBGs operate at ~1570 nm at 20 ◦C, temperature response of polymer FBGs can be simulated as shown in Figure 12, in contrast to that of silica FBGs with $\alpha = 0.55 \times 10^{-6}$ and ζ =-6.3×10^{-6}, described mathematically as [31]:

$$\lambda_B = 1570(1 - 9 \times 10^{-5} T) \text{ (For SM polymer FBG)} \tag{9}$$

$$\lambda_B = 1570(1 + 6.85 \times 10^{-6} T) \text{ (For SM silica FBGs)} \tag{10}$$

For polymer FBGs, the slope is negative, indicating that the Bragg wavelength will decline as the temperature rises due to the large and negative thermo-optic coefficient. The changes in the effective refractive index and thermal expansion would enable the Bragg wavelength to move contradirectionally as temperature changes. Especially, the thermo-optic coefficient is 1 order of magnitude higher than the thermal expansion coefficient for polymer FBGs. Differently, this slope is positive for silica FBGs as both factors are positive and make the Bragg wavelength shift toward the same direction. The slope of polymer FBGs is much bigger than that of silica FBG, showing higher temperature sensitivity of polymer FBGs [31].

4.1.1.3. Thermal Response of SM BPOF Gratings

The thermal response of BPOF gratings is performed with the experimental setup shown in Figure 13 [33].

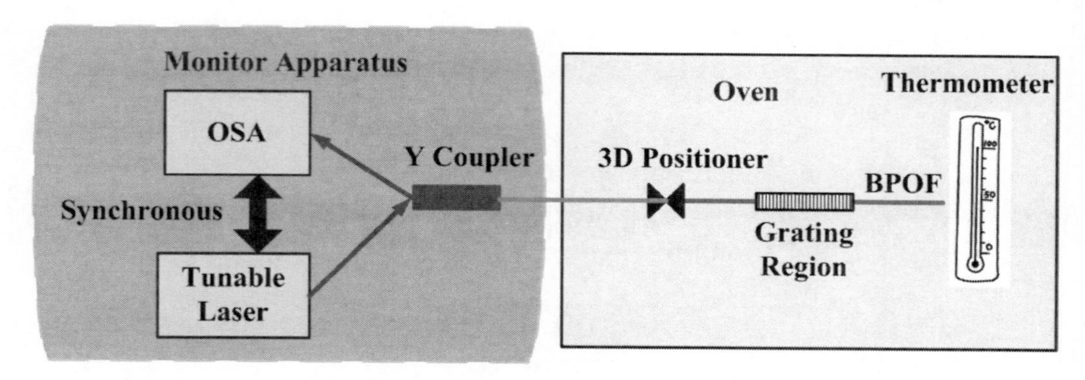

Figure 13. The experiment setup for the thermal response of Grating D.

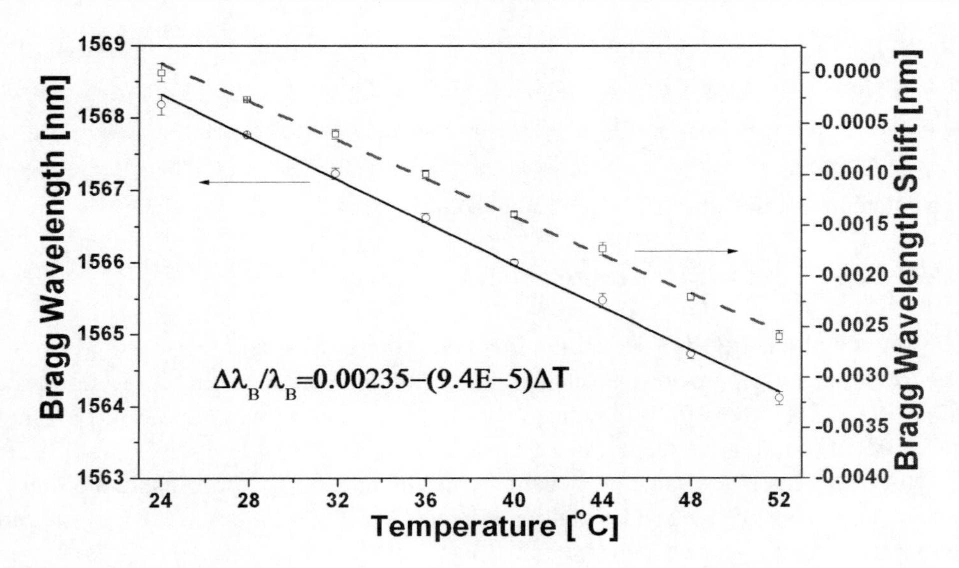

Figure 14. Bragg wavelength and the relative shift of Bragg wavelength of SM BPOF gratings vs temperature.

Both ends of BPOF gratings are sticked using epoxy glue vertically on a metal panel. Then the sensing part of BPOF gratings is put into the oven, whose temperature is controlled by the PID method. The accuracy of the temperature measurement is ±1 °C, and it takes about 2 minutes for the heating system to reach the desired temperature and 5 minutes for the stabilizing of the temperature. The reflection spectra of the BPOF gratings are recorded by an OSA when the thermal test is performed. A tunable laser source is adopted for the BPOF grating characterization, launched into the POF through a 3dB Y-type silica fiber coupler.

With the setup in Figure 13, the thermal response of SM BPOF gratings (**Grating D)** from 24 °C to 52 °C is measured as shown in Figure 14. It can be seen that BPOF gratings display a negative wavelength shift with rising temperature, fitted with the following equation [33]:

$$\lambda_B = 1571.846 - 0.147T \tag{11}$$

According to Eq. (11), the temperature sensitivity can be obtained to be around -147 pm/°C, 10 times larger than the silica gratings, similar to the previous report (-149 pm/°C) [31]. In addition, blueshift of ~4 nm is achieved only with the temperature variation of 28 °C, larger than that in silica fiber gratings [67], indicating higher thermal sensitivity of SM BPOF gratings than that of silica FBGs and showing great potential used as a temperature sensor [33].

Furthermore, the relative shift of Bragg wavelength of BPOF gratings vs temperature could be linearly fitted with the following equation [33]:

$$\frac{\Delta\lambda_B}{\lambda_B} = 0.00235 - 9.4 \times 10^{-5} \Delta T \tag{12}$$

Therefore, the thermal coefficient of the BPOF gratings is about -9.4×10^{-5} /°C, far larger than that silica fiber gratings(7.5×10^{-6} /° C) [68]. In addition, due to the thermal expansion coefficient of the core is 78.51×10^{-6} /°C [33], the thermal optic coefficient can be derived as -1.73×10^{-4} /°C from Eq. (8), which is close to the previous value [69] but larger than the value calculated from the empirical equation(absolute value).

4.1.2. MM BPOF Gratings for Temperature Sensing

4.1.2.1. Theory of MM BPOF Gratings for Temperature Sensing

Due to the characteristics of the interfacial-gel polymerization at the fabrication process of BPOF [30, 70, 71], the BPOF fabricated should be a graded index fiber. For our graded-index MM BPOF(Grating E), the number of propagating modes is 26 at 1.57 μm [32]. Some of them will have almost the same propagation constant and satisfy the Bragg condition [72], which means that some of the modes can be reflected when they satisfy the Bragg condition. Furthermore, the temperature dependence of Bragg wavelengths(corresponding to N[th] mode) of the graded-index MM BPOF gratings can be given by [32, 72]:

$$\frac{d\lambda}{dT}=\frac{\lambda^2}{2n_1\Lambda^2}\frac{d\Lambda}{dT}+\left[\frac{\lambda^2}{2n_1^2\Lambda}-\frac{\lambda^2(N+1)(3n_2-2n_1)}{2\pi a n_1^2\sqrt{2n_1(n_1-n_2)}}\right]\frac{dn_1}{dT}+\frac{\lambda^2(N+1)}{2\pi a n_1\sqrt{2n_1(n_1-n_2)}}\frac{dn_2}{dT} \tag{14}$$

where n_1 and n_2 are refractive indices of the MM BPOF core and cladding, a is core radius and k is wavenumber. As list in Table 2, $\frac{dn_1}{dT}$ for BPOF core and $\frac{dn_2}{dT}$ for BPOF cladding are almost the same value of -1×10^{-4} /° C, so Eq.(14) can be simplified into [32]:

$$\frac{d\lambda}{dT}=\frac{\lambda^2}{2n_1\Lambda^2}\frac{d\Lambda}{dT}+\frac{\lambda^2}{2n_1^2\Lambda}\frac{dn_1}{dT}+\frac{3\lambda^2(N+1)(n_1-n_2)}{2\pi a n_1^2\sqrt{2n_1(n_1-n_2)}}\frac{dn_1}{dT} \tag{15}$$

where n_1 is larger than n_2. According to Eq. (15), with order mode increasing, the varying speed of Bragg wavelength upon temperature change will increase linearly, that is to say, with the order mode increasing, the temperature sensitivity of MM BPOF gratings will increase linearly. According to Eq. (15), the temperature sensitivity of MM silica and MM polymer FBGs in theory can be given by [32]:

$$\frac{d\lambda}{dT}=0.01613+4.6\times10^{-5}(N+1) \tag{16}$$

and

$$\frac{d\lambda}{dT} = -0.07862 - 4.2\times10^{-4}(N+1) \tag{17}$$

respectively, where the positive and negative represents the shift direction of Bragg wavelength. Seen from Eqs. (16) and (17) and Figure 15, the temperature sensitivity of MM polymer FBGs is negative value while that of MM silica FBGs is positive, indicating with the increase of temperature, the Bragg wavelength of MM polymer FBGs blueshifts while that of MM silica FBGs redshifts. For the same order of reflected mode, the temperature sensitivity of MM POF is about 5 times that of MM silica FBGs. In addition, the order influence upon the temperature sensitivity of MM polymer FBGs is almost 1 order larger than that of MM silica FBGs [32]. Furthermore, with the order of the reflected mode increases from 0th to 60th order, for MM silica FBGs the temperature sensitivity will increase linearly from 16.7 pm/°C to 44.4 pm/°C, while for MM polymer FBGs the temperature sensitivity will increase linearly from -79.0 pm/°C to -104.3 pm/°C, showing higher temperature sensitivity compared with MM silica FBGs.

Table 2. Relevant parameters of MM silica FBGs and MM polymer FBGs [32]

Parameter	MM silica FBGs	Reference	MM polymer FBGs	Reference
Core diameter	50μm	[72]	23μm	Our work
Core index	1.4709	[72]	1.477	Our work
Grating Period	534nm	[72]	530.7nm	Our work
Operating Wavelength	1.55μm	[72]	1.57 μm	Our work
Index difference	0.0137	[72]	0.011	Our work
Thermal optic coefficient	1×10^{-5} /°C	[72]	-10×10^{-5} /°C	[7]
Thermal expansion coefficient	0.55×10^{-6}	[72]	5×10^{-5}	[7]

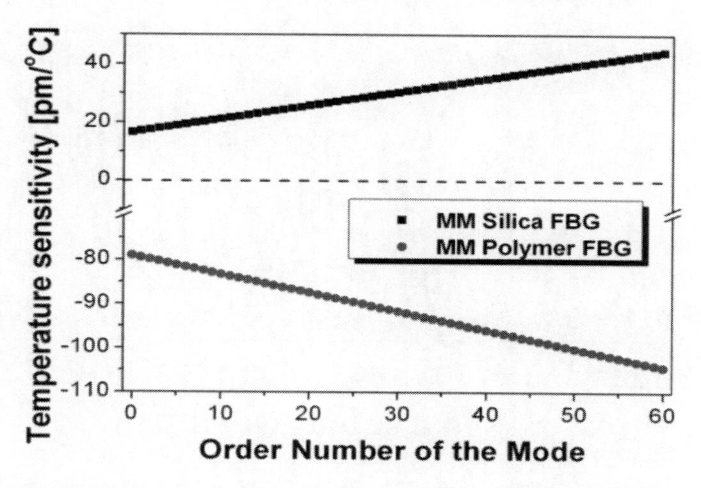

Figure 15. The relationship between the temperature sensitivity of MM silica FBGs and MM polymer FBGs, and the order of the reflected mode.

4.1.2.2. Thermal Response of MM BPOF Gratings

Similarly, the thermal response of MM BPOF gratings is performed with the experimental setup in Figure 16 [31]. With temperature increasing, the Bragg grating spectra of MM BPOF gratings blueshift (Figure 16), similarly to the thermal response of the SM BPOF gratings [31]. When the temperature is increased to 23.4 °C, the spectra will blueshift ~0.6nm. Meanwhile, peaks a, b, c and d change into a', b', c' and d', showing that not only the Bragg wavelength shifts but also the peak intensity changes. The former change is mainly due to the change of refractive index and grating period [73], while the latter changes in the amplitude of different reflected modes is due to the displacement of the peak wavelength of the reflected modes [74, 75] and the temperature and mechanical drift. Due to the latter change, the reflection spectrum looks a bit messy. Each mode will be tracked according to its key features, such as the shape, the peak order, etc. In order to assure the repeatability from grating to grating or fiber to fiber, several techniques, such as the main peak tracking (only identify the main peak), the recalibration according to its key feature, the calculation of the average shift, the construction of a suitable algorithm to modulate and demodulate the shift, etc, will be used to solve this issue.

Furthermore, it can be seen from Figure 17 that with the temperature increased from 16 to 40 °C, the reflection intensity for the mode of a, b, c and d fluctuate from 7.0 to 12.4 dB, 6.4 to 15.3 dB, 8.0 to 16 dB and 2.9 to 13 dB due to the displacement of the peak wavelength of the reflected modes [74, 75] as well as the temperature and mechanical drift, while the Bragg wavelength decreases linearly [32].

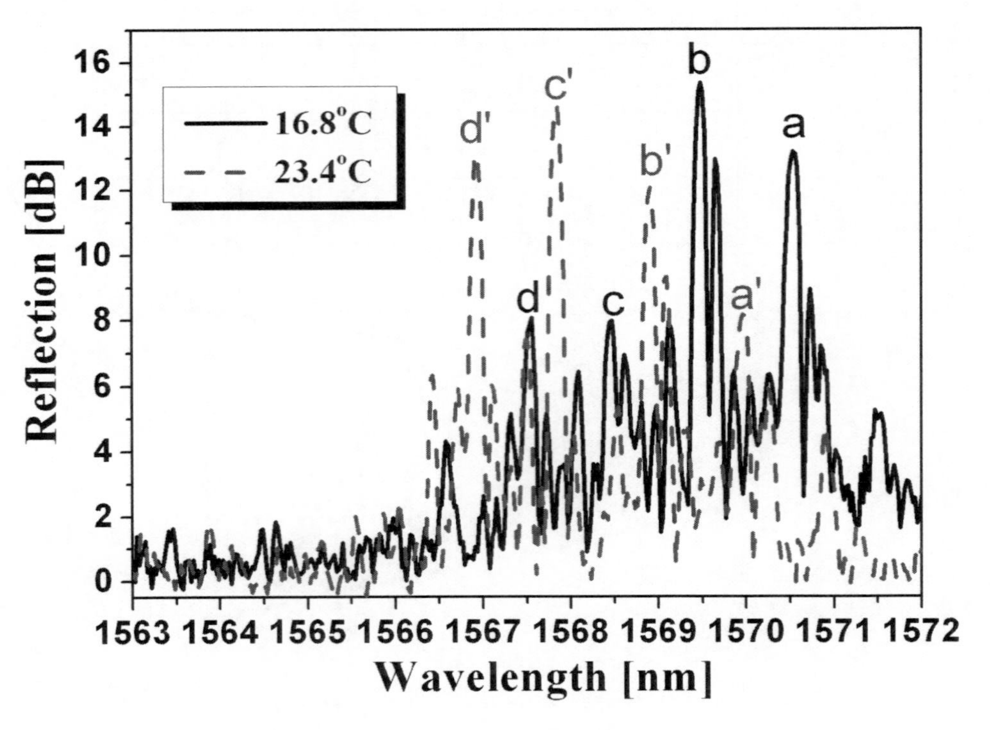

Figure 16. Reflection spectra of MM polymer FBGs(**Grating E**) at 16.8 °C and 23.4 °C, where four peaks a(1570.537 nm), b(1569.485 nm), c(1568.463 nm) and d(1567.547 nm) are chosen for comparison, respectively [32].

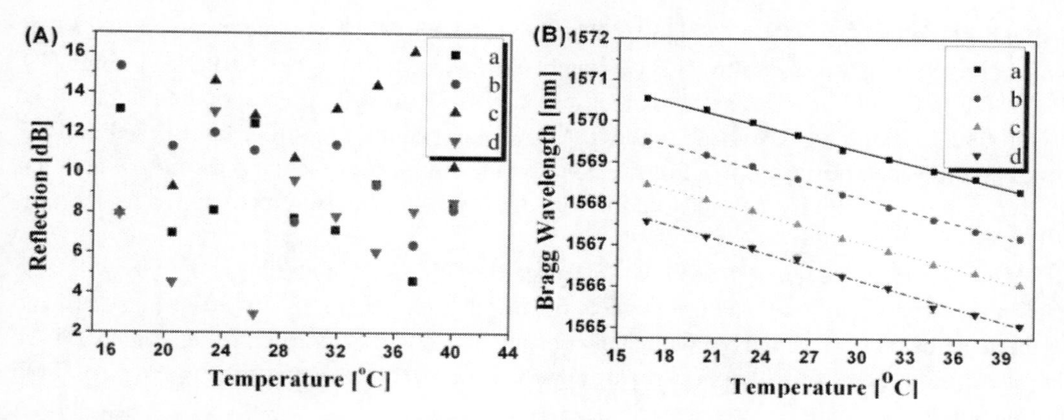

Figure 17. (A) Peak intensity vs temperature for the peak a, b, c, and d. (B) The relationship between the Bragg wavelength of a, b, c and d modes, and temperature, and its corresponding linear fitting curve [32].

With linear regression it can be expressed as [32]:

$$\begin{aligned}
\lambda_b &= -0.097T + 1572.210 \text{ (for mode a)} \\
\lambda_b &= -0.104T + 1571.286 \text{ (for mode b)} \\
\lambda_b &= -0.105T + 1570.247 \text{ (for mode c)} \\
\lambda_b &= -0.111T + 1569.488 \text{ (for mode d)}
\end{aligned} \tag{18}$$

where λ_b refers to the Bragg wavelengths. From Eq. (18), the temperature sensitivity for the mode a, b, c and d are -0.097, -0.104, -0.105 and -0.111, respectively, which is more than 8 times that of MM silica FBGs [32, 72]. In addition, for the higher the mode, the higher temperature sensitivity it is [32]. Although temperature sensitivity for different modes is different, the difference is very small, less than 13%. Therefore, for the reflection of the neighboring modes, the temperature sensitivity is close to each other [32], which is the base of some tracking methods mentioned above in the temperature sensing with MM BPOF gratings.

4.2. Stress & Strain Sensing

Due to the low Young's modulus (about 25 times lower than silica) and high elastic limit of over 10% (about 10 times higher than silica), POF gratings are attractive for fiber-optical strain sensing [76], making that optical devices based on them will allows device properties to be controlled with large dynamic range [9]. POFs are also clinically acceptable, flexible and non-brittle, which makes POF gratings the candidate for in-vivo biomedical sensing applications [2, 77-79]. FBGs have been reported in both step index POFs [1, 12, 49, 80] and mPOFs [12, 15, 57, 81, 82] for sensing application.

The strain sensitivity of POF gratings has been studied previously, revealing a strain coefficient of between 0.64 pm/ με and 2.7 pm/ με [3, 6, 7, 9, 10, 12, 20, 30, 49, 55, 81, 83-86], which has almost the same order as that for conventional silica fiber gratings [87]. In addition, the absolute strain coefficient of polymer FBGs working at 1500 nm is almost twice that at 800 nm [1, 3, 6, 7, 9, 12, 19, 30, 49, 55, 81, 83-86, 88]. Dynamic range up to 270 nm with corresponding strain up to 21.33% has been realized in long period POF gratings [19, 85], while in the polymer FBGs, dynamic range up to 73 nm with corresponding strain up to 5% around 1550 nm has also been reported [50, 83]. Especially, the fiber laser based on POF Bragg gratings can be easily tuned over 35 nm by the simple axial tension method, which has the high strain sensitivity of 1.48 pm/με with the dynamic measurement range as large as 2.37% [9].

Meanwhile, as polymer materials are typical viscoelastic materials, so POF will also exhibit evident viscoelastic nature, which means that the relationship between stress and strain depends on time, and even with step constant stress will cause increasing strain [89]. So far, there have been some reports regarding to the evident viscoelastic properties of POF gratings [3, 25, 30, 50].

M. C. J. Large et al. find that for the mPOF LPG sensor, these viscoelastic effects are small till the sensor is intermittently strained up to 2%. The effect of stress relaxation has been shown experimentally to have only a small effect on the change in the wavelength of the loss features used in the measurement of strain. However, such time dependent effects become significant for a practical sensor [25]. Recently, Z. F. Zhang et al. also find that this shift is notably affected by the time-dependent stress relaxation in the fiber, especially when the FBG is subject to a relatively higher strain (~2%) [50].

4.2.1. SM BPOF Gratings for Stress & Strain Sensing

4.2.1.1. Stress Sensing with SM BPOF Gratings

When BPOF grating is strained, the Bragg wavelength varies due to the change of the grating period and the refractive index. The shift of the Bragg wavelength due to the strain change is given by [90, 91]:

$$\Delta\lambda_B = 2(\Lambda\frac{\partial n_{eff}}{\partial L} + n_{eff}\frac{\partial \Lambda}{\partial L})\Delta L \tag{19}$$

where n_{eff} is the effective refractive index of the fiber core, L is the length of stressed length, and ΔL is the axial displacement. The first term in Eq. (19) represents the elastic stress induced index change (the photoelastic effect) and the second term is about the change of grating period. Furthermore, Bragg wavelength shift induced by the transversal strain (ε_t) and axial strain (ε_z) can be expressed as [91, 92]:

$$\Delta\lambda_B = \lambda_B\{\varepsilon_z - \frac{n_{eff}^2}{2}[p_{11}\cdot\varepsilon_t + p_{12}\cdot(\varepsilon_t + \varepsilon_z)]\} \tag{20}$$

where p_{ij} are the photoelastic constants of the strain optic tensor. The axial strain ε_z is defined as [91]:

$$\varepsilon_z = \frac{\Delta L_Z}{L_Z} \tag{21}$$

While the transversal strain ε_t is expressed by Poisson ratio ν as [91]:

$$\varepsilon_t = -\nu \cdot \varepsilon_z \tag{22}$$

By further simplifying of Eq. (20), the Bragg wavelength shift ratio can be found as [67, 91]:

$$\frac{\Delta\lambda_B}{\lambda_B} = (1 - p_{\text{eff}}) \cdot \varepsilon_z \tag{23}$$

where p_{eff} is an effective photoelastic constant defined as [91]:

$$p_{\text{eff}} = \frac{n_{eff}^2}{2}[p_{12} - \nu(p_{11} + p_{12})] \tag{24}$$

For classic elastomers, the stress (σ) and strain (ε_z) satisfies the following equation [30]:

$$\sigma = \text{E} \cdot \varepsilon_z \tag{25}$$

where E is the Young's modulus. If only the elasticity of BPOF is considered while neglecting the viscoelasticity, the relationship between stress (σ) and axial strain (ε_z) of BPOF can also be described using Eq. (25).

Employing the SM BPOF gratings (**Grating B**), the stress sensing is performed using the setup in Figure 18 [3, 30, 36]. One end of BPOF grating is bonded using epoxy glue vertically on a metal panel. Then ASE source is launched into the BPOF from a 3 dB SM silica fiber coupler.

Afterwards different weights are put on tray connected with the other end of BPOF, so different stress is applied on BPOF grating, of which the interval is about 1 min. After the loading experiment, the unloading experiment begins as followed and different weight is taken away one by one, the interval of which is also about 1min. Meanwhile, the Bragg grating spectra are monitored as different stress is applied in the stress loading and unloading process as shown in Figure 19.

When different stress is applied to BPOF gratings, it would bring about the deformation (tensile strain) of BPOF gratings, resulting in the shift of the reflection spectra of BPOF gratings (Figure 19A) [3].

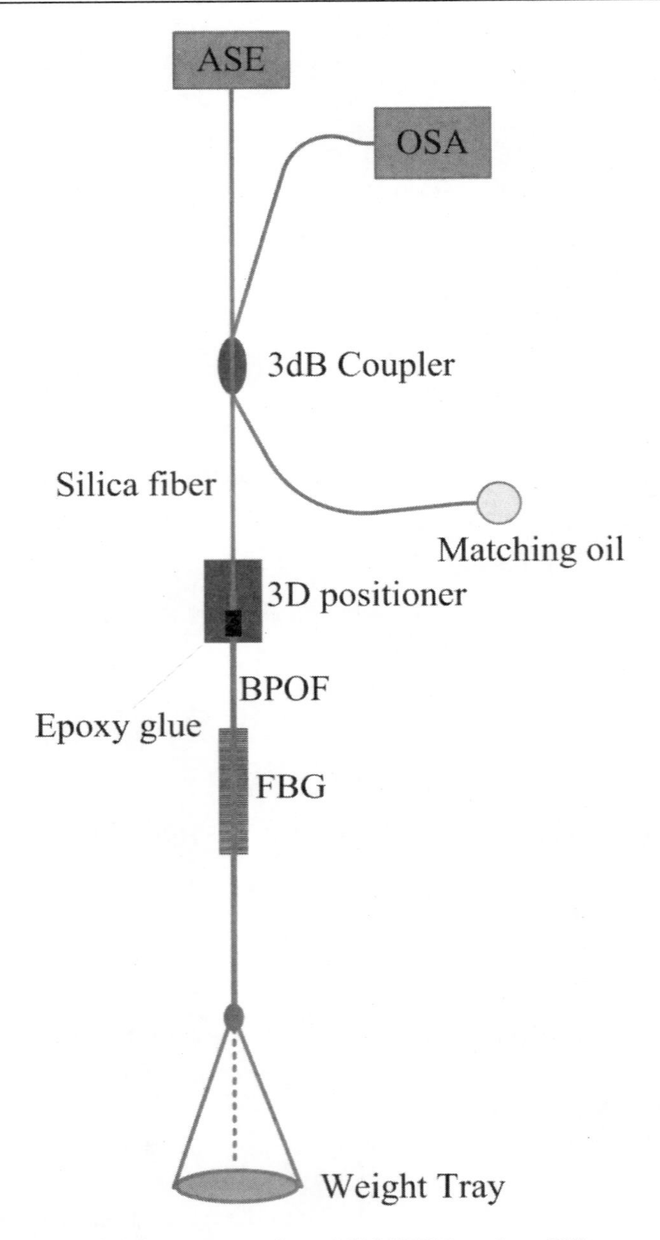

Figure 18. Experiment setup for the stress sensing of SM BPOF gratings [30].

From Figure 19A, it can be seen that the reflection spectra of BPOF gratings will redshift in the loading process while the reflection spectra blueshift in the unloading process.

However, the reflection spectrum in the loading process does not totally overlap with that in the unloading process. Especially, seen from Figure 19B, such difference becomes larger as the stress decreasing from 15.2 MPa. When the stress has completely been unloaded, the Bragg wavelength returns to 1570.8 nm but there is still ~0.4 nm difference compared with the original Bragg wavelength (1570.4 nm), resulted from the viscoelasticity of BPOF [3]. Furthermore, the relationship between Bragg wavelength and stress for BPOF gratings in the loading process can be obtained by the linear fitting [3]:

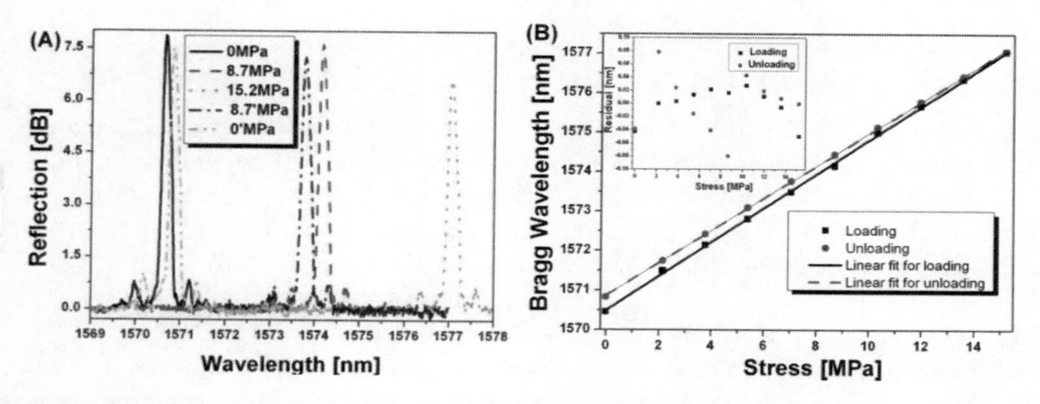

Figure 19. (A) Reflection spectra of SM BPOF gratings after different stress loaded; (B) Bragg wavelength of SM BPOF gratings varying with different stress and the inset is the residual of the linear fitting.

$$\lambda_B = 1570.495 + 0.432\sigma \tag{26}$$

with low residual, seen from the inset of Figure 19B.

From Eq. (26), the stress sensitivity of BPOF gratings in the loading process could be deduced to be 432 pm/MPa. Similarly, the relationship between Bragg wavelength and stress for BPOF gratings in the unloading process can also be linearly fitted as [3]:

$$\lambda_B = 1570.872 + 0.410\sigma \tag{27}$$

with low residual, also seen from the inset of Figure 19B. The stress sensitivity of BPOF gratings in the unloading process could be deduced to be 410 pm/MPa. So the averaged stress sensitivity of SM BPOF gratings is ~421 pm/MPa, 28 times higher than that of conventional silica fiber gratings (15 pm/MPa) [93], because of the lower Young's module of POF(~3 GPa) than that of silica fiber (~100 GPa). Therefore, BPOF grating has great potential for the stress sensing, especially for the condition with small stress sensing.

4.2.1.2. Strain Sensing with SM BPOF Gratings

The strain sensing of SM BPOF gratings (**Grating B**) is performed using the setup in Figure 20 [3, 30]. Both ends of BPOF are glued and fixed on tow steel of blocks connected to two 3D micropositioners. One micropositioner is fixed and the other one can be shifted longitudinally so that an axial tensile strain will be applied to BPOF gratings. The tensile strain values are estimated by dividing the fiber longitudinal extension displacement by the total length between the 3D positioner (about 2.5 cm). The longitudinal displacement accuracy of the moving 3D micropositioner is 0.01 mm. The applied strain is applied by manually moving the 3D micropostioner, where the loading speed is only about 1 με/min.

When the unfixed 3D micropositioner moves to the right side, BPOF will be stretched and induced strain ε_z applied to BPOF gratings, causing redshift of the reflection spectra of BPOF gratings (Figure 21A). According to Eq. (23), with the linear fitting in Figure 21B the relationship between the Bragg wavelength and tensile strain can be given by [3]:

$$\lambda_B = 1569.85 + 12.1\varepsilon \tag{28}$$

with low residual(less than 0.1 nm as shown in the inset of Figure 21B). From Eq. (28), the strain sensitivity of BPOF gratings is deduced to be 12.1 nm/% (1.21 pm/με), similar to that of silica FBGs, basing its great potential as a strain sensor [94]. Furthermore, according to Eq. (23), p_{eff} of the SM BPOF gratings is about 0.23 and Young's module E of the SM BPOF gratings can further be deduced to be 2.87 GPa, close to the value of PMMA [94].

4.2.1.3. Viscoelasticity of BPOF Gratings

As mentioned above, BPOF gratings will exhibit the viscoelasticity, which is the nature of the polymer materials [3, 25, 30, 50]. This means that the relationship between stress and strain depends on time, and even with step constant stress will cause increasing strain [89]. Therefore, the reflection spectra of BPOF gratings (**Grating B**) will evolve with time under 42.5 MPa tensile stress as shown in Figure 22 [34]. When 42.5 MPa stress is loaded, the spectrum has an immediate response with its peak shifted to 1586.83 nm. Then with the time increasing, the spectrum continues redshifting till it reaches relative equilibrium after 110 min with its peak at 1594.905 nm, where the spectrum has already redshifted 23.94 nm since the stress loaded.

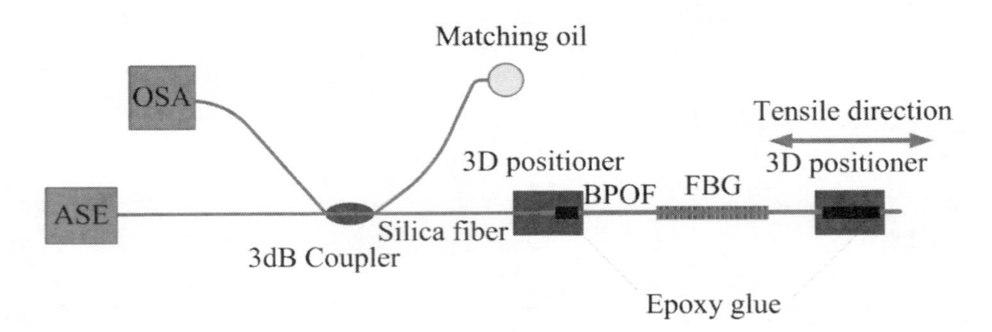

Figure 20. Experiment setup for the strain sensing of SM BPOF gratings [30].

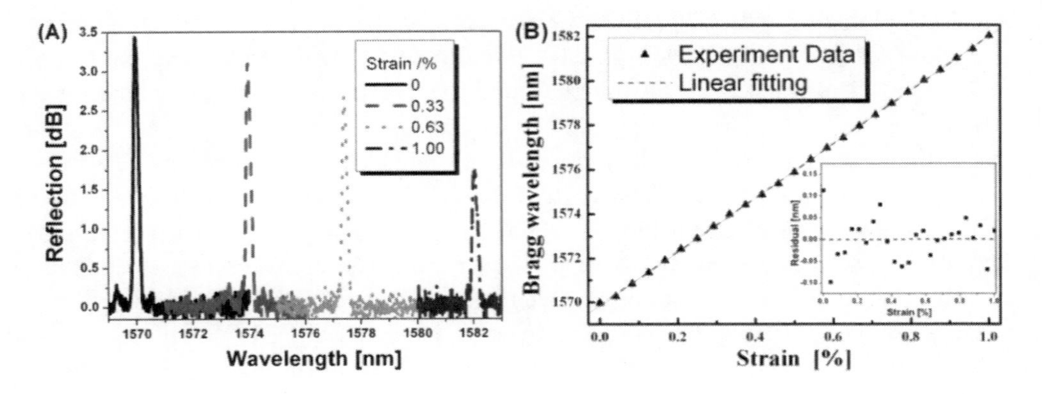

Figure 21. (A) Reflection spectra shifted with strain; (B) Bragg wavelength of BPOF gratings varying with strain, the inset is the residual of the linear fitting.

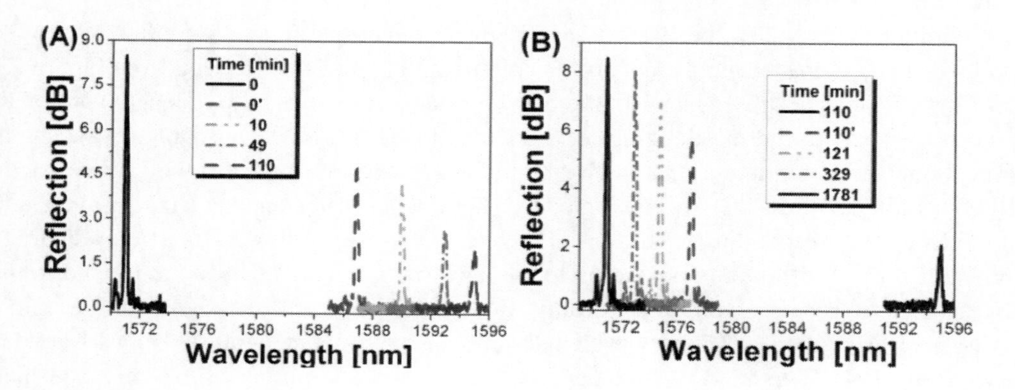

Figure 22. Evolution of reflection spectra for BPOF gratings (**Grating B**) at different time after 42.5 MPa tensile stress was loaded (A) (0-110 min) and unloaded (B) (110-1781 min).

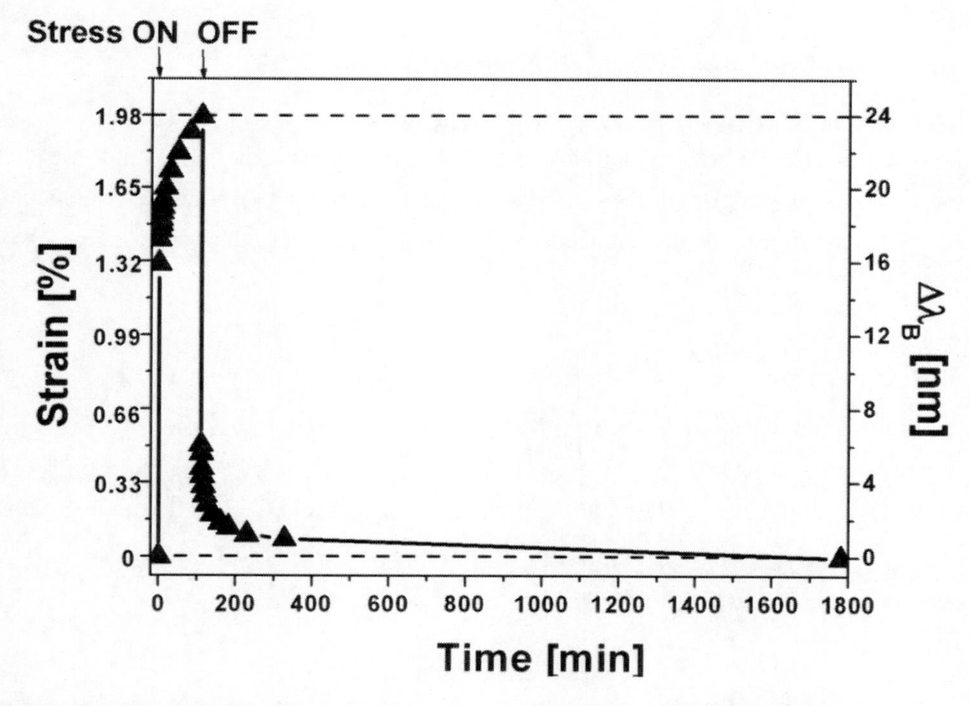

Figure 23. Bragg wavelength shift and strain vs time after loading and unloading 42.5 MPa stress.

When 42.5 MPa stress is unloaded at 110 min, the spectrum has an immediate recovery with its peak returns to 1577.02 nm. Then with the time increasing, the spectrum continues recovering till it returns to its original position (1570.920 nm). The immediate response and recovery is attributed to the elastic effect of the BPOF gratings to stress while the delayed response and recovery is due to the viscoelastic response of the BPOF gratings [3, 30]. In addition, it can also be seen from Figure 22 that since 42.5 MPa stress loaded, the reflection has dropped from 8.5 dB to 2.1 dB, while since 42.5 MPa stress unloaded, the reflection will start to recover, from 2.1 dB to 8.4 dB, due to the lower intensity of the light source at the longer wavelength.

According to Eq. (23), the Bragg wavelength shift and strain vs time after 42.5 MPa stress loaded and unloaded can be obtained as plotted in Figure 23 [34]. Seen from Figure 23, when 42.5 MPa tensile stress is loaded, there is an instantaneous Bragg wavelength shift of 16 nm (corresponding to the elastic strain of 1.3%), followed by a delayed Bragg wavelength shift of 8 nm (corresponding to the creep strain of 0.7% [95]) attributed to the viscoelasticity of BPOF gratings. When the stress is unloaded, there is an instantaneous Bragg wavelength shift of -18 nm (corresponding to the elastic recovery of 1.5%), followed by a delayed Bragg wavelength shift of -5 nm (corresponding to the anelastic recovery of 0.4%) attributed to the viscoelasticity of BPOF gratings. Especially, the final part (0.08% strain) after 219 min's recovery (corresponding to Bragg wavelength shift of only -1 nm) requires almost one day recover completely. This kind of time dependent effects is due to the viscoelasticity, which will bring complexity for the use of BPOF gratings in mechanical sensing [25]. To solve the viscoelastic issue, one method is to embed the FBG region into an elastic material with a lower viscosity to form a device [50] or directly fabricate BPOF gratings with elastic optic materials.

4.2.2. MM BPOF Gratings for Stress & Strain Sensing

4.2.2.1. Theory of MM BPOF Gratings for Stress & Strain Sensing

When an axial tensile strain is applied to MM BPOF gratings, the wavelength of the ^{i}th reflection mode (λ_B^i) will also shift by an amount $\Delta\lambda_B^i$ due to the periodicity change and the effective refractive index change induced by the strain. Similarly, the strain effect can be expressed as [30, 75, 96]:

$$\Delta\lambda_B^i = \lambda_B^i(1 - p_{eff})\varepsilon_z \tag{29}$$

where ε_z is the axial strain suffered by MM BPOF gratings and p_{eff} is the effective strain-optic constant.

Take BPOF gratings as the elastomer, combining Eqs. (25) and (29), $\Delta\lambda_B^i$ can be given by [30]:

$$\Delta\lambda_B^i = \lambda_B^i \frac{(1 - p_{eff})}{E}\sigma \tag{30}$$

However, as BPOF is a viscoelastic material so the measured Young's modulus will change when the stress is applied. When the stress is loaded gradually, the strain increases non-linearly with the stress loaded [94]. When the stress is removed, recovery occurs and strain relaxes non-linearly with the stress unloaded [25]. So Eq. (30) is hardly to be satisfied, and there will be [30]:

$$\Delta\lambda_B^i = \sum_{n=1}^{k} K_n \sigma^n \tag{31}$$

where K_n is the polynomial fitting coefficient and k is the best fitting polynomial order used.

4.2.2.2. Stress Sensing with MM BPOF Gratings

The stress sensing of MM BPOF gratings (**Grating C**) is performed with the same experiment setup in Figure 18 [30]. When the stress is applied, the Bragg grating spectra will change as shown in Figure 24 [30]. It can be seen that as the stress increases, the spectrum with the first (λ_1) and second (λ_2) maximum peak at 1571.383 and 1572.373 nm redshifts. When the stress is 9.59 MPa, the spectrum has already redshifted 2.8 nm without change of shape.

However, the amplitudes of the reflected modes vary a little, due to a major effect of the displacing peak wavelength of the reflected modes [74, 75]. When the stress is unloaded gradually, the spectrum returns. But it can not return its original position when the stress is reduced to 1.35 MPa, where there exists ~0.1 nm difference [30].

The first and second maximum Bragg resonant wavelengths of MM BPOF gratings vs different tensile stress are plotted in Figure 25A and B, respectively, fitted with the linear regression and second order polynomial regression.

Taking into account the linear fit to experimentally measured data, the stress sensitivity varies from 349 to 352 pm/MPa, and the average stress sensitivity is ~351 pm/MPa [30]. In addition, there are ~0.1 nm difference between intercepts of the loading and unloading process due to the viscoelasticity of BPOF gratings [25].

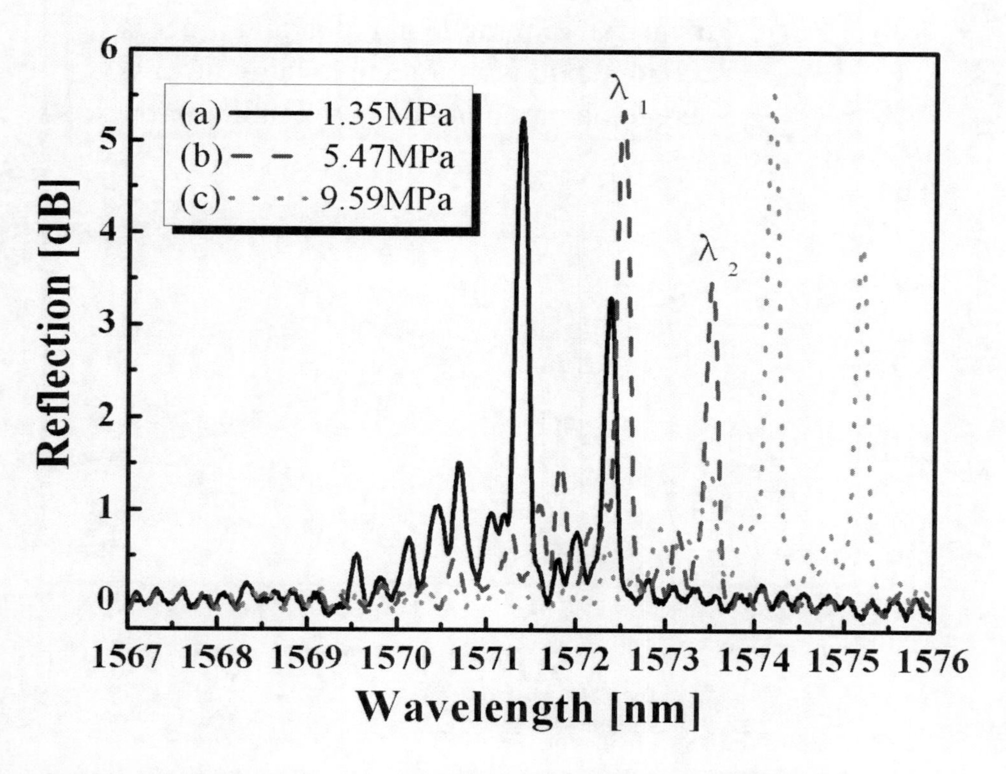

Figure 24. Shift in Reflection spectra of MM BPOF gratings (**Grating C**) with different stress. The values of the stress are: (a) 1.35MPa, (b) 5.47MPa and (c) 9.59MPa, respectively [30].

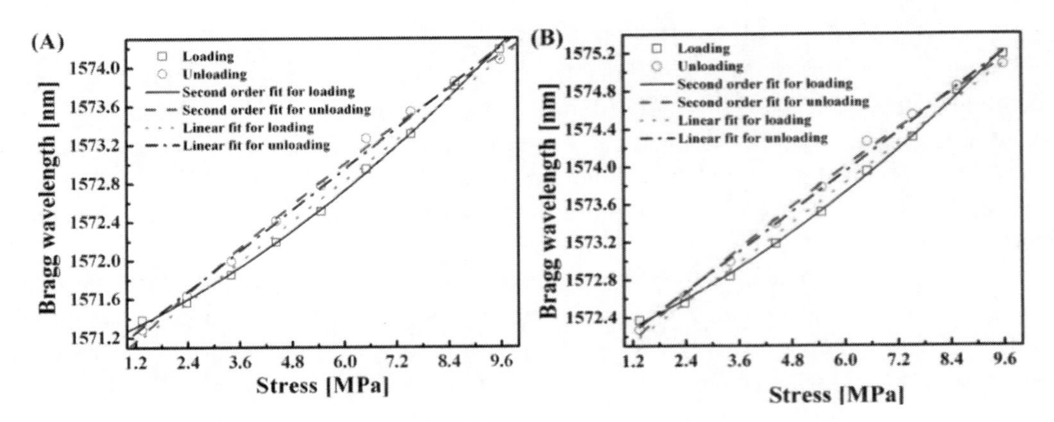

Figure 25. The first (A) and second (B) maximum Bragg resonant wavelength and their corresponding linear and second order fit with the axial loading and unloading stress process for MM BPOF gratings. The dot and dash-dot lines on all the graphs represent the linear fit to the experimental data, whereas the solid and dash lines represent the best second order polynomial fit [30].

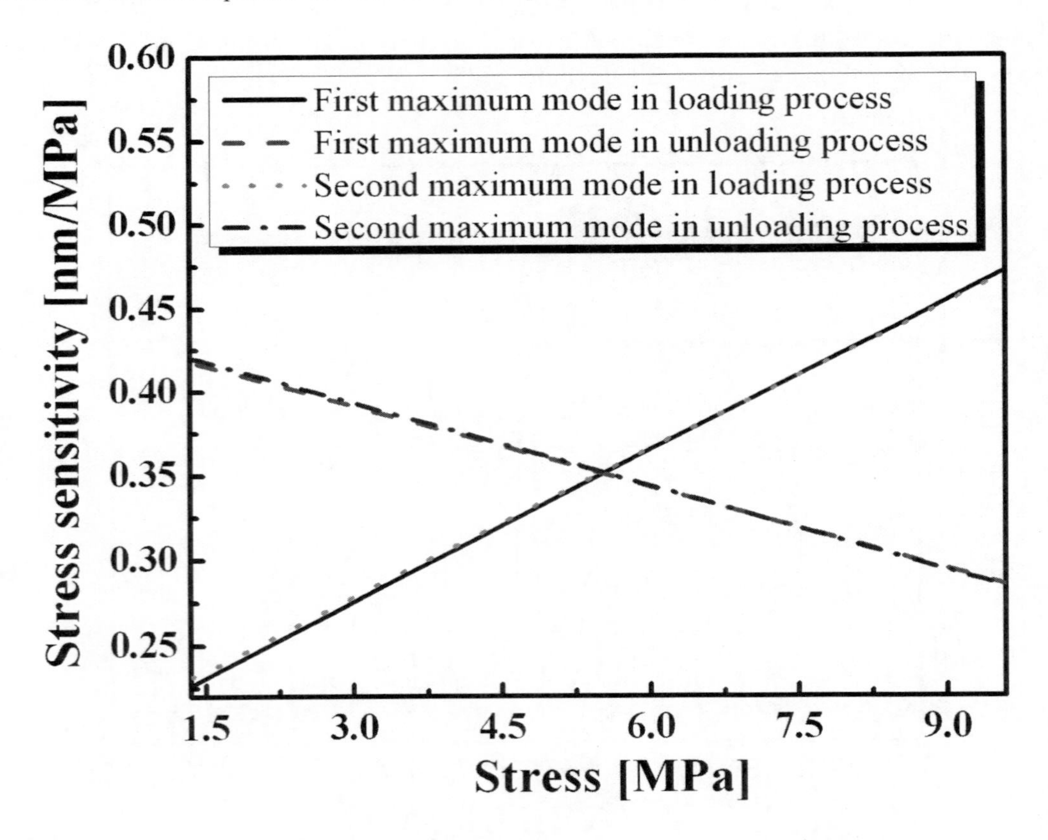

Figure 26. Variations of the stress sensitivities of MM BPOF grating for stress sensing [30].

Compared with the linear regression, the best fit is the second order polynomial fit, which shows higher adjusted R-squared and much smaller deviations. The stress response of the Bragg wavelengths for the second order polynomial regression, can be expressed as [30]:

$$\lambda_b = \begin{cases} 1571.068 + 0.18638\sigma + 0.01488\times\sigma^2 & \text{(for first maximum mode under axial loading stress)} \\ 1572.050 + 0.19111\sigma + 0.01454\times\sigma^2 & \text{(for second maximum mode under axial loading stress)} \\ 1570.653 + 0.43957\sigma - 0.00804\times\sigma^2 & \text{(for first maximum mode in under unloading stress)} \\ 1571.639 + 0.44279\sigma - 0.00827\times\sigma^2 & \text{(for second maximum mode under axial unloading stress)} \end{cases} \quad (32)$$

where λ_b is the corresponding Bragg wavelength with different stress, σ MPa.

Considering the second order polynomial fit, the stress sensitivities of MM BPOF gratings are stress dependent shown in Figure 26, which lie in the range from 226 to 420 pm/MPa with 1.35 MPa stress, whereas the corresponding stress sensitivities with 9.59 MPa stress lie in the range 285–472 pm/MPa for MM BPOF gratings [30]. It can also be seen that the stress sensitivities are almost the same for different modes, but the changing trend of the stress sensitivities in the stress loading and unloading process are totally different, attributed to the viscoelasticity of BPOF gratings [25].

4.2.2.3. Strain Sensing with MM BPOF Gratings

Similarly, strain sensing with MM BPOF gratings is also performed with the experiment setup in Figure 20. The first and second maximum Bragg resonant wavelengths of MM BPOF gratings vs different tensile strain are plotted in Figure 27A and B, respectively, fitted with the linear regression and second order polynomial regression [30]. Taking into account the linear fit, the strain sensitivity varies from 1.16 to 1.19 pm/με, and the average strain sensitivity is about 1.18 pm/με, almost the same as that of conventional silica FBGs [97], but a little lower than the previous result [6, 30].

Compared with the linear regression, the second order polynomial fit is also considered and shows almost the same results as that of the linear fit (Figure 27) [30]. The adjusted R-squared has no evident improving, showing that it is suitable to be fitted with the linear regression different from the stress sensing. The strain response of the Bragg wavelengths for the second order polynomial regression can be expressed as [30]:

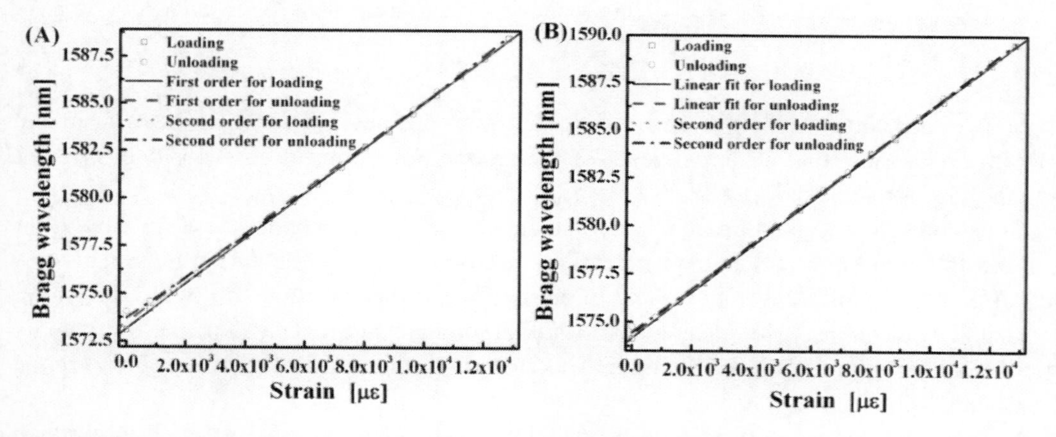

Figure 27. The first (A) and second (B) maximum Bragg resonant wavelengths and their corresponding linear and second order fit with the axial loading and unloading strain process for MM BPOF gratings. The solid and dash lines on all the graphs represent the linear fit to the experimental data, whereas the dot and dash-dot lines represent the second order polynomial fit [30].

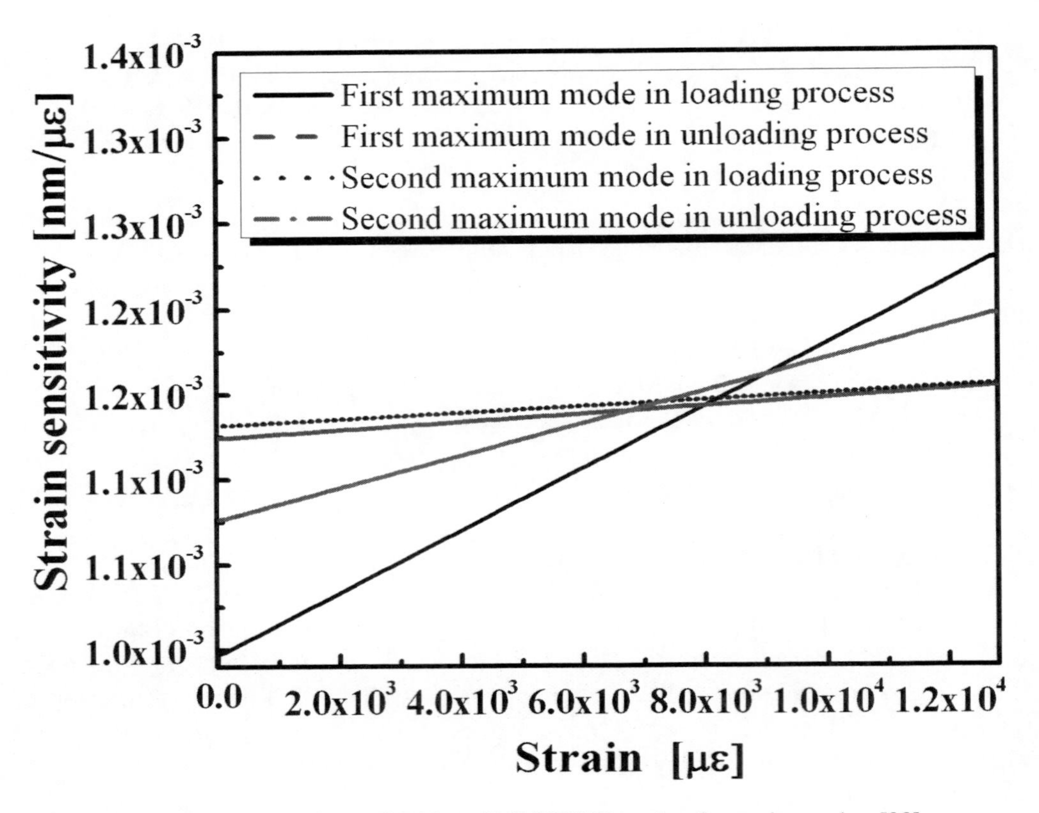

Figure 28. Variations of the strain sensitivities of MM BPOF grating for strain sensing [30].

$$\lambda_b = \begin{cases} 1573.181 + 1.17\times10^{-3}\varepsilon + 1.17\times10^{-9}\varepsilon^2 & \text{(for first maximum mode under axial loading strain)} \\ 1574.182 + 1.18\times10^{-3}\varepsilon + 9.33\times10^{-10}\varepsilon^2 & \text{(for second maximum mode under axial loading strain} \\ 1573.720 + 1.05\times10^{-3}\varepsilon + 9.15\times10^{-9}\varepsilon^2 & \text{(for first maximum mode in under unloading strain)} \\ 1574.444 + 1.13\times10^{-3}\varepsilon + 4.75\times10^{-9}\varepsilon^2 & \text{(for second maximum mode under axial unloading Strain)} \end{cases} \quad (33)$$

where λ_b is the corresponding Bragg wavelength with different strain, ε με. Seen from Eq. (33), it is noticeable that all coefficients of the second polynomial fit are positive different from the case for stress sensing.

Considering the second order polynomial fit, the strain dependent strain sensitivity has been illustrated in Figure 28. It can be clearly seen from this figure that the strain sensitivities are different for different modes and processes, and increase linearly with the strain increasing. The strain sensitivities of MM BPOF grating are strain dependent and found to lie over the range from 1.05 to 1.28 pm/με. The changing range is very narrow showing good consistency [30].

Furthermore, the value of p_{eff} can be calculated to be ~0.25, which is a little larger than that of the silica fiber (0.22), which means that the strain sensitivity of MM BPOF gratings is a little smaller than that of the silica fiber [30, 91]. And the value of E for such MM BPOF gratings can be estimated to be about 3.368GPa, close to the value of PMMA [94]. These results are very important for the measurement of strain/stress with MM BPOF gratings [30].

5. Remark

Both SM and MM BPOF gratings have higher temperature sensitivity, higher stress sensitivity, higher yield limit and wider dynamic range compared with their counterparts in silica fiber. All these results demonstrate that BPOF gratings have great potential for the temperature, stress and strain sensing. Although the photosensitivity of BPOF has improved, a number of issues: the poor stability of the BPOF, relatively low index change, etc, remain to be addressed. Moreover the viscoelasticity brings some complexity for the application of the BPOF gratings. In this regard the materials may need to be modified and optimized for enhanced photosensitivity and stability and reduced viscoelasticity.

Acknowledgments

The authors gratefully acknowledge the National Natural Science Foundation of China (NSFC) (21074123, 91027024, and 50973101), Open Fund of State Key Laboratory of Information Photonics and Optical Communications (Beijing University of Posts and Telecommunications) and an International Science Linkages (ISL) project (CG130013) from the Department of Industry, Innovation, Science and Research (DIISR), Australia.

References

[1] Xiong, Z.; Peng, G. D.; Wu, B.; Chu, P. L., *IEEE Photon. Technol. Lett.* 1999, 11, 352-354.

[2] Emiliyanov, G.; Jensen, J. B.; Bang, O.; Hoiby, P. E.; Pedersen, L. H.; Kjær, E. M.; Lindvold, L., *Opt. Lett.* 2007, 32, 460-462.

[3] Wang, T.; Luo, Y.; Peng, G.-D.; Zhang, Q., *Proc. SPIE* 2012, 8351, 83510M-1-8.

[4] Webb, D. J.; Kalli, K.; Zhang, C.; Komodromos, M.; Argyros, A.; Large, M.; Emiliyanov, G.; Bang, O.; Kjaer, E., *Proc. SPIE* 2008, 6990, 69900L-1-10.

[5] Dobb, H.; Carroll, K.; Webb, D. J.; Kalli, K.; Komodromos, M.; Themistos, C.; Peng, G. D.; Argyros, A.; Large, M. C. J.; Eijkelenborg, M. A. v.; Fang, Q.; Boyd, I. W., *Proc. SPIE* 2006, 6189, 618901-1-12.

[6] Liu, H. B.; Liu, H. Y.; Peng, G. D.; Chu, P. L., *Opt. Commun.* 2003, 219, 139–142.

[7] Peng, G.-D.; Chu, P. L., *Proc. SPIE* 2002, 4929 303-311.

[8] Yu, J. M.; Tao, X. M.; Tam, H. Y., *Opt. Lett.* 2004, 29, 156-158.

[9] Liu, H. Y.; Liu, H. B.; Peng, G. D.; Chu, P. L., *Opt. Commun.* 2006, 266, 132-135.

[10] Peng, G. D.; Xiong, Z.; Chu, P. L., *Opt. Fiber Technol.* 1999, *5*, 242-251.

[11] Sáez-Rodríguez, D.; Munoz, J. L. C.; Johnson, I.; Webb, D. J.; Large, M. C. J.; Argyros, A., *Proc. SPIE* 2009, 7357, 73570L-1-8.

[12] Stefani, A.; Yuan, W.; Markos, C.; Bang, O., *IEEE Photon. Technol. Lett.* 2011, 23, 660-662.

[13] Liu, H. Y.; Liu, H. B.; Peng, G. D.; Chu, P. L., *Opt. Commun.* 2003, 220, 337-343.

[14] Luo, Y. H.; Zhou, J. L.; Yan, Q.; Su, W.; Li, Z. C.; Zhang, Q. J.; Huang, J. T.; Wang, K. Y., *Appl. Phys. Lett.* 2007, 91, 071110.

[15] Dobb, H.; Webb, D. J.; Kalli, K.; Argyros, A.; Large, M. C. J.; Eijkelenborg, M. A. v., *Opt. Lett.* 2005, 30, 3296-3298.
[16] Li, Z.; Tam, H. Y.; Xu, L.; Zhang, Q., *Opt. Lett.* 2005, 30, 1117-1119.
[17] Hiscocks, M. P.; Eijkelenborg, M. A. v.; Argyros, A.; Large, M. C. J., *Opt. Express* 2006, 14, 4644-4649.
[18] Luo, Y.; Zhang, Q.; Liu, H.; Peng, G.-D., *Opt. Lett.* 2010, 35, 751-753.
[19] Blacket, D. R.; Large, M.; Argyros, A., *Proc. 18th Int. Conf. Polymer Opt. Fiber* 2009, 52.
[20] Johnson, I. P.; Webb, D. J.; Kalli, K.; Large, M. C. J.; Argyros, A., *Proc. SPIE* 2010, 7714, 77140D-1-10.
[21] Webb, D. J.; Chen, X.; Kalli, K.; Saez-Rodriguez, D.; Zhang, C.; Barton, J.; Johnson, I.; Ye, C., *Proc. 18th Int. Conf. Polymer Opt. Fiber* 2009, 26.
[22] Webb, D. J.; Kalli, K.; Carroll, K.; Zhang, C.; Komodromos, M.; Argyros, A.; Large, M.; Emiliyanov, G.; Bange, O.; Kjaer, E., *Proc. SPIE* 2007, 6830, 683002-1-9.
[23] Chu, P. L., *Optics and Photonics News* 2005, *16*, 52-56.
[24] Carroll, K. E.; Zhang, C.; Webb, D. J.; Kalli, K.; Argyros, A.; Large, M. C., *Opt. Express* 2007, 15, 8844-8850.
[25] Large, M. C. J.; Moran, J.; Ye, L., *Meas. Sci. Technol.* 2009, 20, 034014-1.
[26] Kuang, K. S. C.; Quek, S. T.; Koh, C. G.; Cantwell, W. J.; Scully, P. J., *J. Sensors* 2009, 2009, 1-13.
[27] Chen, X.; Zhang, C.; Webb, D. J.; Suo, R.; Peng, G. D.; Kalli, K., *Proc. SPIE* 2009, 7503, 750327-1-4.
[28] Ye, C. C.; Dulieu-Barton, J. M.; Webb, D. J.; Zhang, C.; Peng, G.-D.; Chambers, A. R.; Lennard, F. J.; Eastop, D. D., *Proc. SPIE* 2009, 7503, 75030M-1-4.
[29] Wu, W.; Luo, Y.; Cheng, X.; Tian, X.; Qiu, W.; Zhu, B.; Peng, G.-D.; Zhang, Q., *J. Optoelectron. Adv. Mater.* 2010, 12, 1652-1659.
[30] Luo, Y.; Yan, B.; Li, M.; Zhang, X.; Wu, W.; Zhang, Q.; Peng, G.-D., *Opt. Fiber Technol.* 2011, 17, 201-209.
[31] Cheng, X.; Qiu, W.; Wu, W.; Luo, Y.; Tian, X.; Zhang, Q.; Zhu, B., *Chin. Opt. Lett.* 2011, 9, 020602.
[32] Luo, Y.; Wu, W.; Wang, T.; Cheng, X.; Zhang, Q.; Peng, G.-D.; Zhu, B., *Opt. Commun.* 2012, in press, DOI: 10.1016/j.optcom.2012.06.051.
[33] Luo, Y.; Wu, W.; Cheng, X.; Wang, X.; Tian, X.; Wang, T.; Zhang, Q.; Peng, G.-D.; Zhu, B., *J. Optoelectron. Adv. Mater.* 2012, submitted.
[34] Luo, Y.; Wang, T.; Wu, W.; Wang, X.; Yan, B.; Zhang, Q.; Peng, G.-D., *the 37th Australian Conference on Optical Fibre Technology* 2012, submitted.
[35] Luo, Y.; Liu, H.; Zhang, Q.; Peng, G., Grating writing with 355nm wavelength in polymer optical fiber doped with benzildimethylketal, *the 2009 Australasian Conference on Optics, Lasers and Spectroscopy and Australian Conference on Optical Fibre Technology* 2009, 8, 48.
[36] Luo, Y.; Yan, B.; Li, M.; Zhang, X.; Zhang, Q.; Peng, G., Stress and strain sensing with multimode POF Bragg gratings. In PIERS Proceedings, Cambridge USA, 2010; pp 696 - 699.
[37] Tomlinson, W. J.; Kaminow, I. P.; Chandross, E. A.; Fork, R. L.; Silfvast, W. T., *Appl. Phys. Lett.* 1970, 16, 486-489.

[38] Zhang, C.; Carroll, K.; Webb, D. J.; Bennion, I.; Kalli, K.; Emiliyanov, G.; Bang, O.; Kjaer, E.; Peng, G. D., *Proc. SPIE* 2008, 7004, 70044G-1-5.
[39] Li, Z.; Ma, H.; Zhang, Q.; Ming, H., *J. Optoelectron. Adv. Mater.* 2005, 7, 1039-1046.
[40] Luo, Y.; Li, Z.; Zheng, R.; Chen, R.; Yan, Q.; Zhang, Q.; Peng, G.; Zou, G.; Ming, H.; Zhu, B., *Opt. Commun.* 2009, 282, 2348-2353.
[41] Franke, H., *Appl. Opt.* 1984, 23, 2729-2733.
[42] Park, O.-H.; Jung, J.-I.; Bae, B.-S., *J. Mater. Res.* 2001, 16, 2143-2148.
[43] Seidl, B.; Kalinyaprak-Icten, K.; Fuß, N.; Hoefer, M.; Liska, R., *J. Polym. Sci., Part A: Polym. Chem.* 2007, 46, 289-301.
[44] Colom, X.; García, T.; Suñol, J. J.; Saurina, J.; Carrasco, F., *J. Non-Crystal. Solids* 2001, 287, 308-312.
[45] Peng, G. D.; Chu, P. L.; Xiong, Z. J.; Whitbread, T. W.; Chaplin, R. P., *J. Lightwave Technol.* 1996, 14, 2215-2223.
[46] Ming, H.; Zhang, G.; Xie, J., Optoelectronic Technology. Publishing company of USTC: Hefei, 1998.
[47] Ankiewicz, A.; Wang, Z. H.; Peng, G. D., *Opt. Commun.* 1998, 156, 27-31.
[48] Saekeang, C.; Chu, P. L.; Whitbread, T. W., *Appl. Opt.* 1980, 19, 2025-2030.
[49] Yuan, W.; Stefani, A.; Bache, M.; Jacobsen, T.; Rose, B.; Herholdt-Rasmussen, N.; Nielsen, F. K.; Andresen, S.; Sørensen, O. B.; Hansen, K. S.; Bang, O., *Opt. Commun.* 2011, 284, 176-182.
[50] Zhang, Z. F.; Zhang, C.; Tao, X. M.; Wang, G. F.; Peng, G. D., *IEEE Photon. Technol. Lett.* 2010, 22, 1562-1564.
[51] Hill, K. O.; Malo, B.; Bilodeau, F.; Johnson, D. C.; Albert, J., *Appl. Phys. Lett.* 1993, 62, 1035-1037.
[52] Yu, H.-G.; Wang, Y.; Xu, Q.-Y.; Xu, C.-Q., *J. Lightwave Technol.* 2006, 24, 1903-1912.
[53] Liu, H. Y.; Peng, G. D.; Chu, P. L., Highly reflective polymer fiber Bragg gratings and its growth dynamics. In Optical Fiber Communication Conference Sawchuk, A., Ed. Optical Society of America: Anaheim, California, 2002; Vol. 70 pThGG33.
[54] Liu, H. Y.; Peng, G. D.; Chu, P. L., *IEEE Photon. Technol. Lett.* 2001, 13, 824-826.
[55] Chen, X.; Zhang, C.; Webb, D. J.; Peng, G.-D.; Kalli, K., *Meas. Sci. Technol.* 2010, 21, 094005.
[56] Liu, H. Y.; Peng, G. D.; Chu, P. L., *Opt. Commun.* 2002, 204 151–156.
[57] Johnson, I. P.; Yuan, W.; Stefani, A.; Nielsen, K.; Rasmussen, H. K.; Khan, L.; Webb, D. J.; Kalli, K.; Bang, O., *Electron. Lett.* 2011, 47, 271–272.
[58] Dobb, H.; Carroll, K.; Webb, D. J.; Kalli, K.; Komodromos, M.; Themistos, C.; Peng, G. D.; Argyros, A.; Large, M. C. J.; Eijkelenborg, M. A. v.; Arresy, M.; Kukureka, S., *Proc. SPIE* 2006, 6193 61930Q-1-12.
[59] Ishigure, T.; Hirai, M.; Sato, M.; Koike, Y., *J. Appl. Polym. Sci.* 2004, 91, 404–409.
[60] Sato, M.; Hirai, M.; Ishigure, T.; Koike, Y., *J. Lightwave Technol.* 2000, 18, 2139-2145.
[61] Yang, D. X.; Xu, J.; Tao, X.; Tam, H., *Mater. Sci. Eng. A* 2004, 364, 256-259.
[62] E. E.Shafee, *Polym. Degrad. Stabil.* 1996, 53, 57-61.
[63] Zhang, Z.; Zhao, P.; Lin, P.; Sun, F., *Polymer* 2006, 47, 4893–4896.
[64] V, G.; RO, N.; S, S.; BJ., S., *Physical properties of polymers handbook*. AIP Press: Woodbury, 1996; p p. 535.

[65] Jing, Y.; Baojin, P.; Jianwen, F.; Tiefei, H.; Liao, Y.-b.; Zhang, M., *Proc. SPIE* 2007, 6279, 627961-1-6.
[66] Cariou, J. M.; Dugas, J.; Martin, L.; Michel, P., *Appl. Opt.* 1986, 25, 334-336.
[67] Hill, K. O.; Meltz, G., *J. Lightwave Technol.* 1997, 15, 1263-1276.
[68] Jian, H. Z., Study of theory and technique of fiber gratings sensors, Xi'an Institute of Optics and Precision Mechanics of CAS, 2000, Xi'an, PhD thesis.
[69] Kang, E.-S.; Lee, T.-H.; Bae, B.-S., *Appl. Phys. Lett.* 2002, 81, 1438-1440.
[70] Zhang, F.; Wang, X.; Zhang, Q., *Polymer* 2000, 41, 9155–9161.
[71] Zhang, Q.; Wang, P.; Zhai, Y., *Macromolecules* 1997, 30, 7874-7879.
[72] Mizunami, T.; Djambova, T. V.; Niiho, T.; Gupta, S., *J. Lightwave Technol.* 2000, 18, 230-235.
[73] Diemeer, M. B. J., *Opt. Mater.* 1998, 9, 192-200.
[74] Djambova, T. V.; Mizunami, T., *Jpn. J. Appl. Phys.* 2000, 39, 1566-1570.
[75] Cazo, R. M.; Lisbôa, O.; Hattori, H. T.; Schneider, V. M.; Barbosa, C. L.; Rabelo, R. C.; Ferreira, J. L. S., *Microw. Opt. Technol. Lett.* 2001, 28, 4-8.
[76] Webb, D. J.; Kalli, K., *Polymer fiber Bragg gratings*. Bentham Science Publishers: IL, 2009; p 1-20.
[77] Jensen, J.; Hoiby, P.; Emiliyanov, G.; Bang, O.; Pedersen, L.; Bjarklev, A., *Opt. Express* 2005, 13, 5883-5889.
[78] Emiliyanov, G.; Jensen, J. B.; Bang, O.; Hoiby, P. E.; Pedersen, L. H.; Kjær, E. M., *Opt. Lett.* 2007, 32, 1059-1059.
[79] Markos, C.; Yuan, W.; Vlachos, K.; Town, G. E.; Bang, O., *Opt. Express* 2011, 19, 7790-7798.
[80] Zhang, C.; Zhang, W.; Webb, D. J.; Peng, G. D., *Electron. Lett.* 2010, 46, 643–644.
[81] Johnson, I. P.; Kalli, K.; Webb, D. J., *Electron. Lett.* 2010, 46, 1217-1218.
[82] Yuan, W.; Khan, L.; Webb, D. J.; Kalli, K.; Rasmussen, H. K.; Stefani, A.; Bang, O., *Opt. Express* 2011, 19, 19731-19739.
[83] Peng, G. D.; Chu, P. L., *Proc. SPIE* 2000, 4110, 123-138.
[84] Liu, H. Y.; Liu, H. B.; Peng, G. D., *Proc. SPIE* 2005, 5855, 663-666.
[85] Blacket, D. R.; Large, M.; Argyros, A., High Strain sensing with mPOF and Long Period Gratings, Proc. 3rd International Workshop Miscrostructure Polymer Optical Fibre 2009, 52.
[86] Yuan, W.; Stefani, A.; Bang, O., *IEEE Photon. Technol. Lett.*, 2012, 20, 401-403.
[87] Jin, W.; Michie, W. C.; Thursby, G.; Konstantaki, M.; Culshaw, B., *Opt. Eng.* 1997, 36, 598–609.
[88] Johnson, I. P.; Webb, D. J.; Kalli, K.; Large, M. C. J.; Argyros, A., *Proc. SPIE* 2010, 7714, 77140D-1-10.
[89] http://en.wikipedia.org/wiki/Viscoelasticity.
[90] Othonos, A., *Rev. Sci. Instrum.* 1997, 68, 4309-4341.
[91] Liu, H. Polymer Optical fiber Bragg Gratings. University of New South Wales, Sydney, 2003.
[92] Melle, S. M.; Liu, K.; Measures, R. M., *Appl. Opt.* 1993, *32*, 3601-3609.
[93] Inoue, A.; Shigehara, M.; Ito, M.; Inai, M.; Hattori, Y.; Mizunami, T., *Optoelectr. Dev. Tech.* 1995, 10, 119-130.
[94] Ishiyama, C.; Higo, Y., *J. Polym. Sci. Part B: Polym. Phys.* 2002, *40*, 460–465.
[95] Kelly, P., Solid Mechanics Part I An Introduction to Solid Mechanics. 2008.

[96] Zhao, W.; Claus, R. O., *Smart Mater. Struct.* 2000, 9, 212–214.
[97] Jin, W.; Michie, W. C.; Thursby, G.; Konstantaki, M.; Culshaw, B., *Opt. Eng.* 1997, 36, 598–609.

In: Optical Fibers: New Developments
Editor: Marco Pisco

ISBN: 978-1-62808-425-2

Chapter 4

MICROFABRICATION ON THE SURFACE OF OPTICAL FIBERS FOR THE NEXT GENERATION OF MEMS DEVICES

Harutaka Mekaru*

Senior Research Scientist, Green Nano Device Research Team, Research Center for Ubiquitous MEMS and Micro Engineering (UMEMSME), National Institute of Advanced Industrial Science and Technology (AIST)

ABSTRACT

As one of the techniques for overcoming the size limitations of micro-electro-mechanical-systems (MEMS) devices, we chose to fabricate MEMS structure and electric circuits on fibrous substrates instead of using the Si chips; and thus we are now witnessing ordinary fibers evolving into "smart fibers" embedded with functions such as sensors and actuators. In addition, it is planned that these smart fibers be woven into e-textiles with these new functions which work as a large-area display and a wearable health checker. To transfer continuously MEMS patterns and weaving guides that can determine the positions for fixing smart fibers on the surface of optical fibers, we selected thermal imprinting and focused-ion-beam (FIB) etching technologies. We developed two kinds of reel-to-reel thermal imprint system. In the first system, a fiber is sandwiched by two plane molds and rolls under the traction force of sliding molds traveling in opposite directions. With this method, 5-μm-width square and circular dotted patterns were successfully transferred onto the entire curved surface of a 250-μm-diameter plastic optical fiber (POF) using a reel-to-reel feeder as a batch processing operation by a repetition of roller-imprinting. In the second system, we used a cylindrical mold with hybrid-layered microstructures consisting of 260-μm-wide macro patterns and several-10-μm-wide patterns fabricated by precise cutting with a forming tool and 3-D photolithography using a flexible mask. The cylindrical mold heated up to 50 °C with an infrared lamp, was pushed into the surface of POF to a depth of 50 μm. As a result, we succeeded in the transfer of weaving guides and various micro-patterns onto the surface

* Corresponding Author address: 1-2-1 Namiki, Tsukuba, Ibaraki 305-8564, JAPAN. Tel: +81-29-861-7100; Fax: +81-29-861-7225; E-mail: h-mekaru@aist.go.jp.

of the POF at a sending speed of 20 m/min. On the other hand, a mold for high-temperature imprinting on a quartz optical fiber (QOF) was fabricated with a glass-like carbon (GC) substrate. Mold patterns with high accuracies were processed by applying MEMS fabrication techniques. Precise patterns with 5-μm-linewidths were transferred on the front surface of a 200-μm-square QOF by thermal imprinting at a temperature of 1350 °C. In another work, a 3-D processing technology using a FIB etching was developed to fabricate a spiral coil on the surface of QOFs so that a knot may not exist. Using a prototype rotation stage, we succeeded in the seamless patterning of 1-μm-wide features on the full-circumference surface of a 250-μm-diameter QOF. The micro-coils can be applied to receiver coils of nuclear magnetic resonance imaging systems used in imaging blood vessels.

Keywords: Optical fiber, MEMS, Hot embossing, Nanoimprint, FIB etching

INTRODUCTION

Micro-electromechanical-systems (MEMS) are widely used in optical, electrical, chemical, and biological fields such as in digital micro-mirror devices of projectors, precise nozzles installed in the head of inkjet printers, and in many kinds of sensors that measure various physical conditions (Kaajakari 2009). These MEMS devices are processed by the technology that is employed in mechatronics and in the manufacturing of semiconductor devices. In a conventional electronic device, their functions are provided by the assembling and wiring of numbers of electronic components (e.g. ICs, resistors, and capacitors) that are formed on the surface of a printed-circuit board (PCB). Typically, PCBs are made of materials such as paper/phenol and glass/epoxy resins that are not flexible enough to be of use in certain applications. MEMS devices, unlike semiconductor integrated circuit devices, may comprise movable 3-dimensional (3-D) structures with various machine elements and electronic circuits built on substrates of silicon, glass, and organic materials. However, continued increase in Si wafer size has its limitations and other alternatives for the development of large size MEMS devices need to be explored. In our works, we are departing from industry's conventional dependence on large-size Si wafers, and we are by passing the constraints of MEMS manufacturing equipment that require high vacuum environment for their operations. As one of the techniques for overcoming the size limitations of MEMS devices, we chose to fabricate MEMS structure on fibrous substrates instead of doing it on Si chips. We are now witnessing an evolution of ordinary fibers into ''smart fibers'' laid out with functions such as sensors and actuators as shown in Figure 1(a). These smart fibers equipped with new functions would eventually be woven into e-textiles. The woven fibers would easily conform to the anatomy of human body like any other clothing does, and would fit on the substrate of many wearable devices, such as health checkers and safety jackets, and thus replacing the need for plastic films currently used for wearing such devices.

To form micro-/nanoscale structures on curved surfaces, special techniques must be employed. It is very difficult to process a seamless pattern similar to a spiral structure onto a cylindrical surface such as the surface of fibers, especially those with small diameters of 1 mm or less. To process a precise pattern on a cylindrical surface, various techniques have been proposed. The patterning technologies so far reported fall under three main classifications. The first one is a direct-write scheme using: (a) laser (Kikuchi et al. 2007; di

Benedetto et al. 2008); (b) focused-ultraviolet (UV) lights (Siewert and Löffler 2005; Lee et al. 2010); and (c) electron beam (Chand et al. 2009; Taniguchi and Asatani 2009). The second one is a projection technique using: (a) hard photomasks (Lu et al. 2010; Mineta et al. 2011), and (b) X-ray masks (Katoh et al. 2001; Mekaru et al. 2004). And the last one is an exposure technique using a flexible mask lapped around the cylindrical surface (Li et al. 1999). Although all the above methods are innovative and unique but none of them, because of their high cost and low throughput, are suitable for high-volume manufacturing processes. In order to fabricate woven-MEMS devices at a high throughput, novel technologies such as continuous manufacturing of smart fibers would be necessary. We have also looked into the applications of hot embossing (Worgull 2009) and nanoimprinting (Sotomayor Torres 2003) as a technology that can directly fabricate microstructures on the cylindrical surface of fibers. And next, we plan to manufacture a smart fiber after metallization by inkjet printing to fill up the cavities of the imprinted patterns with various materials.

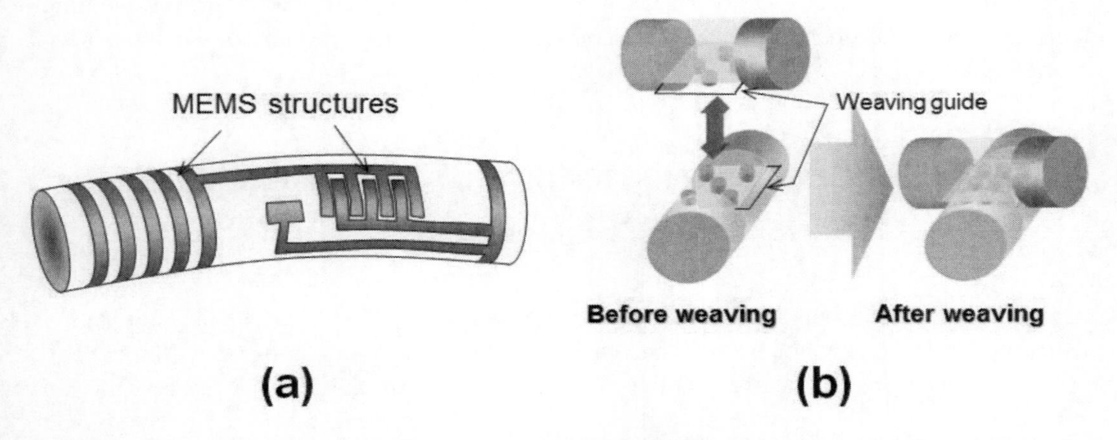

Figure 1. Illustrations of; (a) smart fiber with MEMS structures on its surface, and (b) contact point of smart fibers using weaving guides.

On the other hand, the substrates made of e-textile are quite flexible and can be worn by humans where the devices can conform to any surface of human anatomy. This flexibility in the substrate led to the development of smart clothes that can be worn by humans, such that, by proper positioning of the physical and chemical sensors on the clothes, many physical and biological conditions of the wearer's body can be monitored (Nugent et al. 2003; Engin et al. 2005). In one case, simple computational elements and light emitting-diodes were assembled into a woven substrate, using which a wearable computer with monitoring systems was fabricated (Marculescu et al. 1995; Buechley and Eisenberg 2009). Another technique for making e-textile was proposed where fibers with specific functions were woven into one another. For instance, in one case, a sensor was made by weaving functional fibers coated with piezo-resistive materials (Huang et al. 2008), and in another case, an information routing device was developed using smart fibers with organic transistors formed on the fibers' surface (Lee and Subramanian 2005). In a fabric where warps and wefts intersections are designed to function as mechanical and electrical contact points for its e-textile device, the uniformity of the weaving density becomes very critical. In one case, a solar cell embedded ribbon fiber with a rectangular cross-section was employed that could preserve the cell's positions because such a ribbon-shaped structure could not easily be twisted (Wagner et al. 2002). However, in case of fibers with circular cross-sections, the task of maintaining their assigned positions becomes difficult. In one case in order to make inverter circuits, a fabric of conductive fibers was used where the warps and wefts of the fibers were connected using contact pads (Bonderover and Wagner 2004). In another case, conducting-polymer-coated woven fibers with electrical contacts were used taking advantage of their globe-shaped contact points (Hamedi et al. 2007). For the processing of weaving guides, deep concave structures in fiber were created resulting in significant deformation of the fiber's cross-section as shown in Figure 1(b). Since a weaving guide has to fix several-hundred-μm diameter woofs, a comparatively large structure of the order of hundreds of microns wide was required. On the other hand, since cantilevers and electric circuits are currently assumed as main components in MEMS structures, we presume that the pattern width would be tens of μm. Such structures because of the differences in their pattern sizes, would naturally differ in their imprint conditions. One involves processing on the cylindrical surface of the fiber, and the other is to purposefully induce a deformation in the bulk material of the fiber. However, the molding characteristics of the fiber are not well understood at the present time and not much has been reported on the subject.

MOLDING CHARACTERISTICS OF PLASTIC OPTICAL FIBER

Motivation

One of the thermoplastic materials widely used in MEMS field is polymethyl methacrylate (PMMA) which has also been used in the field of X-ray lithography (Soper et al. 2000). Moreover, mechanical properties and the physics of the PMMA process in hot embossing and thermal nanoimprint lithography have already been addressed (Heckele and Schomburg 2003; Hirai et al. 2003; Hirai 2010). However, the shapes of the PMMA substrates in the previous molding experiments used to be in form of plane surface. Now we deal with curved surfaces of plastic optical fibers (POF) that are used quite often in the field of MEMS (Keiser 2003; Grattan and Meggitt 2003). We investigated the relationship between the imprinted depth on POF and the heating temperature of molds with various press depths. In our thermal imprinting work, four kinds of buffer materials were studied by placing them under the POF, followed by an application of pressures as dictated by thermal imprinting operation. Then after each imprinting the deformation of POF was studied.

Thermal Imprinting Experimental Setup

Figure 2 shows a setup for thermal imprinting that uses a desktop nanoimprint system NI-1075 (Nano Craft Technologies) (Takahashi et al. 2006a). In this system, ceramic heaters were installed on its upper and bottom stages, where on those stages, mold and molding materials of 30 mm^2 or less were set. The mold that served as a master structure was fabricated by Ni electroforming after patterning a photoresist film on it by employing UV lithography. The size of the Ni mold was $15 \times 15 \times 2$ mm^3. A buffer material of $30 \times 30 \times 3$ mm^3 was placed on the bottom stage. For buffer, four kinds of materials with different hardness values were employed as shown in Table 1. The electroformed-Ni mold and the buffer material were fixed to the upper and bottom stages with polyimide double-faced tapes. A POF CK-10 (Mitsubishi Rayon) made of a 240-μm-diameter polymethylmetacrylate (PMMA) core covered with a 5-μm-thick fluorine clad material [Figure 3] was stretched and fixed on the flange of a bellows-type vacuum chamber with a double-faced tape as shown in Figure 2. The upper stage was movable along the vertical direction with a servo motor, while the bottom stage was set to be stationary. The upper stage was connected to a load cell to monitor the contact force during its press operation. After a buffer material was set up, the upper stage was lowered down toward the POF while monitoring the contact force that did not register any value until the stage made its first contact with the fiber, where that position was recorded and referred as a press beginning point. From that point on, and in reference to the press beginning point, the press depth was changed to 25, 75, and 125 μm. The maximum press depth of 125 μm equaled half the size of the diameter of POF CK-10. The value of the press depth was kept within a range of 25–125 μm. The reason was that on the low end side, a threshold value of 25 μm was defined; and below that point no imprinted pattern could be observed, or where any imprinted depth and the press depth did not exhibit any proportionality. On the other hand, in case of a press depth above 125 μm, the POF sank into the buffer material by an excessive pressure, where the experiment could not be repeated using the same buffer. The speed of the pressing and de-molding was fixed to 300 μm/s. The glass transition temperatures of PMMA core and of fluorine clad of POF CK-10 were 110 °C and below room temperature (or less than 25 °C). While the temperature of the bottom stage was maintained at room temperature (25 °C), the temperature of the upper stage was varied from 25 to 125 °C in steps of 25 °C intervals. After the heated upper stage was pressed on POF by a preset press depth and held there for 1 s, the stage was raised while pulling the Ni mold and the POF apart, without allowing any cooling of the materials to occur.

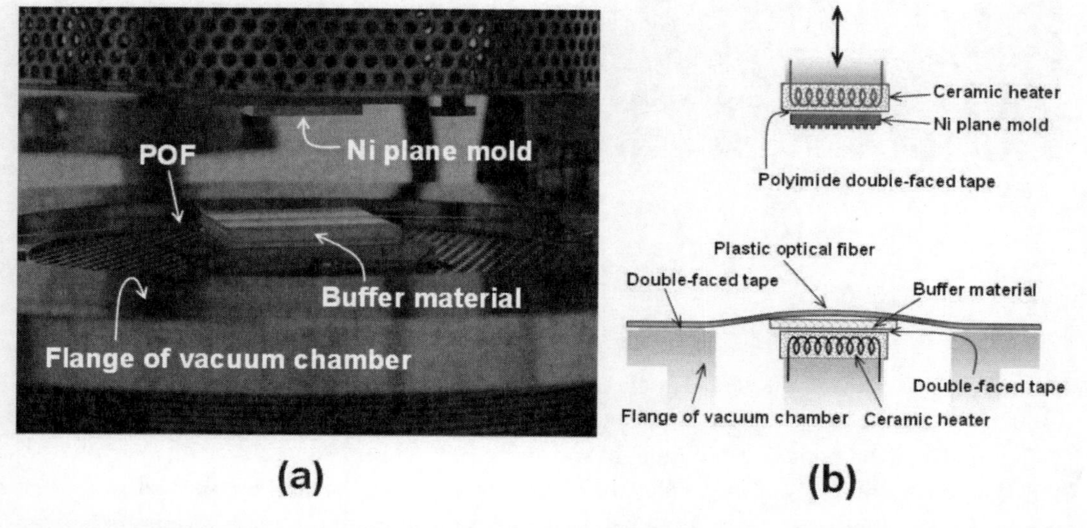

Figure 2. (a) Photograph and (b) schematic diagram of setup for thermal imprint experiments on POFs.

Table 1. Hardness and heat-proof limit temperature of four kinds of buffer materials for thermal imprint on POF

Elastic material	Hardness (ASKER C)	Heat-proof limit (°C)
Silicone rubber	73.8	120
Ethylene-propylene terpolymer (EPT) rubber	84.8	180
Polymethylmethacrylate (PMMA) sheet	99.0	100
Ni plate	—	300

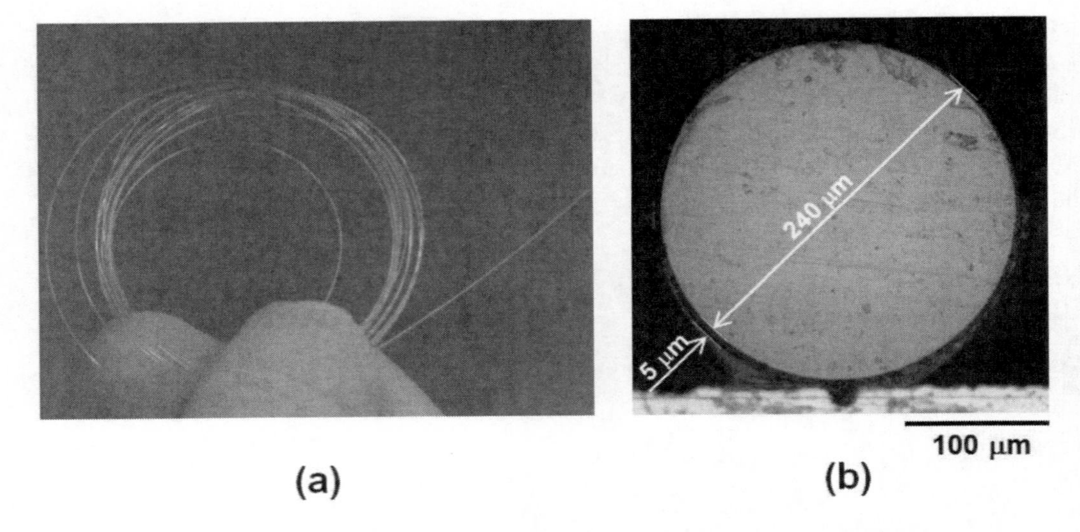

Figure 3. Photographs of (a) 250-μm-diameter POF and (b) cross-sectional shape of POF.

Measurement of Imprinted Depth on POFs

Figure 4(a) is an optical micrograph of a convex 10×10 μm^2 square dotted pattern of the electroformed-Ni mold used as a standard pattern to evaluate the molding accuracy in this experiment. The square dots 3.2 μm in height were arranged in 20 μm pitch forming a lattice. Figures 4(b) show concave patterns imprinted on the surface of POF. In Figure 4(b-1), the imprint result shows the case of using ethylene-propylene terpolymer (EPT) rubber as the buffer material at a heating temperature of 125 °C, and a press depth of 125 μm. Figure 4(b-2) on the other hand, shows the imprint result where Ni plate was used as buffer material under the same heating temperature and press depth as used in the previous case. A comparison between the photographs from the two cases shows that the transferred mold patterns were quite wide in the case where a buffer material with high hardness was used. The reason for this was that the cross-sectional shape of POF changed from a circular arc to a rectangular trough, which will be described later in detail.

Figure 6 shows the measurement results of the imprinted depths using the four kinds of buffer materials. A trend of increasing imprinted depth with a temperature rise was apparent in all cases. However, the range of the imprinted depths seemed to be dependent on the kind of buffer material used. When the buffer material with low hardness (in comparison to PMMA), like silicone and EPT rubbers, was used, the imprinted depth roughly increased in direct proportion to the increase of heating temperature. In case of silicone rubber at a heating temperature of 125 °C the imprinted depth reached 3.05 μm for all press depths. In case of EPT rubber at a heating temperature of 125 °C, and press depth of 125 μm, the imprinted depth reached 3.11 μm. However, in case of PMMA sheet and Ni plate at a heating temperature of 25 °C, pattern with imprinted depth of 2 μm or more could be processed. In case of PMMA sheet, at a heating temperature of 75 °C or more and press depth of 75 μm or more the imprinted depth became 3.0 or more. When the heating temperature was 125 °C and the press depths was 125 μm the imprinted depth exceeded 3.25 μm. The imprinted depth also exceeded 3.20 μm in the case of using Ni plate (hardest among the four kinds of buffer materials) when the heating temperature was 100 °C or more and the press depth was 25 μm. Naturally, when the press depth increased, the depth of the imprinted pattern also increased. The differences between the imprinted depths with increasing of press depths became small at high heating temperature. From the above-mentioned results, it was clarified that mold patterns were transferred onto the surface of POF almost completely when a hard buffer material was used; when the mold was heated at a comparatively high temperature and large press depth, the depth of the imprinted pattern increased.

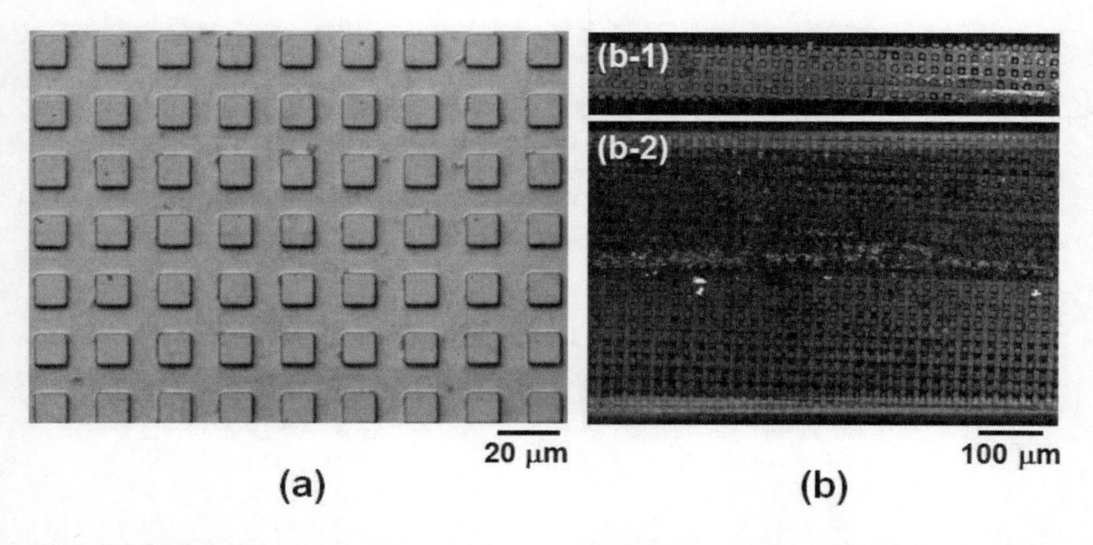

Figure 4. Optical micrographs of 10 μm×10 μm square dotted pattern: (a) in electroformed-Ni mold, and (b) on the surface of imprinted POF.

To measure the depth of the imprinted pattern on the surface of POF, we mounted an imprinted fiber under the lens of a five-line confocal microscope Optelics S130 (Lasertec) for observation [Figure 5]. The two ends of the POF were held by fiber rotators F264N-1 (Suruga Seiki) on the setup. The fiber was rotated about its axis by turning the rotators until the imprinted surface of the POF faced the microscope lens through which the pattern could be observed and measured.

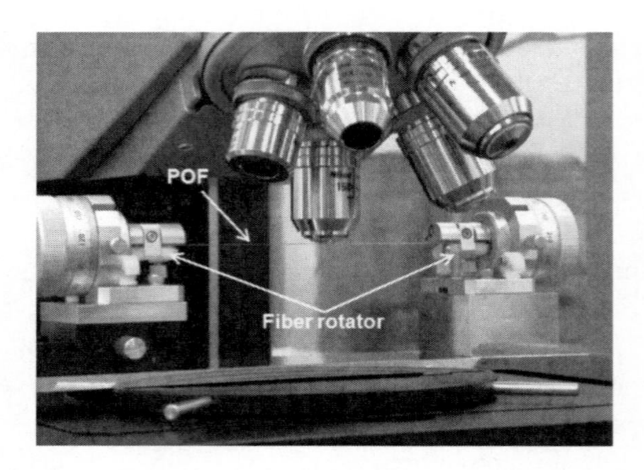

Figure 5. Photograph of the setup for measurement of imprinted depth on the surface of POF using fiber rotators and confocal microscope.

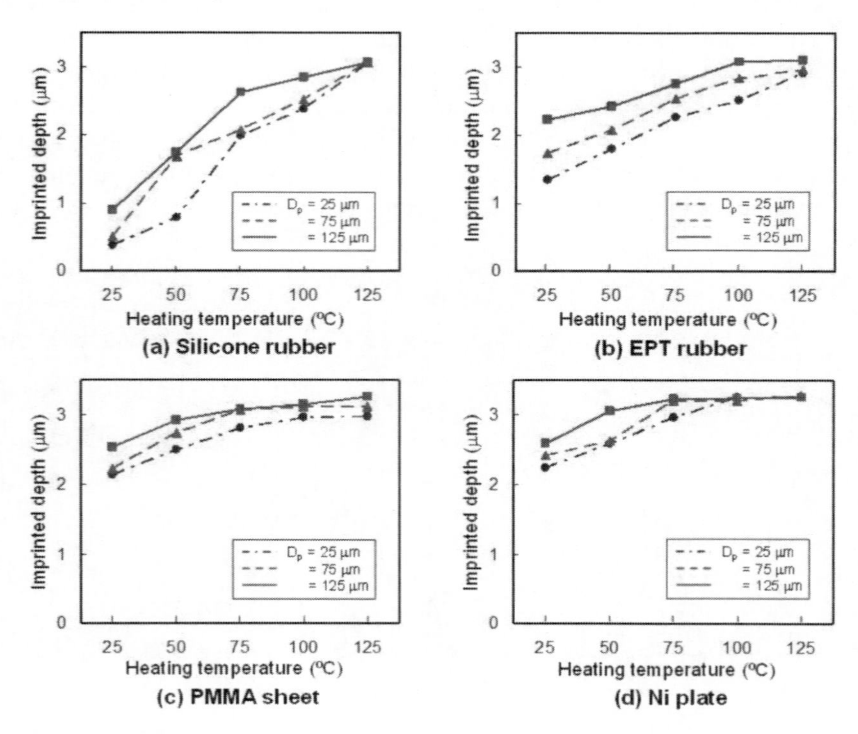

Figure 6. Relationship between the imprinted depth on POF and heating temperature of mold. The press depth (D_p) was changed to 25, 75, and 125 μm. Buffer materials used: (a) silicone rubber, (b) EPT rubber, (c) PMMA sheet, and (d) Ni plate.

Observation of Deformed Core of the POFs

POF's cross-sectional shapes were observed with an optical microscope after cutting the imprinted POFs by a dicing saw. At first, POFs after imprinting were secured on a 4-inch Si wafer with a double-faced tape as shown in Figure 7, and this wafer was then set on a working stage of a wafer dicer DAD 522 (Disco). POFs were then covered with a polyimide tape so that the POFs do not get loose and fall off from the Si wafer during the insertion of distilled water that accompanies the dicing and cutting operation. However, the fluorine clad coated on the PMMA core might peel off with the polyimide tape because there is a strong adhesive force that exists between the polyimide tape and POFs. Hence the cutting part using the dicing saw was covered by a mending tape with a comparatively weak adhesive force to prevent POFs coming off and to ascertain not to deform POFs during the dicing and cutting processes. POFs and the Si wafer were cut by dicing while moving the working stage at a speed of 0.4 mm/s and the saw's rotational speed of 30,000 rpm.

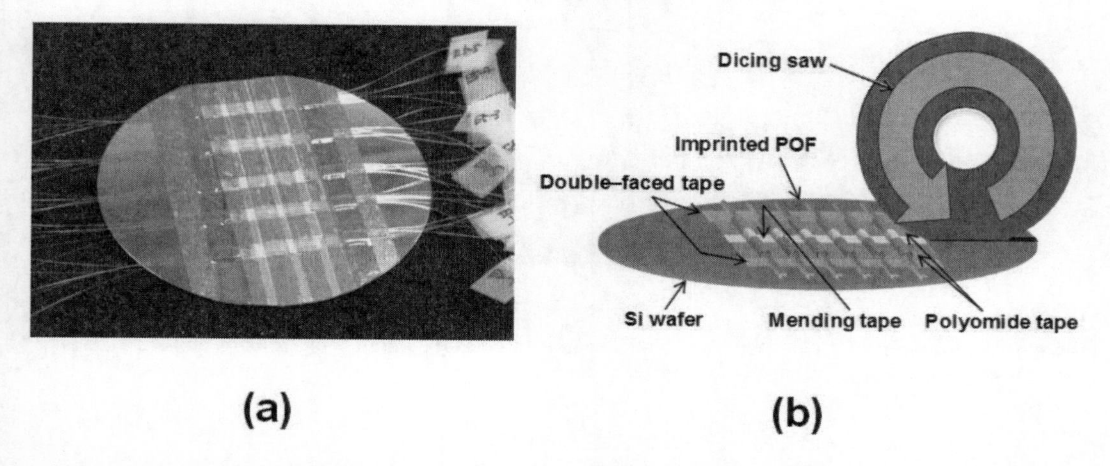

Figure 7. (a) Photograph of imprinted POFs fixed on a 4-inch Si wafer by tapes, and (b) illustration of cutting of the POFs by dicing.

Figure 8 shows cross-sectional optical micrographs of POFs after they were thermal imprinted at the press depths of 25 and 125 μm. At a comparatively high hardness buffer material, the circular cross-section of the PMMA core began to deform into a ribbon shape with rising temperature. When this phenomenon was compared with the results from press depths of 25 and 125 μm, it became obvious that at deeper press depths, the cross-sectional shape of the PMMA core was deformed considerably. To quantify this deformation a scheme is presented in Figure 9 and results from these measurements are shown in Figure 10. The results show that when silicone and EPT rubbers were used as the buffer material, the diameter of PMMA core was not affected by thermal imprinting. However, in the case of the buffer material being equal to or harder than PMMA, the PMMA core exhibited deformation even at a room temperature of 25 °C. And moreover, the deformed diameter of PMMA cores decreased as the press depth, and the heating temperature increased. This means that the contact force impressed on all materials with different hardness during the imprint process caused thermal deformation of the material with comparatively low hardness. Therefore, in case of buffer material with hardness less than that of PMMA, any contact force applied to PMMA could be warded off.

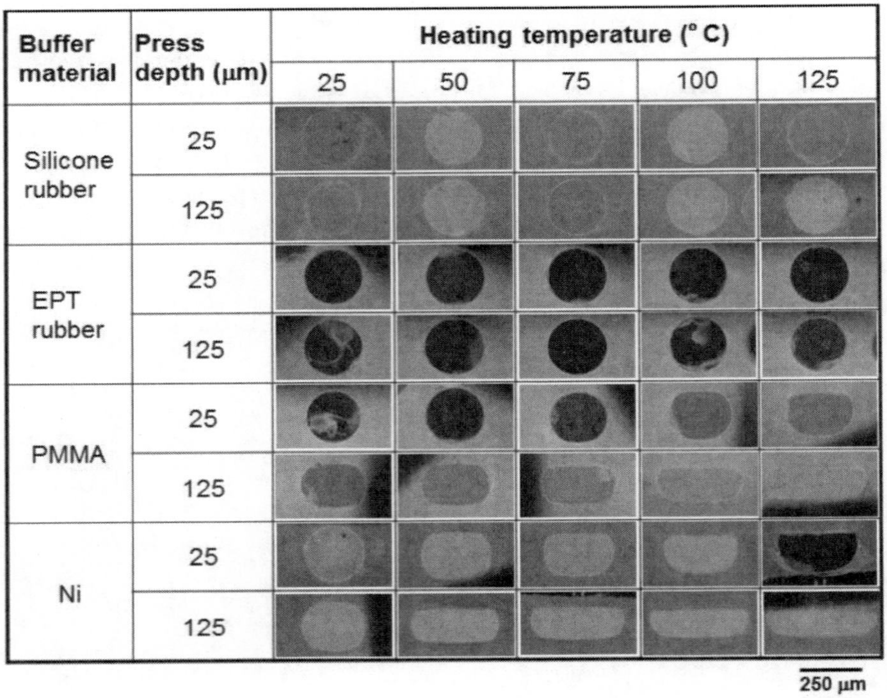

Buffer material
Press depth (μm)
Heating temperature (° C)
25
50
75
100
125
Silicone rubber
25
125
EPT rubber
25
125
PMMA
25
125
Ni
25
125
250 μm

Figure 8. Photographs of cross-sectional shapes of imprinted POFs using four kinds of buffer materials.

Contact force
Clad material
Deformed diameter of core
Original cross-sectional shape of core

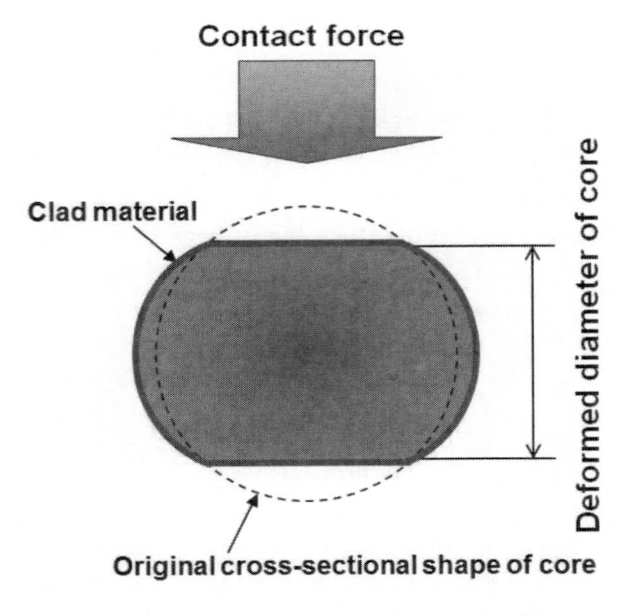

Figure 9. Deformation process of cross-sectional shape of POF by thermal-imprinting.

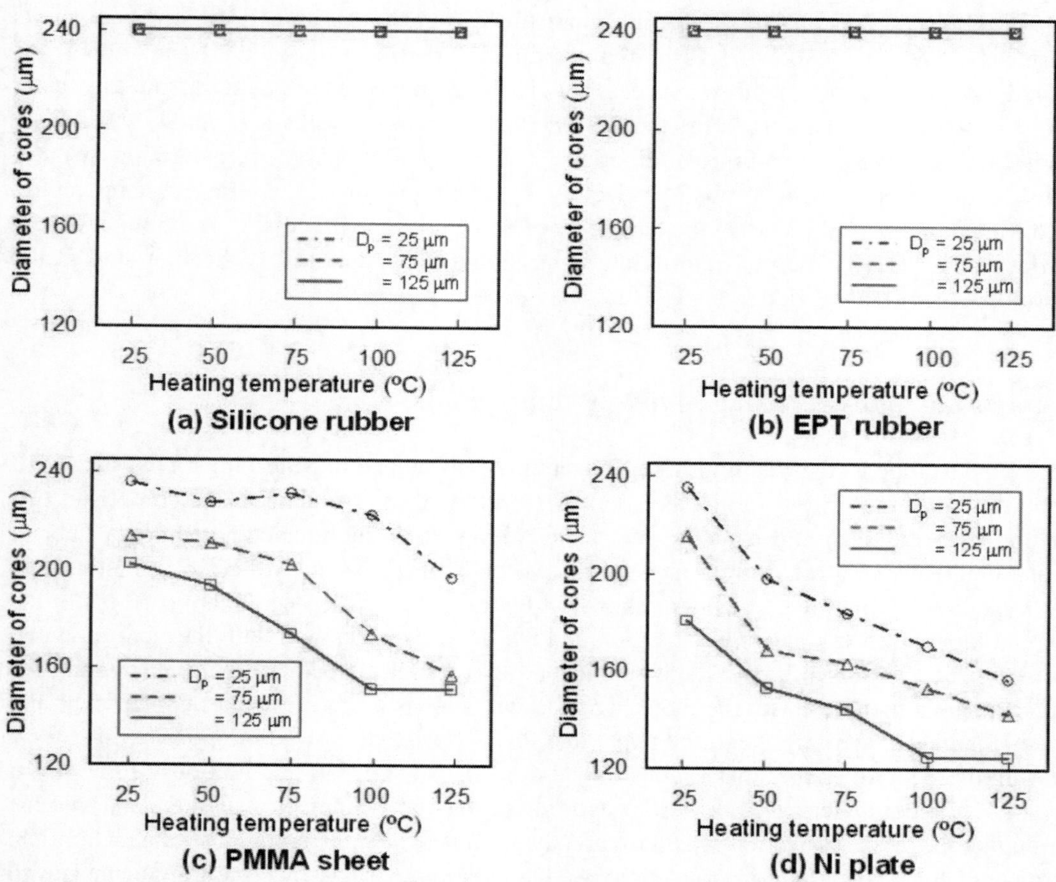

On this subchapter, Springer and the original publisher (Microsystem Technologies, 19, 2013, 325-333, Effect of buffer materials on thermal imprint on plastic optical fiber, Harutaka Mekaru, Akihiro Ohtomo, and Hideki Takagi, original copyright notice) is given to the publication in which the material was originally published, by adding: With 2012/10/03 kind permission from Springer Science and Business Media.

Figure 10. Relationship between the diameter of POF's core and heating temperature of mold. The press depth (D_p) was changed to 25, 75, and 125 µm. Buffer material was: (a) silicone rubber, (b) EPT rubber, (c) PMMA sheet, and (d) Ni plate.

SURFACE PATTERNING ON PLASTIC OPTICAL FIBER

Motivation

Roller hot embossing and roller nanoimprinting using a cylinder mold are very attractive technologies because of their mass producing capabilities. A roller imprinting is planned to be applied as a technique with which we can process smart fibers at high speed. However, it is very difficult to apply the process on arbitrarily shaped patterns such as MEMS structures and electronic circuits of nano-micro size on the cylindrical surface of roller mold. We are developing a flexible display for one of the applications. In this device, we plan to use inlet tubes and plastic optical fibers to form a smart fibrous substrate with MEMS structures on its surface. In previous works, a 1/16-in.-diameter Teflon perfluoroalkoxy (PFA) inlet tube was rolled onto a planar mold, and we successfully transferred the convex pattern from the planar mold to the cylindrical surface of the Teflon-PFA inlet tube (Mekaru et al. 2009a). In the technique that we used, only a cylindrical surface could impart soft patterning without crushing a Teflon-PFA inlet tube by selecting a suitable press force, an imprint temperature, and a rotational speed. We then began the development of the processing technology on the surface of a POF by thermal imprinting using sliding planar molds as the first step in the development of smart fibers.

Imprinting on POFs using Sliding Planar Molds

We developed a thermal imprinting system that specialized in patterning on the surface of a fibrous substrate. The system was essentially a thermal imprint device that comprised two sliding planar molds, and a reel-to-reel feeder designed to automatically send the fiber at preset intervals. In a reel-to-reel feeder, the two ends of a fiber are rolled in a sending reel station and a winding reel station, as shown in Figure 11 (step 1). The pattern surfaces of the two molds face each other (one facing up and the other facing down) with the fiber mounted at their center. The two planar molds are heated to a preset temperature beforehand, and the upper stage is mounted with the mold facing down is lowered until it makes contact with the fiber [Figure 11 (step 2)]. Next, the upper and lower molds are moved in opposite directions. During these motions, the fiber is constrained to roll against the sliding motions of the planar molds. The fiber undergoes a twist in proportion to the distance traveled by the planar molds; through this process a precise pattern of planar molds is transferred onto the cylindrical surface of the fiber. Then, in order to avoid the twisting of fibers, we built a mechanism to rotate both ends of the fiber by rotating the two reel stations in synchronization with the sliding motion of the two planar molds [Figure 11 (step 3)]. After the fiber and reel stations have reached their half-way rotation point, the upper mold is pulled away from the fiber [Figure 11 (step 4)], and then the two upper and lower molds and both reel stations are moved back to their initial positions and orientations [Figure 11 (step 5)]. At that time the imprinted part of the fiber is then wound up into the winding reel station, and the next part of the fiber is set between the planar molds using the reel-to-reel feeder [Figure 11 (step 7)]. The upper right photograph in Figure 11 shows the loading stage for thermal imprinting using sliding planar molds. Electroformed-Ni molds with a size of 15×15×2 mm^3 were mounted on Aluminum loading stages. A self-assembled monolayer (SAM) of a release agent Optool HD-2101TH (Motoji 2008) (Daikin Industries) was formed on the surface of the electroformed- Ni mold.

A plastic fiber was placed on the center of the mold pattern area. The maximum press force and imprint temperature were designed to be 300 N and 250 °C, respectively. Moreover, the other pictures in Figure 11 are a sequential photograph where the sending reel station is rotated halfway to prevent the fiber from twisting. The rotational speed of the reel stations is calculated by a computer to synchronize it with the speed of the sliding movement of the planar molds, and is automatically set. The maximum feeding speed of the fiber has been designed to be 40 m/min. The tension was controlled between 0 and 100 N using a several weights. The size of this system is 3.6 m in length, 1 m in width, and 1.75 m in height.

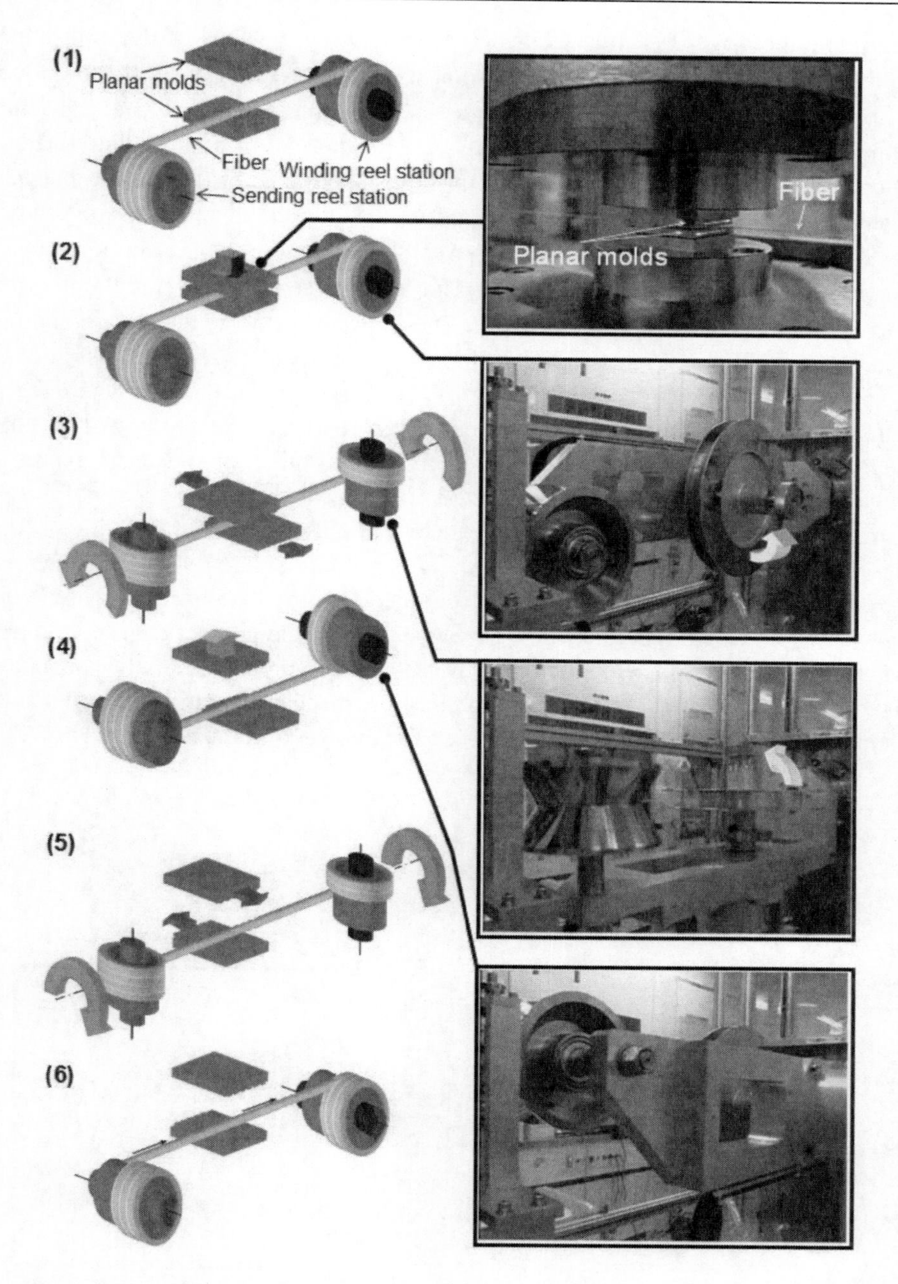

Figure 11. Patterning procedure on the cylindrical surface of POFs by a combination of sliding planar-mold stages and 180°-rotatable reel stations. (Right) Photographs of thermal imprint stages during pressing, and reel station during 180° rotation.

Optimization of Sliding Speed of Planar Molds

In this experiment, we selected the POF CK-10. Based on the experimental data in molding characteristics of the POF (Mekaru et al. 2013a), the contact force, heating temperature, and contact time were set at 12 N, 50 °C, and 0.5 s, respectively. The width of the sliding plane mold was 0.4 mm. Considering the sliding distance of the both upper and lower molds, it became possible to imprint a total width of 0.8 mm. The 250-μm-diameter of the fiber approximately amounted to 785 μm as its circumference. Therefore, a plastic fiber can be imprinted covering its entire circumference. Our thermal imprint method differs from the traditional hot embossing and thermal imprinting techniques in the sense that here the sliding speed of the planar molds exists as one of the important molding conditions. The maximum sliding speed of our imprint system was 800 μm/s. We then examined four sliding speeds that were 800, 400, 200, and 40 μm/s. The heating temperatures and the press depths were set to be 50 °C and 20 μm, respectively. The imprint results of the square and circular dotted patterns are shown in Figures 12(a) and 12(b), respectively. An inset illustration shows the cross-sectional drawing of the imprinted POF. In these figures, the part where the planar molds came into contact with the POFs was chosen to represent the contact edge (point A), and the part where the planar molds separated from the POF was chosen to represent the release edge (point B). The rotation angle in the x-axis means the turning angle of the POF to indicate a relative position of the imprinted patterns that exist between the positions A and B. The scale of the turning angle referred to the fiber rotators F264N-1 that were used to maintain both ends of the imprinted POF in confocal microscope observations. When the sliding speed was 40 μm/s, the pattern depth was deeper at the contact edge than that at the release edge in both figures. The deviation of the pattern depth became smaller in accordance with an increase of the sliding speed. In thermoplastics, mechanical properties such as tensile strength and the elasticity modulus, etc. decrease with the rise of temperature.

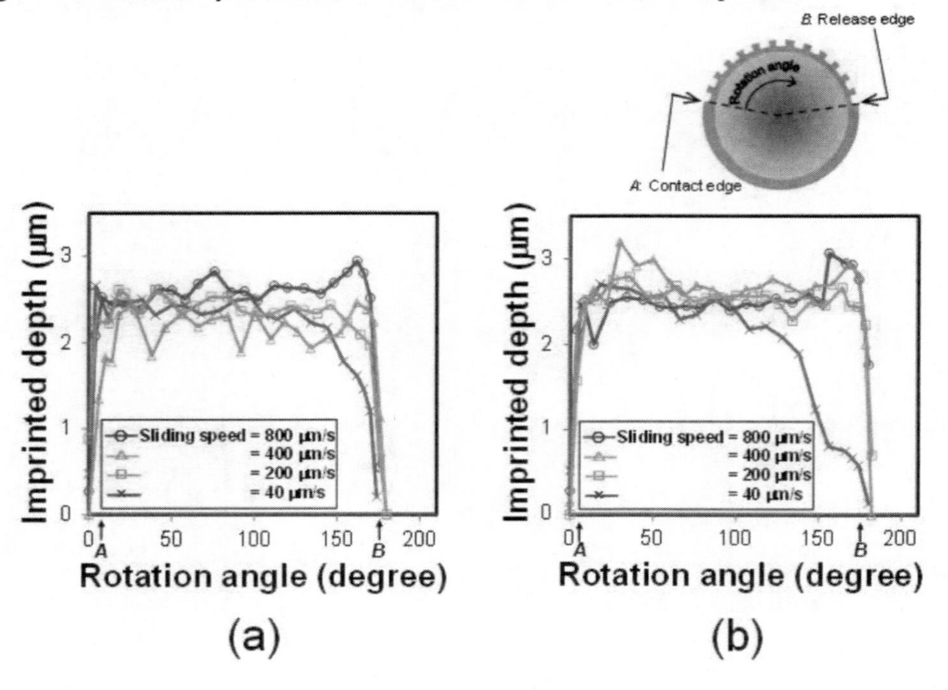

Figure 12. Relationship between the rotation angle and measured depth of the imprinted patterns on the POFs using (a) the upper mold with a 10-μm-width square dotted pattern, and (b) the lower mold with a 10-μm-diameter circular dotted pattern. The inset illustration shows the contact and release edges of the imprinted patterns on the POFs.

When the sliding speed slows down, the heat becomes easier to transmit to the PMMA core from the molds, since at the slow speed the planar molds remain in contact with a POF for a longer period of time. Therefore, the elasticity modulus of the PMMA core gradually decreased while the POF was rolling on the planar molds at a slow sliding speed, and the PMMA core thermally deformed from a circular to a flattened shape. As a result, it can be seen that the contact area between the POF and molds increases, and the contact pressure impressed onto the fluoroplastics-clad layer on the surface of the POF decreases. However, a decrease in the imprinted depth according to the rise of the sliding speed was not observed. From these results, we chose 800 μm/s as an optimized sliding speed.

Imprinting Results on the POF

Figure 13(a) is a photograph of an imprinted plastic fiber. The discolored sections in the figure (white arrows) are the patterned sections as shown in Figure 13(b). Figure 13(c) shows an observed optical micrograph of the imprinted patterns using the confocal microscope. It is thus confirmed that a fine pattern was completely transferred on the surface of POF. Each square dotted pattern of 5 μm in width and pitch of 10 μm, and circle dotted pattern of 5 μm in diameter, and pitch of 10 μm were selected as a pattern of upper and lower plane molds respectively in thermal roller-imprint experiments. Figure 14 shows an optical micrograph of square and circle mold patterns, and imprinted patterns after the first and 100th shots. Regardless of the kind of pattern or the shot frequency, 5-μm-width square and 5-μm-diameter circular dotted patterns with 10 μm pitch were successfully transferred.

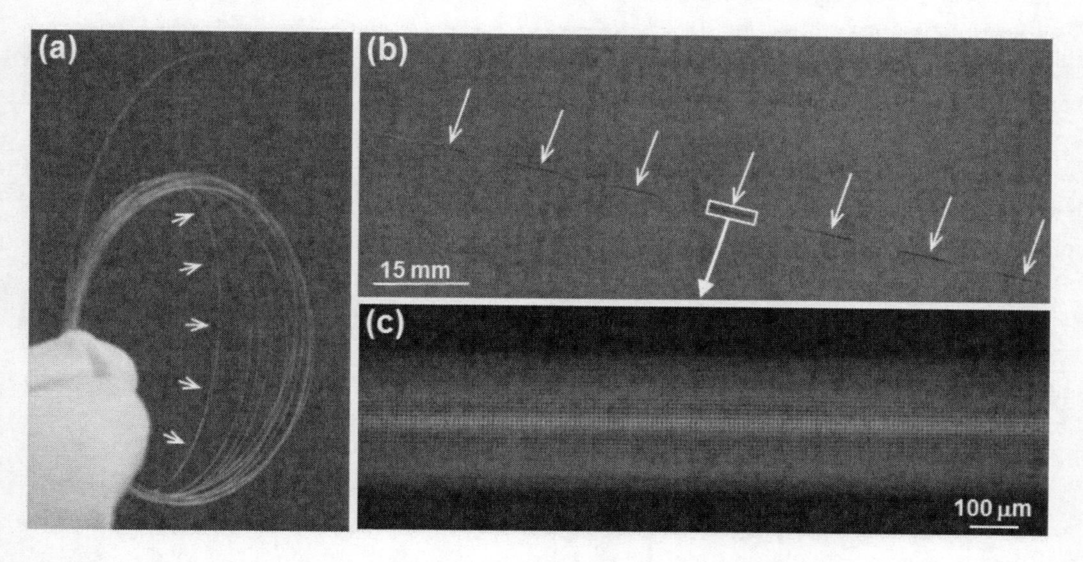

Figure 13. (a) Photograph of 250-μm-diameter POF after imprinting, (b) Photograph of POF patterned intermittently, (c) Optical micrograph of 5-μm dotted pattern on imprinted POF.

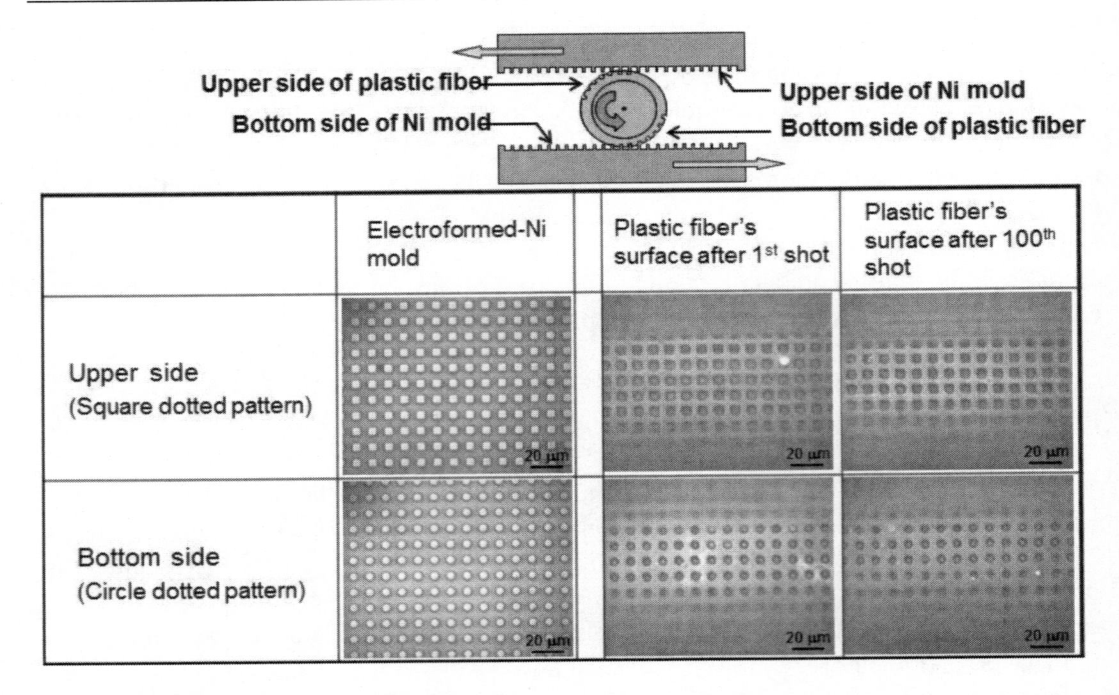

Figure 14. Cross-sectional schematic image of sliding roller-imprint and optical micrographs of mold patterns and imprinted patterns on the surface of POF after the 1st and 100th shots.

Figure 15 shows measured results of imprinted depths for every ten shots on the POF continuously roller-imprinted from the first through the 100th shots, as obtained by the confocal microscope.

The upper mold pattern was a 3.22 μm high convex square dotted pattern. On the other hand, the lower mold pattern was a 3.28 μm high convex circle dotted pattern. When the imprinted depths of 11 areas picked from the square dotted patterns on the surface of POF were compared among each other, the minimum value was 2.99 μm (at the 50th shot), and the maximum value was 3.20 μm (at the 10th shot). In the case of the circle dotted patterns, the minimum depth was 2.83 μm (at the 40th shot) and the maximum depth was 3.20 μm (at the 60th shot).

Considering the heights of convex mold patterns, the molding rate (imprinted depth/mold height) amounted to approximately 88–99%. The differences between the minimum values and the maximum values of both patterns were found to be 0.21 and 0.37 μm, respectively. Originally, the irregularity on the surface of the plastic fiber that had existed before imprinting was also observed.

Therefore, these differences show that a considerably uniform and continuous imprinting could be executed. The processing time for single roller-imprinting was 14.2 s. This reel-to-reel process was repeated 100 times by establishing a pitch of 16 mm. The total processing time was roughly 24 min. No significant differences were seen among a set of 11 imprinted patterns chosen between the first and 100th imprinting operations.

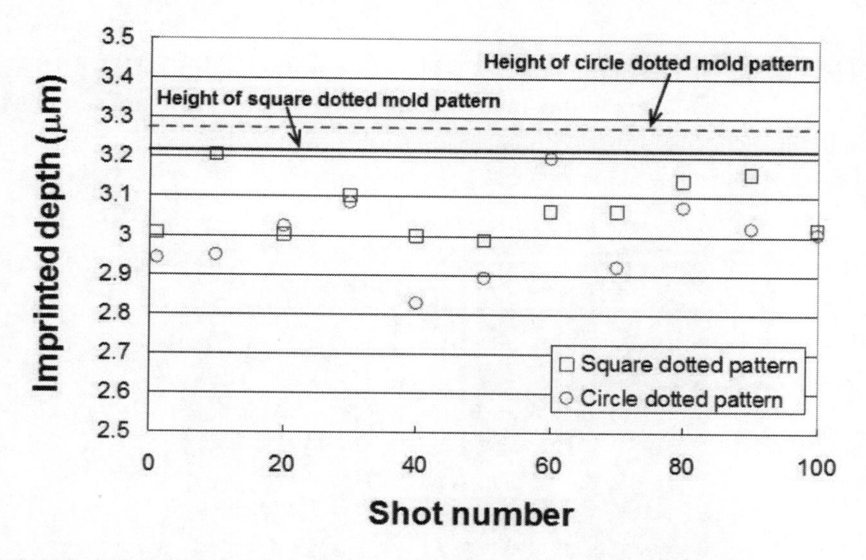

Figure 15. Relationship between shot number and imprinted depth on the surface of POF.

Patterning of Diffraction Grating Structures on the POF

As mentioned in the previous paragraphs, the conditions used to deeply imprint the precise pattern were a heating temperature of 50 °C, a press depth of 20 μm, and a sliding speed of 800 μm/s without deforming the cross-sectional shape of the 250-μm-diameter POF by thermal-imprinting using sliding planar molds. To demonstrate this technology under these imprinting conditions, we employed a special electroformed-Ni mold with a mirror image of a string of characters forming the word "MACROBEANS" (BEANS means bio-electro-mechanical-autonomous-nano-systems. It is a Japanese national project on the development of regarded as a next generation MEMS that has incorporated into nano- and bio-technologies from 2007 to 2013) engraved like a lattice with a vertical pitch of 200 μm, where the individual characters were composed of diffraction grating structures with linewidths of 1 and 2 μm [Figure 16]. The size of the character "M" was 180×136 μm^2. The height of the convex mold pattern was 1.1 μm. Figure 16 shows optical micrographs of the characters "MACROBEANS" composed of the diffraction grating transferred onto the cylindrical surface of the POF. The upper picture shows that the character pattern composed of diffraction grating structures with a 1 μm linewidth was quite bright. The lower picture of the figure confirms that only the cylindrical surface was patterned without causing any damage to the POF. The cross-sectional circumference of the POF is approximately 785 μm. Therefore, four character lines of "MACROBEANS" were carved onto the cylindrical surface of the POF. We fine-tuned the setting position of the upper and lower molds so that the characters imprinted by both molds should not overlap each other. The depth of the concave imprinted pattern was measured to be 1.0 μm. Figure 17(a) shows that the color of each character string changed slightly to cause different reflective properties, depending on the kind of diffraction gratings in the Ni mold. Figure 17(b) shows a mirror image of characters "MA" in the Ni mold observed by a field-emission scanning-electron microscope (FE-SEM). Figure 17(c) is a FE-SEM image that covered up the characters "BEANS" on the imprinted POF fiber after sputter-deposition of a 10-nm-thick Pt layer. Figure 17(d) shows another FE-SEM image of the characters "MA" on the imprinted POF. The pattern of the diffraction grating can be clearly recognized as characters in this image. To further investigate the transfer condition of an individual diffraction grating with linewidths of 1 and 2 μm, close examinations of expanded FE-SEM images were carried out.

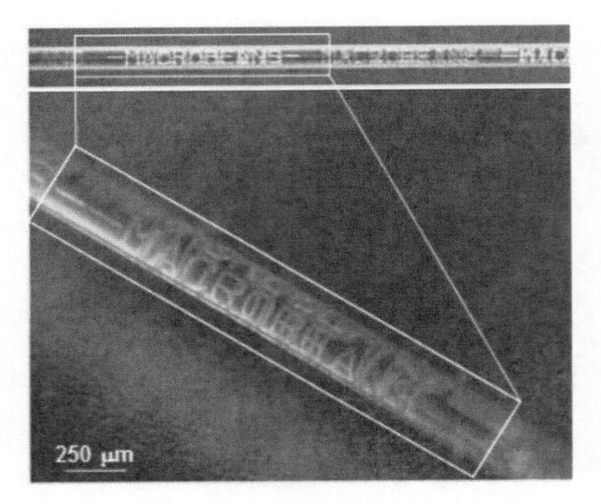

Figure 16. Optical micrographs of the character patterns with diffraction grating structures transferred from the planar electroformed-Ni molds to the cylindrical surface of the POF.

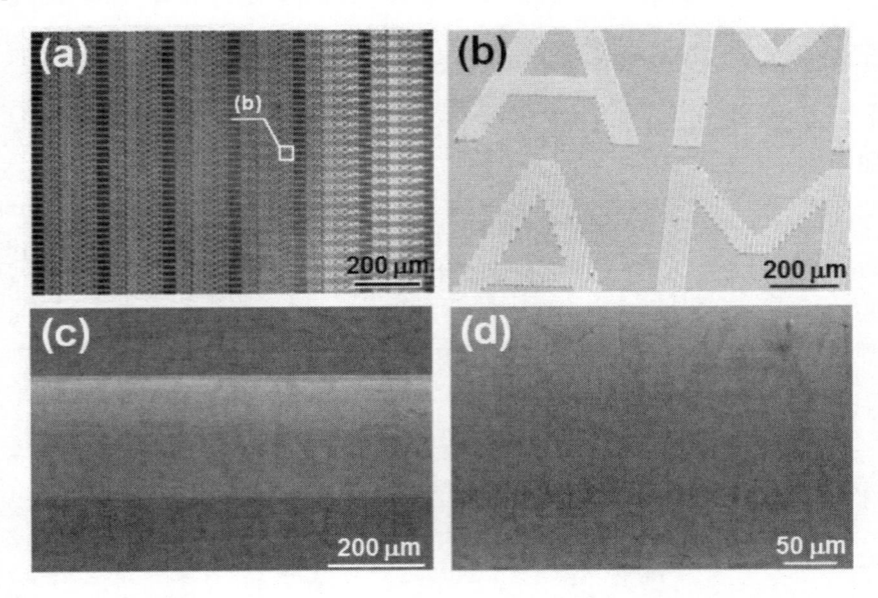

Figure 17. (a) Optical micrograph of the character patterns in the planar electroformed-Ni mold. (b) FE-SEM image of diffraction grating structures in the characters "AM" in the mold, observed at an inclined angle of 45°. (c) FE-SEM images of the character patterns on the POF after 10-nm-thick Pt was sputter-deposited. (d) FE-SEM image of the diffraction grating structures in the characters "MA" on the POF, observed at an inclined angle of 0°.

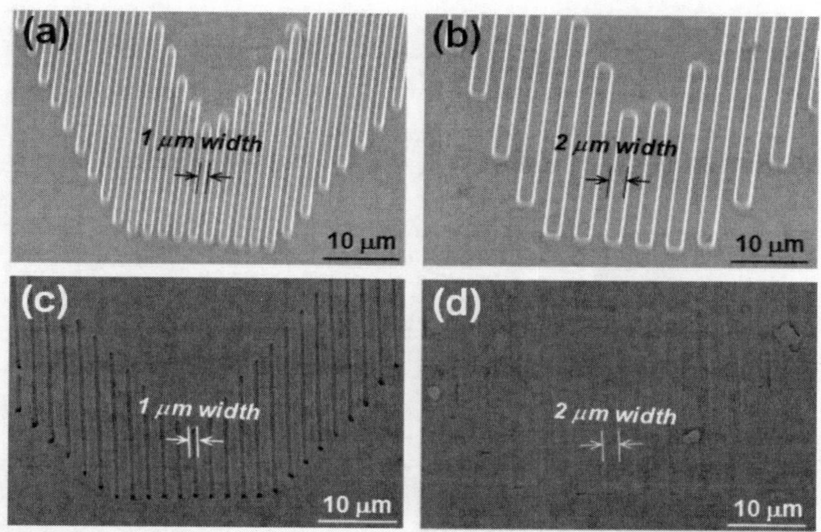

Figure 18. (a) and (b) Expanded FE-SEM images of diffraction grating structures in the character "M" with linewidths of 1 and 2 μm in the mold, respectively, observed at an inclined angle of 45°. (c) and (d) Expanded FE-SEM images of the diffraction grating structures in the character "M" with linewidths of 1 and 2 μm, on the POF after 10-nm-thich Pt was sputter-deposited, respectively, observed at an inclined angle of 45°.

The characters in Figure 18(a) were composed of the diffraction grating pattern with the linewidth of 1 μm and the pitch of 2 μm on the Ni mold. On the contrary, the linewidth and the pitch of the diffraction grating were 2 μm and 4 μm, respectively [Figure 18(b)]. Figures 18(c) and 18(d) show the results of observing the cylindrical surface of the POF to correspond to the mold patterns of the planar electroformed-Ni molds shown in Figures 18(a) and 18(b). It was confirmed that the diffraction grating pattern with linewidths of 1 and 2 μm was clearly transferred from the flat surface of the Ni molds to the cylindrical surface of the POFs. As previously mentioned, we succeeded in soft patterning on the surface of POFs with a diameter of 250 μm without crushing the fiber.

BULK DEFORMATION TO FABRICATE WEAVING GUIDES

Motivation

A fibrous substrate used in a traditional weaving process does not support a structure for securing the relative positions of its warps and wefts. And hence any shift in their contact positions, caused by any applied forces of tension, or by any insertions to the warp and weft in the central part of a fabric can be different from that of its surrounding region [Figure 19(a)].

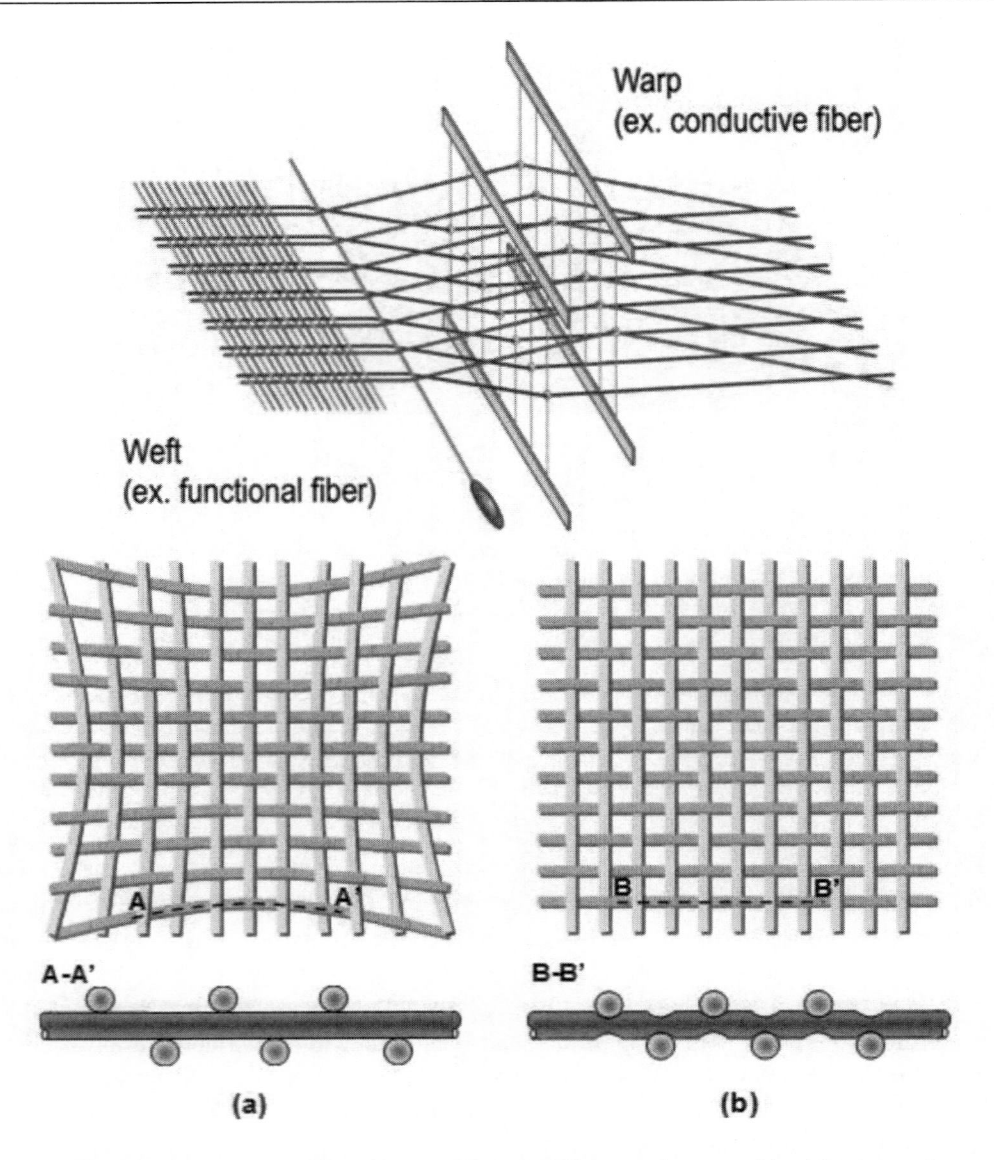

Figure 19. Illustrations of the weaving process and woven fibrous substrates: (a) without weaving guides, and (b) with weaving guides. Lower figures show cross-sectional images taken along a dotted line in the upper figures.

We then engineered weaving guides on the top and bottom surfaces of a fibrous substrate to secure the relative positions of the warps and wefts [Figure 19(b)]. A thermal imprint technology was employed for a continuous processing of the weaving guide structures on the surface of a fibrous substrate. We have previously reported on our success in the processing of weaving guides by hot-pressing of the surface of a nylon fiber using a planar mold with convex patterns (Mekaru et al. 2010; Mekaru and Takahashi 2010). However, that technique could not continuously process long fibers because of the patching requirement involved in the process without a feeding system. A case was reported where diffraction grating patterns were processed onto the surface of a polyester fiber by a roll-to-roll imprint using cylinders that were wrapped with electroplated-Ni molds (Schift 2006). In this work, a similar technique using seamless cylindrical molds was employed to form a weaving guide for e-textiles.

Reel-to-Reel Imprint System using Cylindrical Molds

Figure 20 shows a special system developed to process the weaving guides on the fibrous substrate that combines a reel-to-reel feeder with a roller-imprint device.

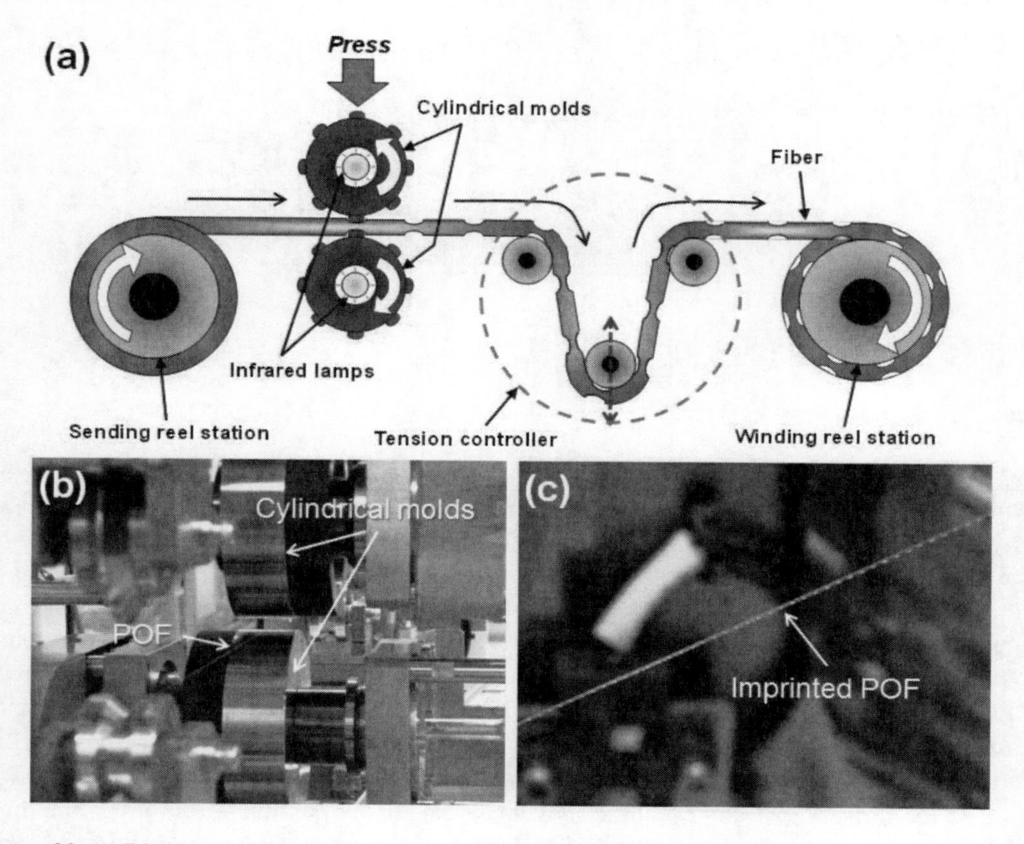

Figure 20. (a) Diagrammatic illustration of a reel-to-reel imprint system. Photographs of: (b) roller-imprint device with the cylindrical molds, and (c) imprinted POF before winding.

Here, the fibrous substrate is sent off from a sending reel station, and then is winded up on a receiving reel station after being pressed between the two cylindrical molds rotating in synchronization with the sending speed. The maximum sending speed was designed to be 30 m/min. To prevent any breakage of the fibrous substrate during its movement, the substrate was run through a tension controller comprising two fixed pulleys, and a vertically moveable pulley. The cylindrical molds were designed to reach a temperature up to 250 °C by infrared lamps installed inside the cylinders' center cores. The molds were made by precision cutting on a several-hundred-µm-thick Ni–P alloy layer electroless-plated on a cylindrical base material (100 mm in diameter, and 30 mm in width). The mold pattern on the cylindrical surface comprised convex rectangle shaped structures 260 μm wide and 46 μm high, and 145 μm radius convex arc shaped structures with a pitch of 1 mm.

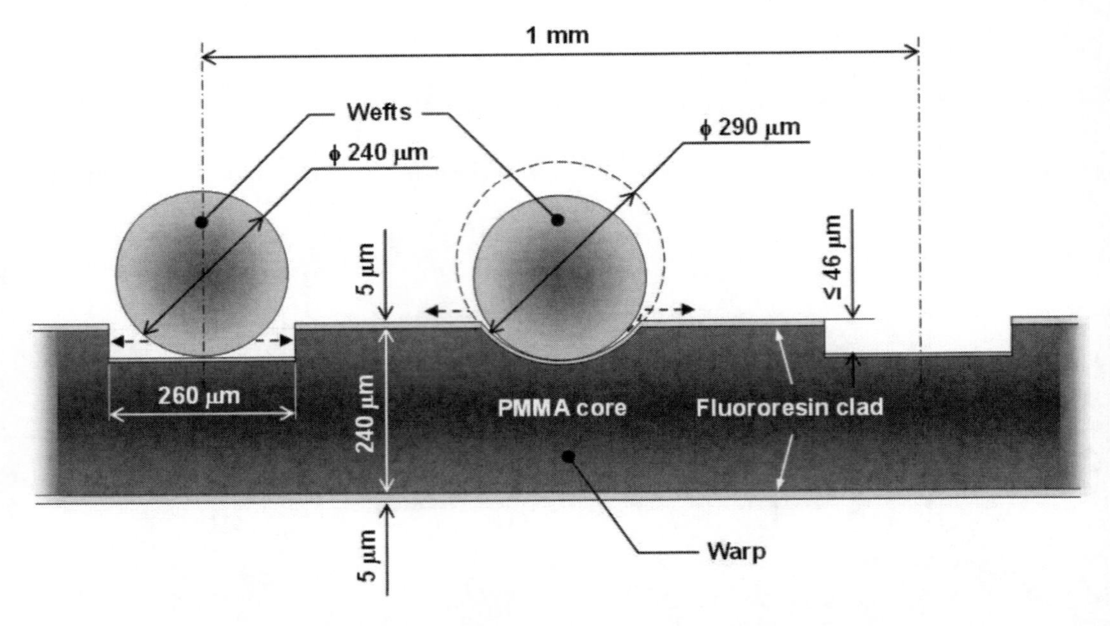

Figure 21. Schematic cross-section of weaving guides is formed on the surface of POF.

Figure 21 shows the design of the weaving guides of 250-μm-diameter fibrous substrate imprinted using this cylindrical mold. In the weaving scheme, when one set of imprinted fibers are used as warps and another set with the same diameter are used as wefts, then a weft fiber falling in a rectangle-shaped weaving guide with vertical sidewalls is rigidly fixed, and a weft fiber in the arc-shaped weaving guide with curved sidewalls can be moved comparatively freely. The rectangle-shaped weaving guide is used as a fixed contact point which maintains the lattice structure of e-textiles, and the arc-shaped weaving guide is operated as a moving contact point which supports the flexibility of e-textiles (Mekaru and Takahashi 2010). The upper cylindrical mold could be moved up and down by using a servomotor; and the stage holding the upper cylindrical mold was connected to a load cell. To minimize the influence of any eccentric movement caused by possible errors in assembling the cylindrical molds, an automatic press-force correction mechanism was added to the press device. The vertical position of the upper cylindrical mold was dynamically adjusted according to the difference between the measured press-force, and a pre-set value. Figure 22 shows the measured press-force and vertical position of the upper mold at a sending speed of the fibrous substrate as 1 m/min. In the uncorrected case, the measured press-force fluctuated periodically within a range of 33 N or less; and this phenomenon was synchronous to the rotation operation of the cylindrical molds. However, when the correction mechanism was used, the changes in the press-force could be controlled to within 3 N by adjusting the vertical position of the upper cylindrical mold within a range of ±37 μm from its center position. This result concludes that by using the correction mechanism the fluctuation range of the press-force could be decreased to one-eleventh of that found in the case of without controlling. By this function, a press operation under a constant press-force became possible.

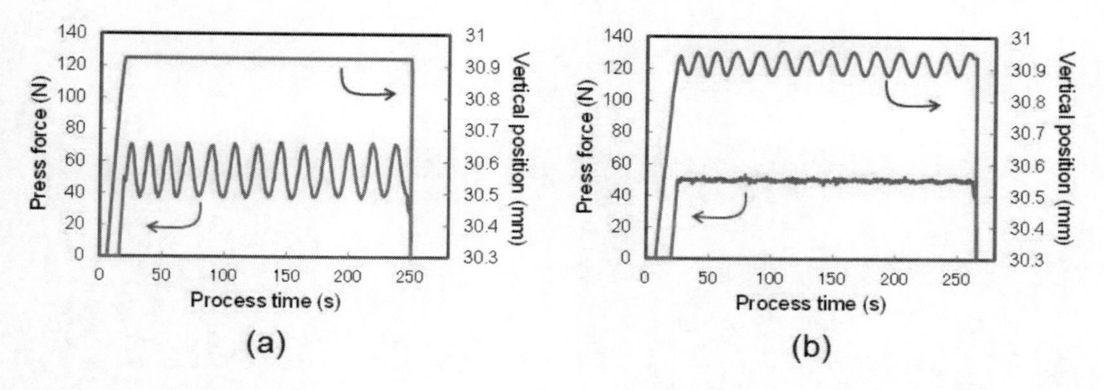

Figure 22. Relationship between process time and press-force measured by load cell/vertical position of the cylindrical mold: (a) without, and (b) with the correcting of press-force.

Processing of Weaving Guides on the POF

In the next step, to validate the functioning of our reel-to-reel imprint system, a weaving guide structure was processed on the surface of the POF CK-10. Figure 23(a) and 23(b) show the photographs of the imprinted POFs at the sending speed (v_s) of 1 and 5 m/min, respectively. The imprinted patterns were checked on the surface of the POFs, and the white dotted-line frame in these photographs of Figure 23(a) and 23(b) were observed in detail. Figure 23(c)–(f) show the top views and side views, respectively, of the imprinted weaving guides where the imprint temperature, and the press-force were 50 °C and 20 N. In terms of the sending speeds, the Figure 23(c) and 23(e) show the results of imprinting at a sending speed (v_s) of 1 m/min. And Figure 23(d) and 23(f) show the results of imprinting at an increased speed (v_s) of 5 m/min. It was confirmed that both, rectangle shaped and arc shaped weaving guides were clearly transferred on the surface of the POFs. Figure 24 shows the result of measuring the depths of the imprinted pattern using the confocal microscope. When comparing the imprinted depths under the two sending speeds, it appeared that the slower speed resulted in the formation of deeper weaving guides. Because the diameter of the cylindrical mold was 100 mm, the POF of a length of approximately 314 mm was sent during the time the cylindrical mold completed one full rotation. Measurement and investigation of each imprinted depth at 12-divided points on the 314-mm-length POF showed that the weaving guides on the POF in form of depths appeared at every 30° rotation of the cylindrical mold.

When the sending speed was 1 m/min, the maximum depths in case of rectangle shaped and arc shaped weaving guides were found to be 9.87 and 10.00 μm, respectively; and the minimum depths in case of the rectangle shaped and arc shaped weaving guides were 9.29 and 9.01 μm, respectively. The mean values were calculated to be 9.56 and 9.64 μm. And the values of the standard deviations (σ_s) in the two cases were calculated to be 0.19 and 0.26 μm. When the sending speed was raised to 5 m/min, then 1.79 μm in the maximum depth and 1.51 μm in minimum depth were measured in the rectangle shaped weaving guides. The mean value and standard deviation were calculated to be 1.61 and 0.07 μm. On the other hand, in the arc shaped weaving guides the maximum and minimum depths were measured to be 2.19

and 2.00 μm, respectively. The mean value and standard deviation in this case were calculated to be 2.07 and 0.07 μm.

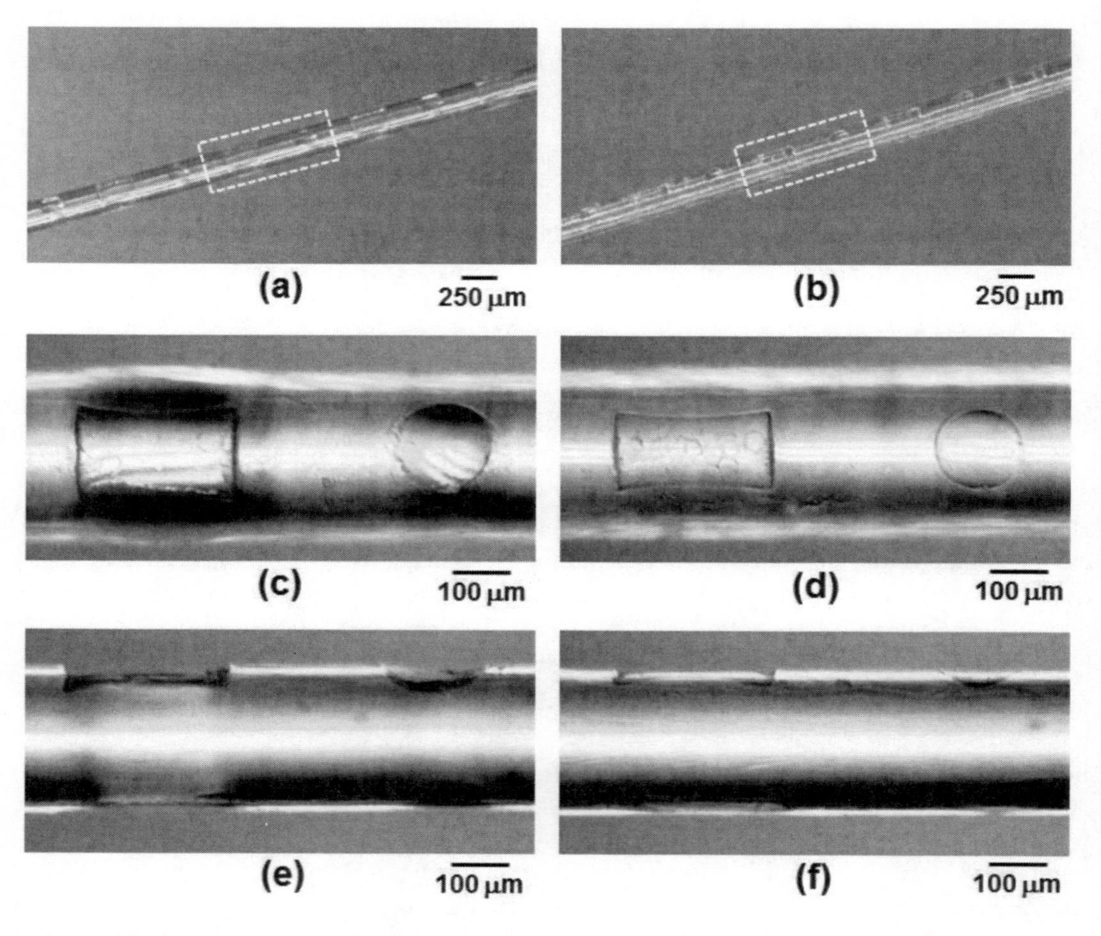

Figure 23. Photographs of imprinted POFs at the sending speed of: (a) 1 m/min and (b) 5 m/min. Optical micrographs of the imprinted patterns of the white dotted-line frames in Figs (a) and (b) on the surface of POFs at a sending speed of; (c and e) 1 and (d and f) 5 m/min.

With the standard deviations of 26.7–36.7% as compared to the case of 1 m/min of the sending speed, the imprinted depths in case of 5 m/min were remarkably decreased. The explanation of this behavior is that at high sending speed, the press operation reaches its completion before enough heat is transmitted from the heated cylindrical mold to the POF at the room-temperature, and that the de-molding is executed after an insufficient thermal deformation. An error range of 2.9 μm inevitably occurred because the diameter of POF was measured to be in the range of 247.5–250.4 μm. This value happens to be much larger than that of the imprinted depth. Therefore, the observation that the differences between the imprinted depths on POFs with varying diameters were very small, confirms the excellence of the transfer accuracy. Thus, using our reel-to-reel imprint system, it was confirmed that the surface of a fibrous substrate could be processed with a high degree of uniformity. To achieve a faster imprinting, the speed-up of the response speed of the control sequence will be addressed in the next step.

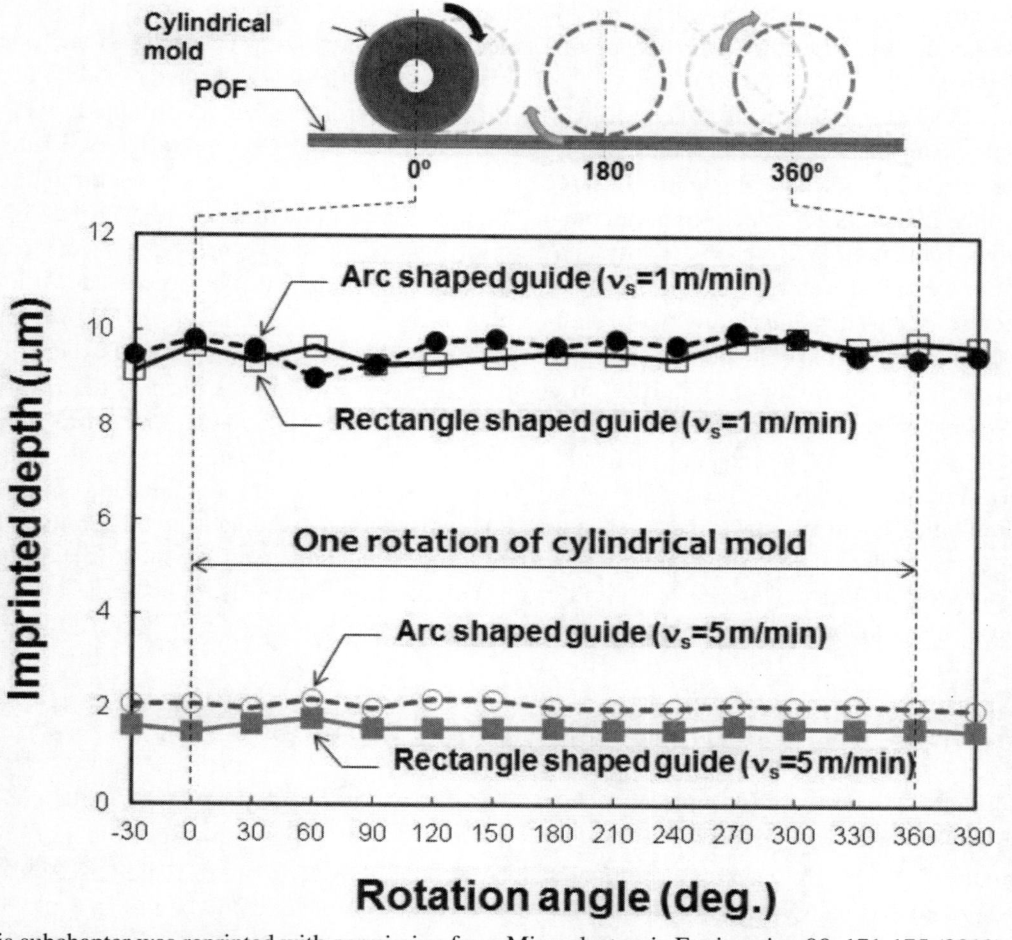

Figure 24. Relationship between the rotation angle and imprinted depth on the surface of POF. Here, "v_s" in the figure stands for the sending speed of POF.

Reel-to-Reel Imprinting of Maluti-Layered Microstructures on Plastic Optical Fiber

Motivation

Among the various nanoimprint technologies, the roller imprint technology (Tan et al. 1998) using a cylindrical mold is a suitable way to continuously fabricate micro-/nano-structures on the surface of films and fibrous substrates without putting any limits on the length of the substrate (Wiley et al. 2010). Especially when a molding material happens to be a large-size film, a roll-to-roll nanoimprinting (Ahn and Guo 2008) where a pattern is continuously being transferred by a cylindrical mold onto a film that is conveyed from a sending roll to a winding roll, is seen as an ultimate mass-production method of micro-/nanostructures. Since UV roll-to-roll nanoimprinting (without causing any thermal expansion or contraction) can transfer micro-/nanostructures with high accuracy (Watts 2008), the process has been used to fabricate optical devices for displays (Ahn et al. 2006) and anti-reflective films (Ting et al. 2008), as well as has been used in flexible electronics (Maury et al. 2011). On the other hand, thermal roll-to-roll nanoimprinting has contributed to the development of chemical /bio-chips like microfluidics (Yeo et al. 2009) and lab-on-a-chips (Vig et al. 2011) with comparatively large-size patterns because it offers a wide-range of molding materials to select from. In our technique, various patterns are transferred on a fibrous substrate by a reel-to-reel thermal imprint method. In above-mentioned works, by taking special precaution that a fibrous substrate be not damaged, we processed MEMS structures by roller imprinting with sliding planar molds (Mekaru et al. 2011a; 2011b; Ohtomo et al. 2011). On the other hand, weaving guides had been processed where the convex mold patterns of several 100-μm width were pushed into the surface of the fibrous substrates, where the fibrous substrates were partially damaged (Mekaru and Takahashi 2010; Mekaru et al. 2012; Ohtomo et al. 2012). However, if such structures were fabricated by some other imprint methods, the result would have been of low throughput and imprint technologies could not have met the criteria for high-volume manufacturing. Therefore, in this experiment, MEMS structures and weaving guides are processed on the surface of a cylindrical mold, and we tried to imprint both patterns simultaneously by a single shot of roller-imprinting on a fibrous substrate.

Cylindrical Mold with Hybrid-Layered Microstructures

Figure 25 shows an example of a smart fiber that we designed. A weaving guide and a deep hollow which patterns MEMS structures are arranged in 1 mm of pitches on a 250-μm-diameter POF. A circular cross-sectional shaped hollow with a curvature radius of 290 μm was formed on the POF so that another POF with the same diameter could be inserted into the hollow as warps. Moreover, in order to insert the conducting material into the cavity of a concave imprinted pattern, obtained by ink-jet printing, a flat surface rather than a cylindrical surface would be the right shape for holding the dropped material. Therefore, a rectangular cross-sectional-shaped groove with a width of 260 μm was fabricated on the POF, and a several 10-μm-wide MEMS patterns were transferred on the flat bottom surface of the rectangular groove of the POF. In order to process such structures on POFs by a single shot of the roller thermal imprinting, we developed a seamless cylindrical mold 100 mm in diameter and 30 mm in width as shown in Figure 26(a). The polished surface of the cylinder was covered with electroless-plated Ni-P, and convex structures with rectangular cross-sections and arc shaped convex structures were cut with a forming tool. Next, the surface of the cylinder was coated with a positive-tone photoresist by a dipping process. UV lights were then irradiated on the photoresist through a micro-patterned flexible photomask wrapped around the cylinder. In this way, the micro-patterns from the mask were transferred onto the cylinder surface. Next, following a Cu electroplating, hybrid-layered patterns comprising 260-μm-wide macro and several 10-μm-wide microstructures were made; and the photoresist was then removed. T. Huang, *et al.* had also reported on a fabrication method of a cylindrical mold using a plane photomask (Huang et al. 2009). If there were no uneven structure on the cylindrical surface, a standard 3-D UV lithography using a plane photomask could be used. However, like in our case, where an uneven structure with a width of several 100 μm and a depth of 50 μm does exist on the cylindrical surface, a flexible photomask which can conform to the cylindrical surface is considered apt for the task. In the manufacture process of the cylindrical mold used here, the tasks of electroplating followed by a 3-D photolithography using a flexible photomask were carried out by Optnics precision Co., Ltd., which is a manufacturer of X-ray masks with polyimide membrane. Figure 26(b) shows an optical micrograph of line/space and dotted patterns on a rectangular convex structure and weaving guides arranged alternately on the cylindrical surface. Figure 26(c) shows another optical micrograph to which the edge of the mold pattern area was expanded, and where the rectangular and cylindrical cross-sectional-shaped convex structures can be clearly observed.

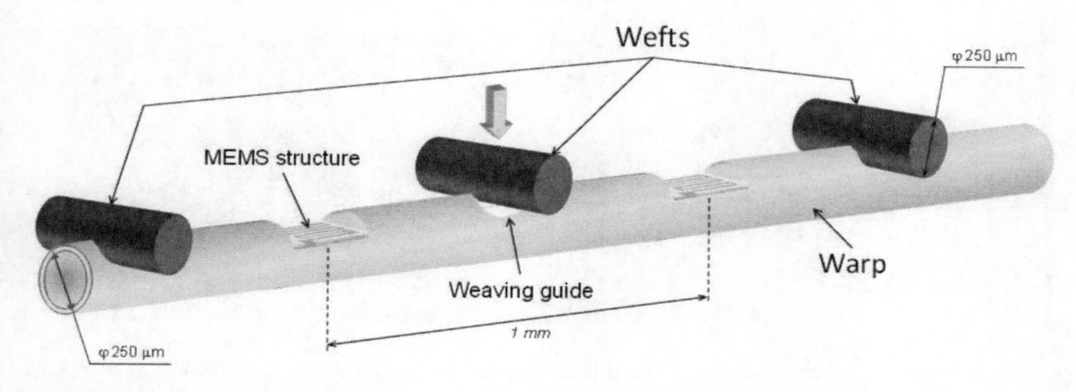

Figure 25. Illustration of MEMS structures and weaving guides on the surface of a smart fiber.

Re-Modelling of Reel-to-Reel Imprint System

Previously, using a cylindrical mold with weaving guides, we had developed a reel-to-reel imprint system, and succeeded in a continuous imprinting at a speed of 5 m/min (Mekaru et al. 2012; Ohtomo et al. 2012). Our new target of thermal imprinting on POF at an imprint speed of 20 m/min was twice the maximum speed of the commercial roll-to-roll systems to do the job. In order to realize this improvement in the imprint speed, a load cell unit, which was set above the upper cylindrical mold, was then moved down to its new position under the lower cylindrical mold, which then raised the mechanical rigidity of a press device as shown in Figure 27. A guide way was fixed on beside the load cell to reduce any horizontal blur; and then vibration of the transverse direction in regard to the sending direction of a fibrous substrate was restrained. Moreover, a feedback system, that controlled the fluctuation of the distance between the upper and lower cylindrical molds by a load measurement, was changed into a feed-forward control device to imprint at a high speed, while referring to the load change data measured at a low rotation speed. Furthermore, the information was read into the database; and a program that predicted the vertical position of the upper cylindrical mold was also developed, while considering the delay time taken to actually move the position of the cylindrical mold by a pulse motor drive. After re-modelling, the standard deviation of the press force was reduced from 5.86 to 2.80 N. In addition, improvement in the imprint speed of the reel-to-reel imprint system by a feed-forward controlling has been explained in the reference (Ohtomo et al. 2012) in detail.

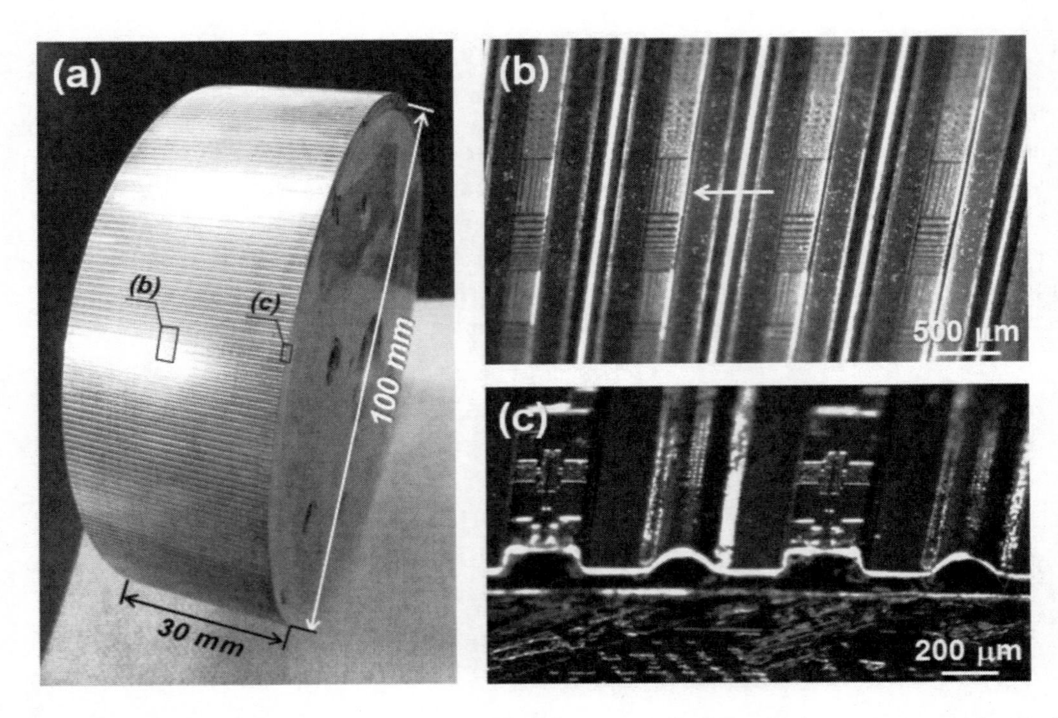

Figure 26. (a) Photograph of a cylindrical mold with convex hybrid-layered microstructures and optical micrographs of mold patterns at: (b) center and (c) edge of the surface of the cylindrical mold.

High-speed Imprinting on the POF

An imprinting result on the 250-μm-diameter POF CK-10 are shown in Figure 28 under conditions of imprint temperature, press force, and sending speed being 50 °C, 30 N, and 20 m/min. Based on the experimental data in the previous work (Mekaru et al. 2013a), conditions that could be used for deep imprinting without crushing the POFs, were selected. On the surface of the fiber, the method successfully fabricated microstructures comprising concave rectangle of a maximum depth of 6.7 μm, and an arc structure of a maximum depth of 5.0 μm, as shown in Figure 28. Furthermore, line/space patterns were transferred onto the bottom surface of the concave rectangle grooves (refer to the white arrow in Figure 26(b)). In order to evaluate the patterns transfer accuracy on POFs from the cylindrical mold, a square dot with its designed width 20 μm was observed.

Figures 29(a) and 29(b) also show the relative setups of the mold and the POF in the imprint step and an optical micrograph of concave-dotted mold patterns, respectively. Figures 29(c) and 29(d) also show an illustration of the imprint position of the POF on the mold pattern surface and an optical micrograph of a convex dotted pattern, respectively. The thickness of Cu top layer varied around 2-3 μm depending upon the position of observation as measured by a high-precision non-contact depth measuring microscope Hisomet-II DH II (Union Optical).

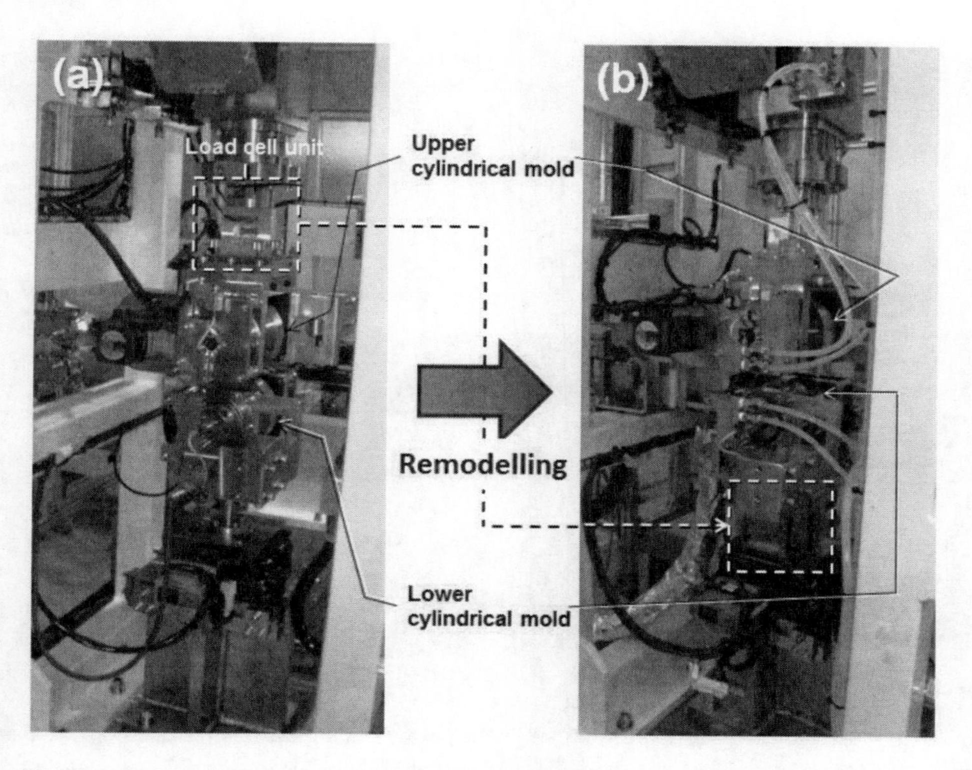

Figure 27. Photographs of press device included upper and lower cylindrical molds in a reel-to-reel imprint system: (a) before, and (b) after re-modelling.

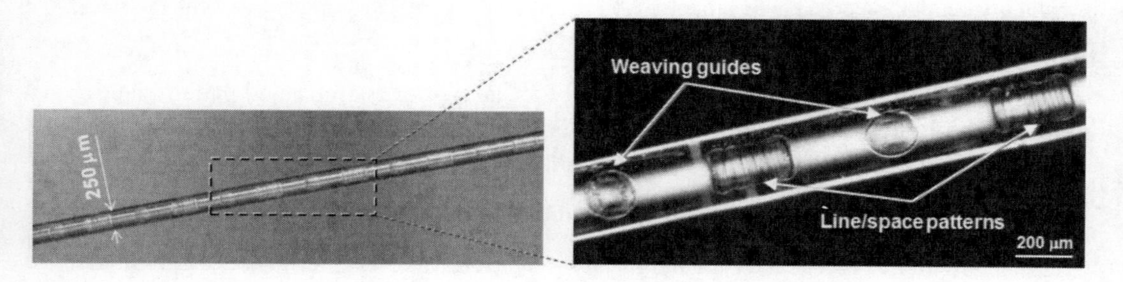

Figure 28. Optical micrograph of imprinted POF, and expanded optical micrograph of weaving guides and hybrid-layered microstructures on the surface of the POF.

From the enlarged optical micrographs inserted at the upper-right corner of Figures 29(b) and 29(d), the average width of the concave and convex square-dotted patterns on the mold was found to be 13.0 and 25.3 μm, respectively. These values are estimated to be -35% and +26.5% as compared to the ideal width of 20 μm which was a designed value in the photomask. And it was observed that as the widths became large, the error became small. This behavior can result in the shrinkage of the photoresist structures in a 3-D photolithography process, and in the expansion of the Cu structures in an electroplating process.

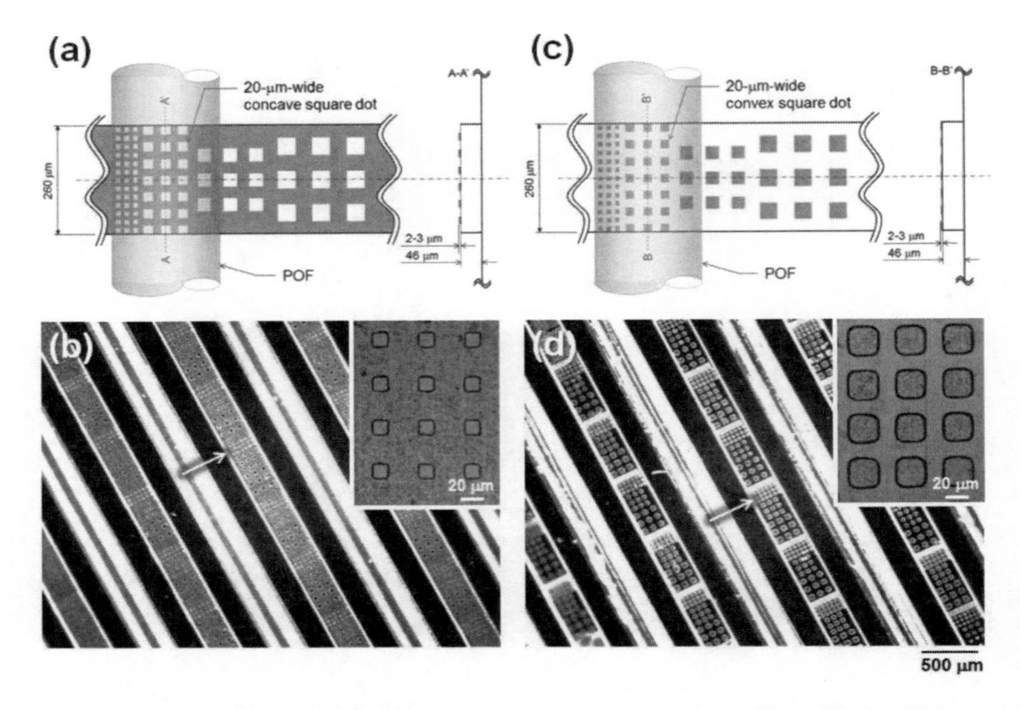

Figure 29. (a) Illustration of relative positions between concave-square-dot mold patterns and POF in the imprint process, and (b) optical micrograph of concave-dots patterns on the cylindrical mold. Enlarged optical photograph inserted in the upper- right corner shows concave square dots with a designed pattern width of 20 μm. (c) Illustration of relative position between the convex-square-dot mold patterns and POF in the imprint process, and (d) optical micrograph of convex-dots patterns on the cylindrical mold. Enlarged optical photograph inserted in upper- right corner shows convex square dots with a designed pattern width of 20 μm.

Figure 30 shows a cross-sectional profile of the square-dotted imprinted pattern measured by the confocal microscope. Here, after considering the disparity between the size of a mold pattern and its actual design value, we now address the difference between the size of an imprinted pattern and that of the mold pattern. The bottom surface of POFs where dotted patterns were transferred was not deformed into a perfect flat surface. In convex square dotted patterns of Figure 30(a), the average height of the pattern was 547 nm. Moreover, the average pattern width was 14.2 μm and the error in reference to the concave mold pattern width was +9%. Next, in concave square-dotted patterns of Figure 30(b), the average value of the pattern depth and width was 662 nm and 28.4 μm, respectively. The imprinted pattern width became larger than 12% as compared with the convex mold pattern, and the error in reference to the pattern width became large in comparison to that of the concave mold pattern. In general, when we compare the vertical movement of a convex mold pattern on a plane surface of mold with a convex mold pattern on a convex surface of a rotating cylindrical mold, we can find that during the imprinting and de-molding process the later setup may cause some scooping out of material (Wu et al. 2009). Therefore, the patterns which are transferred from the cylindrical mold to a molding material result in the imprinting of relatively large widths.

Moreover, the width and depth of the rectangular groove were measured to be 267.5 and 6.78 μm in Figure 30(a), and 271.1 and 4.58 μm in Figure 30(b), respectively. Since the design value of rectangle width was 260 μm, the error of the rectangular groove width estimates as 2.9 and 4.3%, respectively. In the reference (Yeo et al. 2009), when a PMMA film was processed by a roll-to-roll embossing using a cylindrical mold at a feeding speed of 0.6-0.8 m/min, it can be estimated that an error in the imprinted pattern width was nearly 15% compared to that of the bottom width of a mold pattern. The transfer accuracy of the reel-to-reel imprinting on POFs does not show great difference as compared with the result of embossing on the PMMA film. From these results, although the manufacture process of a seamless cylindrical mold needs to be improved, its practicality was checked, and it was proven that a high-speed of imprinting at 20 m/min was possible without causing any damage to the POFs.

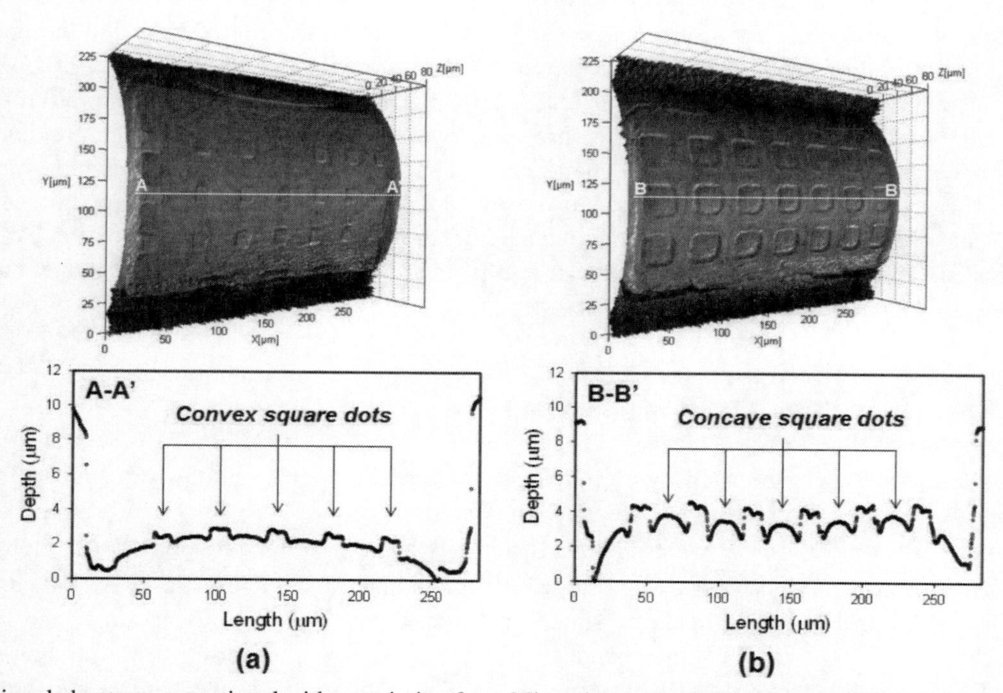

This subchapter was reprinted with permission from Microelectronic Engineering, (2012), *in press*.

Figure 30. (a) 3-D image of convex square-dot patterns imprinted on the surface of POF and cross-sectional profile along a line A-A' of convex square patterns. (b) 3-D image of concave square-dot patterns imprinted on the surface of POF, and cross-sectional profile along a line B-B' of concave square patterns.

THERMAL IMPRINTING ON QUARTZ OPTICAL FIBER

Motivation

For the current work, depending on the type of device under consideration, plastic or glass fiber is selected as the base material. The glass transition temperature of PMMA and polycarbonate (PC) used as core material for optical fibers are typically 105 and 145 °C (Gauthier 1995). Patterning on the surfaces of plastic fibers to make MEMS devices is not feasible where high temperature annealing is required. On the other hand, a glass fiber is stable at high temperatures. Therefore, we are pursuing a technology for processing MEMS on the surface of quartz fibers. In general, the glass transition temperature of quartz is found to be 1175 °C (Mazurin et al. 1993), and therefore, for thermal nanoimprinting on quartz plates the temperature of mold has to be heated up to 1300 °C or higher (Takahashi and Maeda 2006). We have developed a high speed thermal nanoimprint system that can be heated up to 1400 °C (Takahashi et al. 2006a), and used it to replicate micro-parts of quartz. In the devices made here, many ceramic and molybdenum components were used as refractory materials. Glasslike carbon (GC), one of the stable materials to stand high temperatures, was used as a mold material for imprinting on quartz. Mold patterns were processed on GC substrates by focused-ion-beam (FIB) etching (Takahashi et al. 2006b). Using this nanoimprint system, line/space patterns with the linewidth of 500 nm and the line depth of 100 nm on a quartz plate were imprinted and reported (Takahashi et al. 2005). We demonstrated pattern transfer on the surface of a quartz fiber by using a thermal imprinting technology where deep patterns could be processed without much difficulty. In the previous experiment that reported on imprinted quartz plates, a GC mold patterned by FIB etching was used. However, the FIB etching process takes very long time even though the process does produce precise patterns with arbitrary shapes. The mold used in our experiment was fabricated using MEMS technology that can also form arbitrary patterns on GC substrate, but it does so in a considerably short time.

Fabrication of Glass-Like Carbon Mold

A GC mold was fabricated by employing MEMS technologies as shown in Figure 31. Using a conventional process, MEMS structures were fabricated by using a Si wafer. The thickness of a typical 4 in. Si wafer is 400±1 μm, with average surface roughness of 1 nm or less. However, the surface of a GC substrate was rough and the thickness was 2 mm±40 μm because GC substrates were made by sintering. When a wafer exhibits a curvature, and its surface happens to be rough, it then becomes difficult to fabricate uniform MEMS structures on it. Hence, both sides of the GC substrate were polished by chemical-mechanical polishing (CMP), where the final substrate thickness amounted to 2 mm±0.5 μm, with average surface roughness of 5 nm. These values matched the specifications of the sample holder of the ultraviolet (UV) stepper used in the patterning process. Figure 32(1) shows a GC wafer that cut an orientation flat after polishing on its both sides. At first, a 900-nm-thick Au film was sputter deposited on the GC wafer to serve as a masking layer. Prior to this, a 10 nm thick Ti layer was sputter deposited as an interlayer to increase the adhesive force between the Au layer and the GC wafer. Next, a positive-negative inversion photoresist AZ 5200NJ (AZ Photoresist Products) in the thickness of 4 μm was spin coated and then heated up to 110 °C for 90 s using a hot plate. After letting it cool down to the room temperature, UV lights with 200 mJ exposure energy were irradiated on the resist using a UV stepper 1500 MVS R-PC system (Ultratech). In order to reverse the pattern transferred from a reticle, the substrate was heated up to 125 °C for 125 s. In the develop process, a 25% AZ 400K developer (AZ

Photoresist Products) was used. After developing for 2 min, the Au coated GC substrate was washed with distilled water for 1 min, and then spin dried [Figure 32(2)]. After that, the Au layer was etched for 13 min using a reactive-ion-etching (RIE) system Multiplex ASE (Surface Technology Systems) with an Ar gas [Figure 32(3)]. Next, 12 square shaped chips in the size of 20 mm were cut out from the 4 in. GC wafer by dicing [Figure 32(4)]. After patterning, the Au layer was used as a masking layer, and the GC substrate was etched using a mixture of O_2 and SF_6 gases in another RIE system model RIE-10NRS (Samco). By controlling the etching time to 20, 40, 50, and 100 min, the depths of mold patterns were obtained as 2.4, 3.5, 7.0, and 13.5 μm. The remaining Au layer was then chemically removed using an AURUM-301 etchant (Kanto Chemical) [Figure 32(5)]. The GC plate was soaked for several minutes in a 20% buffered hydrofluoric acid (Stella Chemifa), and then the Ti interlayer was removed. After washing with acetone, isopropyl alcohol, and distilled water, a GC mold was thus completed [Figure 32(6)]. SEM images of the mold pattern are shown in Figure 33. Figure 33(a) shows a line/space pattern 14.7 μm in depth. The minimum linewidth was 5 μm, the maximum linewidth was 100 μm, and the line/space widths ratio was 1:1. Figure 33(b) shows a square dotted pattern in the sizes of 5 and 10 μm. In general, the width and depth of patterns that are suitable for electrical circuits in MEMS devices are assumed to be 5–50 and 5–10 μm, respectively. These figures result in a range of aspect ratios from 0.5 and 10. Moreover, the beam widths of cantilevers and comb actuators happen to be in the range of 50–100 μm. GC molds fabricated for our experiment do take these sizes into consideration. The edges of the patterns were well defined and pattern transfer was very successful.

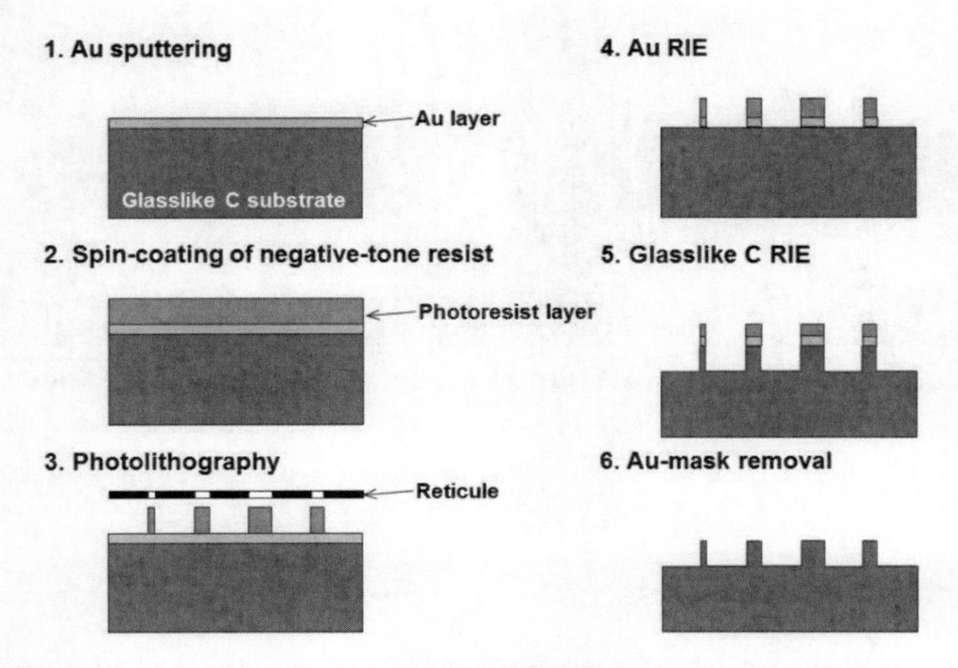

Figure 31. Process flow to fabricate a GC mold by MEMS fabrication technologies. Mold patterns were fabricated on the GC substrate by using the conventional processes such as sputtering, spin coating, photolithography, and RIE.

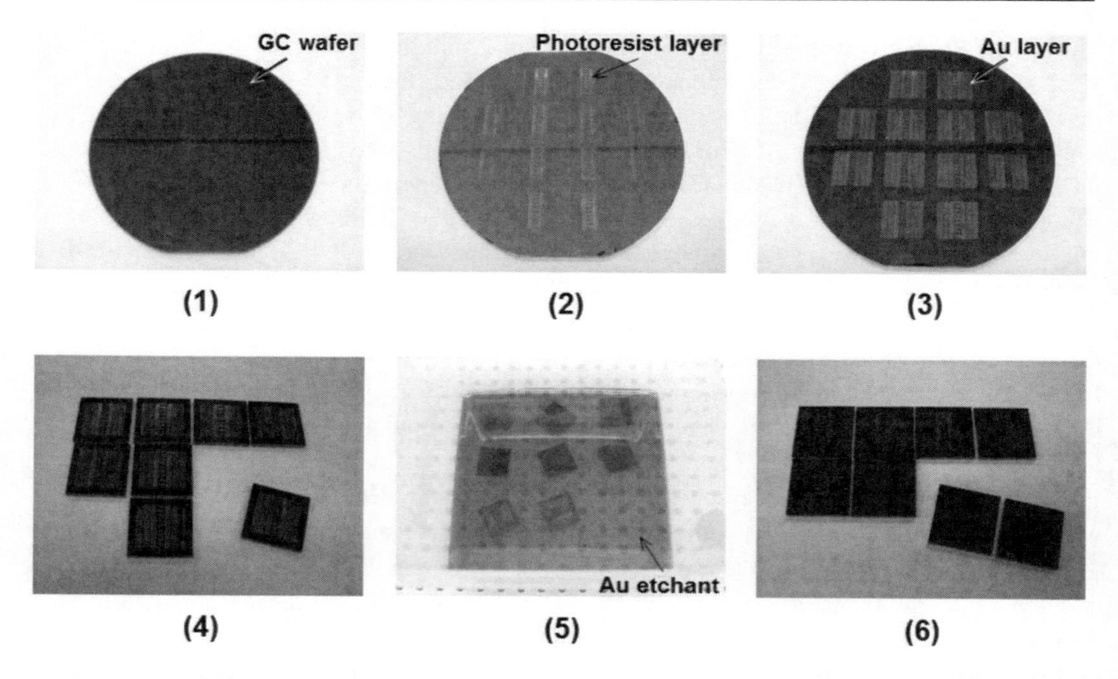

Figure 32. Photographs of the intermediate stages in the fabrication of GC mold: (1) GC wafer, (2) after photolithography, (3) after Au RIE, (4) after dicing, (5) after Au removal, and (6) completed GC molds.

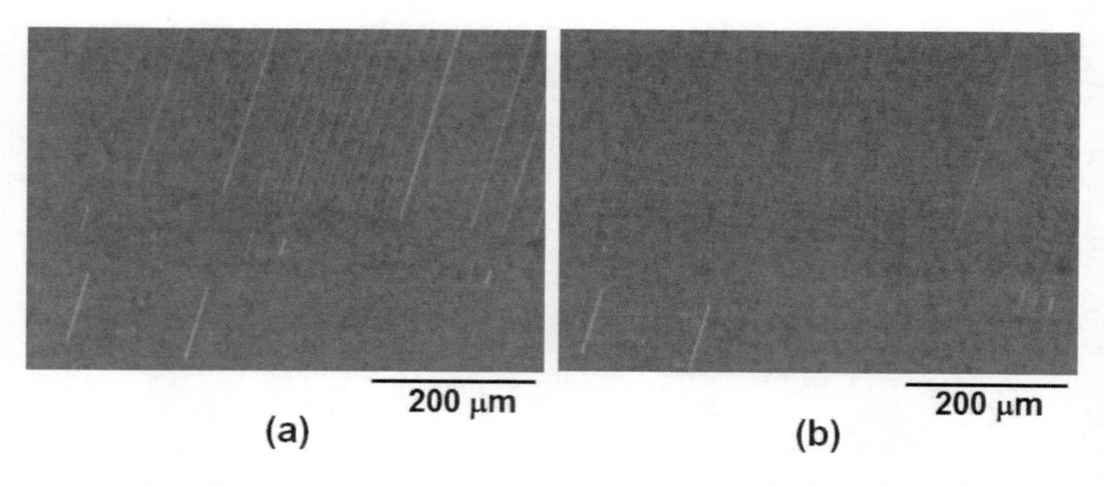

Figure 33. SEM images tilted at 45° of the surface on a GC mold with: (a) line/space with linewidths of 5–20 μm and (b) convex dotted patterns with 5 and 10 μm size squares. The depth of all patterns in the figure is 14.7 μm.

Experimental Procedure

We used a quartz fiber with 200 μm size square shaped cross section. The surface of the fiber was coated with a 10 nm thick protective layer of UV cured resist. In a previous experiment, the protection film remained strongly adhered to the surface of the quartz fiber when we attempted to remove it by ultrasonic cleaning with acetone. The UV cured resist was finally removed by H_2SO_4 etching as described here. The quartz fiber coated with the UV cured resist was cut into 40 mm long pellets with a nipper, and the pellets were then fixed into a Teflon sample holder. The Teflon sample holder with the pellets was then soaked for 10 min into a mixture of 18% H_2SO_4 and 30% H_2O_2 heated up to 100 °C. The UV cured resist coated on the quartz fiber was thus completely removed by this method. The pellets were then soaked in distilled water then dried by spraying N_2 gas. A thermal nanoimprint system ASHE0201 (Takahashi et al. 2006a) was used for imprinting on the quartz fiber. The maximum heating temperature of this system is prescribed to be 1400 °C, and the ramp time to reach this temperature is 15 min. An upper loading stage is removable by a servo motor, and the permissible loading force is 9.8 kN. The upper and bottom loading stages were stored inside a bellows-type chamber where a pressure in the vacuum chamber could be decompressed down to 0.07 Pa with a turbo molecular pump.

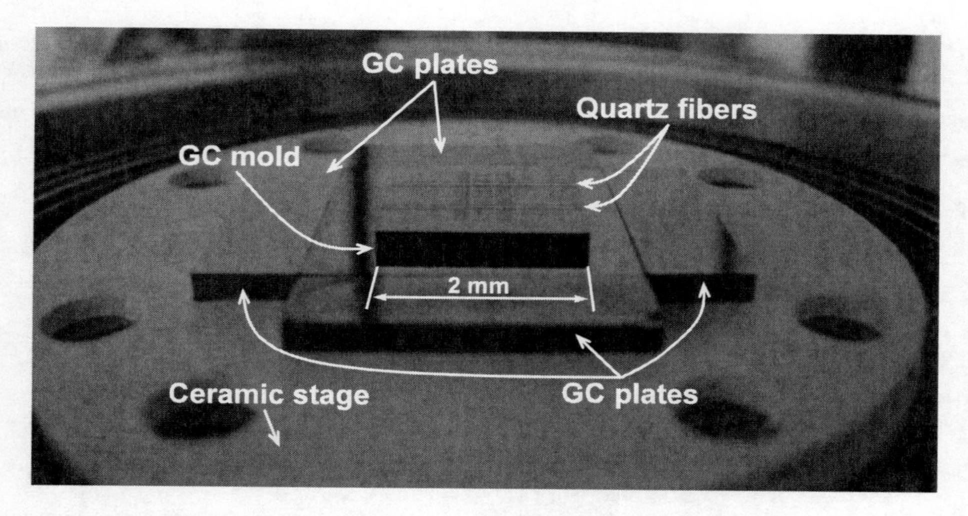

Figure 34. Photograph of experimental setup in ASHE0201 nanoimprint system after the quartz fibers were arranged on the GC mold. The size of the GC mold was $20 \times 20 \times 2$ mm^3.

Figure 34 shows a photograph of the setup ready for imprinting on the quartz fiber with a GC mold. The bottom loading stage made of ceramic was processed into the shape of a discoid with a diameter of 60 mm. A 30 mm size square GC plate (3 mm thick) was placed on the center of the bottom loading stage and held secured by four smaller CG plates placed at the four sides of the center CG plate to make it immovable during the decompression process. A 20-mm-size square GC mold was set facing up on the center GC plate. Two quartz fibers were put on the GC mold crossing line/space patterns. Another 30 mm size square GC plate in the thickness of 3 mm was put on top of quartz fibers. Because the mold and fibers were small in this experiment, it was not possible to fix separately on the upper and bottom loading stages. Moreover, the loading stages were made of ceramic material, which happens to exhibit a certain degree of roughness on its surface. When the upper loading stage is pressed against the surface of the quartz fiber, small ruggedness on the surface of the ceramic stage is transferred onto the surface of the quartz fiber. This ruggedness can result in an increase in the release force during the de-molding process. Therefore, the GC plate with a smooth surface was inserted between the quartz fiber and the upper loading stage. The upper loading stage was approached until it reached a distance of 1 mm (or less) from the back side of the uppermost GC plate. After that, the bellows-type chamber was closed and decompressed, and the upper and bottom loading stages were heated up to 1350 °C. Experiments on thermal imprint were executed in a vacuum, and the GC mold was pressed on the quartz fibers for 2 min using a press force of 300 N. Finally, the GC mold and quartz fibers were let to cool down to the room temperature for about an hour.

Imprinting Results on the QOF

Figure 35 shows the results of the imprinted patterns on the surface of the quartz fiber as seen through an optical microscope. Figure 35(a) shows the result of the imprinted pattern on the quartz fiber that was coated with the UV cured resist. Figure 35(b) shows the result of imprinted pattern on the quartz fiber from which the UV cured resist was removed by H_2O_2+H_2SO_4 etching. When the pattern was imprinted on the protective film of the UV cured resist, cracks were observed in the resist layer, although the patterns still got transferred on the quartz. Because of these crack formations, the imprinted quartz fibers could not be used for subsequent processing.

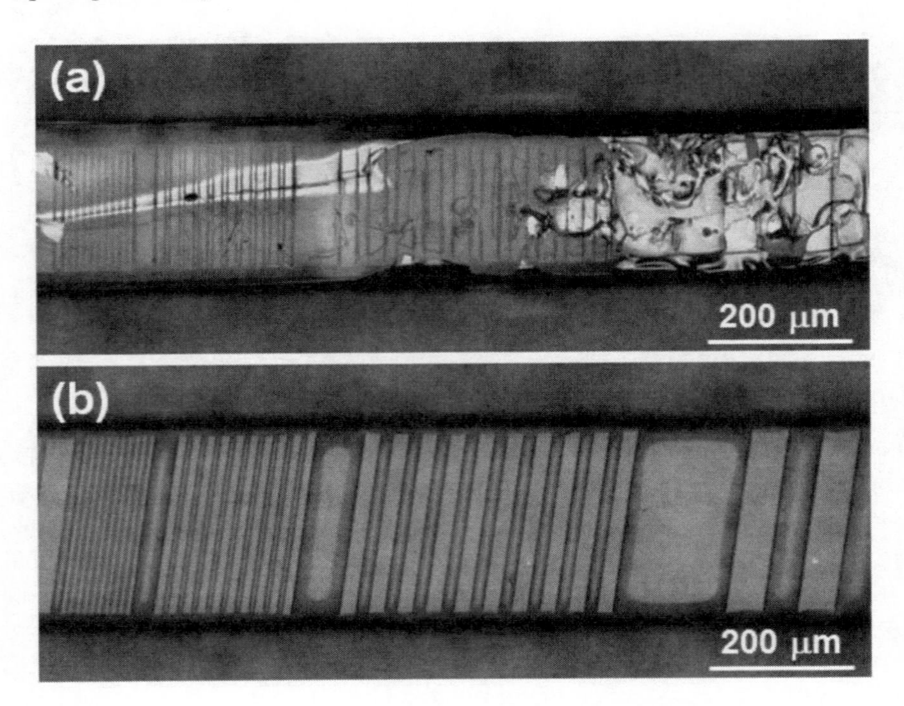

Figure 35. Optical micrograph of line/space patterns imprinted on the surface of a quartz fiber: (a) a quartz fiber was coated by UV cured resin and (b) a quartz fiber without any coating. When the pattern was imprinted on the protective film of the UV-cured resist, cracks were observed in the resist layer coated on the quartz fiber. Molding conditions where the heating temperature, press force, and press time were 1350 °C, 300 N, and 2 min, respectively.

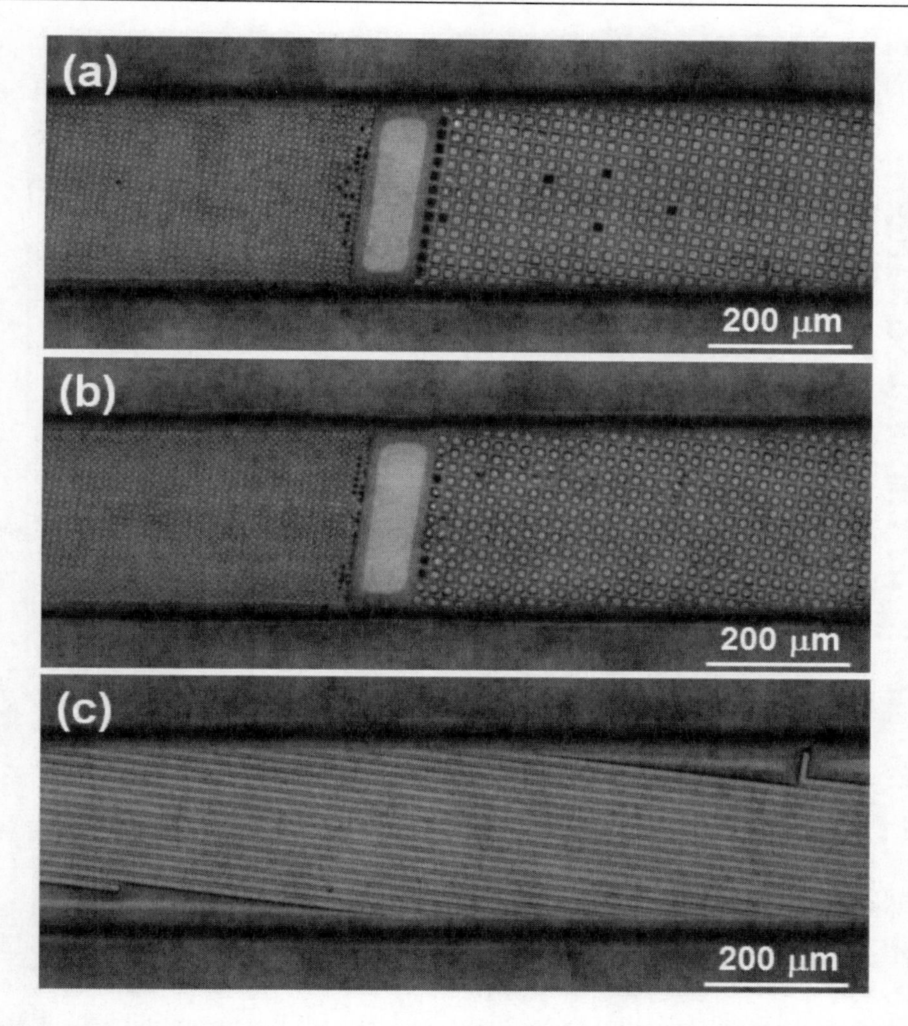

Figure 36. Optical micrograph of imprinted surfaces on a quartz fiber with (a) square dotted patterns, (b) circle dotted patterns, and (c) antenna patterns for demonstration. The 14.7 μm deep GC mold was pressed against a quartz fiber without any coating, under conditions where the heating temperature, press force, and press time were 1350 °C, 300 N, and 2 min, respectively.

On the other hand, when the clean pure quartz was imprinted, then the line/space patterns on the GC mold transferred very well. In the GC mold, besides line/space and square dotted patterns, there were circle dotted patterns and MEMS structures, which also transferred well. Results of imprinting on the quartz fiber using a 14.7 μm deep GC mold are shown in Figure 36. Figure 36(a) shows an optical micrograph of imprinted square-dotted pattern with the sizes of 5 and 10 μm appearing on the left and right sides of the picture. Figure 36(b) shows an optical micrograph of imprinted circle-dotted pattern with the diameters of 5 and 10 μm appearing on the left and right sides of the picture. In all these pictures, some black spots were also observed. It is suspected that during the imprinting process some impurities in the mold might have gotten transplanted onto the quartz fibers. Figure 36(c) is another example where an antenna pattern with the linewidth of 20 μm in MEMS structures was transferred for the purpose of demonstration. All mold patterns were transferred undistorted, and no defect in imprinting was noticeable.

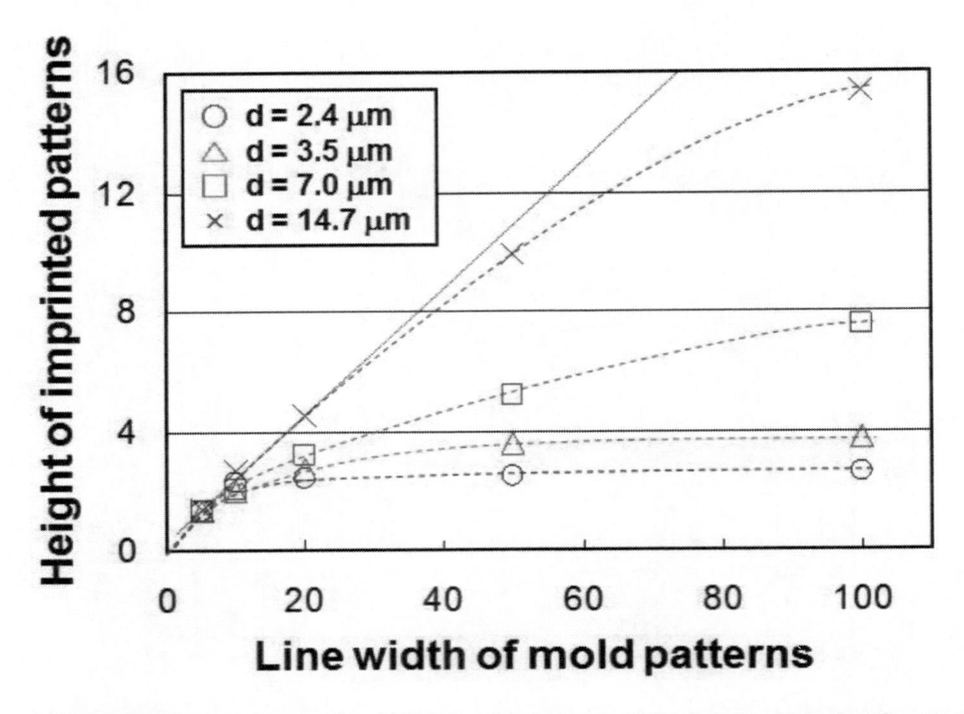

Figure 37. Relationship between the linewidth of mold patterns and the height of imprinted patterns on quartz fibers. "*d*" **means the depth** of mold patterns in the figure. The height was measured by a 3D optical profiler.

Figure 37 shows the heights of the imprinted line/space patterns on the quartz fiber as functions of the linewidths of the GC mold pattern. The study involved four different kinds of GC molds, with pattern depths (*d*) of 2.4, 3.5, 7.0, and 14.7 μm. The imprint conditions were same in all cases, and the measurements were made with 3-D optical profiler NEWVIEW 5000 (Zygo). The widths of the line patterns on the CG mold were 5, 10, 20, 50, and 100 μm. Because the width of the quartz fiber (on which the lines were imprinted) was 200 μm, a convex area of line/space patterns on CG mold making physical contact with quartz fiber was calculated and found to be 0.185 mm^2. Therefore, a pressure of 1.62×10^9 Pa was estimated on the melted quartz fiber when it was imprinted with a press force of 300 N. Therefore, the height of imprinted patterns on quartz fiber against the linewidths on the CG mold increased in proportion to the linewidth on the CG mold. When the depth (*d*) of the mold pattern was 14.7 μm, the relationship between the imprinted pattern on the quartz fiber and linewidth of the mold pattern was found to be linear until the linewidth of the molted pattern reached 20 μm in size. After this, the imprinted height began to saturate as its value approached the depth of the mold patterns [Figure 37]. When GC molds with comparatively low depth values (d = 2.4 and 3.5 μm) were used, the imprinted height on quartz fiber did not increase with mold pattern linewidths it was the case even where the pattern widths were small. When the GC mold with the pattern depth equal to 7.0 μm was used, the range where the increase in imprinted pattern height was proportional to the increase in mold's linewidth was still very limited. We believe this reason is that the shape of the molding material was not plate but fiber influences although we cannot be certain about this contention. As for the utilization of loading force during imprinting, all power was not channeled toward the imprinting of patterns.

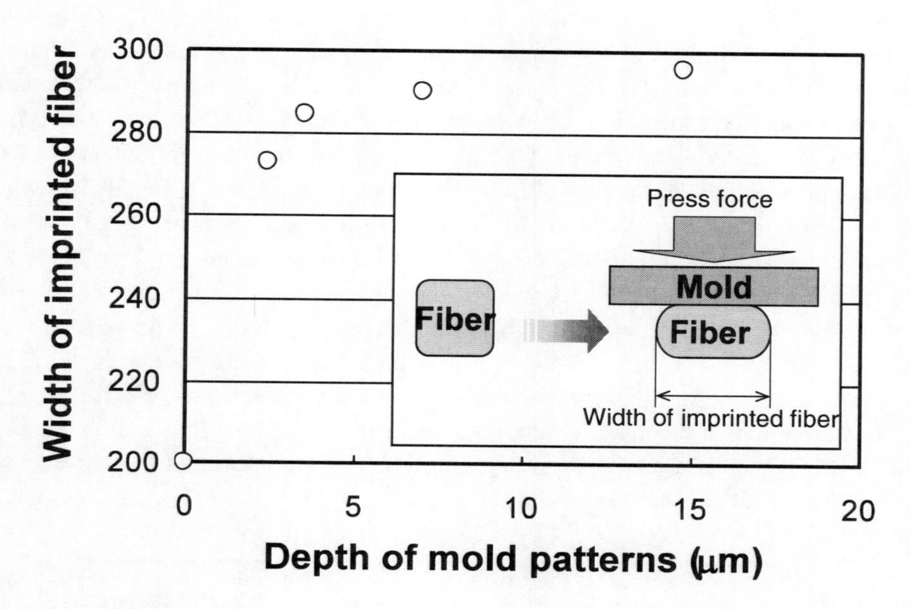

Figure 38. Relationship between the depth of mold patterns and the width of imprinted quartz fibers. The insertion figure shows the location marked as "Width of imprinted fiber." A quartz fiber was crushed by a press force during an imprint process that resulted in extending the width of the quartz fiber.

The quartz fiber was flattened by the press force where its cross section changed its shape from square to a rectangle with rounded corners as shown in the inserted illustration of Figure 38. The width of quartz fiber was measured using an optical microscope. Figure 38 shows that the width of imprinted quartz fiber increased to 270 μm or above. Thus, a part of press force applied to quartz fiber was used up in deforming the quartz fiber, and the imprinted height on quartz fiber seemed smaller than when the imprinting is done on quartz plates. Figure 39 shows a relationship between the aspect ratio of mold patterns and the filling rate. Here an aspect ratio is defined as the ratio of depth to linewidth in the mold pattern, and filling rate is defined by dividing the height of the imprinted pattern by the depth of the mold pattern. It becomes difficult for softened quartz to enter a cavity of mold patterns when the aspect ratio is high, and in that case complete filling becomes difficult. A dotted curve in Figure 39 shows a trend of the relationship between the filling rate and aspect ratio of the mold pattern. The filling rate is found to be inversely proportional to the aspect ratio of mold patterns. If the aspect ratio happens to be same then the filling rate becomes near, even if the depths of mold patterns were different. The filling rate changed within a range of 0 to 1, where "1" stands for complete filling. The reason why some filling rates seemed larger than 1 can be attributed to the depressurization during the cooling process. In this process, carbon heaters installed on the upper and bottom loading stages were turned off before the start of the cooling process. Because the GC mold and quartz fibers were cooled down under a vacuum, their temperatures approached the room temperature by a very slow process of thermal transmission through the upper and bottom loading stages. In this experiment, the press force during the cooling process was not altered, while the position of the upper loading stage remained fixed. Therefore, the GC mold and quartz fibers shrank during the cooling process, and the press force was decreased gradually. On the other hand, imprinted patterns was expansion by residual heat at the same time as decreasing the press force because the contact part, where the GC mold touched the quartz fibers was cooled at the end. Therefore, some filling rates exceeded 1. Hirai *et al.* reported on the relationship between the height of imprinted patterns and the aspect ratio of mold patterns in the case of thermoplastics (Hirai et al. 2004). They compared simulations with the experiment results of patterns with the linewidths of approximately 250 nm and 1 μm. They took into account a mechanism where a melted resin filled into a cavity of the mold. As a result, the molding accuracy showed that the pattern is not dependent on the absolute size, but is rather dependent on the relative initial thickness of the resin and on the aspect ratio of the pattern. Figure 39 shows that the same tendency was observed in the imprinting of quartz fiber, as mentioned earlier. Moreover, we are planning to fabricate an electrical circuit by inserting an electro-conductive material into a thermally imprinted concave pattern on quartz fibers by inkjet and electroless-plating processes. Therefore, when the electric circuit pattern with the same cross-sectional area is to be fabricated, it will be taken into account that the width of the imprinted pattern is made as wide as possible to in order to minimize the aspect ratio.

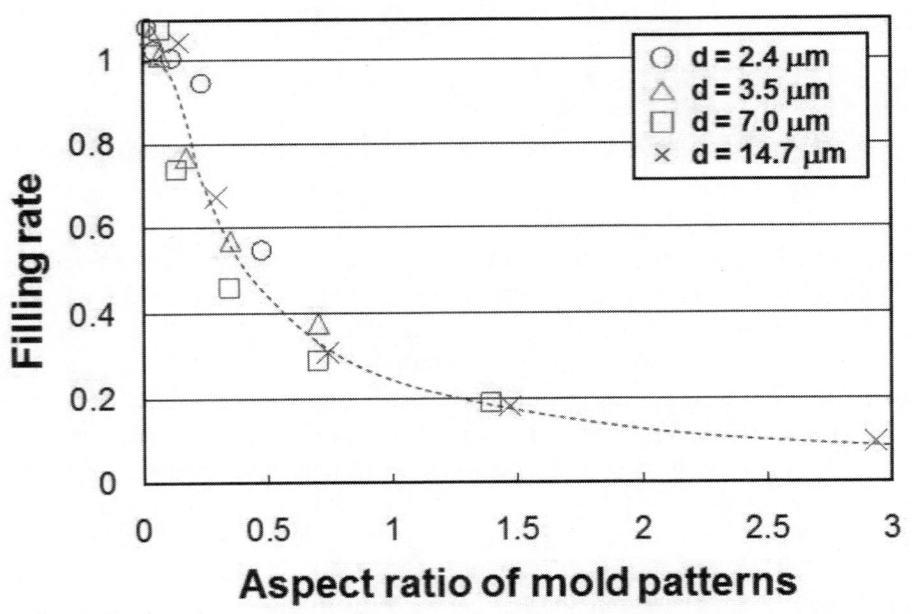

Figure 39. Relationship between the aspect ratio of mold patterns and filling rate. The filling rate was defined by dividing the height of the imprinted pattern by the depth of the mold pattern. *d* means the depth of mold patterns in the figure.

3-D SEAMLESS PATTERNING ON QURATZ OPTICAL FIBER BY FOCUSED-ION BEAM ETCHING

Motivation

Nanoimprint technologies are faced serious difficulties in stitching patterns. We then tried processing a precise spiral structure by FIB (Meingailis 1987) etching of a processing area that was larger than that of electron-beam lithography. We took advantage of the feature of seamless processing by direct etching without employing any development cycle. FIB technologies were applied to nanofabrication by direct etching on a solid substrate (Tseng 2004) and in combination with chemical-vapor deposition (CVD) (Matsui et al. 2000). Moreover, the applicable field was expanded from sample preparation techniques for material analysis (Ishitani and Yaguchi 1996; Heaney et al. 2001; Wirth 2004) to semiconductor manufacture (Morimoto et al. 1986) and micro-/nanodevice development (Matsui and Ochiai 1996; Reyntjens and Puers 2001). We used FIB as one of the fabrication tools for precise molds to thermal imprinting on glass materials (Mekaru et al. 2007; 2008; 2009b). For the work presented in this paper, we fabricated a simple rotation stage for trial purposes. We also reported our experimental results of processing a spiral line pattern on a cylindrical surface of a 250-μm-diameter QOF by FIB etching. Our technique can be applied to the fabrication of the capillaries of micro-fluid devices, as well as of receiver coils connected to a catheter and an endoscope of nuclear magnetic resonance imaging (MRI) systems to be used in imaging blood vessels, by forming a spiral coil on the surface of a cylindrical quartz material.

Fabrication of Fiber Rotation Stage

A mechanism for rotating the fiber being etched with FIB is necessary for the continuous processing of the full circumference surface of the QOF. However, the cost of incorporating a rotating element into a conventional FIB etching system is too high. Also, in such a case, a technical problem is encountered when using the prototype rotation stage developed before finalizing the large-scale remodeling of our FIB etching system. For the etching process, the FIB system EIP-5400 (Elionix) was used. Figure 40 shows a drawing and a photograph of a simple rotation stage fabricated for FIB etching. The rotation stage was compact in design with the dimensions of $163 \times 120 \times 55$ mm^3. The fiber was held by a four-jaw chuck connected to the rotation system, and the height of the rotation axis of the chuck from the bottom of the simple rotation stage was 32.5 mm, which is the same as the height of a standard surface when the Si wafer is normally etched by FIB. An LR3 battery was connected to the direct-current (DC) magnetic motor FP030-KN (Standard Motor) through a variable resistor. A plastic gear mechanism for changing the speed (5402:1) was used because the typical rotational speed of the motor was 9600 rpm, but in our case, a speed as low as 0.17 rpm was necessary. To rotate the gears smoothly, silicone grease was injected into the extra space between the gears. EIP-5400 is generally run under a vacuum of less than 1.0×10^{-4} Pa. Normally, the motor, gear changer, and battery used in this experiment cannot function under such a vacuum. We used a small motor typically employed to drive a plastic model because it enabled a simple rotation stage to be fabricated at a very low cost. Therefore, the pressure and temperature when the motor was driven in another vacuum chamber with an exhaust unit were measured in an exploratory experiment. From Figure 41, it can be seen that the pressure was decompressed to 2.8×10^{-4} Pa for 1 h. The temperature remained steady although it was 8 °C higher than that of the atmosphere. Therefore, it seemed that this motor could be used under vacuum only for a brief period. Moreover, the LR3 battery was inserted into the vacuum chamber of the FIB etching system containing the motor because there was no preliminary flange in the vacuum chamber, although the electric power was generally supplied from an outside source via a current introduction terminal. In the exploratory work, the battery did not explode, and the phenomenon of electrobath leakage was not encountered.

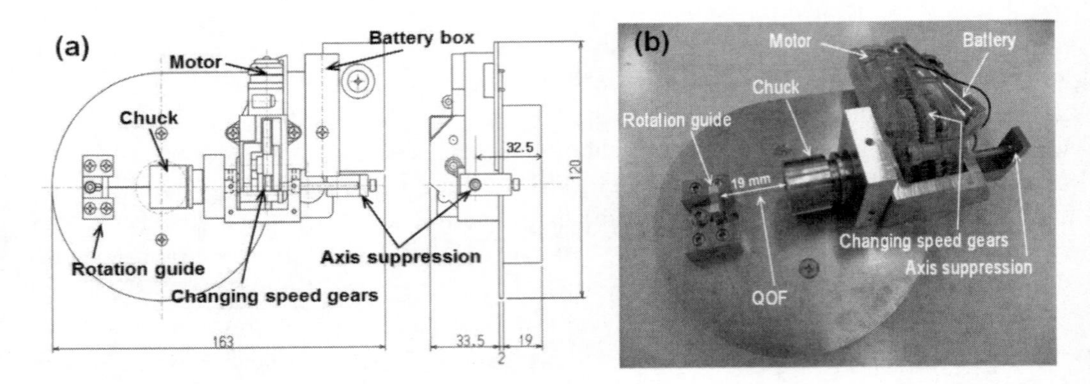

Figure 40. (a) Drawing and (b) photograph of simple rotation stage for etching QOF using FIB. The unit in the figure is mm.

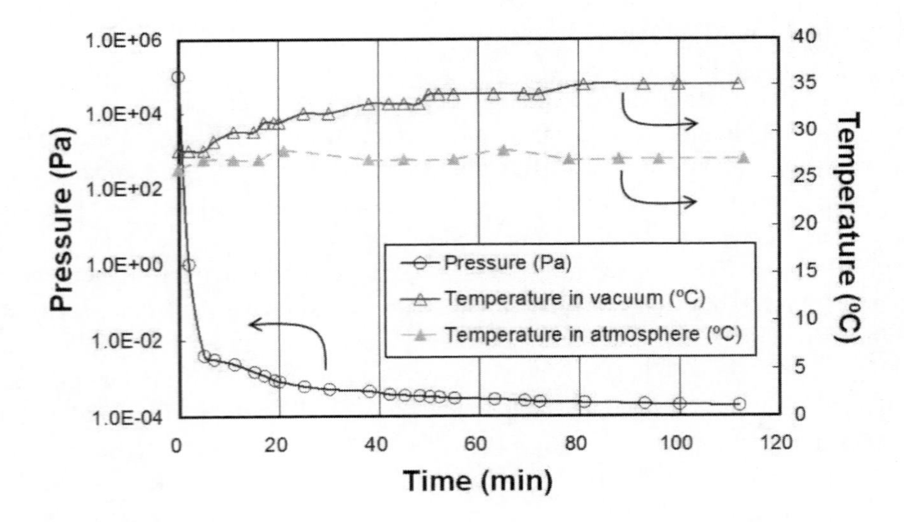

Figure 41. Pressure and temperature when dc motor was driven in vacuum.

Improvement of Eccentric Movement of the QOF

The test sample was a 250-μm-diameter QOF made of a 125-μm-diameter quartz core coated with a UV-cured resin cladding. A 10-nm-thick Pt layer was sputter-deposited on the cylindrical surface of the fiber to prevent any charge buildup during FIB etching. At the initial stage of the experiment, the line-shape etched on the UV-cured resin appeared to be zigzag, and the line pitch was not stable [Figure 42(a)]. This behavior was related to an eccentric spinning (or wobbling) of the rotating fiber, and the sliding movement of the rotation axis in the gear mechanism [Figure 42(b)].

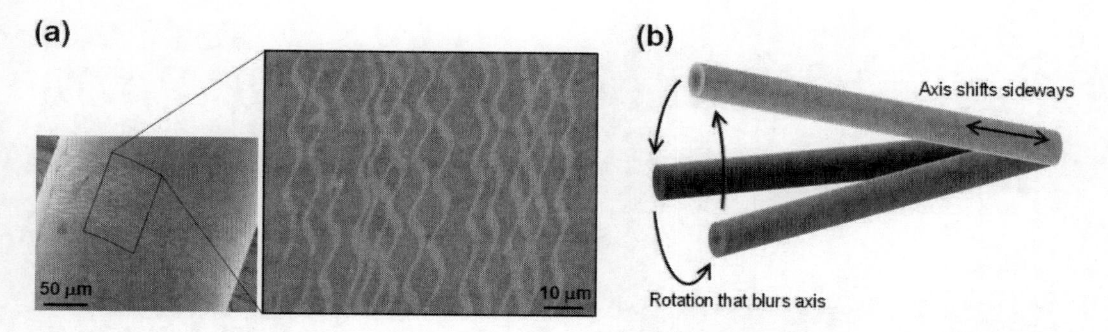

Figure 42. (a) FE-SEM images of spiral line before improvement. (b) Schematic image of blurring and sliding fiber in rotation stage.

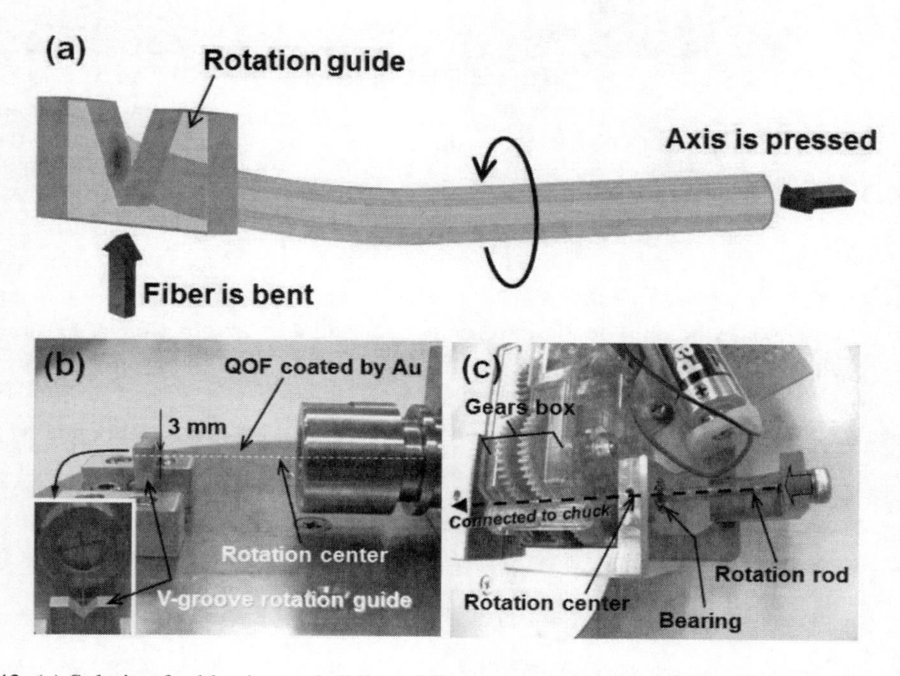

Figure 43. (a) Solution for blurring and sliding of fiber in rotation stage. Enlarged photographs of (b) V-groove rotation guide in simple rotation stage for controlling blurring effect from fiber, and (c) axis suppression device in rotation stage for controlling sliding of fiber.

The problem was solved by installing a rotation guide and an axis suppression device onto the rotation stage, as shown in Figure 43. With this setup, the free end of the QOF was bent to prevent its eccentric movement, which also stopped its horizontal sliding motion because the rotation rod that was connected to the chuck with a bolt was pushed from its back end. Figures 43(b) and 43(c) show enlarged photographs of the rotation guide and the axis suppression device of the simple rotation stage, respectively. The cross-sectional shape of the rotation guide was a V-groove with an intersection angle of 45°, where the V-groove structure was fine-processed by electric-discharge machining. The surface roughness of the V-groove was 1 μm or less. The 250-μm-diameter QOF was held by the V-groove, while in contact with its two surfaces allowing the horizontal axis blur to be controlled.

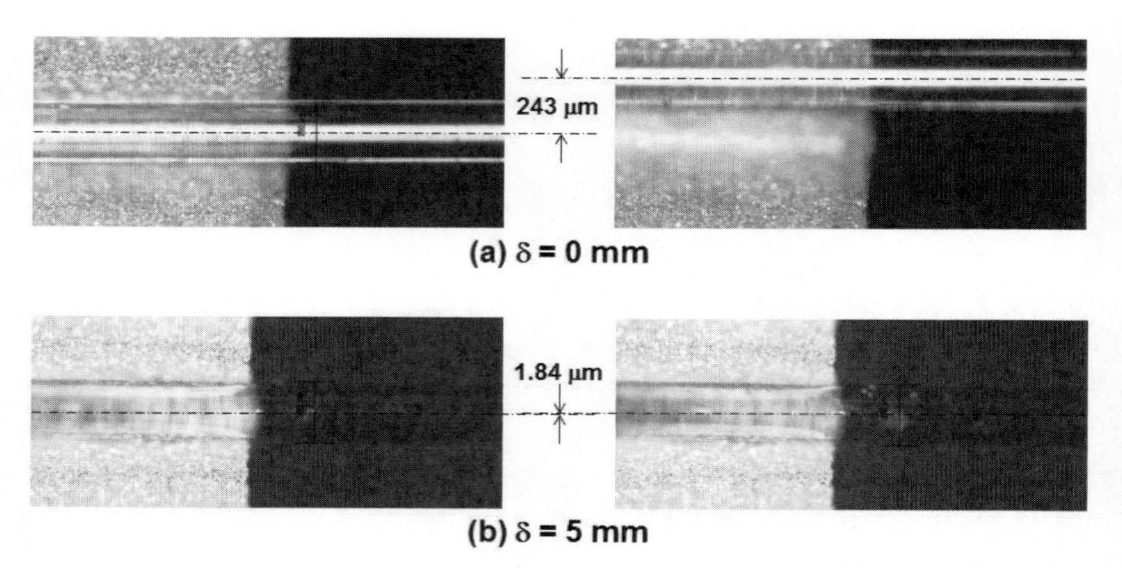

Figure 44. Optical micrographs of QOF on V-groove rotation guide with bending heights of (a) 0 mm (standard position) and (b) 5 mm.

Figure 44 shows the result of our investigation of the central of the eccentric exercise when the V-groove rotation guide is raised higher than the rotation axis of the fiber. Figure 44(a) shows the result when the height of the V-groove rotation guide was kept the same as that of the rotation axis, where no bending of fibers occurred. On the other hand, Figure 44(b) shows the result for a fiber bent by 5 mm by moving the V-groove rotation guide upward, where the distance of the V-groove rotation guide from the chuck was 19 mm. Bending force was calculated from the bending height of the QOF using a well-known formula related to the cantilever with a circular cross section. The geometric moment of inertia I_q of the quartz core is denoted by (The Society of Polymer Science, Japan 1986)

$$I_q = \frac{\pi D_q}{64}, \tag{1}$$

where D_q is the diameter of the quartz core. On the other hand, the geometric moment of inertia I_r of a UV-cured resin, which covered the quartz core as a clad layer, is shown by (The Society of Polymer Science, Japan 1986)

$$I_r = \frac{\pi}{64}\left(D_r - D_q\right), \tag{2}$$

where D_r is the outer diameter of the UV-cured resin and is the same as the diameter of the quartz core. In eq. (3) (The Society of Polymer Science, Japan 1986), the bending force W is described. Here, W is needed to bend the fiber with a displacement of δ while the front end of the fiber remains fixed:

$$W = \frac{3\delta}{l^3}\left(E_q I_q + E_r I_r\right). \qquad (3)$$

Here, l is the length of the cantilever beam, and E_q and E_r are Young's moduli of quartz and the UV-cured resin, respectively. Because the details of the UV-cured resin were unknown, the bending force was calculated by substituting the values for epoxy resin. D_q = 125 μm, D_r = 250 μm, l = 19 mm, E_q = 73:1 GPa (National Astronomical Observatory of Japan 2011), and E_r = 2:36 GPa (The Japan Society of Mechanical Engineers 1987) were substituted in eq. (3), and each bending height in Figure 44 was converted into its corresponding bending force and plotted in relation to the distance of axis blur, resulting in the eccentric movement observed in Figure 45. When the front end of the QOF was forcibly pushed up away from the axis of rotation by a bending force of 28.7 mN, the distance of axis blur decreased from 243 to 1.84 μm and the eccentric movement was suppressed.

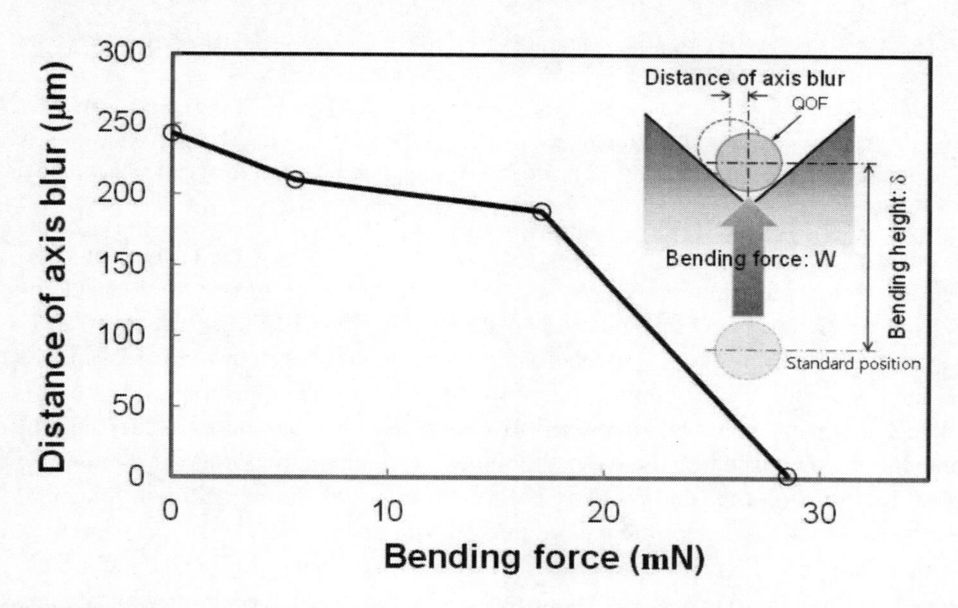

Figure 45 Relationships between bending force on QOF using V-groove rotation guide and distance of axis blur.

3-D Etching on the QOF by FIB

In FIB etching, the rotational speed of the QOF was fixed to 0.5 rpm, where the rotation stage was placed inside the vacuum chamber of the FIB etching system. After the pressure in the chamber reached 5×10^{-4} Pa, the focus and stage position of FIB were adjusted. FIB was scanned at a speed of 37 nm/s along a line defined by the axis of rotation, as shown in Figure

46(a). The line length programmed in the FIB system was 60 μm. Figure 46(b) shows the etched surface of QOF observed using a FE-SEM. Figure 45(c) shows a magnified FE-SEM image of the area within the white dotted square in Figure 45(b), and a spiral line pattern processed on the cylindrical surface of QOF by FIB etching. The white arrows in the figure indicate the spiral line pattern. Because the beam diameter of FIB was 0.9 μm, a trench with a width of approximately 1 μm and an average pitch of 3.75 μm could be processed.

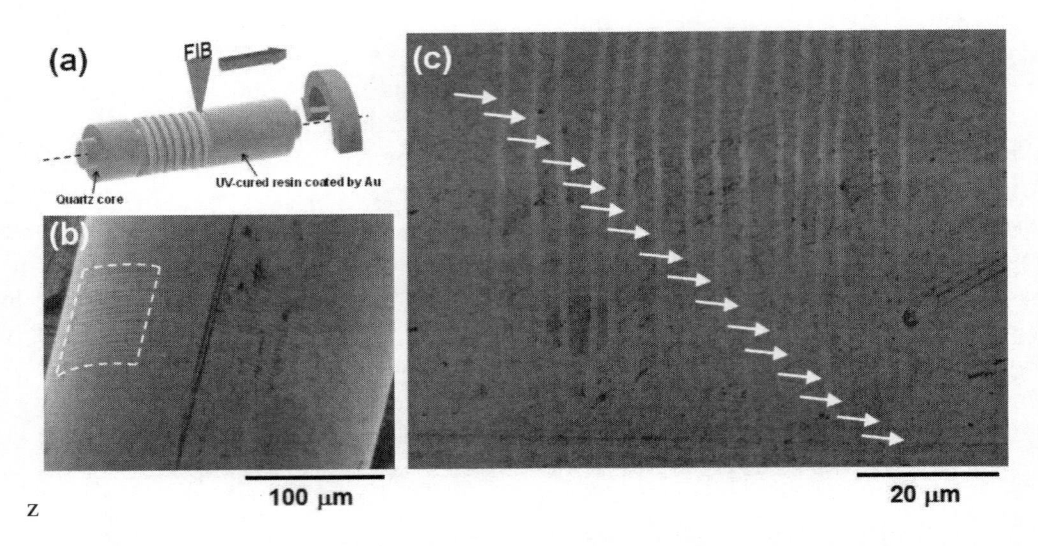

z

Figure 46. (a) Illustration of 3-D FIB etching for patterning spiral line on fiber. (b) FE-SEM image of spiral line on QOF after improvement. (c) The inset is a magnified FE-SEM image of area within dotted square in Figure (b).

To measure the width, depth, and pitch of the etched trenches, the surface of QOF was observed using the confocal microscope. A 3-mm-diameter rotation rod was connected to the last gear in the gear box and to the chuck to fix the QOF via bearings, as shown in Figure 43(c). During axis suppression, the rotation rod was deliberately pushed using an M3 bolt with a sharpened cone tip to suppress the horizontal sliding phenomenon. At the end of the rotation rod, a conical hole was processed to ensure that the rotation axis does not shift its position while being pushed. In the suppression device, a maximum torque of 50 mN·m could be applied by thrusting the M3 bolt. To observe the effect of torque addition, the spiral line pattern etched without added torque was compared with that etched at a suppression torque of 50 mN·m, as shown in Figures 47(a) and 47(b). A cross-sectional profile taken along the dot-dash line in the upper figure is shown in the lower figure. In the absence of suppression torque, individual line patterns were jaggedly drawn with a line-edge roughness of 0.98 μm. The pattern pitch was not uniform and the variation in pitch was also large. Moreover, the etched area width of the spiral pattern in Figure 47(a) spread farther than 60 μm, which was the designed value. However, when the maximum torque was added to the rotation rod, lines were drawn smoothly on the cylindrical surface, and it was confirmed that the line-edge roughness of an individual spiral line improved to 0.37 μm. The peak-to-peak value of the pattern depths between the dotted lines in the cross-sectional profile of Figure 47(b) was measured as 252 nm. Moreover, to numerically compare the phenomenon of reduction of the variation in pattern pitch with torque by axis suppression, the standard deviation σ regarding the pitch of the spiral lines etched on the cylindrical surface of QOF at three torques (0, 10, and 50 mN·m) was estimated by calculation. The relationship between suppression torque and σ is shown in Figure 48. The 10mN·m torque means that the M3 bolt was tightly screwed manually. σ, which was 2.4 μm before adjustment, decreased rapidly to 0.96 μm upon adding a suppression torque of 10mN·m, and improved to 0.81 μm upon adding a suppression torque of 50 mN·m. In summary, the spiral line pattern was processed to the planned precision.

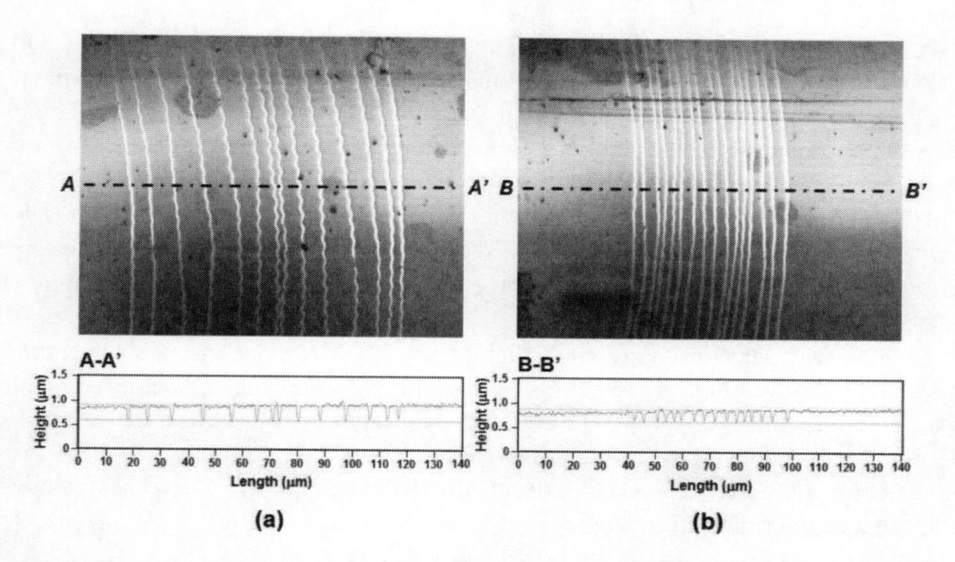

Figure 47. All in-focus images observed by confocal microscope and cross-sectional profiles on white dotted line in upper figure: (a) without and (b) with suppression of torque of 50 mN·m.

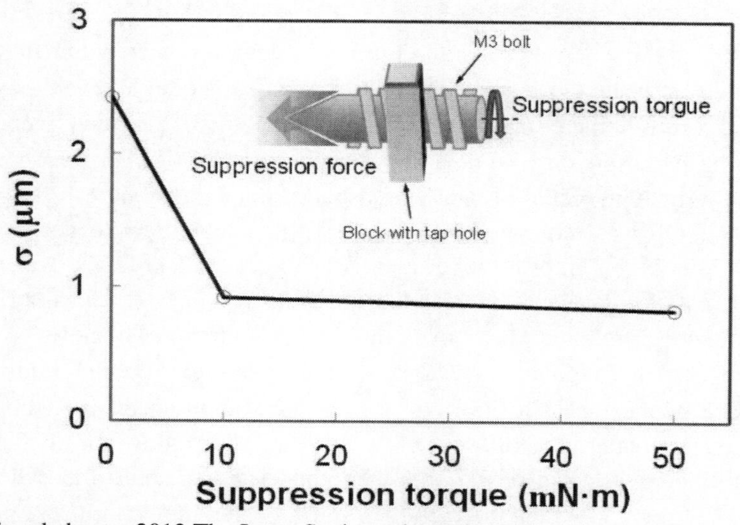

Figure 48. Relationships between suppression torque applied to rotation rod and standard deviation of pattern pitch on cylindrical surface of QOF.

CONCLUSION

Mold Characteristics of Plastic Optical Fiber

We experimentally determined the molding characteristics of POF that provided valuable information when POF was used as a molding material for hot embossing, and for thermal nanoimprinting. We paid special attention to heating temperatures, press depths, and buffer materials with different hardness values. The depths of the imprinted patterns and the deformation of PMMA cores were also investigated under various conditions. We found that the depth of the imprinted pattern increased with the increase in press depth. This trend became more pronounced in case of buffered materials with higher hardness values. From the observation of POFs cut segments obtained from the dicing operation, we learned that the shape of PMMA core varied significantly with the different hardness values of the buffer material. The buffer material, which was softer than the PMMA core, was elastically deformed to ease off any excessive contact force on PMMA core during the imprinting which protected the PMMA core from any damage. On the other hand, when the hardness of the buffer material was same or more than that of the PMMA core, the PMMA core was deformed significantly. Thus, by selecting an appropriate buffer material in reference to its hardness value, the processing depth in POF can be easily controlled from the fluorine clad layer to the PMMA core. We believe that the basic data for the imprinting on POFs is useful in the development of smart fibers and woven MEMS devices, and will become a strong tool for the advancement of our future work.

Surface Patterning of Plastic Optical Fiber

We developed a new-type roller nanoimprinting system to precisely transfer fine patterns from a plane mold onto the curved surface of a fibrous substrate. With this system, 5-μm-width square and 5-μm-diameter circular dotted patterns with 10 μm pitch were successfully imprinted covering the entire surface of a 250-μm-diameter POF. Imprinted conditions such as press force, heating temperature of plane molds, and contact time were 12 N, 50 °C and 0.5 s, respectively. A POF was moved intermittently with a reel-to-reel feeder under the same condition, and a continuous imprint experiment was executed 100 times over a pitch of 16 mm. As a result, a molding rate of 88% or more was achieved. Excellent imprinted patterns were obtained and compared with the convex mold pattern of approximately 3.2 μm in height. In the next step, we investigated a suitable sliding speed for thermal imprinting on the cylindrical surface of the POF. In the final design, the sliding speed was set to 800 μm/s, where the character pattern composed of a diffraction grating on a flat surface of electroformed-Ni molds was transferred onto the cylindrical surface of the POFs by thermal-imprinting. It was observed that the diffraction gratings of the linewidth of 1 μm and the pitch of 2 μm, or the linewidth of 2 μm and the pitch of 4 μm were plainly imprinted on the surface of the POFs. From this demonstration, it was confirmed that the thermal imprinting technology using sliding planar molds was effective for soft patterning on the cylindrical surface of fibrous substrates without damaging them.

Bulk Deformation to Fabricate Weaving Guides

We developed a reel-to-reel thermal imprint system using a cylindrical mold to process weaving guide structures with uniform imprinted depths onto the surface of a fibrous material. A continuous molding with a constant press-force was enabled by building a press-force correction mechanism into the imprint system; and by suppressing the installation error margin of a cylindrical mold, the fluctuation of the press-force decreased to one-11th as compared to the case of without controlling at a sending speed of 1 m/min. In addition, it succeeded in the continuous processing of a rectangle shaped and arc shaped cross-section weaving guide structures onto the surface of a 250-μm-diameter POF using a cylindrical mold of 100 mm in the diameter and 30 mm in width where the Ni–P alloy surface layer had been cut by precision machining. In imprint experiments, POFs were sent at two different speeds of 1 and 5 m/min, and the imprinted depths were then compared. As a result, the variations of the imprinted depths were evaluated in terms of their mean values and $2\sigma_s$, they fell below 35% of the variation of the diameters of the POFs, and exceptional processing accuracy could be proven. From these results, it was confirmed that our imprint system that combined the roller-imprint and the reel-to-reel sending mechanisms using the cylindrical mold was effective as a technique for continuous processing weaving guides onto the surface of the fiber.

Reel-to-Reel Imprinting of Multi-Layered Microstructures on Plastic Optical Fiber

We manufactured the cylindrical mold with hybrid-layered patterns in order to fabricate weaving guides and MEMS structures on a fibrous substrate by a single shot of imprinting to develop a smart fiber. Circular and rectangular cross-sectional shaped convex structures with a width of 260 μm were processed on the cylindrical mold by precise cutting using a forming tool. Then MEMS the structure that was patterned by a 3-D photolithography using the flexible photomask and Cu electroforming was formed as a hybrid-layered pattern. Moreover, in order to make high-speed imprinting possible at a feeding speed of 20 m/min, the control system of press force was changed from a feedback method to a feed-forward method in consideration of the delay time, and then the standard deviation of the press force was reduced to half the value obtained from a conventional method using feed-back controlling. The cylindrical mold was included in the reel-to-reel thermal imprint system after re-modelling, and experiment was carried out using the POF CK-10 as a fibrous substrate for a demonstration.

Then, we succeeded in high-speed imprinting of circular shaped weaving guides, and rectangular shaped grooves which contained MEMS patterns on its bottom surface. These results show that the reel-to-reel imprint technology using the cylindrical mold is quite a useful tool to manufacture a smart fiber.

Thermal Imprinting on Quartz Optical Fiber

We succeeded in imprinting various patterns on 200 μm wide quartz fibers using GC molds fabricated by MEMS technologies. A GC substrate was polished by CMP until the thickness and the average surface roughness became 2 mm±0.5 μm and 5 nm, respectively. Polished GC substrates were processed same way by sputtering, photolithography, and RIE, as done in Si wafers. GC molds with four different pattern depths were fabricated and were set in a thermal nanoimprint system that had developed to imprint quartz. The GC molds were pressed against the quartz fibers in vacuum under prescribed conditions: heating temperature = 1350 °C, press force=300 N, and press time = 2 min. As a result, when the depth of mold patterns was sufficient, the height of imprinted patterns was found to be proportional to the linewidth of mold patterns. However, when the height of imprinted patterns came close to the depth of mold patterns, it reached a saturation point. The width of the imprinted quartz fibers grew from 200 to 296 μm by the press force, but part of the press force was also used up in deforming the quartz fibers. Moreover, the filling rate was inversely proportional to the aspect ratio of mold patterns. We proved experimentally that the molding accuracy is not totally dependent on the pattern size, although it is relatively dependent on the aspect ratio of mold patterns as same as thermoplastics.

3-D Seamless Patterning on Quartz Optical Fiber by Focused-Ion Beam Etching

We succeeded in a seamless patterning with 1-μm-wide features on the surface of a 250-μm-diameter QOF by FIB etching. For this purpose, a simple rotation stage system comprising a DC motor and an LR3 battery was fabricated. In this process, a technical problem resulting in axis blur was encountered prior to the final development in a large-scale remodeling of our FIB etching system. It turned out that the eccentric spinning (or wobbling) of the axis fiber of the system caused by the unrestrained free end of the fiber was the cause of axis blur. The problem was addressed by installing a rotation guide and an axis suppression device onto the rotation stage. Moreover, we successfully used the dc motor and LR3 battery for the plastic model that could be used in vacuum under a pressure of 10^{-5} Pa for a short duration. We plan to develop high-precision FIB processing technology by assembling a rotating mechanism for a high-vacuum chamber of the FIB etching system, as well as various MEMS devices using the base material that may not necessarily be a Si substrate but can be other materials, such as a 3-D shaped optical fiber, in the future.

ACKNOWLEDGMENTS

Technical support in part of the works was provided by Akihiro Ohtomo of Toshiba Machine Co., Ltd., Akihisa Ueno of Nano Craft Technologies Co., Chieko Okuyama of the National Institute of Advanced Industrial Science and Technology (AIST), and Takayuki Yano of the Institute for Molecular Science (IMS), National Institutes of Natural Sciences. The author thanks them for their cooperation and advice.

The development of thermal-imprint technologies on optical fibers was supported in part by the Bio-Electromechanical-Autonomous-Nano-Systems (BEANS) Project of the New Energy and Industrial Technology Development Organization (NEDO), Japan.

The work concerning FIB etching on QOF was supported by the Joint Studies Program (2009 and 2010) of the Institute for Molecular Science (IMS), National Institutes of Natural Sciences.

REFERENCES

Ahn S., Cha J., Myung H., Kim S., Kang S., (2006). Continuous ultraviolet roll nanoimprinting process for replicating large-scale nano- and micropatterns. *Appl. Phys. Lett.* 89, 213101.

Ahn S.H., Guo L. J., (2008). High-speed roll-to-roll nanoimprint lithography on flexible plastic substrates. *Adv. Mater.* 20, 2044-2049.

Bonderover E., Wagner S., (2004). A woven inverter circuit for e-textile applications. IEEE *Electron Device Lett.* 25, 295–297.

Buechley L., Eisenberg M., (2009). Fabric PCBs, electronic sequins, and socket buttons: Techniques for e-textile craft. *Pers. Ubiquit. Comput*. 13, 133–150.

Chand T., Lai W., Luo S., Chen C., Yang H., Chou T., Tsai J., (2009). Electron-beam lithography for roller mold in nanoimprint process. *Proceedings of the 2nd Asian Symposium. Nanoimprint Lithography (ASNIL 2009), Taipei, Taiwan*, 32–133.

di Benedetto F., Camposeo A., Pagliara S., Mele E., Persano L., Stabile R., Cingolani R., Pisignano D., (2008). Patterning of light-emitting conjugated polymer nanofibers. *Nature Nanotech*. 3, 614–619.

Engin M., Demirel A., Engin E.Z., Fedakar M., (2005). Recent developments and trends in biomedical sensors. *Measurement* 37, 173–188.

Gauthier M.M., (1995). Engineered Materials Handbook. *ASM International Handbook Committee, Metals Park, USA*.

Grattan L.S., Meggitt B.T., (2003). Optical fiber sensor technology: Advanced applications—Bragg gratings and distributed sensors. *Kluwer Academic Publishers, Dordrecht, Netherlands*.

Hamedi M., Forchheimer R., Inganas O., (2007). Toward woven logic from organic electronic fibers. *Nature Mater.* 6, 357–362.

Heaney P.J., Vicenzi E.P., Giannuzzi L.A., Livi K.J.T, (2011). Focused ion beam milling: A method of site-specific sample extraction for microanalysis of Earth and planetary materials. *Am. Mineral*. 86. 1094-1099.

Heckele M., Schomburg W.K.. (2003). Review on micro molding of thermoplastic polymers. *J. Micromech. Microeng.* 14, R1–R14.

Hirai Y., (2010). Process physics in thermal nanoimprint. *J. Jpn. Soc. for Prec. Eng.* 76, 143–147.

Hirai Y., Konishi T., Yoshikawa T., Yoshida S., (2004). Simulation and experimental study of polymer deformation in nanoimprint lithography. *J. Vac. Sci. Technol.* B 22, 3288-3293.

Hirai Y., Yoshikawa T., Takagi N., Yoshida S., Yamamoto K., (2003). Mechanical properties of poly-methyl methacrylate (PMMA) for nano imprint lithography. *J. Photopolym. Sci. Technol.* 16, 615–620.

Huang C., Shen C., Tang C., Chang S., (2008). A wearable yarn-based piezo-resistive sensor. *Sens. Actuators* A 141, 396–403.

Huang T., Wu J., Yang S., Huang P., Chang S., (2009). Direct fabrication of microstructures on metal roller using stepped rotating lithography and electroless nickel plating. *Microelectron. Eng.* 86, 615-618.

Ishitani T., Yaguchi T., (1996). Cross-sectional sample preparation by focused ion beam: A review of ion-sample interaction. Microsc. Res. Technol. 35, 320-333.

Kaajakari, V., (2009). Practical MEMS: Design of Microsystems, Accelerometers, Gyroscopes, RF MEMS, Optical MEMS, and Microfluidic Systems. *Small Gear Publishing, Las Vegas, USA*.

Katoh T., Nishi N., Fukagawa M., Ueno H., Sugiyama S., (2001). Direct writing for three-dimensional microfabrication using synchrotron radiation etching. *Sens. Actuators* A 89, 10–15.

Keiser G., (2003). Optical fiber communications. *Wiley, Hoboken, USA*.

Kikuchi T., Takahashi H., Maruko T., (2007). Fabrication of three-dimensional platinum microstructures with laser irradiation and electrochemical technique. *Electrochim. Acta* 52, 2352–2358.

Lee D., Hiroshi H., Zhang Y., Itoh T., Maeda R., (2010). Cylindrical projection lithography for microcoil structures. *Microelectron. Eng.* 88, 2625–2628.

Lee J.B., Subramanian V., (2005). Weave patterned organic transistors on fiber for e-textiles. *IEEE Trans. Electron Dev.* 52, 269–275.

Li W.J., Mai J.D., Ho C., (1999). Sensors and actuators on non-planar substrates. *Sens. Actuators* A 73, 80–88.

Lu Y., Zhang Y., Lu J., Mimura A., Matsumoto S., Itoh T., (2010). Three-dimensional photolithography technology for a fiber substrate using a microfabricated exposure module. *J. Micromech. Microeng.* 20, 125013.

Marculescu D., Marculescu R., Zamora N.H., Stanley-Marbell P., Khosla P.K., Park S., Jayaraman S., Jung S., Lauterbach C., Weber W., Kirstein T., Cottet D., Grzyb J., Troster G., Jones M., Martin T., Nakad Z., (1995). Electronic textiles: A platform for pervasive computing. *Proc. IEEE.* 91, 1995-2018.

Matsui S., Kaito T., Fujita J., Komuro M., Kanda K., Haruyama Y., (2000). Three-dimensional nanostructure fabrication by focused-ion-beam chemical vapor deposition. *J. Vac. Sci. Technol.* B 18, 3181-3184.

Matsui S., Ochiai Y., (1996). Focused ion beam applications to solid state devices. Nanotechnology 7, 247-258.

Maury P., Turkenburg D., Stroeks N., Giesen P., Barbu I., Meinders E., van Bremen A., Iosad N., van der Werf R., Onvlee H., (2011). Roll-to-roll UV imprint lithography for flexible electronics. Microelectron. Eng. 88, (2011) 2052-2055.

Mazurin O.V., Streltsina M.V., Shvaiko-Shvaikovskaya T.P., (1993). Handbook of Glass Data. *Elsevier, Amsterdam, Netherlands*.

Meingailis J. (1987). Focused ion beam technology and applications. *J. Vac. Sci. Technol.* B 5, 469-495.

Mekaru H., Fukushima E., Hiyama Y., Takahashi M., (2009a). Thermal roller imprint on surface on Teflon perfluoroalkoxy inlet tube. *J. Vac. Sci. Technol.* B 27, 2814-2819.

Mekaru H., Kitadani T., Yamashita M., Takahashi M., (2007). Glass nanoimprint using amorphous Ni-P mold etched by focused-ion beam. *J. Vac. Sci. Technol.* A 25, 1025-1028.

Mekaru H., Koizumi O., Ueno A., Takahashi M., (2010). Guide structure with pole arrays imprinted on nylon fiber. *Microelectron. Eng.* 87, 922–926.

Mekaru H., Kusumi S., Sato N., Yamashita M., Shimada O., Hattori T., (2004). Fabrication of mold master for spiral microcoil utilizing x-ray lithography of synchrotron radiation. *Jpn. J. Appl. Phys.* 43, 4036–4040.

Mekaru H., Ohtomo A., Takagi H., (2013a). Effect of buffer materials on thermal imprint on plastic optical fiber. *Microsyst. Technol.* 19, 325-333.

Mekaru H., Ohtomo A., Takagi H., Kokubo M., Goto H., (2011a). Development of reel-to-reel process system for roller-imprint on plastic fibers. *Microelectron. Eng.* 88, 2059-2062.

Mekaru H., Ohtomo A., Takagi H., Kokubo M., Goto H., (2012). Reel-to-reel imprint system to form weaving guides on fibers. *Microelectron. Eng.* 98, 171-175.

Mekaru H., Ohtomo A., Takagi H., Kokubo M., Goto H., (2013b). High-speed imprinting on plastic optical fibers using cylindrical mold with hybrid microstructures. *Microelectron. Eng. in press.*

Mekaru H., Okuyama C., Tsuchida T., Yasui M., Kitadani T., Yamashita M., Uegaki, J., Takahashi M., (2009b). Development of Ni-P-plated Inconel alloy mold for imprinting on Pyrex glass. *Jpn. J. Appl. Phys.* 48, 06FH06.

Mekaru H., Okuyama C., Ueno A., Takahashi M., (2009c). Thermal imprinting on quartz fiber using glasslike carbon mold. *J. Vac. Sci. Technol.* B 27, 2820-2825.

Mekaru H., Takagi H., Ohtomo A., Kokubo M., Goto H., (2011b). Soft patterning on cylindrical surface of plastic optical fiber. *J. Vac. Sci. Technol.* B 29, 06FC07.

Mekaru H., Takahashi M., (2010). Imprinting of guide structure to weave nylon fibers. *J. Vac. Sci. Technol.* A 28, 706–712.

Mekaru H., Tsuchida T., Uegaki J., Yasui M., Yamashita M., Takahashi M. (2008). Micro lens imprinted on Pyrex glass by using amorphous Ni-P alloy mold. *Microelectron. Eng.* 85, 873-876.

Mekaru H., Yano T., (2012). Patterning of spiral structure on optical fiber by focused-ion-beam etching. *Jpn. J. Appl. Phys.* 51, 06FB01.

Mineta T., Deguchi T., Makino E., Kawashima T., Shibata T., (2011). Fabrication of cylindrical micro actuator by etching of TiNiCu shape memory alloy tube. *Sens. Actuators* A 165, 392–398.

Morimoto H., Sasaki Y., Saitoh K., Watakabe Y., Kato T., (1986). Focused ion beam lithography and its application to submicron. *Microelectron. Eng.* 4, 163-179.

Motoji T., (2008) Technological trends of fluorine release agents for precise molds. *Tribology* 249, 54-56 [in Japanese].

National Astronomical Observatory of Japan, (2011). Rika Nenpyo, *Maruzen, Tokyo, Japan* [in Japanese].

Nugent C.D., McCullagh P.J., McAdams E.T., Lymberis A., (2003). Personalised Health Management Systems: The Integration of Innovative Sensing, Textile, Information and Communication Technologies, *IOS Press, Amsterdam, Netherlands.*

Ohtomo A., Kokubo M., Goto H., Mekaru H., Takagi H., (2012). Fast and continuous patterning on the surface of plastic fiber by using thermal roller imprint. *J. Vac. Sci. Technol.* B 29, 06FB01.

Ohtomo A., Mekaru H., Takagi H., Kokubo M., Goto H., (2011). A sliding-press-type reel-to-reel thermal imprint system for fiber substrates. *IEEJ Trans. on Sensors and Micromachines* 131, 240-245 [in Japanese].

Ohtomo A., Mekaru H., Takagi H., Kokubo M., Goto H., (2012). Continuous forming of weaving guides using a reel-to-reel thermal roller imprint. *IEEJ Trans. on Sensors and Micromachines* 132, 37-41 [in Japanese].

Reyntjens S., Puers R., (2001). A review of focused ion beam applications in microsystem technology. *J. Micromech. Microeng.* 11, 287-300.

Schift H., Halbeisen M., Schütz U., Delahoche B., Vogelsang K., Gobrecht J., (2006). Surface strucuring of textile fibers using roll embossing. *Microelectron. Eng.* 83, 855-858.

Siewert C., Lo¨ffler F., (2005). Photolithographic structuring of a thin metal film coil on a Zerodur cylinder. *Surf. Coat. Technol.* 200, 1061–1064.

Soper S.A., Ford S.M., Qi S., McCarley R.L., Kelly K., Murphy M.C., (2000). Polymeric microelectromechanical systems. *Anal. Chem.* 72, 642A–651A.

Sotomayor-Torres C.M., (2003). Alternative Lithography, Unleashing the potentials of nanotechnology, nanostructure science and technology. *Springer, New York. USA.*

Takahashi M., Goto H., Maeda R., Maruyama O., (2006a). Desktop nanoimprint system—prototype and performance. *Proc. Japan Soc. for Preci. Eng. Conf., Chiba, Japan*, 737–738 [in Japanese].

Takahashi M., Maeda R., (2006). Large-area micro-hot embossing of glass materials with glassy carbon mold machined by dicing. *J. Jpn. Soc. Technol. Plast.* 47, 963-667.

Takahashi M., Maeda R., Sugimoto K., (2006b). Micro/nano-hot embossing of quartz glass materials with glassy carbon mold prepared by focused ion beam. *J. Jpn. Soc. Technol. Plast.* 47, 958-962.

Takahashi M., Sugimoto K., Maeda R., (2005). Nanoimprint of glass materials with glassy carbon molds fabricated by focused-ion-beam etching. *Jpn. J. Appl. Phys.* 44, 5600-5605.

Tan H., Gilbertson A., Chou S.Y., (1998). Roller nanoimprint lithography. *J. Vac. Sci. Technol.* B 16, 3926-3928.

Taniguchi J., Asatani M., (2009). Fabrication of a seamless roll mold by direct writing with an electron beam on a rotating cylindrical substrate. *J. Vac. Sci. Technol.* B 27, 2841–2845.

The Japan Society of Mechanical Engineers, (1987). A. Kisohen and B. Ohyohen: Kikai Kogaku Binran (JSME Mechanical Engineer's Handbook), *Maruzen, Tokyo, Japan* [in Japanese].

The Society of Polymer Science, Japan, (1986). B. Ohyohen: Kobunshi Data Handbook, *Baifukan, Tokyo, Japan* [in Japanese].

Ting C., Chang F., Chen C., Chou C.P., (2008). Fabrication of an antireflective polymer optical film with subwavelength structures using a roll-to-roll micro-replication process. *J. Micromech. Microeng.* 18, 075001.

Tseng A.A., (2004). Recent developments in micromilling using focused ion beam technology. *J. Micromech. Microeng.* 14, R15-R34.

Vig A.L., Mäkelä T., Majander P., Lambertini V., Ahopelto J., Kristensen A., (2011). Roll-to-roll fabricated lab-on-a-chip devices. *J. Micromech. Microeng.* 21, 035006.

Wagner S., Bonderover E., Jordan W.B., Sturm J.C., (2002). Electrotextiles: Concepts and challenges. *Int. J. High Speed Electron. Syst.* 12, 391–399.

Watts M.P.C., (2008). Advanced in roll to roll processing of optics. *Proc. SPIE* 6883, 688305.

Wiley B.J., Qin D., Xia Y., (2010). Nanofabrication at high throughput and low coast. *ACS Nano* 4, 3554-3559.

Wirth R., (2004). Focused ion beam (FIB): A novel technology for advanced application of micro- and nanoanalysis in geosciences and applied mineralogy. *Eur. J. Mineral.* 16, 863-876.

Worgull M., (2009). Hot embossing. Theory and technology of microreplication, micro and nano technologies. *William Andrew, Oxford, UK.*

Wu C., Lin J., Fang T., (2009). Molecular Dynamics simulations of the roller nanoimprint process: Adhesion and other mechanical characteristics. *Nanoscale Res. Lett.* 4, 913-920.

Yeo L.P., Ng S.H., Wang Z., Wang Z., de Rooij N.F., (2009). Micro-fabrication of polymeric devices using hot roller embossing. *Microelectron. Eng.* 86, 933-936.

In: Optical Fibers: New Developments
Editor: Marco Pisco

ISBN: 978-1-62808-425-2

Chapter 5

DYNAMICS OF BRIGHT SOLITONS AND THEIR COLLISIONS FOR THE INHOMOGENEOUS COUPLED NONLINEAR SCHRÖDINGER-MAXWELL-BLOCH EQUATIONS

***Camus Gaston Latchio Tiofack*[1*], *Alidou Mohamadou*[2,3], *Timoléon Crépin Kofané*[1,3] and *Kuppuswamy Porsezian*[4]**

[1] Laboratory of Mechanics, Department of Physics, Faculty of Science, University of Yaounde I, Yaounde, Cameroon

[2] Condensed Matter Laboratory, Department of Physics, Faculty of Science, University of Douala, Douala, Cameroon

[3] The Abdus Salam,International Center For Theoretical Physics, Strada Costiera, Trieste, Italy

[4] Department of Physics, Pondicherry University, Pudhucherry, India

ABSTRACT

We consider the coupled nonlinear Schrödinger and Maxwell-Bloch equations with variable dispersion and nonlinearity management functions. These equations describe the propagation of an optical soliton in an inhomogeneous nonlinear waveguide doped with two-level resonant atoms. Based on the linear eigenvalue problem, the complete integrability of such nonlinear equations is identified by admitting an infinite number of conservation laws. The first three constants of the motions from the obtained conservation laws, namely, energy, momentum and

Hamiltonian are presented. Using the Darboux transformation method, we obtain some explicit bright multi-soliton solutions in a recursive manner. As an example, the one- and two-soliton solutions in explicit forms are generated. Then, we explain the soliton control by the effect of distributed amplification with varying group velocity dispersion, nonlinearity and gain/loss functions. The propagation characteristics of the solitons and their collision dynamics are discussed via analytic solutions and graphical

[*] E-mail: glatchio@yahoo.fr.

illustration. It is shown that the self-induced transparency induces an overall phase shift of the soliton during the propagation.

1. INTRODUCTION

Communications with the aid of solitons have renowned interest and deserve the attention because of their potential applications in high-bit-rate signal transmission over trans-oceanic distances [1-6]. In recent years, long haul optical communication through fibers has attracted considerable research activities among scientists all over the world. Especially, soliton-type pulse propagation plays a vital role in these long haul communication systems. They are considered to be the futuristic tools in achieving low-loss, cost-effective high speed communication throughout the world. Soliton-type pulse propagation through nonlinear optical fibers is realized by means of the exact counterbalance between the major constraints of the fiber, i.e, group velocity dispersion (linear effect) which broadens the pulse and the self-phase modulation (nonlinear effect) which contracts the pulse. The propagation of optical pulses through a nonlinear fiber in the picosecond regime is described by the well-known nonlinear Schrödinger (NLS) equation, which was first proposed by Hasegawa and Tappert in 1973. Considering the coupling between copropagating fiber modes through nonlinearities and nonuniformities of different types and the interactions between two waves of different frequencies/polarizations, the coupled NLS (CNLS) systems have been proposed to support stable optical solitons [7-10]. The CNLS systems are also used in the theory of soliton wavelength division multiplexing [11], multi-channel bit parallel-wavelength optical fiber network [12] and so on. In recent years, there has been a growing interest in investigating the vector solitons governed by the CNLS systems which can be used as carriers of the switched information, when they collide with other vector solitons [13-15]. Furthermore, they can retain their shapes and energies invariant during the propagation process. However, communication grade optical fibers or as a matter of fact, any optical transmitting medium does posses finite attenuation coefficient, thus optical loss is inevitable and the pulse is often deteriorated by this loss. To make the soliton based communication systems highly competitive, reliable and economical when compared to the conventional linear systems, attenuation in a fiber must be avoided. Therefore, optical amplifiers have to be employed to compensate for this loss. The erbium-doped fiber amplifiers are widely use for this purpose. Erbium is selected because the energy difference between the two levels is nearly equal to that of the frequency at which present day optical signals are transmitted. In 1967, McCall and Hahn [16] described a special type of lossless pulse propagation in two-level resonant media. They showed that if the energy difference between the two levels of media coincides with the frequency of the optical signal, then coherent absorption and re-emission of light takes place and the medium becomes optically transparent to that particular wavelength; this is called self-induced transparency (SIT).

In erbium-doped fibers, the resultant solitons are collectively called nonlinear-Schrödinger-Maxwell-Bloch (NLS-MB) solitons. This type of soliton pulse propagation was theoretically shown for the first time by Mamitsov and Manykin [17], and experimentally demonstrated by Nakazawa et al. [18]. In [19], the possibility of coexistence of NLS and SIT solitons with higher order linear and nonlinear effects has been investigated. However, the above model is based on ideal system. Among the most important models of modern nonlinear sciences are the variable-coefficient Schrödinger-typed ones, which describe such situations more realistically than their constant-coefficient counterparts, in plasma physics, arterial mechanics and long-distance optical communications [20-25]. The problem of nonlinear wave propagation in inhomogeneous media has been found to be of great interest, which has a wide range of applications.

The inhomogeneity in the fiber mainly arises due to two factors: (i) the variation in the lattice parameters of the fiber medium and (ii) variation of the fiber geometry. Of late, the effect of these inhomogeneities on the propagation of solitary wave pulses in an optical fiber has produced considerable activity among researchers. In particular, for the theoreticians, the question is to analyze the way in which the behavior of solitons is affected and to find out whether these inhomogeneous systems are still integrable like their homogeneous counterparts.

Although the NLS-MB system has been widely investigated, we find that there has not been much discussion about the conservation laws and the collision between two solitons of the generalized inhomogeneous CNLS-MB equations. Being motivated by the above aspects, in the present work, we make a detailed study of the integrability aspects and relevant soliton structures of the inhomogeneous CNLS-MB system. Particularly, the generation and stability of multisolitary pulses in a system of inhomogeneous CNLS-MB equations by means of the simple, straightforward Darboux transformation method based on the Lax pair is obtained. These solutions include the one- and two-soliton solutions. The paper is organized as follows. In section 2, the generalized inhomogeneous coupled nonlinear Schrodinger-Maxwell-Bloch system will be proposed.

In section 3, the Lax pair and conservation laws of the above model will be obtained. In section 4, we will construct the Darboux transformation of the generalized inhomogeneous coupled nonlinear Schrodinger-Maxwell-Bloch system. In addition, we will present some analytical exact solutions of the system understudy in section 5. Finally, the conclusions will be presented in section 6.

2. The Generalized Inhomogeneous Coupled Nonlinear Schrödinger-Maxwell- Bloch Model

The solution of NLS and CNLS equations in an inhomogeneous medium is of great importance for investigating wave propagation in various types of physical situations such as plasma physics, nonlinear optics, condensed matter, and so on. Serkin et al. [26] introduced the inhomogeneous NLS equation and obtained the one- and two-soliton solution through the Lax pair technique [27].

Very recently, we have extended the study in the inhomogeneous CNLS equation [28]. In this work, we have modified the inhomogeneous CNLS equation to suit with the pulse propagation in erbium-doped fibers, wherein the effect of SIT should be included and the governing equation is now called the generalized inhomogeneous CNLS-MB equation of the following forms:

$$\begin{aligned}
&i\frac{\partial q_1}{\partial z}-\frac{D(z)}{2}\frac{\partial^2 q_1}{\partial t^2}+R(z)\left[|q_1|^2+B|q_2|^2\right]q_1+i\frac{\Gamma(z)}{2}q_1+\langle P_1\rangle=0,\\
&i\frac{\partial q_2}{\partial z}-\frac{D(z)}{2}\frac{\partial^2 q_2}{\partial t^2}+R(z)\left[B|q_1|^2+|q_2|^2\right]q_1+i\frac{\Gamma(z)}{2}q_2+\langle P_2\rangle=0,\\
&\frac{\partial P_1}{\partial t}-2i\omega P_1=f(z)\left(-Nq_1+M_{11}q_1+M_{21}q_2\right),\\
&\frac{\partial P_2}{\partial t}-2i\omega P_2=f(z)\left(-Nq_2+M_{12}q_1+M_{22}q_2\right),\\
&\frac{\partial M_{11}}{\partial t}=-f(z)\left(q_1P_1^*+q_1^*P_1\right),\\
&\frac{\partial M_{12}}{\partial t}=-f(z)\left(q_1^*P_2+q_2P_1^*\right),\\
&\frac{\partial M_{21}}{\partial t}=-f(z)\left(q_1P_2^*+q_2^*P_1\right),\\
&\frac{\partial M_{22}}{\partial t}=-f(z)\left(q_2P_2^*+q_2^*P_2\right),\\
&\frac{\partial N}{\partial t}=-f(z)\left(-q_1P_1^*+q_1^*P-q_2P_2^*+q_2^*P_2\right),
\end{aligned} \tag{2.1}$$

where q_1 and q_2 are the slowly varying amplitudes of the signal, $P_j(z,t)$ is the measure of the polarization of the resonant medium, $M_{i,j}(z,t)\,(i,j=1,2)$ denotes the extent of population inversion, $D(z)$ represents the GVD, $R(z)$ is the nonlinearity parameter, $\Gamma(z)$ corresponds to gain or loss and $f(z)$ is a parameter describing the interaction between the propagating field and the energy levels of the erbium atoms, $*$ represents the complex conjugate, $\langle ...\rangle$ represents the averaging function over the entire frequency range. For example,

$$\langle P(z,t)\rangle=\int_{-\infty}^{+\infty}P(z,t;\omega)d\omega, \tag{2.2}$$

such that

$$\int_{-\infty}^{+\infty}g(\omega)d\omega=1, \tag{2.3}$$

$g(\omega)$ being the distribution function which represents the uncertainty in the energy level of the resonant atoms. The cross-phase modulation coefficient B is a function of ellipticity angle θ, and is given by [29]

$$B=\frac{2+2\sin^2(\theta)}{2+\cos^2(\theta)}. \tag{2.4}$$

This give $B=2/3$ for linear birefringent fiber $\theta=0$; $B=2$ for circular birefringent fiber $\theta=\pi/2$, and $B=1$ for the ideal birefringence case $\theta\approx 35,3^{\circ}$. The latter case used in the rest of the paper is of particular interest because Eq. (2.1), with $B=1$ correspond to the Manakov equation, which can be solved by the Darboux transformation method. Eq. (1) also describes the amplification or absorption of pulses propagating in a coupled optical fibers with distributed dispersion and nonlinearity. In practical applications, the model is of primary interest not only for the amplification and compression of optical solitons in inhomogeneous systems, but also for the stable transmission of dispersion and nonlinear management soliton.

3. Lax Pair and Infinite Conservation Laws

To ensure the complete integrability of a nonlinear system, the Lax pair plays a vital role in studying the integrable properties of nonlinear evolution equations [30]. By employing the Ablowitz-Kaup-Newell-Segur (AKNS) technique, one can construct the linear eigenvalue problem for Eq. (2.1) as follows:

$$\phi_t = U\phi = (\lambda U_0 + U_1)\phi, \quad \phi_z = V\phi = \left(\lambda^2 V_0 + \lambda V_1 + \frac{1}{2}V_2 + V_{-1}\right)\phi, \tag{3.1}$$

where $\phi=(\phi_1,\phi_2,\phi_3)^T$, T denotes the transpose of the matrix and λ is a isospectral parameter, while matrices U_0, U_1, V_0, V_1, V_2, and V_{-1} are presented in the form

$$U_0 = \begin{pmatrix} -i & 0 & 0 \\ 0 & i & 0 \\ 0 & 0 & i \end{pmatrix}, \quad U_1 = \sqrt{\frac{-R(z)}{D(z)}} \begin{pmatrix} 0 & q_1 & q_2 \\ -q_1^* & 0 & 0 \\ -q_2^* & 0 & 0 \end{pmatrix}, \tag{3.2}$$

$$V_0 = D(z)U_0, \quad V_1 = D(z)U_1, \tag{3.3}$$

$$V_2 = \begin{pmatrix} iR(z)\left(|q_1|^2+|q_2|^2\right) & -i\sqrt{-D(z)R(z)}q_{1t} & -i\sqrt{-D(z)R(z)}q_{2t} \\ -i\sqrt{-D(z)R(z)}q_{1t}^* & -iR(z)|q_1|^2 & -iR(z)q_1^*q_2 \\ -i\sqrt{-D(z)R(z)}q_{2t}^* & -iR(z)q_1q_2^* & -iR(z)|q_2|^2 \end{pmatrix}, \tag{3.4}$$

$$V_{-1} = \begin{pmatrix} N & P_1 & P_2 \\ P_1^* & M_{11} & M_{12} \\ P_2^* & M_{21} & M_{22} \end{pmatrix}. \tag{3.5}$$

It is easy to prove that the compatibility condition $U_z - V_t + [U,V] = 0$, gives rise to Eq. (2.1) under the following restriction condition [31, 32]:

$$\Gamma(z)=\frac{1}{2}\frac{W\left[R(z),D(z)\right]}{R(z)D(z)},\quad f(z)=\left(\frac{R(z)}{D(z)}\right)^{\frac{1}{2}}, \tag{3.6}$$

where $W\left[R(z),D(z)\right]=RD_z-DR_z$.

According to [33], we know that an infinite number of conservation laws have a close relationship to the Lax pair. We will prove the existence of infinitely many independent conservation laws as further support of the integrability for equation (2.1). Introducing two new variables

$$\Gamma_1=\frac{\phi_2}{\phi_1},\quad \Gamma_2=\frac{\phi_3}{\phi_1}, \tag{3.7}$$

and taking the derivative of $\Gamma_j\quad (j=1,2)$ with respect to t by the use of equation (3.1) gives rise to the following two Riccati-type equations:

$$\Gamma_{1t}=-f(z)q_1^*+2i\lambda\Gamma_1-f(z)q_1\Gamma_1^2-f(z)q_2\Gamma_1\Gamma_2. \tag{3.8}$$

$$\Gamma_{2t}=-f(z)q_2^*+2i\lambda\Gamma_2-f(z)q_2\Gamma_2^2-f(z)q_1\Gamma_1\Gamma_2. \tag{3.9}$$

Multiplying equations (3.8) and (3.9) respectively by q_1 and q_2 and expanding $q_1\Gamma_1$ and $q_2\Gamma_2$ in power series of $1/\lambda$,

$$q_1\Gamma_1=\sum_{m=1}^{\infty}\lambda^{-m}\Gamma_{1m}(z,t),\quad q_2\Gamma_2=\sum_{m=1}^{\infty}\lambda^{-m}\Gamma_{2m}(z,t), \tag{3.10}$$

the recursion formulae for Γ_{1m} and Γ_{2m} $(m=1,2,...)$ can be determined as

$$\Gamma_{11}=-\frac{i}{2}f(z)|q_1|^2,\quad \Gamma_{21}=-\frac{i}{2}f(z)|q_2|^2, \tag{3.11}$$

$$\Gamma_{12}=-\frac{1}{4}f(z)q_1q_{1t}^*,\quad \Gamma_{22}=-\frac{1}{4}f(z)q_2q_{2t}^*, \tag{3.12}$$

$$\begin{aligned}\Gamma_{1m+1}&=-\frac{i}{2}\left(f(z)\sum_{k=1}^{m-1}\Gamma_{1m-k}\Gamma_{1k}+f(z)\sum_{k=1}^{m-1}\Gamma_{1m-k}\Gamma_{2k}+\left(\frac{\Gamma_{1k}}{q_1}\right)_t q_1\right),\\ \Gamma_{2m+1}&=-\frac{i}{2}\left(f(z)\sum_{k=1}^{m-1}\Gamma_{2m-k}\Gamma_{2k}+f(z)\sum_{k=1}^{m-1}\Gamma_{1m-k}\Gamma_{2k}+\left(\frac{\Gamma_{2k}}{q_2}\right)_t q_2\right),\quad m>2.\end{aligned} \tag{3.13}$$

By virtue of the compatibility condition $\left(\ln\phi_1\right)_{zt}=\left(\ln\phi_1\right)_{tz}$, we obtain the following conservation form

$$i\frac{\partial}{\partial t}\rho_k(z,t)+\frac{\partial}{\partial z}J_k(z,t)=0, \tag{3.14}$$

where $\rho_k(z,t)$ and $J_k(z,t)$ $(k=1,2,...)$ are called conserved densities and conserved fluxes, respectively. The first three significant physical conservation laws are presented as

$$\begin{aligned}
&\rho_1(z,t)=f(z)\left(|q_1|^2+|q_2|^2\right),\quad \rho_2(z,t)=f(z)\left(q_1q_{1t}^*+q_2q_{2t}^*\right),\\
&\rho_3(z,t)=f(z)\left(q_1q_{1tt}^*+q_2q_{2tt}^*+f^2(z)\left(|q_1|^4+2|q_1|^2|q_2|^2+|q_2|^4\right)\right),\\
&J_1(z,t)=\frac{D(z)}{2}\left(q_1q_{1t}^*-q_{1t}q_1^*+q_2q_{2t}^*-q_{2t}q_2^*\right)+N,\\
&J_2(z,t)=\frac{D(z)}{2}\left(q_{1t}q_{1t}^*-q_1q_{1tt}^*+q_{2t}q_{2t}^*-q_2q_{2tt}^*\right)+\frac{1}{2}R(z)\left(|q_1|^2+|q_2|^2\right)^2+\left(p_1q_1^*+p_2q_2^*\right)-2i\omega N,\\
&J_3(z,t)=\frac{D(z)}{2}\left(q_{1t}q_{1tt}^*-q_1q_{1ttt}^*+q_{2t}q_{2tt}^*-q_2q_{2ttt}^*\right)+R(z)\left(\left(|q_1|^2+|q_2|^2\right)_t\left(|q_1|^2+|q_2|^2\right)+\left(|q_1|^2+|q_2|^2\right)\left(q_1q_{1t}^*-q_{1t}q_1^*+q_2q_{2t}^*-q_{2t}q_2^*\right)\right)+\\
&\left(p_1q_{1t}^*+p_2q_{2t}^*\right)-2i\omega\left(p_1q_1^*+p_2q_2^*\right)-4\omega^2N.
\end{aligned} \tag{3.15}$$

Using the vanishing boundary condition, we can gain the first three constants of the motions from the obtained conservation laws,

$$\begin{aligned}
&H_1=\int_{-\infty}^{+\infty}\left(|q_1|^2+|q_2|^2\right)dt,\quad H_2=\int_{-\infty}^{+\infty}\left(q_1q_{1t}^*+q_2q_{2t}^*\right)dt,\\
&H_3=\int_{-\infty}^{+\infty}\left(q_1q_{1tt}^*+q_2q_{2tt}^*+f^2(z)\left(|q_1|^4+2|q_1|^2|q_2|^2+|q_2|^4\right)\right)dt,
\end{aligned} \tag{3.16}$$

where H_1, H_2, and H_3 may be related to the energy, momentum and Hamiltonian for System (2.1).

4. Darboux Transformation Method

In order to reveal the analytic soliton-like solutions of Eq. (2.1) under constraints (3.6), we will employ the Darboux transformation method, which is an effective and computerizable procedure and has been widely used to construct soliton-like solutions for a class of variable coefficient nonlinear evolution equations [31, 32]. The Darboux transformation can give rises to a general procedure in order to recursively generate a series of analytic solutions including multisoliton solutions from a seed solution [34]. On the basis of the 3×3 Lax pair and with constraints (3.6), we construct the Darboux transformation for Eq. (2.1) in the following form [35]

$$\phi'=(\lambda I+S)\phi \tag{4.1}$$

where I is the 3×3 identity matrix, S is a nonsingular matrix and its entries $S_{ij}\,(1\le i,j\le 3)$ are all parameters to be determined. It requires that ϕ' should also satisfy the linear eigenvalue problem (3.1), i.e.

$$\phi_t' = \left(\lambda U_0' + U_1'\right)\phi, \quad \phi_z' = \left(\lambda^2 V_0' + \lambda V_1' + \frac{1}{2}V_2' + V_{-1}'\right)\phi, \tag{4.2}$$

where $U_0', U_1', V_0', V_1', V_2'$, and V_{-1}' have the same form as U_0, U_1, V_0, V_1, V_2 and V_{-1} except that $q_1(z,t)$ is replaced by $q_1'(z,t)$, $q_2(z,t)$ by $q_2'(z,t)$, and should also satisfy the following set of equations:

$$U_0' = U_0, \quad V_0' = V_0, \tag{4.3}$$

$$U_1' - U_1 + U_0 S - S U_0 = 0, \tag{4.4}$$

$$U_1' S - S U_1 - S_t = 0, \tag{4.5}$$

$$V_1' - V_1 + V_0\left(S - i\omega_0 I_3\right)S - \left(S - i\omega_0 I_3\right)V_0 = 0, \tag{4.6}$$

$$V_2' - V_2 + V_1\left(S - i\omega_0 I_3\right)S - \left(S - i\omega_0 I_3\right)V_1 = 0, \tag{4.7}$$

$$V_2'\left(S - i\omega_0 I_3\right) - \left(S - i\omega_0 I_3\right)V_2 - S_z = 0, \tag{4.8}$$

where Eqs. (4.3)-(4.8) are actually identical and are satisfied if and only if

$$q_1' = q_1 + 2i\sqrt{\frac{-D(z)}{R(z)}}S_{12}, \quad q_2' = q_2 + 2i\sqrt{\frac{-D(z)}{R(z)}}S_{13}, \tag{4.9}$$

$$S_{12} = -S_{21}^*, \quad S_{13} = -S_{31}^*. \tag{4.10}$$

Then, based on the investigation of [36], we can specially define

$$S = -H\Lambda H^{-1}, \tag{4.11}$$

with

$$H = \begin{pmatrix} \phi_{11}(\lambda_1) & \phi_{21}(\lambda_1) & \phi_{31}(\lambda_1) \\ \phi_{21}(\lambda_1) & -\phi_{11}(\lambda_1) & 0 \\ \phi_{31}(\lambda_1) & 0 & -\phi_{11}(\lambda_1) \end{pmatrix}, \quad \Lambda = \begin{pmatrix} \lambda_1 & 0 & 0 \\ 0 & \lambda_1^* & 0 \\ 0 & 0 & \lambda_1^* \end{pmatrix}. \tag{4.12}$$

From expression (4.9), we can get the relation between the new potentials $q_1^{'}(z,t)$ and $q_2^{'}(z,t)$ and old potentials $q_1(z,t)$ and $q_2(z,t)$ as given below

$$q_1^{'}(z,t) = q_1(z,t) + 4i\sqrt{\frac{-D(z)}{R(z)}}\frac{\mathrm{Im}(\lambda_1)\phi_{1,1}(\lambda_1)\phi_{2,1}^{*}(\lambda_1)}{|\phi_{1,1}|^2+|\phi_{2,1}|^2+|\phi_{3,1}|^2},$$
$$q_2^{'}(z,t) = q_2(z,t) + 4i\sqrt{\frac{-D(z)}{R(z)}}\frac{\mathrm{Im}(\lambda_1)\phi_{1,1}(\lambda_1)\phi_{3,1}^{*}(\lambda_1)}{|\phi_{1,1}|^2+|\phi_{2,1}|^2+|\phi_{3,1}|^2}. \tag{4.13}$$

Analogous to this procedure and using the Darboux transformation n times, we find the following nth-iterated potential transformation formula:

$$q_1^{'}[n] = q_1[n] + 4i\sqrt{\frac{-D(z)}{R(z)}}\sum_{j=1}^{n}\frac{\mathrm{Im}(\lambda_j)\phi_{1,j}(\lambda_j)\phi_{2,j}^{*}(\lambda_j)}{A_j},$$
$$q_2^{'}[n] = q_2[n] + 4i\sqrt{\frac{-D(z)}{R(z)}}\sum_{j=1}^{n}\frac{\mathrm{Im}(\lambda_j)\phi_{1,j}(\lambda_j)\phi_{3,j}^{*}(\lambda_j)}{A_j}. \tag{4.14}$$

where

$$\phi_{m,j+1}(\lambda_{j+1}) = \left(\lambda_{j+1}-\lambda_j^{*}\right)\phi_{m,j}(\lambda_{j+1}) - \frac{B_j}{A_j}\left(\lambda_j-\lambda_j^{*}\right)\phi_{m,j}(\lambda_j), \tag{4.15}$$

$$A_j = |\phi_{1,j}(\lambda_j)|^2 + |\phi_{2,j}(\lambda_j)|^2 + |\phi_{3,j}(\lambda_j)|^2, \tag{4.16}$$

$$B_j = \phi_{1,j}(\lambda_j)\phi_{1,j}^{*}(\lambda_{j+1}) + \phi_{2,j}(\lambda_j)\phi_{2,j}^{*}(\lambda_{j+1}) + \phi_{3,j}(\lambda_j)\phi_{3,j}^{*}(\lambda_{j+1}), \tag{4.17}$$

with $m = 1, 2, 3,$ $j = 1, 2,$ and $\left[\phi_{1,j}, \phi_{2,j}, \phi_{3,j}\right]^T$, the vector solution of the Lax pair corresponding to $\lambda = \lambda_j$.

5. Bright Soliton Solutions of The Generalized Inhomogeneous Coupled Nonlinear Schrödinger-Maxwell-Bloch Equation

5.1. Bright One-Soliton Solution

In order to obtain the bright one-soliton solution of system (2.1), we take the zero seed solutions and choose $\phi_{1,1}(\lambda_1), \phi_{2,1}(\lambda_1)$ and $\phi_{3,1}(\lambda_1)$ as follows

$$\phi_{1,1}(\lambda_1)=c_1\exp(\mu_1),\ \phi_{2,1}(\lambda_1)=c_1\exp(-\mu_1),\ \phi_{3,1}(\lambda_1)=c_1\exp(-\mu_1), \tag{5.1}$$

where $\mu_1=-i\lambda_1 t-i\lambda_1^2\int D(z)dz$, and c_1, c_2 and c_3 are both complex constants. By taking the complex spectral parameter $\lambda_1=\frac{1}{2}(\eta_1+i\xi_1)$, the one-soliton solution of system (3.1) can be derived as:

$$\begin{aligned} q_1 &= 2\xi_1\frac{c_2^*}{c_1^*}\exp(-2i\phi_1)\sqrt{\frac{-D(z)}{R(z)}}\operatorname{sech}(2\theta_1),\\ q_2 &= 2\xi_1\frac{c_3^*}{c_1^*}\exp(-2i\phi_1)\sqrt{\frac{-D(z)}{R(z)}}\operatorname{sech}(2\theta_1), \end{aligned} \tag{5.2}$$

where

$$\theta_1=\xi_1 t+\xi_1\eta_1\int_0^z D(z)dz+\frac{\eta_1}{\frac{1}{4}\eta_1^2+\left(\frac{1}{2}\xi_1-\omega_0\right)^2}\int_0^z D(z)dz+\theta_{10},$$

$$\phi_1=\eta_1 t+\frac{1}{2}(\eta_1^2-\xi_1^2)\int_0^z D(z)dz+\frac{\xi_1-2\omega_0}{\frac{1}{4}\eta_1^2+\left(\frac{1}{2}\xi_1-\omega_0\right)^2}\int_0^z D(z)dz+\phi_{10},$$

which $\left|\frac{c_2}{c_1}\right|^2+\left|\frac{c_3}{c_1}\right|^2=1$. θ_{10} and ϕ_{10} are the arbitrary real constants. The real part of the spectral parameter is related to the velocity of the soliton and its imaginary part corresponds to the amplitude of the soliton. Because solutions (5.2) include two arbitrary distributed functions $D(z)$ and $R(z)$, thus by choosing the different form, one can explain the various soliton controls. Here, as an example, we consider a periodic distributed amplification system [37] with the varying group velocity dispersion parameter

$$D(z)=D_0\exp(\sigma z)R(z), \tag{5.3}$$

and the nonlinearity parameter

$$R(z)=R_0\left[1+\varepsilon\sin(gz)\right], \tag{5.4}$$

where D_0 is the parameter related to the initial peak power in the system and R_0 is the parameter describing Kerr nonlinearity. ε is a small quantity and g is related to the varying period of fiber parameters. For convenience, we take parameters $D_0=-1$, $R_0=1$, $\varepsilon=0.05$ and $g=1$. In this situation, the gain/loss distributed function (3.6) is of the constant form

$\Gamma(z)=\dfrac{\sigma}{2}$, which represents the dispersion decreasing (increasing) fiber media for $\sigma<0\,(\sigma>0)$.

Figures 1, 2 and 3 present the evolution of the solution (5.2) with $D(z)$ and $R(z)$ given by Eqs. (5.3) and (5.4) for different values of the parameter σ. These figures show that the intensity of the solitary wave decreases (Figure 1 for $\sigma=-0.05$) or increases (Figure 2 for $\sigma=0.05$) while propagating through optical medium. For $\sigma=0$, the profile of the solitary wave is constant (Figure 3). To understand the effect of erbium doped atoms, we have represented the evolution of the one soliton solution in absence of the MB part obtained in [28]. By comparing the results of Figures 1(a), 2(a), 3(a) with those of Figures 1(b), 2(b) and 3(b), it is clear that the SIT phenomenon is responsible for introducing a phase shift. In all these figures, the time shift and the group velocity of the solitary wave are changing while the solitary wave keeps its shape in propagating along the fiber. This is one of the important properties of solitary waves.

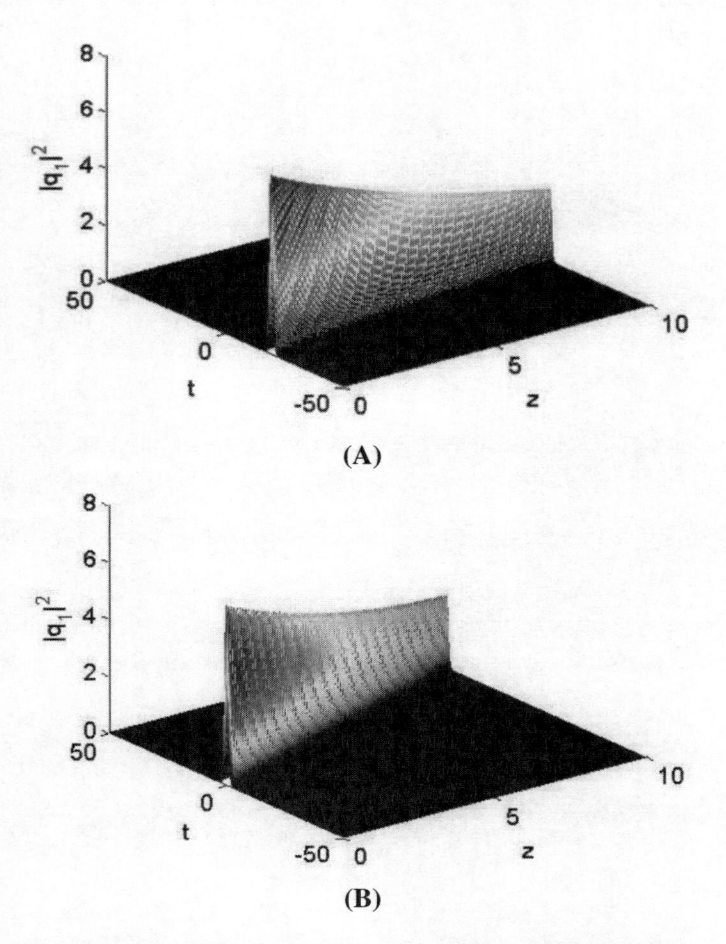

Figure 1. The evolution plot of solution given by Eq. (5.2) with periodic influence for $\sigma=-0.05, \eta_1=-1.75, \xi_1=1.25, c_1=1, c_2=1, \theta_{10}=0, \phi_{10}=0, \omega_0=0.8$; (a) without the MB term; (b) with the MB term.

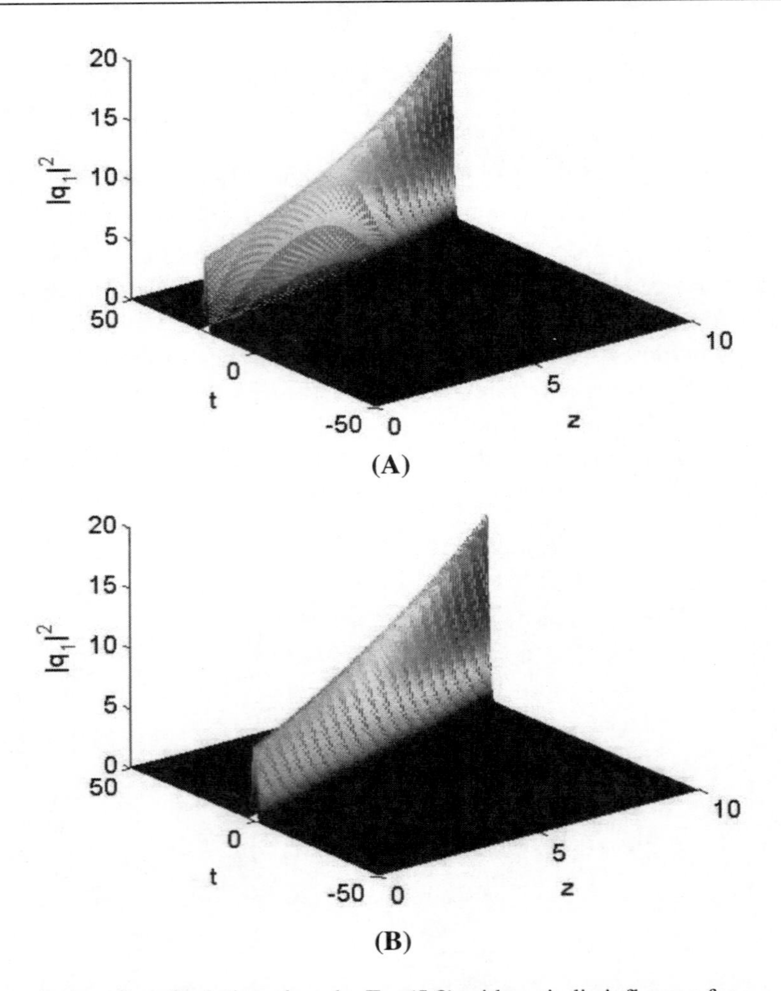

Figure 2. The evolution plot of solution given by Eq. (5.2) with periodic influence for $\sigma = 0.05, \eta_1 = -1.75, \xi_1 = 1.25, c_1 = 1, c_2 = 1, \theta_{10} = 0, \phi_{10} = 0, \omega_0 = 0.8$; (a) without the MB term; (b) with the MB term.

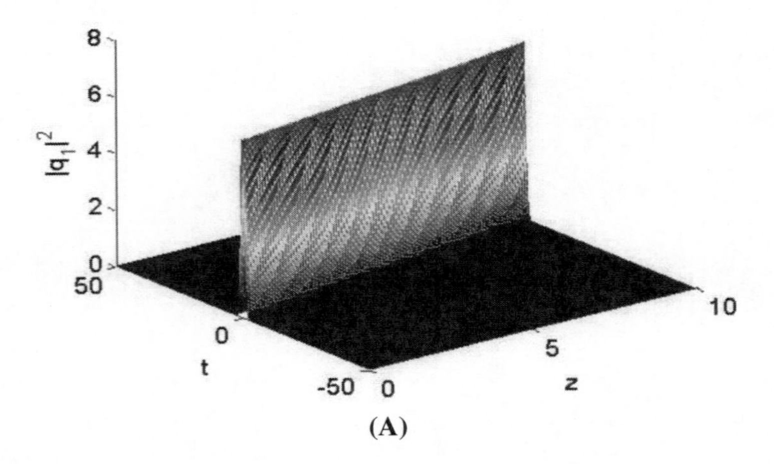

Figure 3. (Continued).

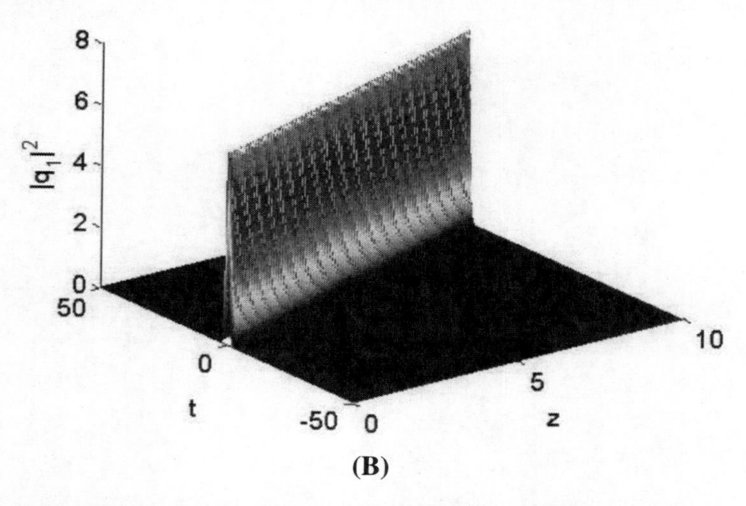

(B)

Figure 3. The evolution plot of solution given by Eq. (5.2) with periodic influence for $\sigma = 0, \eta_1 = -1.75, \xi_1 = 1.25, c_1 = 1, c_2 = 1, \theta_{10} = 0, \phi_{10} = 0, \omega_0 = 0.8$; (a) without the MB term; (b) with the MB term.

5.2. Bright Two-Soliton Solution

Similarly, setting $n = 2$ in expression (4.14) and choosing

$$\phi_{1,1}(\lambda_2) = c_4 \exp(\mu_2), \phi_{2,1}(\lambda_2) = c_1 \exp(-\mu_2), \phi_{3,1}(\lambda_2) = c_1 \exp(-\mu_2), \tag{5.5}$$

where $\mu_2 = -i\lambda_2 t - i\lambda_2^2 \int D(z)dz$, and c_4, c_5 and c_6 being complex constants, the two-soliton solution can be written in an explicit form as follows:

$$q_1[2] = \frac{c_4^*}{c_6^*}\sqrt{\frac{-D(z)}{R(z)}}\frac{G_1}{F}, \quad q_2[2] = \frac{c_5^*}{c_6^*}\sqrt{\frac{-D(z)}{R(z)}}\frac{G_1}{F}, \tag{5.6}$$

where

$$\begin{aligned} G_1 &= a_1 \cosh(2\theta_2)\exp(2i\phi_1) + a_2 \cosh(2\theta_1)\exp(2i\phi_2) + ia_3\left[\sinh(2\theta_2)\exp(2i\phi_1) - \sinh(2\theta_1)\exp(2i\phi_2)\right] \\ F &= b_1 \cosh(2\theta_1 + 2\theta_2) + b_2 \cosh(2\theta_1 - 2\theta_2) + b_3 \cos(2\phi_2 - 2\phi_1), \end{aligned} \tag{5.7}$$

with

$$
\begin{aligned}
a_1 &= \frac{\eta_1}{2}\left[\eta_1^2 - \eta_2^2 + (\xi_1 - \xi_2)^2\right], \\
a_2 &= \frac{\eta_2}{2}\left[\eta_2^2 - \eta_1^2 + (\xi_1 - \xi_2)^2\right], \\
a_3 &= \eta_1\eta_2(\xi_1 - \xi_2), \\
b_1 &= \frac{1}{4}\left[(\eta_1 - \eta_2)^2 + (\xi_1 - \xi_2)^2\right], \\
b_2 &= \frac{1}{4}\left[(\eta_1 + \eta_2)^2 + (\xi_1 - \xi_2)^2\right], \\
b_3 &= -\eta_1\eta_2,
\end{aligned}
\tag{5.8}
$$

$$
\begin{aligned}
\theta_j &= \xi_j t + \xi_j\eta_j \int_0^z D(z)dz + \frac{\eta_j}{\frac{1}{4}\eta_j^2 + \left(\frac{1}{2}\xi_j - \omega_0\right)^2}\int_0^z D(z)dz + \theta_{j0}, \\
\phi_j &= \eta_j t + \frac{1}{2}\left(\eta_j^2 - \xi_j^2\right)\int_0^z D(z)dz + \frac{\xi_j - 2\omega_0}{\frac{1}{4}\eta_j^2 + \left(\frac{1}{2}\xi_j - \omega_0\right)^2}\int_0^z D(z)dz + \phi_{j0},
\end{aligned}
\tag{5.9}
$$

where $\left|\frac{c_4}{c_1}\right|^2 + \left|\frac{c_6}{c_1}\right|^2 = 1$, θ_{j0} and ϕ_{j0} are the arbitrary real constants, and we have used $\lambda_j = \frac{1}{2}(\eta_j + i\xi_j)$, $j = 1, 2$.

From the expression of θ_j, one can clearly see that if one manages to control the parameters η_j and ξ_j, it is possible to form a separating evolution behavior of solitons. Although the fundamental soliton propagation cannot be obtained in standard fibers, pulse propagation over relatively long distances (and even transoceanic distance) can still be obtained through an appropriate combination of dispersion management and optical amplification (now mostly based on erbium-doped fiber amplifiers and Raman amplifiers) [38]. Figure 4 presents the evolution of the two soliton solutions without the MB term (Figure 4 (a)) and with the MB term (Figure 4 (b)) for $\sigma = 0$. $D(z)$ and $R(z)$ are given by Eqs. (5.3) and (5.4). It is clearly seen from the figure that the presence of doped atoms induces an overall phase shift. One can also see that the separation between the two solitons can decreases or increases if one adjusts the initial separation. From Figure 4 (a), we observe that the separation between the two soliton decreases. The decreasing initial separation between adjacent pulses may result in the strong interaction and pulse distortion [39]. Obviously, this is disadvantageous for the transmission of the information. For comparison, we observe in Figure 4(b) that the separation between the two soliton increases progressively with the propagation distance. This case is advantageous to increase the information bit rate in optical soliton communication. These results will be very useful in increasing the bit rate in ultrahigh and long-distance optical communication links.

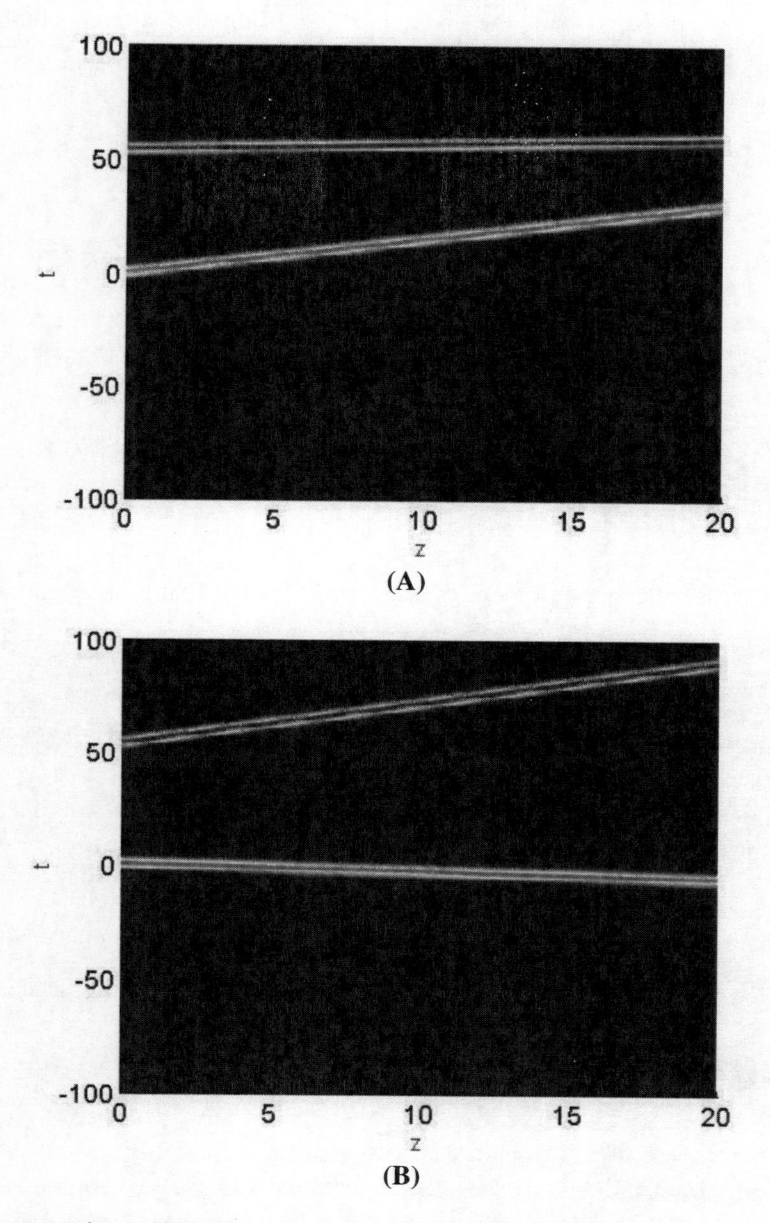

Figure 4. The separating evolution plot of solution given by Eq. (33) with $\sigma = 0, \eta_1 = 2.25, \eta_2 = -0.5, \xi_1 = 0.5, \xi_2 = -0.5, c_4 = 1, c_5 = 1, c_6 = 1, \theta_{01} = 0, \theta_{02} = 0, \phi_{01} = 0, \phi_{02} = 0, \omega_0 = 0.8$; (a) without the MB term; (b) with the MB term.

In order to find whether the elastic interactions between the two solitons along the erbium-doped fiber is maintained in the presence of inhomogeneities, and better understand the collision dynamics between the two solitons, we perform the asymptotic analysis [40] to investigate expression (43) as follows:

- Before collision $(z \to -\infty)$.

a) When we set $(\theta_2 \to +\infty)$, the first component of the two-soliton solution has the form

$$q_j^{1+} \to \frac{\xi_j}{\sqrt{b_1 b_2}} \sqrt{\frac{-D(z)}{R(z)}} \frac{(a_1 + ia_2)\exp(2i\varphi_1)}{\cosh(2\theta_1 + R)}, \tag{5.10}$$

where $j = 1, 2$ and $\exp(R) = \sqrt{\frac{b_1}{b_2}}$.

b) When we set $(\theta_2 \to -\infty)$, the first component of the two-soliton solution has the form

$$q_j^{1-} \to \frac{\xi_j}{\sqrt{b_1 b_2}} \sqrt{\frac{-D(z)}{R(z)}} \frac{(a_1 + ia_3)\exp(2i\varphi_1)}{\cosh(2\theta_1 - R)}. \tag{5.11}$$

Thus we can define the amplitude's transition matrix for the first component of the two-soliton solution before the collision as

$$T_j^1 = \frac{a_1 + ia_2}{a_1 - ia_2}. \tag{5.12}$$

- After collision $(z \to +\infty)$

a) When we set $(\theta_1 \to +\infty)$, the second component of the two-soliton solution has the form

$$q_j^{2+} \to \frac{\xi_j}{\sqrt{b_1 b_2}} \sqrt{\frac{-D(z)}{R(z)}} \frac{(a_3 - ia_2)\exp(2i\varphi_2)}{\cosh(2\theta_2 + R)}, \tag{5.13}$$

b) When we set $(\theta_1 \to -\infty)$, the second component of the two-soliton solution has the form

$$q_j^{2-} \to \frac{\xi_j}{\sqrt{b_1 b_2}} \sqrt{\frac{-D(z)}{R(z)}} \frac{(a_3 + ia_2)\exp(2i\varphi_2)}{\cosh(2\theta_2 - R)}. \tag{5.14}$$

Also we can define the amplitude's transition matrix for the second component of the two-soliton solution after the collision as

$$T_j^1 = \frac{a_3 - ia_2}{a_3 + ia_2}. \tag{5.15}$$

One can easily find from equations (5.12) and (5.15) that $|T_j^l| = 1, \quad (l = 1, 2)$.

This result shows that the collision of two inhomogeneous solitons in erbium doped fiber is elastic. The solitons will pass through each other without being affected in their shapes and sizes when the collision happens. However, the phase shift as a result of collision may be obtained as $2R = \ln\left(\frac{b_1}{b_2}\right)$. Figures 5, 6, and 7 display the collision process of the two solitons propagating in erbium-doped fiber with the MB term for different values of the parameter σ. We observe that the solitons can undergo elastic collision (shape preserving). Two solitons start separately and walk through each other unaffectedly and keep their shapes and velocities invariant after the collision.

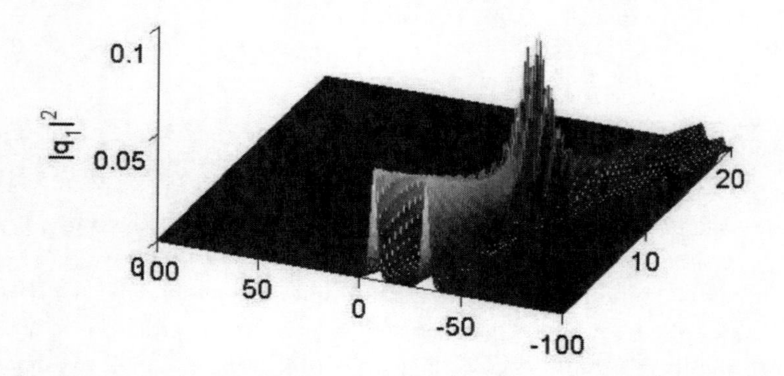

Figure 5. The evolution plot of the solution given by Eq. (5.6) describing the collision of two solitons with opposite velocities in the case $\sigma = -0.09$. The others parameters are $\eta_1 = 3.5, \eta_2 = 0.25, \xi_1 = 0.5, \xi_2 = 0.5, c_4 = 1, c_5 = 1, c_6 = 1, \theta_{01} = 0, \theta_{02} = 0, \phi_{01} = 0, \phi_{02} = 0, \omega_0 = 0.8.$

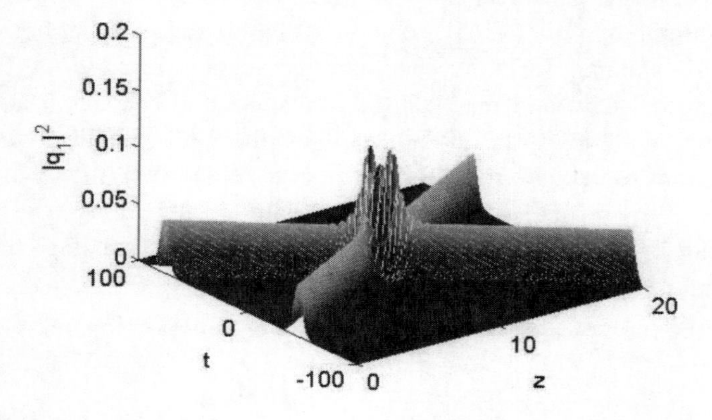

Figure 6. The evolution plot of the solution given by Eq. (5.6) describing the collision of two solitons with opposite velocities in the case $\sigma = 0$. The others parameters are $\eta_1 = 1.56, \eta_2 = -0.56, \xi_1 = 0.2, \xi_2 = 0.2, c_4 = 1, c_5 = 1, c_6 = 1, \theta_{01} = -8, \theta_{02} = 5, \phi_{01} = 0, \phi_{02} = 0, \omega_0 = 0.8.$

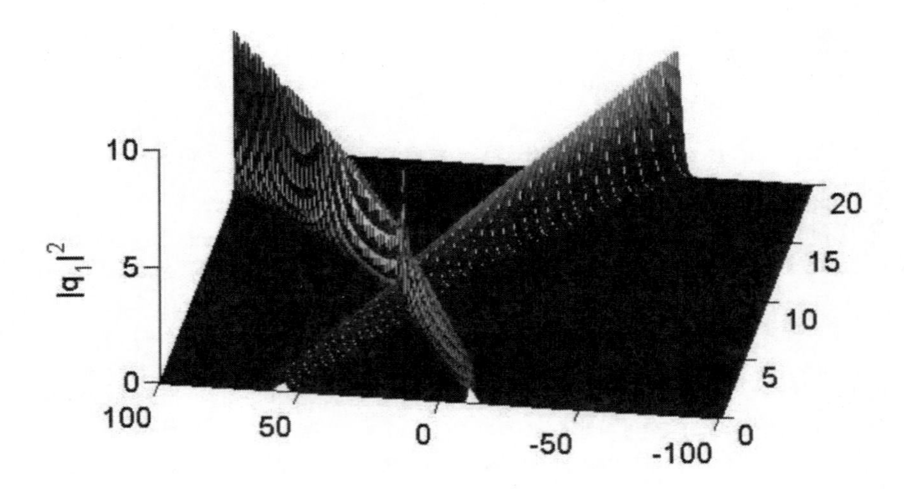

Figure 7. The evolution plot of the solution given by Eq. (5.6) describing the collision of two solitons with opposite velocities in the case $\sigma = 0.09$. The others parameters are $\eta_1 = 1.25, \eta_2 = -2.5, \xi_1 = 0.6, \xi_2 = 0.8, c_4 = 1, c_5 = 1, c_6 = 1, \theta_{01} = 0, \theta_{02} = 0, \phi_{01} = 0, \phi_{02} = 0, \omega_0 = 0.8.$

CONCLUSION

Optical solitons and their interactions have their potential applicability in the optical communication systems. In this work, with symbolic computation, we have studied the interaction properties of solitons in the generalized inhomogeneous CNLS-MB system. We have presented the explicit 3×3 Lax pair for such a system. In order to better explain the stable pulse propagation in erbium-doped fiber communication systems, we have given the first three conservation laws and the first three conserved quantities of the system. Via the Darboux transformation method, a simple procedure to derive the N-soliton solutions has been presented. For instance, the explicit one- and two-soliton solutions have been generated. The solutions' characteristics and collisions are discussed analytically under a given soliton control system. Soliton based optical fiber communication systems using erbium-doped fiber amplifiers are more suitable for long haul communication because of their very high information carrying capacity and repeater less transmission. Then, the application of the results obtained here would be useful in the design of fiber optic amplifiers or and in optical soliton communications to increase the information bit rate. However, it should be noted that, in realistic situations, it is difficult to produce exact gain/loss functions as given by equation (3.6). So, it is our belief that, in future, the difficulties in realizing such fibers with inhomogeneous parameters could be overcome.

ACKNOWLEDGMENTS

KP wishes to thank the IFCPAR (No. IFC/3504- F/2005/2064), CSIR, DST-DFG and DST Ramanna Fellowship, Government of India, for the financial support through major projects.

REFERENCES

[1] Ablowitz MJ, Segur H. Solitons and Inverse Scattering Transform. Philaladelphia: SIAM; 1981.

[2] Hasegawa A, Tappert F. Nonlinear Optical Pulses in Dispersive Dielectric Fibers. *Applied Physics Letter* 1973; 23: 142-144. DOI: 10.1063/1.1654836.

[3] A. Hasegawa, Y. Kodama. Solitons in Optical Communications. Oxford; 1995.

[4] Mollenauer LF, Stolen RH, Gordon JP. Experimental Observation of Picosecond Pulse Narrowing and Solitons in Optical Fibers. *Physical Review Letter* 1980; 45: 1095-1098. DOI: 10.1103/PhysRevLett.45.1095.

[5] Agrawal GP. Fiber-Optic Communication Systems. New York: Wiley Interscience; 1997.

[6] Kivshar YS, Malomed BA. Dynamics of Solitons in Nearly Integrable System. *Review of Modern Physics* 1989; 61: 763-915. DOI: 10.1103/RevModPhys.61.763.

[7] Kanna T, Lakshmanan M. Exact Soliton Solutions, Shape Changing Collisions, and Partially Coherent Solitons in Coupled Nonlinear Schrödinger Equations. Physical Review Letter 2001; 86: 5043-5046. DOI: 10.1103/PhysRevLett.86.5043.

[8] Y. Barad, Y. Silberberg. Polarization Evolution and Polarization Instability of Soliton in a Birefringent Optical Fiber. *Physical Review Letter* 1997; 78; 3290-3293. DOI: 10.1103/PhysRevLett.78.3290.

[9] Christodoulides DN, Singh SR, Carvalho MI, Segev M. Incoherently Coupled Soliton in Biased Photorefractive Crystals. *Applied Physics Letter* 1996; 68: 1763-1765. DOI: 10.1063/1.116659.

[10] Radhakrishnan R, Lakshmanan M, Hietarinta J. Inelastic Collision and Switching of Coupled Bright Solitons in Optical Fibers. *Physical Review E* 1997; 56: 2213-2216. DOI: 10.1103/PhysRevE.56.2213.

[11] Chakravarty S, Ablowitz MJ, Sauer RJ, Jenkins RB. Multisoliton Interactions and Wavelength Division Multiplexing. *Optics Letter* 1995; 20: 136-138. DOI: 10.1364/OL.20.000136.

[12] Yeh C, Bergman L. Enhanced Pulse Compression in a Nonlinear Fiber by a Wavelength Division Multiplexed Optical Pulse. *Physical Review* E 1998; 57: 2398-2404. DOI: 10.1103/PhysRevE.57.2398.

[13] Zakharov VE, Wabnitz S. Optical Solitons-Theoretical Challenges and Industrial Perspectives. Berlin: Springer; 1998.

[14] Maimistov AI. Reversible Logic Elements as a New Field of Application of Optical Solitons. *Quantum Electron* 1995; 25: 1009-1013. DOI: 10.1070/QE1995v025n10ABEH000520.

[15] Adamatzky A. Collision Based Computing. Berlin: Springer, 2002.

[16] McCall SL, Hahn EL. Self-Induced Transparency by Pulsed Coherent Light. *Physical Review Letter* 1967; 18: 908-911. DOI: 10.1103/PhysRevLett.18.908.

[17] Basharov AM, Maimistov AI. Propagation of Ultrashort Electromagnetic Pulses in a Kerr Medium with Impurity Atoms under Quasi-Resonance Conditions. *Quantum Electronics* 2000; 30: 1014-1018. DOI: 10.1070/QE2000VO30n11ABEH001854.

[18] Nakazawa M, Kurokawa K, Kubota H, Yamada E. Coexistence of a Self-Induced-Transparency Soliton and a Nonlinear Schrödinger Soliton in an Erbium-Doped Fiber. *Physical Review Letter* 1991; 66: 5973-5987. DOI: 10.1103/PhysRevA.44.5973.

[19] Nakkeeran K, Porsezian K. Coexistence of a Self-Induced Transparency Soliton and a Higher Order Nonlinear Schrödinger Soliton in an Erbium Doped Fiber. *Optics Communications* 1996; 123 : 169-174. DOI: 10.1016/0030-4018(95)00477-7.

[20] Xue JK. Propagation of Nonplanar Dust-Acoustic Envelope Solitary Waves in a Two-Ion-Temperature Dusty Plasma. *Physics of Plasmas* 2004; 11:1860. DOI: 10.1063/1.1689355.

[21] Serkin VN, Hasegawa A. Novel Soliton Solutions of the Nonlinear Schrödinger Equation Model. *Physical Review Letter* 2000; 85: 4502-4505. DOI: 10.1103/PhysRevLett.85.4502

[22] Kruglov VI, Peacock AC, Harvey JD. Exact Solutions of the Generalized Nonlinear Schrödinger Equation with Distributed Coefficients. *Physical Review E* 2005; 71: 056619-11. DOI: 10.1103/PhysRevE.71.056619.

[23] Vidhya N, Mahalingam A, Uthayakumar A. Propagation of Solitary Waves in Inhomogeneous Erbium-Doped Fibers with Third-Order Dispersion, Self-Steepening and Gain/Loss. *Journal of Modern Optics* 2009;.56; 607-614. DOI: 10.1080/09500340802711727.

[24] Tian B, Shan WR, Zhang CY, Wei GM, Gao YT. Transformations for a Generalized Variable-Coefficient Nonlinear Schrödinger Model from Plasma Physics, Arterial Mechanics, and Optical Fibers with Symbolic Computation. *European Physical Journal B* 2005; 47:329-332. DOI: 10.1140/epjb/e2005-00348-3.

[25] Mahalingam A, Porsezian K, Mani Rajan MS, Uthayakumar A. Propagation of Dispersion-Nonlinearity-Managed Solitons in an Inhomogeneous Erbium-Doped Fiber System. *Journal of Physics A: Mathematical and* Theoretical 2009. 42, 165101-165112. DOI: 10.1088/1751-8113/42/16/165101.

[26] Serkin VN, Hasegawa A, Belyaeva TL. Nonautonomous Solitons in External Potentials. *Physical Review Letter* 2007 ; 98 : 0714102-4. DOI : 10.1103/PhysRevLett.98.074102.

[27] Hirota R. Exact N-Soliton Solutions of the Wave Equation of Long Waves in Shallow-Water and in Nonlinear Lattices. *Journal of Mathematical Physics* 1973; 14: 810-814. DOI: 10.1063/1.1666400.

[28] Tiofack CGL, Mohamadou A, Kofane TC, Porsezian K. Exact Quasi-Soliton Solutions and Soliton Interaction for the Inhomogeneous Coupled Nonlinear Schrödinger Equations.*Journal of Modern Optics* 2010; 57: 309-320. DOI: 10.1080/09500340903531370.

[29] Menyuk CR. Pulse Propagation in an Elliptically Birefringent Kerr Medium. *IEEE Journal of Quantum Electron* 1989; 25: 2674-2682. DOI: 10.1109/3.40656.

[30] Ablowitz MJ, Clarkson PA, Solitons, Nonlinear Evolution Equations and Inverse Scattering. Cambridge; 1992.

[31] Hao R, Li L, Li Z, Xue W, Zhou G. A New Approach to Exact Soliton Solutions and Soliton Interaction for the Nonlinear Schrödinger Equation with Variable Coefficients. *Optics Communications* 2004; 236: 79-86. DOI: 10.1016/j.optcom.2004.03.005.

[32] Tian J, Zhou G. Exact Bright Soliton Solution for a Family of Coupled Higher-Order Nonlinear Schrödinger Equation in Inhomogeneous Optical Fiber Media. *European Physical Journal D* 2007; 41:171-177. DOI: 10.1140/epjd/e2006-00194-y.

[33] Wadati M, Sanuki H, Konno K. Relationships among Inverse Method, Bäcklund Transformation and an Infinite Number of Conservation Laws. *Prog. Theor. Phys* 1975; 53, 419-436. DOI: 10.1143/PTP.53.419.

[34] Hirota R. Exact Solution of the Korteweg-de Vries Equation for Multiple Collisions of Solitons. Physical Review Letter 1971; 27: 1192. DOI: 10.1103/PhysRevLett.27.1192.

[35] Wright OC, Forest MG. On the Bäcklund-Gauge Transformation and Homoclinic Orbits of a Coupled Nonlinear Schrödinger System. *Physica D* 2000; 141: 104-116. DOI: 10.1016/S0167-2789(00)00021-X.

[36] Weiss J, Tabor M, Carnevale G. The Painlevé Property for Partial Differential Equations: Bäcklund Transformation, Lax Pair, and the Schwarzian Derivative. *Journal of Mathematical Physics* 1983; 24: 522-530. DOI: 10.1063/1.525875.

[37] Hao R, Li L, Li Z, Zhou G. Exact Multisoliton Solutions of the Higher-Order Nonlinear Schrödinger Equation with Variable Coefficients. *Physical Review E* 2004; 70: 066603-6. DOI: 10.1103/PhysRevE.70.066603.

[38] Mollenauer LF, Smith K. Demonstration of Soliton Transmission Over More Than 4000 Kmin Fiber with Loss Periodically Compensated by Raman Gain. *Optics Letter* 1988; 23: 675-677. DOI: 10.1364/OL.13.000675.

[39] Uzunov IM, Stoev VD, Tzoleva TI. N-Soliton Interaction in Trains of Unequal Soliton Pulses in Optical Fibers. *Optics Letter* 1992; 17: 1417-1419. DOI: .1364/OL.17.001417.

[40] Hisakado M, Iizuka T, Wadati M. Coupled Hybrid Nonlinear Schrödinger Equation and Optical Solitons. *Journal of Physical Society of Japan* 1004; 63: 2887-2894. DOI: 10.1143/JPSJ.63.2887.

In: Optical Fibers: New Developments
Editor: Marco Pisco

ISBN: 978-1-62808-425-2

Chapter 6

SOLITONS IN OPTICAL FIBERS WITH POWER LOSS/GAIN AND PHASE MODULATION

M. Idrish Miah[1,2,*]

[1]Department of Physics, University of Chittagong, Chittagong, Bangladesh
[2]Queensland Micro- and Nanotechnology Centre, Griffith University, Nathan, Brisbane, QLD 4111, Australia

ABSTRACT

Solitons are special breed of optical pulses that can propagate through an optical fiber undistorted for very long distances. Solitons are therefore an important development in the field of optical communications. The key to soliton formation is the careful balance of the opposing forces of chromatic dispersion and self-phase modulation (nonlinear phenomena). We give the basic principle of soliton formation and study the nonlinear electromagnetic wave propagation in inhomogeneous fiber cores in the both normal and anomalous dispersive regimes. In order to include the inhomogeneous physical effects, the nonlinear Schrödinger (NLS) equation, which governs the solitary pulse propagation in optical fiber, is modified by adding terms for phase modulation and fiber power loss/gain. Advanced mathematical solution techniques, e.g. Lax pair construction and Hirota transformation are introduced in order to bilinearize and solve the modified NLS equations, which are inhomogeneous and higher-order nonlinear partial differential equations. A general integrability condition is arrived for all the cases. The modified NLS equations in the normal and anomalous dispersive regimes are then Hirota bilinearized, and exact bright and dark solitons solutions are obtained. The analytical soliton solutions are also obtained.

The results show that in both cases the areas of the pulse envelopes remain preserved during the propagation in the fibers, demonstrating that bright/dark solitary wave propagation is maintained in the cores. The results are discussed in details. The discussion, however, highlights some problems in solitonic propagation.

Keywords: Optical soliton; Nonlinear Schrödinger equation; Horota transformation; Lax pair; Self-phase modulation; Group velocity dispersion

---

* Email: m.miah@griffith.edu.au.

1. INTRODUCTION

We are now at the age of high demand of frequent information, in particular, we expect all information we need at the speed of light! In order to respond to this demand, the old analog metal cable communication devices are lagging behind because they are slow and noisy. Modern communications relies on fast digital optical fiber systems, though there is, however, a draw back to using fiber solitons as the current electronic amplifiers and switches just cannot operate quickly enough for the short pulses and high speeds that optical solitons can provide. The idea of optical fiber communications is to inject laser pulses into a fiber. Information is coded within these pulses. In order to increase the amount of information we can send in one second, we can make the pulses as short as possible and inject many pulses per second, but there is a limit of the width of laser pulses. When a pulse propagates along a fiber, its width increases: the narrower the pulses, the more rapid the increase. This effect is called dispersion. We then can imagine that if we inject too many short pulses into a fiber, they will overlap after propagating over some distance. We then will not be able to distinguish between pulses, and information will have been lost. Fortunately, there is also another counter-effect which shortens the width of a pulse. This effect is called nonlinearity. If the intensity of a laser pulse is strong enough, this nonlinear effect can be in an active balance with the dispersion effect. The result is a pulse that can keep its shape for a long propagation distance, even including small disturbance or perturbation. These steady pulses are called optical solitons.

Due to their short pulse duration and high stability, solitons could form the high-speed communications backbone of tomorrow's information super-highway. Solitons are a special breed of optical pulses that can propagate through an optical fiber undistorted for tens of thousands kilometers. The key to soliton formation is the careful balance of the opposing forces of chromatic dispersion and self-phase modulation. Solitons refer to the functional form of the specific solutions of the underlying nonlinear equation that describes light propagation in nonlinear optical media. The fundamental property of solitons that makes them attractive for optical communications is that solitons maintain their shape upon propagation and even exhibit particle behavior by surviving collisions with one another. A wide applicability of the soliton equation implies soliton phenomena which are common in various fields of physics. This is the essence of soliton physics. Solitons appear in almost all branches of physics, such as hydrodynamics, plasma physics, nonlinear optics, condensed matter physics, low temperature physics, particle physics, nuclear physics, biophysics and astrophysics [1].

An optical soliton is a pulse that travels without distortion due to dispersion or other effects. They are a nonlinear phenomenon caused by self-phase modulation (SPM), which means that the electric field of the wave changes the index of refraction seen by the wave (Kerr effect). SPM causes a red shift at the leading edge of the pulse. Solitons occur when this shift is canceled due to the blue shift at the leading edge of a pulse in a region of anomalous dispersion (bright soliton), resulting in a pulse that maintains its shape in both frequency and time. Solitons are therefore an important development in the field of optical communications. At the beginning of this paper a general discussion on optical soliton, its formation and properties and related terms is given. Nonlinear Schrödinger (NLS) equation in anomalous (bright soliton) and normal (dark soliton) dispersive regimes with both gain and loss (damping) has been considered and exactly solved. Bright and dark solitons in both cases have been constructed. In order to solve eigenvalue problem, Lax pair and Hirota transformation have been introduced and used. Challenges or problems in soliton propagation have also been discussed.

2. CHROMATIC DISPERSION

An electromagnetic wave, such as the light sent through an optical fiber, is actually a combination of electric and magnetic fields oscillating perpendicular to each other. When an electromagnetic wave propagates through free space, it travels at the constant speed of c =3.0 x10^8 meters per second, equivalent to about seven trips around the earth every second. However, when light propagates through a material rather than through free space, the electric and magnetic fields of the light induce a polarization in the electron clouds of the material. This polarization makes it more difficult for the light to travel through the material, so the light must slow down to a speed less than its original 3.0 x10^8 meters per second. The degree to which the light is slowed down is given by the material's refractive index n. The speed of light within material is then v = c/n (m/s). This equation shows that a high refractive index means a slow light propagation speed. Higher refractive indices generally occur in materials with higher densities, since a high density implies a high concentration of electron clouds to slow the light.

Since the interaction of the light with the material depends on the frequency of the propagating light, the refractive index is also dependent on the light frequency. This, in turn, dictates that the speed of light in the material depends on the light's frequency, a phenomenon known as chromatic dispersion. In order to understand what chromatic dispersion means for fiber optic communication systems, one must first understand the nature of an optical pulse. Although an optical pulse represents only a single bit of information in a communication system, it is actually composed of many hundreds or even thousands of particles of light known as photons.

Optical pulses are generated by a near-monochromatic light source such as a laser or a LED. If the light source were completely monochromatic, then it would generate photons at a single frequency only, and all of the photons would travel through the fiber at the same speed. In reality, small thermal fluctuations and quantum uncertainties prevent any light source from being truly monochromatic. This means that the photons in an optical pulse actually include a range of different frequencies. Since the speed of a photon in an optical fiber depends on its frequency, the photons within a pulse will travel at slightly different speeds from each other. The result is that the slower photons will lag further and further behind the faster photons, and the pulse will broaden. Optical pulses are often characterized by their shape, and a typical pulse shape is Gaussian, like that shown in Figure 1. In a Gaussian pulse, the constituent photons are concentrated toward the center of the pulse, making it more intense than the outer tails. An example of this Gaussian-type broadening is given in Figure 1.

Chromatic dispersion may be classified into two different regimes: normal and anomalous. With normal dispersion, the lower frequency components of an optical pulse travel faster than the higher frequency components. The opposite is true with anomalous dispersion.

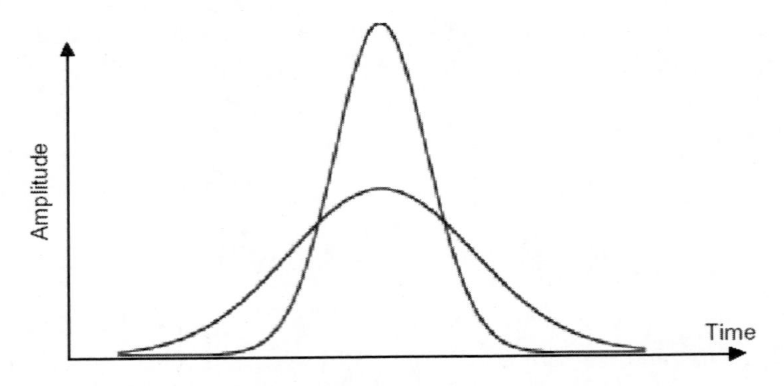

Figure 1. A Gaussian pulse broadens due to chromatic dispersion.

The type of dispersion a pulse experiences depends on its wavelength; a typical fiber optic communication system uses a pulse wavelength of 1.55 μm, which falls within the anomalous dispersion regime of most optical fibers. Pulse broadening, and hence chromatic dispersion, can be a major problem in fiber optic communication systems for obvious reasons. A broadened pulse has much lower peak intensity than the initial pulse launched into the fiber, making it more difficult to detect. Worse yet, the broadening of two neighboring pulses may cause them to overlap, leading to errors at the receiving end of the system. However, chromatic dispersion is not always a harmful occurrence. As we shall see soon, when combined with self-phase modulation, chromatic dispersion in both normal and anomalous dispersive regimes may lead to the formation of optical solitons.

3. Self-Phase Modulation

As we see in the previous section that a material's refractive index is dependent on the frequency of the light traveling through it. In fact, the refractive index is also dependent on the intensity of the light. This is due to the fact that the induced electron cloud polarization in a material is not actually a linear function of the light intensity. The degree of polarization increases nonlinearly with light intensity, so the material exerts greater slowing forces on more intense light. The result is that the refractive index of a material increases with the increasing light intensity. Phenomenological consequences of this intensity dependence of refractive index in fiber optics are known as fiber nonlinearities. There exist many different types of fiber nonlinearities, but the one of most concern to soliton theory is SPM. With SPM, the optical pulse exhibits a phase shift induced by the intensity-dependent refractive index. The most intense regions of the pulse are slowed down the most, so they exhibit the greatest phase shift. Since a phase shift changes the distances between the peaks of an oscillating function, it also changes the oscillation frequency along the horizontal axis.

For simplicity, you may think of a phase shift as stretching out or squishing part of an oscillating function along its horizontal axis. The concepts of phase shift and chirp is given by introducing an optical pulse wave (Gaussian pulse) before undergoing a phase shift (Figure 2, top). This wave is unchirped, meaning that there is no ordered variation in frequency along the length of the wave. In the case of Figure 2 top, this is due to fact that the pulse wave has a constant frequency along its entire length.

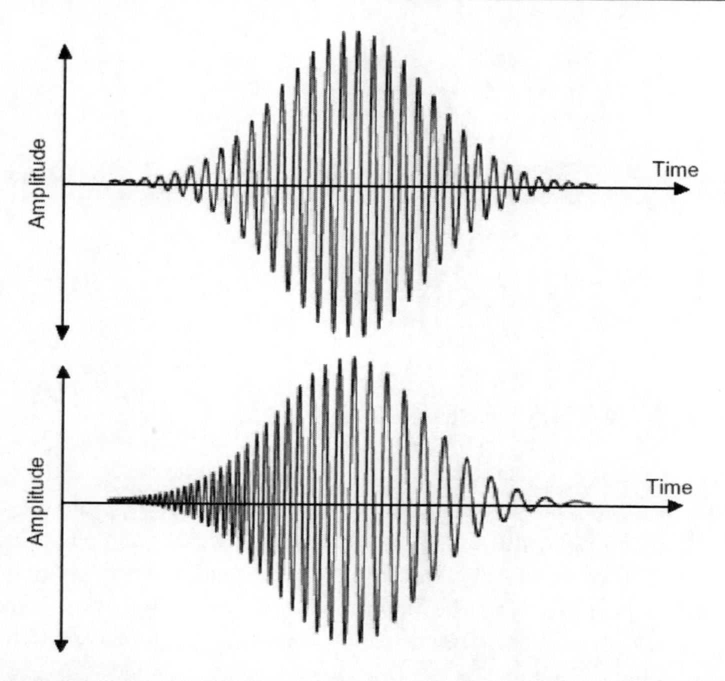

Figure 2. Unchirped ((Top) and chirped (Bottom) Gaussian pulses.

Figure 2 bottom shows the same pulse wave after undergoing a phase shift such that the left-hand side of the wave has a higher frequency than the right-hand side. As a result, the wave is chirped and so there is an ordered variation in frequency along the horizontal axis. This is because SPM only broadens the pulse in the frequency domain, not the time domain.

As in Figure 2 bottom, SPM leads to a chirping with lower frequencies on the leading (right-hand) side and higher frequencies on the trailing (left-hand) side of the pulse. Like dispersion, SPM may lead to errors at the receiving end of a fiber optic communication system. This is particularly true for wavelength-division multiplexed systems, where the frequencies of individual signals need to stay within strict upper and lower bounds to avoid encroaching on the other signals.

4. Soliton Formation

The origins of chromatic dispersion and SPM and how they separately affect the propagation of an optical pulse are introduced in Sections 2 and 3. These phenomena can, individually, create major problems in a fiber optic communication system, but if a system is designed such that the effects of dispersion and SPM perfectly cancel each other out, a pulse may propagate through a fiber without any broadening in the time or frequency domains. Let us now see how this is possible for a pulse with a wavelength in the anomalous dispersion regime of a fiber. SPM leads to lower frequencies at the leading side of the pulse and higher frequencies at the trailing side of the pulse and anomalous dispersion causes lower frequencies to travel slower than higher frequencies. Therefore, anomalous dispersion causes the leading side of the pulse to travel slower than the trailing side, effectively compressing the pulse and undoing the frequency chirp induced by SPM.

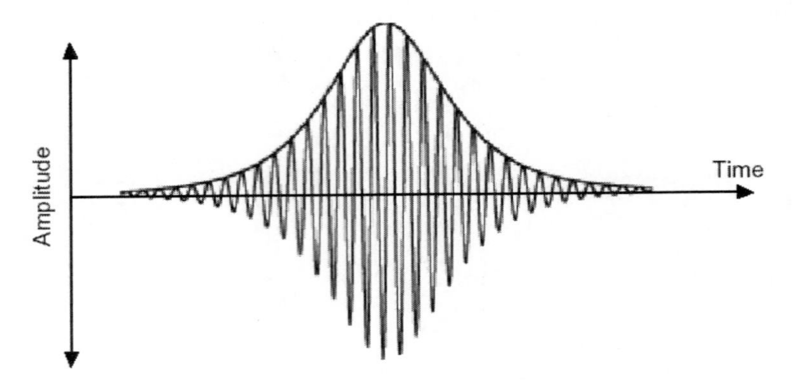

Figure 3. An optical soliton with a hyperbolic-secant envelope.

If the properties of the pulse are just right, the instantaneous effects of SPM and anomalous dispersion cancel each other out completely, and the pulse remains unchirped and retains its initial width along the entire length of the fiber, we can form an optical soliton.

The formation of solitons requires a careful balance among the properties of the pulse and the fiber. In addition, the pulse must be unchirped and have a hyperbolic-secant shape, as shown in Figure 3. If all of these qualifications are met, solitons may be successfully generated and propagated through a fiber.

Theoretical and experimental results show that if the exact power and shape parameters are not met or satisfied, the interplay between dispersion and SPM can actually mold the pulse into a hyperbolic-secant shape with the appropriate peak power. Of course, the initial values must be close enough to the final soliton values in order for the soliton to form, and there are drawbacks when the input parameters vary substantially from the ideal soliton values. Namely, the dispersive effects will cause the pulse to lose some of its energy as it evolves into a soliton, and this energy loss can cause interference with neighboring pulses and therefore adversely affect the performance of a system. Soliton formation is much more sensitive to chirp in the initial pulse than it is to inaccurate pulse shape and power. Ideally, the input pulse should be completely unchirped, but soliton formation may still occur is there is a very small amount of initial chirp. Again, the interplay between dispersion and SPM will eventually remedy the pulse, but it may lose a large percentage of its energy as it evolves into an unchirped soliton. The generation of unchirped pulses close to a hyperbolic-secant shape requires careful attention to the design of the semiconductor laser at the fiber input.

5. Nonlinear Schrödinger Equation

It has been well-known that soliton-type wave propagation is achieved by means of a counterbalance of the major constraints in the optical fibre, namely, the group velocity dispersion (GVD) and SPM. This type of soliton pulse propagation in fibers is governed by the well-known NLS equation [1]

$$i\frac{\partial A}{\partial z} - s\frac{\partial^2 A}{\partial t^2} + 2|A|^2 A = 0\,, \tag{1}$$

where A(z,t) is the slowly varying pulse envelope of electric field (of the electromagnetic wave) and i denotes imaginary. Considering homogeneous fiber core medium, equation (1) has widely been studied by numerous authors (e.g. [2-4]). However, in an actual fiber, the core medium is, in general, not uniform or homogeneous [5]. This inhomogeneity arises mainly due to imperfection or defects in fiber core medium and fluctuations in core/cladding radius. Therefore, the study of nonlinear wave propagation in inhomogeneous media is of great interest in the area of fiber optics communications.

Recently, depending on the physical situations, several modifications to equation (1) has been suggested and the pulse propagation has been discussed (e.g. [3,6-10]). Of them, one of the important modifications, for example, was given by Flach and co-workers [9], who discussed resonant light scattering by solitary waves, by considering the process of light scattering by optical solitons in a planar waveguide with homogeneous and inhomogeneous refractive index cores. Considering certain inhomogeneous effect, nonlinear wave propagation in optical fiber has also been discussed in Ref. [10], where the author considered nonlinear compression of chirped solitary waves with and without phase modulation.

In case of inhomogeneities in optical fiber or fiber medium, the NLS equation [11] can be written in the general form

$$i\frac{\partial A}{\partial z}-s\frac{\partial^2 A}{\partial t^2}+2|A|^2 A=\varphi t^2 A\pm igA \tag{2}$$

where $f(z,t) = \varphi(z)\, t^2 A(z,t) \pm ig(z)\, A(z,t)$ is added in order to include the fiber power gain ($g>0$)/ loss ($g<0$) and phase modulation (φ) are both functions of distance z. φ is the coefficient of phase modulation. Since $s = \mathrm{sign}(\beta) = \pm 1$ the solution of equation (2), either gain or loss case, depends on whether β, the GVD parameter related to the frequency dependence of group velocity (v_g) defined by

$$\beta = \partial/\partial\omega(1/v_g) \approx (\lambda^3/2\pi c^2)(\partial^2 n/\partial\lambda^2), \tag{3}$$

is positive or negative. In all cases, the NLS equation can be solved by the inverse scattering transform method. The pulse-like solutions are found to occur only in the case of anomalous dispersion ($\beta< 0$), and are called bright solitons. In the case of normal dispersion ($\beta >0$), the solitary wave solutions of equation (2) appear as a dip in a constant background, and are called dark solitons. Equation (2) contains arbitrary functions of z, so we need to identify the integrability condition of the equation through linear eigenvalue problem [12]. The Lax pair [2] assures the complete integrability condition of a nonlinear system of equations and is used to achieve N-soliton solutions by means of inverse scattering transformation technique [13].

6. Soliton Solutions, Results and Discussion

6.1. Bright Soliton

6.1.1. Fiber Gain

In case of fiber gain, the NLS equation (Eq. (2)) with phase modulation reduces to

$$i\frac{\partial A}{\partial z}+\frac{\partial^2 A}{\partial t^2}+2|A|^2 A=\varphi t^2 A+igA\cdot \tag{4}$$

To construct the Lax pair for equation (4), we conveniently introduce a variable transformation (For more information about Lax pair construction, see [2]):

$$A(z,t)=a(z,t)\exp(-igt^2/2)\,.$$

The Lax pair associated with equation (4) is derived as

$$\frac{\partial\psi}{\partial t}=P\psi \quad \frac{\partial\psi}{\partial z}=Q\psi \quad \psi=(\psi_1\psi_2)^T \tag{5}$$

$$P=\begin{pmatrix}-i\Delta & a\\ -a & i\Delta\end{pmatrix}$$

$$Q=2i\Delta^2\begin{pmatrix}-1 & 0\\ 0 & -1\end{pmatrix}+2\Delta\begin{pmatrix}-igt & a\\ -a^* & igt\end{pmatrix}+i\begin{pmatrix}|a|^2 & a_t^*-2igta^*\\ a_t^*+2igta^* & -|a|^2\end{pmatrix},$$

where Δ is the variable spectral parameter given by $\Delta=\Delta_0\exp(2\int_0^z gdz)\cdot$ Using the above variable transformation in equation (4), we obtain

$$i\frac{\partial a}{\partial z}+\frac{dg}{dz}t^2\frac{a}{2}+\frac{\partial^2 a}{\partial t^2}+2|a|^2 a-2igt\frac{\partial a}{\partial t}-2iga-(g^2+\varphi)at^2=0\cdot \tag{6}$$

The compatibility condition $\frac{\partial P}{\partial z}-\frac{\partial a}{\partial t}+[P,a]=0$ gives

$$i\frac{\partial a}{\partial z}+\frac{\partial^2 a}{\partial t^2}+2|a|^2 a-2igt\frac{\partial a}{\partial t}-2iga=0\cdot \tag{7}$$

By comparing equations (6) and (7), we find that equation (2) is completely integrable and it thus gives exact solutions, and the integrability condition is $\frac{dg}{dz}-2(\varphi+g^2)=0$.

Now we introduce the Hirota derivative operators. A symbol D_x is called the Hirota derivative with respect to variable x and defined to act on a pair of functions $f(x)$ and $p(x)$ as follows:

$$D_x f(x).p(x)=(\partial_x-\partial_{x'})f(x)p(x')\big|_{x'=x},$$

where ∂_x denotes partial derivative with respect to x. Hirota derivatives are bilinear operators. We define the bilinear operators for our system as

$$D_z^m D_t^n b(z,t).c(z,t)=(\partial_x-\partial_{x'})^m(\partial_t-\partial_{t'})^n b(z,t)c(z',t')\big|_{x'=x,t'=t} \tag{8}$$

with the transformation in the form (for more information about Hirota's transformation, see [6])

$$a = \frac{b}{c}, \tag{9}$$

where $b(z,t)$ and $c(z,t)$ are complex and real functions respectively. Using the above transformation in equation (6), we obtain

$$(iD_z + D_t^2 - 2igtD_t - 2ig)(b.c) = 0, \;\; D_t^2(c.c) = 2b^2. \tag{10}$$

Multi-soliton solutions can be obtained by the following perturbation expansions of b and c:

$$b = \sigma b_1 + \sigma^3 b_3 + \sigma^5 b_5 + \sigma^7 b_7 + ... \tag{11}$$

$$c = 1 + \sigma^2 c_2 + \sigma^4 c_4 + \sigma^6 c_6 + ..., \tag{12}$$

where σ is an expansion coefficient. For one-soliton solution (1SS) only the first term and the first two terms in equations (11) and (12) respectively are needed. As we want 1SS, we will consider here those terms only. Pluging $b = \sigma b_1$ and $c = 1 + \sigma^2 c_2$ in equation (10) and collecting the terms (coefficients of σ, σ^2, σ^3 and σ^4) with same power of σ, we obtain:

$$\left.\begin{array}{l} (iD_z + D_t^2 - 2igtD_t - 2ig)(b_1.1) = 0 \\ D_t^2(1.c_2 + c_2.1) = 2|b_1|^2 \\ (iD_z + D_t^2 - 2igtD_t - 2ig)(b_1.c_2) = 0 \\ D_t^2(c_2.c_2) = 0 \end{array}\right\}. \tag{13}$$

We solve the set of equations (13). In deriving the solutions for b_1 and c_2, we conveniently assume $b_1 = e^{\nu+\mu}$ and $c_2 = e^{\nu+\mu+(\nu+\mu)^*}$, where $(\nu+\mu)^*$ is the complex conjugate of $(\nu+\mu)$. After some algebra, equation (7) yields

$$a(z,t) = 2\varphi_2 e^{\nu} \sec h(\mu), \tag{14}$$

where $\nu = -2i\{\varphi_1 t + 2\int_0^z (\varphi_1^2 - \varphi_2^2)dz\}$ and $\mu = 2(\varphi_2 t + 4\int_0^z \varphi_1\varphi_2 dz) + \Omega$ (Ω is an integration constant) with $\varphi_1 = \varphi_{10}\exp(2\int_0^z g dz)$ and $\varphi_2 = \varphi_{20}\exp(2\int_0^z g dz)$. Using variable transformation, defined earlier ($a \to A$), we get

$$A(z,t) = 2\varphi_2 e^{\nu - igt^2/2} \sec h(\mu). \tag{15}$$

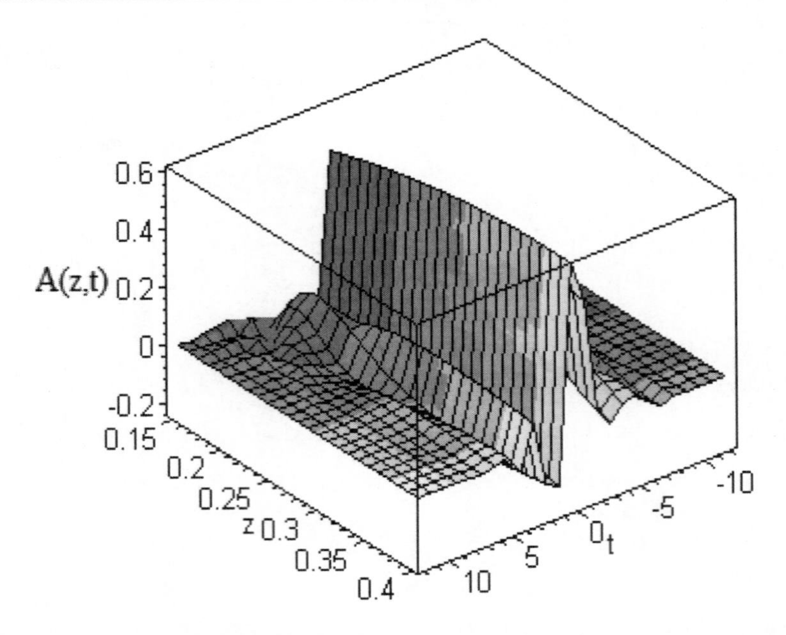

Figure 4. Amplitude profile of A(z,t) for the bright 1SS of equation (15).

Equation (15) is the exact bright 1SS of equation (4), i.e. with fiber gain and phase modulation, which with parameter values $\varphi_{20} = \varphi_{10}/2 = 1/2$, $g = 0.2/z$ and Ω=0 is plotted in Figure 4. We thus have obtained the exact bright soliton solution (Eq. (15)) using Hirota transformation technique.

Figure 4 shows that the pulse amplitude increases as z increases. This is due to the admittance of $\varphi_2(z)$ in the amplitude (equation (15)), which shows that the soliton amplitude grows at the same rate as the respective power amplitude with the inclusion of phase modulation and gain. As a result of chirping, the pulse broadens as it propagates along the fiber. At each stage of propagation, the product of pulse width and pulse amplitude is found to be conserved for which the area, occupied by the pulse envelop, remains preserved with the inclusion of gain and phase modulation. This can be seen in Figure 4.

6.1.2. Fiber Loss

Now we consider Eq. (2) with phase modulation and fiber loss. Proceeding as before, it is found that equation below is integrable with condition $\frac{dg}{dz} + 2(\varphi + g^2) = 0$:

$$i\frac{\partial A}{\partial z} + \frac{\partial^2 A}{\partial t^2} + 2|A|^2 A = \varphi t^2 A - igA \tag{16}$$

The 1SS of equation (16) is obtained by choosing the variable transformation: $A = a(z,t)e^{igt^2/2}$. Following the same procedure, we derive the following soliton solution

$$A(z,t) = 2\varphi_2 e^{\nu + igt^2/2} \sec h(\mu), \tag{17}$$

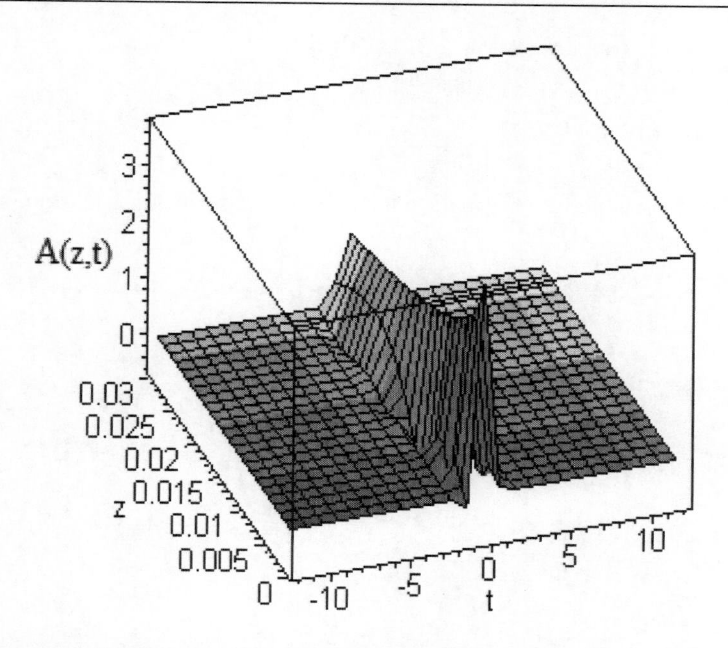

Figure 5. Amplitude profile of A(z,t) for the 1SS of equation (17).

Thus we have obtained the exact bright 1SS for the wave propagation in the inhomogeneous optical fiber with phase modulation and fiber loss. Equation (17) is plotted ($\varphi_{20}=\varphi_{10}/2=1/2$, $g=0.1/z$ and Ω=0) in Figure 5. The same modulating term, as in equation (15) but with opposite sign, is also appeared in the amplitude of soliton solution with fiber loss, i.e. in the amplitude of equation (17), which shows that the amplitude decays at the same rate as the respective power amplitude. This effect can also be seen in Figure 5, where soliton amplitude decreases as z increases. As a result of chirping, the pulse broadens as it propagates along the fiber, and as a result of damping, the amplitude of the pulse reduces. The amplitude of the pulse is found to increase or decrease, depending on the sign of $g(z)$ (gain/ loss), with the same amount of broadening in the pulse width during its propagation such that the area of the pulse remains constant. With the inclusion of phase modulation and fiber loss, Figure 5 clearly depicts the damping effect in solitary pulse propagation in the fiber with an inhomogeneous core and/or inhomogeneous core/cladding radius.

6.2. Dark Soliton

6.2.1. Fiber Gain

Dark solitons have recently been investigated because of their potential in applications, such as, in optical logic devices, wave guide optics as dynamic switches and junctions [14,15], and communication applications due to their inherent stability and less fiber loss [16] than that of bright soliton. Now, in this case (for β>0 and g>0), the NLS equation becomes

$$i\frac{\partial A}{\partial z}-\frac{\partial^2 A}{\partial t^2}+2|A|^2 A=\varphi t^2 A+igA. \tag{18}$$

We make a variable transformation for Eq. (18) as

$$A(z,t) = U(z,t)\exp(igt^2/2). \tag{19}$$

Using the above variable transformation in Eq. (18) the resulting equation is

$$i\frac{\partial U}{\partial z} - \frac{dg}{dz}t^2\frac{U}{2} - \frac{\partial^2 U}{\partial t^2} + 2|U|^2 U - 2igt\frac{\partial U}{\partial t} - 2igU - (g^2 - \varphi)Ut^2 = 0 \cdot \tag{20}$$

The Lax pair associated with Eq. (20) is derived as

$$\frac{\partial \Omega}{\partial t} = P\Omega \quad \frac{\partial \Omega}{\partial z} = Q\Omega \quad \Omega = (\Omega_1 \Omega_2)^T, \tag{21}$$

where

$$P = \begin{pmatrix} -i\Delta/2 & -iU \\ iU^* & i\Delta/2 \end{pmatrix}$$

and

$$Q = 2i\Delta^2\begin{pmatrix} i/2 & 0 \\ 0 & -i/2 \end{pmatrix} + 2\Delta\begin{pmatrix} -igt & iU \\ -iU^* & igt \end{pmatrix} + \begin{pmatrix} i|U|^2 & -\frac{\partial U^*}{\partial t} - 2igtU^* \\ -\frac{\partial U^*}{\partial t} + 2igtU^* & -i|U|^2 \end{pmatrix}$$

Here U^* is the complex conjugate of U and Δ is the variable spectral parameter given by $\Delta = \Delta_0 \exp(2\int_0^z g dz)$. The compatibility condition $\partial P/\partial z - \partial Q/\partial t + [P,Q] = 0$ gives

$$i\frac{\partial U}{\partial z} - \frac{\partial^2 U}{\partial t^2} + 2|U|^2 U = 2igt\frac{\partial U}{\partial t} + 2igU. \tag{22}$$

By comparing Eqs. (20) and (22), we find that equation (18) is completely integrable with the integrability condition $dg/dz + 2(\varphi - g^2) = 0$ and it thus gives exact solutions.

We now define the bilinear operators for our system as

$$D_z^m D_t^n b(z,t).c(z,t) = (\partial_x - \partial_{x'})^m(\partial_t - \partial_{t'})^n b(z,t)c(z',t')\big|_{x'=x,t'=t} \tag{23}$$

with the transformation in the form

$$U = \frac{b}{c}, \tag{24}$$

where b(z,t) and c(z,t) are complex and real functions respectively. Using the above transformation in Eq. (22), we obtain

$$(iD_z - D_t^2 - 2igtD_t - 2ig + f)(b.c) = 0,\ D_t^2(c.c) = 2b^2, \tag{25}$$

where f is an arbitrary function. Dark 1SS can be obtained by the following perturbation expansions of b and c :

$$b = b_0 + \sigma b_0 b_1 \tag{26}$$

$$c = 1 + \sigma c_1, \tag{27}$$

where σ is an expansion coefficient. Pluging $b = b_0 + \sigma b_0 b_1$ and $c = 1 + \sigma c_1$ in Eq. (25) and collecting the coefficient of σ^0, we obtain

$$(iD_z - D_t^2 - 2igtD_t - 2ig + f)(b_0.1) = 0,\ f = 2|b_0|^2. \tag{28}$$

Equations (28) are solved for f and the result gives $f = 8\varphi_{10}^2 \exp(4\int_0^z gdz)$, where $b_0 = 2\varphi_{10}\exp\{\int_0^z (if + 2g)dz\}$. Making use of Eqs. (25)-(27) and then collecting the terms with the same power of σ, i.e. the coefficients of σ and σ^2, we obtain the following equations

$$\left.\begin{aligned} &(iD_z + D_t^2 - 2igtD_t)(b_1.1 + 1.c_1) = 0 \\ &(D_t^2 - \kappa)(1.c_1 + c_1.1) + 4b_1|b_0|^2 = 0 \\ &(iD_z + D_t^2 - 2igtD_t)(b_1.c_2) = 0 \\ &(D_t^2 - \kappa)(1.c_1 + c_1.1) + 2b_1|b_0|^2 = 0 \end{aligned}\right\} \tag{29}$$

We solve the set of equations (29). In deriving the solutions for b_1 and c_1 from equation (29), we conveniently assume $b_1 = -e^{4\varphi_1 t + \delta_1}$ and $b_1 + c_1 = 0$, where δ_1 is a constant. After some algebra, equation (22) yields

$$U(z,t) = 2\varphi_1 e^{\xi_1} \tanh\gamma_1, \tag{30}$$

where $\xi_1 = i\int_0^z (8\varphi_1^2 dz \pm \pi)$ and $\gamma_1 = 1/2(4\varphi_1 t + \delta_1)$ with $\varphi_1 = \varphi_{10}\exp(2\int_0^z gdz)$. Using variable transformation defined earlier (Eq. (19)), we get

$$A(z,t) = 2\varphi_1 e^{\xi_1 + igt^2/2} \tanh\gamma_1. \tag{31}$$

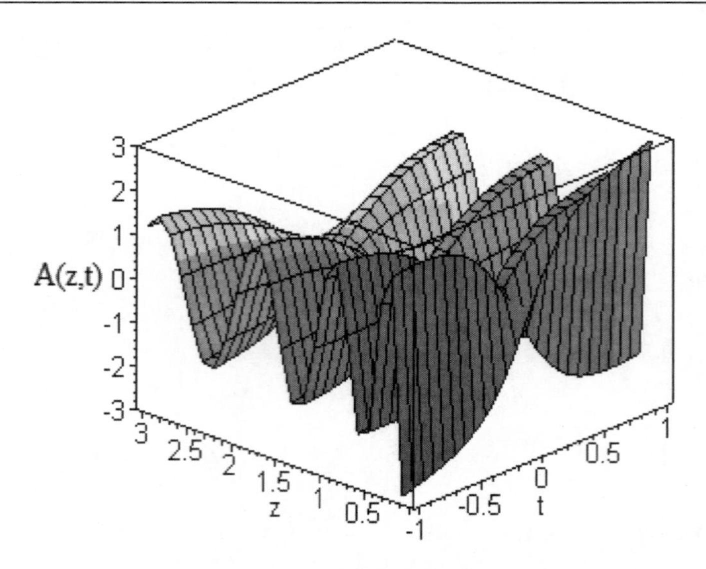

Figure 7. Amplitude profile of A(z,t) for the dark 1SS of equation (31).

Equation (31) is the exact dark 1SS (of Eq. (18)) with fiber gain and phase modulation, which is plotted in Figure 7. As shown in the figure, the depth of the dark soliton increases as z increases while the width decreases. This is due to the admittance of the terms $\varphi_1(z)$ and $\beta>0$ in the Eq. (31), where the soliton amplitude grows at the same rate as the respective power amplitude with the inclusion of phase modulation and gain.

6.2.2. Fiber Loss

Now we consider Eq. (18) with phase modulation and fiber loss ($g<0$) in the normal dispersive regime ($\beta>0$). Proceeding as before, it is found that the appropriate equation for the system below is integrable with condition $dg/dz-2(\varphi-g^2)=0$:

$$i\frac{\partial A}{\partial z}-\frac{\partial^2 A}{\partial t^2}+2|A|^2 A=\varphi t^2 A-igA. \tag{32}$$

The 1SS of Eq. (32) is obtained by choosing the variable transformation: $A(z,t)=U(z,t)\exp(-igt^2/2)$. Following the same procedure, we obtain

$$U(z,t)=2\varphi_2 e^{\xi_2}\tanh\gamma_2, \tag{33}$$

where $\xi_2=i\int_0^z(8\varphi_2^2dz\pm\pi)$, $\gamma_2=2\varphi_2 t+\delta_2$, δ_2 is a constant, $\varphi_2=\varphi_{20}\exp(2\int_0^z gdz)$ and the eigenvalue problem associated with Eq. (32) is

$$P=\begin{pmatrix}-i\Delta/2 & -iU\\ iU^* & i\Delta/2\end{pmatrix},\quad Q=\Delta^2\begin{pmatrix}i/2 & 0\\ 0 & -i/2\end{pmatrix}+\Delta\begin{pmatrix}igt & iU\\ -iU^* & -igt\end{pmatrix}+\begin{pmatrix}i|U|^2 & -\dfrac{\partial U}{\partial t}+2igtU\\ -\dfrac{\partial U^*}{\partial t}-2igtU^* & -i|U|^2\end{pmatrix}.$$

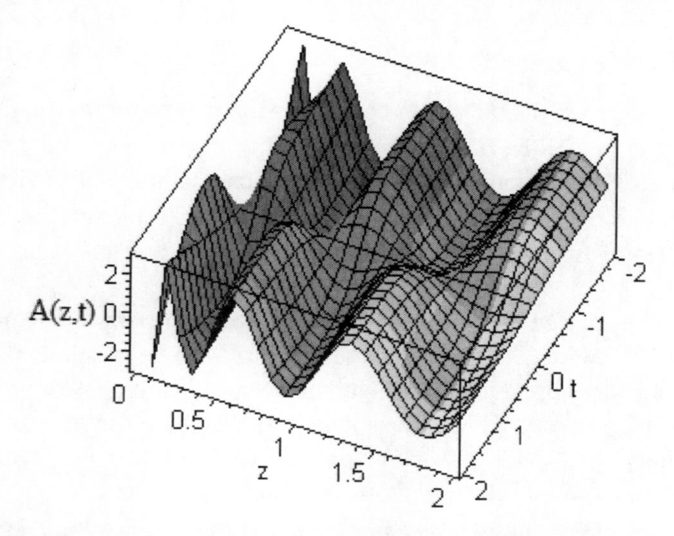

Figure 8. Amplitude profile of A(z,t) for the dark 1SS of equation (34).

The dark soliton solution with fiber loss is then

$$A(z,t) = 2\varphi_2 e^{\xi_2 - igt^2/2)} \tanh \gamma_2 . \tag{34}$$

We thus have obtained the exact dark 1SS for the wave propagation in the inhomogeneous optical fiber with phase modulation and fiber loss. Equation (34) is plotted in Figure 8. The same modulating term, as in equation (31) but with opposite sign, is also appeared in the soliton solution with fiber loss, i.e. in the equation (34), where the depth of the soliton decreases as z increases while the width increase so that the amplitude decays at the same rate as the respective power amplitude. As a result of chirping, the pulse broadens as it propagates along the fiber, and as a result of damping, the amplitude or depth of the pulse reduces. The depth of the solitary pulse is found to increase or decrease, depending on the sign of g, i.e. on fiber gain or loss, with the same amount of broadening in the width during its propagation such that the product of the depth and width, i.e. the area remains constant. With the inclusion of phase modulation and fiber loss, Figure 8 clearly depicts the damping effect in solitary pulse propagation in a fiber with inhomogeneous core and/or inhomogeneous core/cladding radius. The results show that the area of the solitary pulse envelope for either gain or damping (loss) is preserved during propagation, as found for bright solitons in the anomalous dispersive regime shown earlier. The results are also consistent with those obtained in an earlier investigation [11].

7. Problems in Solitonic Propagation

Generating pulses with the appropriate parameters for soliton formation is just the first step in a successful soliton communication system: solitons must also be able to propagate through an optical fiber while maintaining these parameters. In particular, solitons must be able to retain their approximate peak power defined [4, 17] by

$$P = -\beta/(\gamma t_o^2), \tag{35}$$

where P is the peak power of the pulse, β is the dispersion parameter (a negative value since it is in the anomalous regime of the fiber), γ is the nonlinear coefficient, and t_o is the half-width of the pulse at its $1/e$ (36.79%) intensity point. This, of course, means that fiber loss is a huge potential problem. Here, $\gamma = 2\pi n_2 /(\lambda A_{eff})$, n_2 is the nonlinear (refractive) index coefficient, a measure of the degree of fiber nonlinearity, and A_{eff} is the effective core area of the fiber [18, 5].

Fiber loss occurs because the optical fiber is not 100% transparent. The pulse wavelength is chosen in order to minimize loss, but there is always some amount of absorption and scattering of the constituent photons. The result is a decrease in the power of the pulse, and if it is a soliton, the dispersion and SPM effects must broaden the pulse in order to maintain the correct relationship for peak power given above. Soliton broadening in soliton communication systems is harmful for the same reasons of pulse broadening due to chromatic dispersion is harmful in non-soliton systems. This problem may be addressed by periodically passing the solitons through optical amplifiers, which counteract fiber loss by increasing the power of the pulse. The most common type of optical amplifier is the erbium-doped fiber amplifier (EDFA), which amplifies the pulse through a process of stimulated emission similar to what occurs in a laser [1]. However, EDFAs and other amplifiers also introduce noise into the system via spontaneous emission, and this noise may perturb the system by interfering with the soliton signals and changing their parameters. Therefore, high-quality amplifier design is critical to the proper functioning of any soliton communication system.

The bit-rate of a soliton communication system is limited not only by the modulation speed of the laser used to generate the solitons, but also by the interactions that may occur between two neighboring solitons if they are too close together [5]. These interactions, another result of fiber nonlinearity, may cause two adjacent solitons to periodically combine together and then separate from each other. If two neighboring solitons combine, they appear to the detector as just a single pulse; consequently, an error occurs. Soliton-soliton interactions may be minimized by ensuring that adjacent solitons are far enough apart that they do not interfere with each other. However, this may reduce the bit-rate of the system since it limits the number of solitons that may be transmitted in any given time interval. Solitons may encounter many other problems or perturbations as they propagate through an optical fiber due to inhomogeneities in fiber core medium and/or inhomogeneous core/cladding radius.

Conclusion

The more general case of inhomogeneities in fibers was taken in present study. The NLS equation including fiber loss or gain and phase modulation in both anomalous and normal dispersive regimes was considered. A general integrability condition was arrived and the modified NLS was bilinearized and solved using an advanced mathematical solution techniques, namely Lax pair construction and Hirota transformation, for all the cases. We finally constructed the bright and dark solitons with fiber loss/gain. The solutions obtained numerically and analytically explain how one can achieve the pulse compression keeping solitonic property when it is transmitted through the fiber with the addition of fiber loss or gain and phase modulation. In both cases (bright and dark), the amplitude of the phase was found to decrease in an exponential manner for damping (loss) and to increase in an exponential way for gain with the pulse width broadening at each stage of propagation such that the area of the pulse envelope remained preserved. The results were discussed by highlighting present challenges in solitonic propagation in optical fiber communications.

REFERENCES

[1] A. Hasegawa, Fundamentals of optical soliton theory in fibers, In: *Optical solitons: Theoretical challenges and industrial perspectives* (V. E. Zakharov and S. Wabnitz edited), Springer, 1998.

[2] M. Remoissenet, *Waves called solitons*, Springer-Verlag, 1993.

[3] D. E. Pelinovsky, Y. S. Kivshar and V. V. Afanasjev, *Phys. Rev. E 54,* 2015, 1996.

[4] G. P. Agrawal, *Nonlinear fiber optics*, Academic Press, 1989.

[5] L. W. Couch II, *Digital and analog communication systems*, 5th edition, Prentice Hall, 1997.

[6] R. Radhakrishnan and M. Lakshmanan, *J. Phys. A: Math. Gen.* 28, 2683, 1995.

[7] Y. S. Kivshar and B. Luther-Davies, *Phys. Rep.* 298, 81, 1998.

[8] D. P. Caetano and J. M. Hickmann, *J. Opt. Soc. Am. B* 23, 1655, 2006.

[9] S. Flach, V. Fleurov, A. V. Gorbach and A. E. Miroshnichenko, *Phys. Rev. Lett.* 95, 023901, 2005.

[10] J. D. Moores, *Opt. Lett.* 21, 555, 1996.

[11] M. I. Miah, J. *Optoelectron. Adv. Mater.* 9, 2339 (2007).

[12] M. I. Miah, *Solitonic motion and integrable Heisenberg spin equation, in preparation.*

[13] K. Konno and M. Wadati, *Prog. Theor. Phys.* 53, 1652, 1975.

[14] G. A. Swartzlander, *Opt. Lett.* 17, 493, 1992.

[15] B. Luther-Davies and Y. Xiaoping, *Opt. Lett.* 17, 496, 1992.

[16] Y. S. Kivshar and B. Luther-Davies, *Phys. Rep.* 298, 81, 1998.

[17] G. P. Agrawal, *Applications of nonlinear fiber optics*, Academic Press, Inc., 2001.

[18] B. E. A. Saleh and M. C. Teich, *Fundamentals of photonics*, John Wiley and Sons, 1991.

In: Optical Fibers: New Developments
Editor: Marco Pisco

ISBN: 978-1-62808-425-2

Chapter 7

LAST-MILE TECHNOLOGIES: NEW WDM ACCESS PROPOSALS, DEVICES AND EXPERIMENTS

Qian Deniel[1], Fabienne Saliou[1], Philippe Chanclou[1], Lucia Marazzi[2], Paola Parolari[2], Marco Brunero[2], Alberto Gatto[2], Mario Martinelli[2], Giancarlo Gavioli[3], Alessandro Iachelini[3], Antonio Cimmino[3], R. Brenot[4], Sophie Barbet[4], François Lelarge[4], Sean O'Duill[5], Simon Gebrewald[5], Christian Koos[5], Wolfgang Freude[5] and Juerg Leuthold[5]

[1] Orange Labs - France
[2] Politecnico di Milano – Italy
[3] Alcatel Lucent Italia – Italy
[4] III-V Labs – France
[5] Institute of Photonics and Quantum Electronics of the Karlsruhe Institute of Technology – Germany

ABSTRACT

Wavelength division multiplexed passive optical networks (WDM PON) offer a promising solution for a variety of applications in Next Generation Access Network (NGAN). Up to now WDM PON technology has been unsatisfactorily obtained by either costly tuneable transmitters or by exploiting external seeding sources. The key to enabling cost-effective WDM PON architectures is the deployment of colourless Optical Network Unit (ONU) transmitters, so that generic self-tuning transmitters can be employed with an automatic and passive selection by the optical infrastructure of the wavelength for each user. The chapter presents FP7 European project ERMES approach to the provision of colourless ONU transmitters. The breakthrough idea is to use a significant portion of the network to implement an embedded, self-tuning and modulable laser cavity. This idea is based on establishing a very long cavity laser with facility for direct modulation. The goal is achieved by using a dedicated device whose constituents are: the gain chip at the ONU side to create the circulating laser field and to provide modulation functionality; at the remote node only a wavelength multiplexer/

demultiplexer and a reflector is required; the distribution fibre connects the ONU to the RN. Once a steady-state circulating laser field is established, the active chip can be directly modulated enabling up-stream up to 10 Gb/s data-rate per user. The development of a device suitable for this application is the central goal of the ERMES project. The ERMES consortium comprises of leading industrial and academic partners with strong expertize in optical network construction, device fabrication, laboratory experimental trials and theoretical investigations. The intra-consortium collaboration allows for the swift development of such a transmitter. This solution is potentially a highly effective alternative to the existing approaches in terms of cost reduction as it is colourless and obviates the need for wavelength-specific external seeding sources. It is also appealing in terms of achievable performance, with operation to 10 Gb/s transmission and as there are no impairments from Rayleigh back-scattering, longer bridged distances are expected. In the proposed contribution the main applications of the ERMES project approach are initially reviewed together with the requested specifications from component vendors and network operators point of view. The principle of operation is explained and discussed in details with specific attention to issues arisen during the first year project. Experimental results are presented and commented with the help of the first modelling result. Finally the project future goals are illustrated.

Keywords: WDM, PON, FTTH, Backhauling, last mile technologies, seeding, NGAN

INTRODUCTION

For the Next Generation Access Network (NGAN) wavelength division multiplexed passive optical networks (WDM PON) appear a promising and suitable solution offering almost unlimited bandwidth similarly to point-to-point links, while allowing the advantages of fibre sharing. It is widely recognized that WDM PON deployment requires colourless Optical Network Unit (ONU) transmitters, so that each user has the same transmitter. Up to now these characteristics has been unsatisfactorily obtained either with a costly tunable transmitter or by exploiting external seeding sources. This chapter proposes a disruptive approach to the ONU transmitter. The breakthrough idea is to use a significant portion of the network to implement an embedded self-tuning modulable laser cavity.

This idea is based on establishing a very long cavity laser, which can be directly intensity modulated. The goal is achieved by using a dedicated multifunction active chip (MFAC), which acts as the gain medium of the cavity including: the array waveguide grating (AWG) and a reflector at the remote node (RN), and the distribution fibre connecting the ONU to the RN. After cavity set up, the active chip is directly modulated enabling up-stream up to 10 Gb/s data-rate per user. The development of a MFAC suitable for this application is mandatory to bring this approach from the proof-of-principle stage closer to industrial exploitation. This solution is a potential highly effective alternative to the existing approaches in terms of cost reduction as it is colourless and gets rid of the need for external seeding sources. It is also appealing in terms of achievable performance, as it is not impaired by Rayleigh back-scattering, allowing for longer bridged distances.

The success of these last-mile optical technologies will therefore enforce European industrial leadership in the access arena, whose development in terms of capillarity and bandwidth has major social and economical fallouts.

1. WDM PON AS A POSSIBLE SOLUTION FOR SPECIFIC NICHE APPLICATIONS

Passive Optical Network (PON) is considered by all the actors, including regulation authorities and alternative operators, as the least expensive way to deploy FTTH access in the world for residential market. Present PON solutions are based on time division multiplexing & time division multiple access (TDM & TDMA) to achieve a point to multipoint connectivity. G-PON is the technology under deployment following by XG-PON1 and the wavelength stacking of TDM&TDMA PON (named NG-PON2 by FSAN group).

WDM PON based on the definition of one wavelength per user, is not considered as a solution for this mass market. Several reasons could explain this choice but some could become the reason of his adoption by mobile application. We can list here the fact that wavelength per user achieve: a point to point connectivity, an independence of protocol and encapsulation to managed the connectivity and an already capable to achieve about 1Gbit/s and in the future 10Gbit/s per wavelength.

We have now to understand why mobile application needs such fiber network. Radio access networks, GSM (900 MHz, then 1800 MHz) then 3G (2.1 GHz) and LTE (2.5 GHz and 3.5 GHz) use higher frequency bands with higher propagation attenuations which leads to smaller cell sizes and higher sites density. Another reason for sites densification is obviously the increasing traffic demand especially in urban areas. Hence, an increasing number of sites must be acquired or leased by the operator. This results in higher acquisition or rental and maintenance costs. Base station hotelling could be a solution that reduces these costs and enables evolved radio resource management techniques. If the "traditional" base station is located near a tower antenna, in an enclosure (or cabinet) at the base of the tower; the enclosure contains baseband processing, the radio part (power amplifier, frequency conversion, AD/DA converters), network interface and power supply, including backup power source. The electrical cable between the antenna and the base station is lossy, a power amplifier is then installed next to the antenna, to compensate this loss.

Thanks to the recent progress in SDR technology area, it is possible to separate analog and digital processing parts of a base station. The analog processing is performed in the Remote Radio Head (RRH) located near the antenna. The baseband digital processing is performed in the Base Band Unit (BBU). These two units form the "distributed base station". Base station hotelling is the step further. It consists in co-locating many BBUs of different base stations in the same "Base Station Hotel". An emerging strategy to enable cost-effective mobile-coverage improvements is the distributed base station architecture, often referred to as the cloud RAN or C-RAN. Lower-cost, lower-power RRHs can be more easily and pervasively deployed closer to subscribers while the more complex and costly component, the BBU, can be centralized in a data center or base station hotel similar to a cloud computing model. The cloud RAN provides a number of benefits including improved performance, more efficient use of capital and human resources, and energy efficiency.

Going back to our fiber network consideration, these BBUs are connected to the corresponding RRH with optical fiber. Cloud RAN (including BBU pooling & hostelling) creates new fiber network requirements to achieve the FTTA (Fiber To The Antenna). This link between BBU and RRH is also called “fronthaul” in analogy with the standard backhaul of mobile network which correspond to the transport of the aggregated traffic and synchronisation signal. Two standards (Open Base Station Architecture Initiative, OBSAI, and Common Public Radio Interface, CPRI) exist to define the fronthaul interface. CPRI and OBSAI share a number of similarities but CPRI has penetrated the market much more quickly thanks to its more narrow focus (CPRI defines only the interface between RRH and BBU whether OBSAI aims at standardising the whole base station architecture) that leaves more room for flexibility and makes it less complex. Moreover, standardization activities are ongoing for multivendor interoperability in the Open Radio Interface (ORI) working group at ETSI, based on CPRI specifications. CPRI line rates go from 614.4 up to 9830 Mbit/s as a result of the digitisation of the radio signal. As an example, for a 20MHz 2x2 MIMO LTE signal, the corresponding CPRI line rate is 2457 Mbit/s for uplink and downlink. Several vendors are working towards CPRI line rate compression of a factor 3. CPRI and OBSAI support single mode (SMF) and multimode fibres (MMF) with maximum fibre length up to 40km for SMF. The use of transceivers based on Small Form-factor Pluggable (SFP) give a high flexibility. Indeed, SFPs are very low cost and are available at different wavelengths (WDM). The easiest way is then to dedicate a point to point fibre link, but infrastructure mutualisation has to be considered. Due to the fact that for one antenna site, one CPRI link is requested per sector (typically three sectors per antenna), per carriers, per mobile generation system. Multiplexing several CPRIs inside one fiber link could have an economical interest. WDM appear as a candidate multiplexing technology. When you add that CPRI has some stringent requirements such as a low latency (typically round trip times for this segment below 400µs (equivalent to 40km reach) for long term evolution-advanced (LTE-A) and 700µs for LTE), symmetrical bit rates ranging from 0.6 to 10Gbit/s for up- and down-stream, and strong requirement on jitter, time delay calibration concerning synchronisation, the WDM PON become an interesting solution. The fact that WDM PON could do the connectivity without any encapsulation method (CPRI untouched) is also a promising property for front-haul network.

2. Specification for Such Application by Vendors

The pace of fiber network rollouts around the world continues to accelerate. In 2009, there were roughly 20,000 new users connected daily to fiber-based passive optical networks (PON) worldwide; this number will increase by 70 percent in 2010. Governments are making huge investments in fiber initiatives, yet no one can say definitively how all this fiber will be used in the future. In the same way electrical plants were built 70 years ago with enough capacity to support appliances that nobody could have dreamed of back then, today's broadband rollout is laying the foundation for a whole new kind of digital life. Are today's fiber networks future poof? How long will we be able to continue to rely on them? How much of a network operator's investment can be protected over the longer term? The technology innovation looks at the possibilities beyond current-generation PON – and at what operators must do to protect legacy investments while ensuring their networks' readiness for future evolution.

Globally, fiber optic rollouts show no sign of slowing down, although the 'growth profile' varies from country to country, beginning with operators' choice of technology. In Asia, EPON (Ethernet PON) dominates, with Japan and South Korea as the pioneers and

China joining more recently with a mix of EPON and GPON (gigabit PON) deployments. GPON remains the first choice in North America, with a small percentage of point-to-point (P2P) optical deployments driven by municipalities and developers. Europe has a mix of P2P and GPON deployments, the latter being the preferred choice of carriers.

Although current technologies such as GPON will easily meet the mid-term needs of 40 to 60 Mb/s per user, over the longer term they will struggle to answer the requirements of HDTV, 3DTV, multiple image and angle video services, growth in unicast video (versus broadcast), cloud computing, telepresence, multiplayer video gaming and more. Looking into that farther future, operators need to consider today which optical networking platform will allow them to evolve and adapt most cost-effectively and intelligently as conditions change.

Technology Snapshots

EPON

Mass deployments of EPON began in 2004 and today serve more than 30 million subscribers. Of the two PON architectures, the existing version of EPON is most hampered by its capacity, supporting just 1 Gb/s for both uplink and downlink. This is becoming insufficient for the demands of today's IP multimedia services. Recognizing that an update was long overdue, IEEE ratified a new standard for 10G EPON in 2009, with higher-bandwidth products set to ship in the second half of 2010.

GPON

GPON is slightly younger than EPON, with mass rollout starting in 2006. Offering 2.5 Gb/s downlink and 1.25 Gb/s uplink, it does not face the same immediate capacity concerns as EPON. New standards for 10G GPON — also referred to as 'NG-PON' or '10G-PON' for its 10 Gb/s downlink speed — are expected to be ratified in 2010, with initial deployments coming in 2011–12.

What is Next-generation PON?

Operators and equipment vendors are actively collaborating to create the next generation of optical technology. Most research and development has focused on developing and commercializing 10G EPON and 10G GPON, with the goal in both cases being to increase bandwidth without disrupting services or affecting the outside plant. Aware of EPON's limitations, IEEE began working on a 10G EPON white paper in 2007 and ratified the new standard in 2009. IEEE envisioned two tracks for the new standard: one asymmetrical, with 10G downstream and 1G upstream; the other symmetrical at 10G in both directions. ITU/FSAN has taken a slightly different approach to GPON. The FSAN next-generation technology roadmap extends beyond 2015 and identifies two distinct waves. The first, known as NG-PON1, is underway today. It proposes both asymmetrical and symmetrical options: the asymmetrical providing 10G downstream and 2.5G upstream (XG-PON1) and the symmetrical delivering 10G in both (XG-PON2). Next-generation GPON's higher upstream bandwidth compared to EPON) is driven by emerging new services that require greater upstream capacity, such as cloud computing and video sharing.

The second wave, NG-PON2, is expected around 2015 and accounts for technologies such as DWDM and OFDM that will enable even higher bandwidth capabilities. Unlike NG-PON1 — which preserves the existing outside plant — NG-PON2 will likely require changes in the fiber plant.

New Generation PON Requirements

Capacity may be the most conspicuous driver of the shift to next-generation PON, but it's not the only one. Increased reach and split, greater resiliency, improved power consumption and enhanced optical troubleshooting are also among the benefits service providers are seeking from migration. Looking beyond the next decade, even greater network speeds will be required. Given that GPON is more widely distributed than EPON, this next section looks at possible developments post-10G. GPON. Preliminary work is underway to determine how speeds of 40 Gbit/s and even 100 Gbit/s could be achieved.

TDM PON

The TDM PON architecture is very similar to existing GPON systems, with the electronics and optics being much faster. Although this is conceptually simple technology and it allows reuse of the existing fiber plant, the cost of developing and manufacturing optics to support these data rates would be high. Additionally, basic limitations of the fiber start to come into play through factors such as chromatic dispersion — making pure TDM an unlikely approach.

WDM PON

WDM PON is one of the most intriguing potential technologies of the future, combining the best of P2P and PON by creating a logical point-to-point with end users — with no bandwidth limitation — while preserving the economical advantages of PON. In WDM PON, one wavelength is assigned per customer, enabling very high-speed transport. There are, however, challenges. First, it seems likely that the most expensive form of WDM, DWDM, would be needed. Second, ensuring wavelength stability at the user end under extreme temperature conditions would be far from easy. Filtering requirements at the user end would also be complex, and upgrades and changes would be more difficult. It is expected that color-specific ONTs would likely be needed when using DWDM PON; however, there is no clear best solution to this problem at the present.

Dense Self-Seeding WDM-PON Technique (DSS WDM-PON)

A self-seeding DWDM-PON technique can be used in some Optical Distribution Network (ODN) scenarios (pure WDM-PON and overlay with TDM-PON) with the aim to evaluate pros and cons in the context of mobile backhaul and fixed network applications. Low cost, colourless, uncooled, up to 10G scalable baud-rate and photonic/electronic integration technologies are presented to build converged networks leveraging on centralized wireless resource pooling.

The basic configuration of a Dense Self-Seeding WDM-PON (pure DSS WDM-PON) architecture is shown in Figure 1. Fundamentally the architecture uses Wavelength Division Multiplexing to provide point to point logical links over a bidirectional point to multipoint physical infrastructure. These links are merged in the OLT and fanned out at the RN location via Mux/Demux functions. The feeder fiber, the RN Mux/Demux and the distribution fibers constitute the passive bidirectional ODN of the access network.

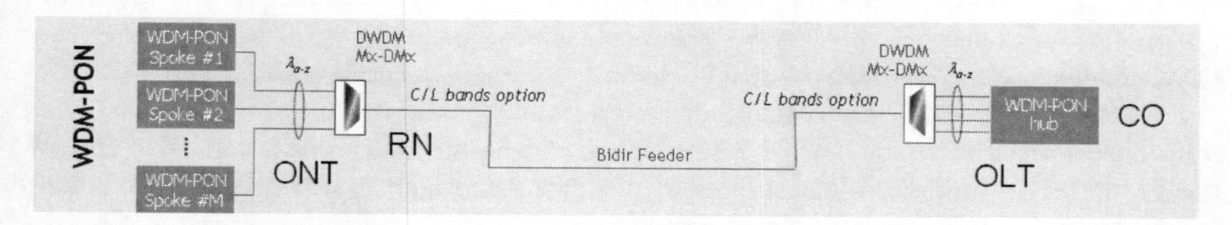

Figure1. Self-seeding based architecture.

The self-seeding technique applies between the ONU/ONT transceiver and related DWDM Mux/Demux filter plus mirror, where there is an unconventional external long cavity instead of a traditional integrated cavity inside the transceiver source. The coloured reflection from the RN allows the ONU/ONT to tune itself without the need for other complex tuning components. The self-seeding concept and performances (power budget, reach, transport capacity, etc.) will be described in the following chapters. For those analysis will be assumed a symmetric architecture, with ONU/ONT and OLT ERMES transmitters technology as a particular case of a more general system development. The proposed wavelength plan uses the S to L band and aims to coexist with other TDM-PON platforms and video technologies (GPON, XG-PON, RF-VIDEO, etc.). Further wavelength plans in other bands will be analysed in the future if there will be request of applications.

Self-seeding DWDM-PON addresses main NGPON2 requirements such as bandwidth, transport capacity and related symmetry, link security, CO consolidation, versatile ONU/ONT. It is important to consider the roadmap of fixed and mobile services requirements. At this time, 1G25 and 2.5G symmetrical bandwidth per user seems to be enough for a 2015 converged NGPON2 requirements (residential services). However 10G could be required in the same time period for future mobile backhauling of CPRI like based BBU hoteling and pooling or further business applications. 10G DSS WDM PON evolution and specification for 10G ERMES transmitter will be the objective of future elaborations along the ERMES project roadmap.

Beyond these basic architectures, other variants can be conceived. For example we can build a two stage mux/demux function at CO or RN side to provide access to several operators with assigned sub-bands on the common fiber plant. To take care of this open access ODN we may need to consider some extra optical budget with respect the basic single-stage mux/demux option described here below. The power budget margin evaluation will help to understand the effects of some extra insertion loss due to different ODN configurations in field (e.g., different amount of connectors, protections and fiber cabling to reach the ONU/ONT).

Characterization of the ODN, Split Ratios, Reach and Differential Reach

As a reference for next performance discussions we assume to start with a pure DSS WDM-PON architecture (Option A1a) able to manage 64/80 x 1G25 bidirectional channels at 50 GHz in C/L bands. Options with 32/40 bidirectional chs, 128/160 bchs and 2G5 bit-rate per channel will be considered to provide more data about this DSS WDM-PON performances.

Target reach and target differential reach is 40km. Main wavelength plan is settled in C+L band with some options for S band. 40km reach will be a benchmark to understand performances and margins with respect other applications.
ONU architecture to meet the basic requirements.

The DSS WDM-PON ONU transmitter consists of a portion of the network itself. The self-tuning laser cavity includes the RSOA active chip in the ONU whose reflective end is one of the cavity mirrors, the distribution fiber, the remote node (RN) array wave-guide (AWG) which acts as the self-tuning element and a passive reflector/mirror (or faraday Rotator Mirror) also placed at the RN. The key element of this approach is an RSOA which has a triple role, as clarified in Figure 2:

- it sustains cavity lasing as the gain element, i.e., it overcomes the overall cavity losses
- it is the modulating element, through the active chip driving current
- it allows modulation cancellation avoiding its recirculation inside the cavity through the self-gain modulation mechanism, which cancels the recirculating residual modulation reflected back to the active chip allowing new signal modulation.

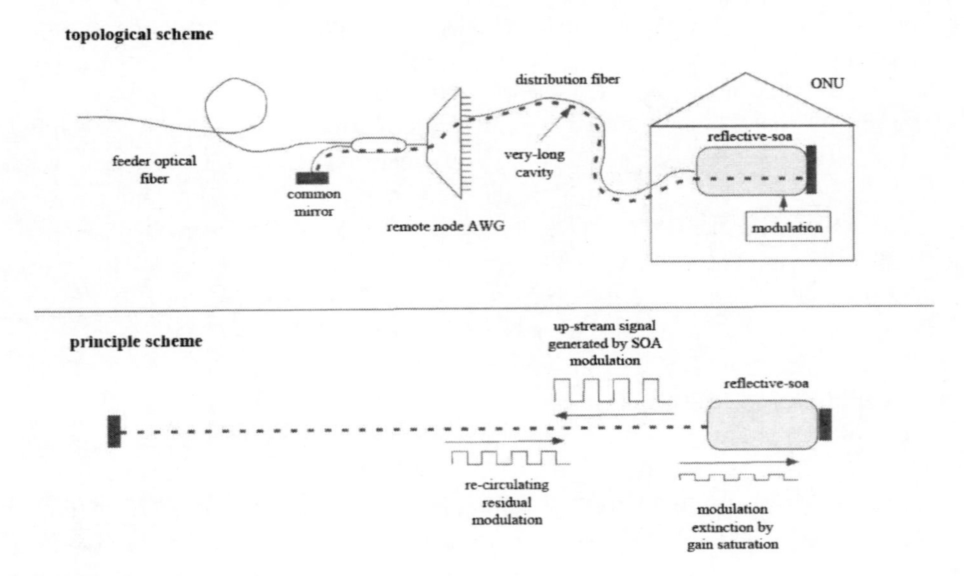

Figure 2. RSOA.

The effectiveness of the self-seeding approach is that it obviates the need for otherwise complex components such as tunable lasers or centralized seeding sources (with associated Rayleigh backscattering impairments which limit reach and bit rate).

ERMES Transmitter Project Specifications in the Range from 1G25 to 2G5

Following the power budget/reach analysis it is possible to define some basic specifications of the ERMES transmitter at R/S (or S/R) interface to cope with the general requirements for residential (NGPON2) and Business & Backhauling applications in the range from 1G25 to 2G5 bit-rate.

The following specifications are defined as transmitter benchmark in the range from 1G25 to 2G5 to support several protocols and related applications (Ethernet, CPRI, FC, SDH, OTH protocols list tbd). The transmitter benchmark for applications in the range from 2G5 to 10G will be defined after the first ERMES project results with the lower bit-rate range (1G25 ÷ 2G5). Final applications (power budget and reach classification vs ODN architectures) and compatibility with the overall range of protocols and bit-rates (e.g., from 1G25 to 10G) will be envisaged during the phases of the ERMES project as guidelines for possible parallel standardization activities.

Proposal of Product Specifications

- L band for downstream and C band for upstream (final wavelength range will be defined following standard activities)
- 100 GHz grid (possible evolution to 50 GHz)
- 40 bidirectional channels (possible evolution to 80 bidir chs)
- EOL PTXmin = +4 dBm
- Sensitivity SRXmin = -30 dBm with Standard RS FEC (1E-4 BER) at 1G25 and Standard RS FEC plus electronic compensation technology (tba) at 2G5 bitrate
- Max Cavity length (SMF) = 5 km (tbc from operator)
- Fiber attenuation (with splices, repair margin) = 0.375 dB/km (tbc from operator)
- Max AWG + FRM insertion loss = 8 dB (9dB if 50GHz / 80chs)
- Max ONU/ONT side distribution cabling insertion loss = 1dB (tbc from operator)
- Max OLT side distribution cabling insertion loss = 1dB (tbc from operator)

3. Principle of Operation

It is widely recognized that WDM PON deployment requires colourless Optical Network Unit (ONU) transmitters, so that each user has the same transmitter. Up to now these characteristics has been unsatisfactorily obtained either with a costly tunable transmitter, which may require wavelength setting algorithm, or by exploiting external seeding sources.

By FP7 European Project ERMES a disruptive approach to the ONU transmitter is proposed. The breakthrough idea is to use a significant portion of the network to implement an embedded self-tuning modulable laser cavity.

This idea is based on establishing a very long cavity laser, which can be directly intensity modulated. The goal is achieved by using a dedicated multifunction active chip (MFAC), which acts as the gain medium of the cavity including: the array waveguide grating (AWG) and a reflector at the remote node (RN), and the distribution fibre connecting the ONU to the RN. After cavity set up, the active chip is directly modulated enabling up-stream up to 10

Gb/s data-rate per user. The development of a MFAC suitable for this application is mandatory to bring ERMES approach from the proof-of-principle stage closer to industrial exploitation.

This solution is a potential highly effective alternative to the existing approaches in terms of cost reduction as it is colourless and gets rid of the need for external seeding sources. It is also appealing in terms of achievable performance, as it is not impaired by Rayleigh back-scattering, allowing for longer bridged distances.

ERMES ONU transmitter physically coincides with a portion of the network itself. The transmitter in fact relies on a laser cavity, which is essentially constituted by an RSOA active chip, placed at the ONU, whose reflective end is one of the cavity mirrors, the distribution fibre, which connects the ONU and the remote node (RN), the RN array wave-guide (AWG) multiplexer and a passive reflector also placed at the RN. The AWG represents the cavity wavelength selective element and every cavity associated to the single AWG channel will emit on that channel, self-tuning itself and thus achieving a colorless transmitter. The RSOA is the cavity active element and it sustains the cavity lasing overcoming the overall cavity losses. The RSOA is also the modulating element, through direct modulation of the active chip current. Finally as for the cavity presence part of the modulated output signal travels through the RN back to the ONU, the RSOA allows recirculating modulation cancellation through the self-gain modulation mechanism, supporting new signal modulation. To sustain this triple role the active device for the embedded self-tuning cavity must show a high gain to overcome the network embedded cavity losses, in deep saturation conditions, which are necessary for recirculating modulation bleaching and to show an electro-optical transfer function fitting the desired data rate.

A major issue with RSOA self-seeded architecture is the polarization evolution within the optical circuit [1], which is established between the optical reflector at the remote node (RN) and the mirror at the reflective semiconductor optical amplifier, placed at the optical network unit (ONU). The distribution fiber, which connects the ONU and the RN WDM multiplexer, namely an arrayed waveguide grating (AWG), can present lengths from a hundred of meters to a few kilometers. The unavoidable birefringence of the fiber modifies in an unpredictable manner the state of polarization (SOP) of the signal returning to the RSOA after the remote mirror reflection and the transit into the AWG.

The problem has been solved for low polarization dependent gain (PDG) RSOA [2, 7] exploiting the properties of the Faraday rotator mirror, known as the universal time-reversal operator, which ensures that the SOP of the signal re-injected in the RSOA is orthogonally aligned to the signal at the RSOA output. 1.25 Gb/s operation of a 32 channels WDM PON exploiting self-seeded transmitters has been demonstrated for low-PDG RSOA [7].

To allow the exploitation of high PDG RSOAs the addition of two simple elements is crucial: a Faraday rotator (FR), located in close proximity to the RSOA output, and a Faraday mirror, shared by all the ONUs and placed at the RN close to the AWG.

The key point of this analysis is the model of the RSOA from the SOP point of view, while the other circuit elements already own a consolidate Jones representation [9]. The RSOA acts both as SOP Generator, when launching the modulated signal into the cavity, and as SOP Analyzer, when intercepting the returning light from the cavity in order to saturate it. As the two functions are on first approximation dependent on the RSOA PDG, they can be both modeled with the same polarization operator, that is with a non-ideal generator of elliptical SOP, or dichroic elliptical analyzer (DEA) for the SOA, followed by the mirror operator (M). The diagonal coefficients of the DEA operator, p_x and p_y, are linked to the SOA gains, G_{TE} and G_{TM}, along the two main Cartesian directions, being TE (TM) parallel (orthogonal) to the epitaxial growth, where $G_i=\exp(\Gamma_i g_i l)$, Γ is the confinement factor, g the material gain and l the active region length [4]. Without any loss of generality, in a principal reference system oriented as the RSOA axes, the SOA operator is:

$$DEA|_{SOA} = \frac{1}{G_{TE}+G_{TM}}\begin{bmatrix} p_x & 0 \\ 0 & p_y \end{bmatrix} \tag{1}$$

where $|p_x|^2 = G_{TE}$, $|p_y|^2 = G_{TM}$. Although in principle all the coefficients in (1) are time-dependent on the bit time scale, as far as the SOP evolution is concerned, their average values are to be considered over the characteristic times of the birefringence phenomena, which are thermally and mechanically driven.

The generic STCT circuit (Figure3a) can be schematized in terms of polarization evolution by the SOP equivalent circuit shown in Figure 4a: both the rear mirror at the RSOA and the remote mirror at the AWG are represented by the ideal mirror operator M [9]; all the fibers in the circuit are represented by an equivalent generic retarder wave-plate (RWP), that is a linearly birefringent waveplate with retardation Δ degrees, with its fast axis making an angle θ degrees with the x-axis of the chosen reference system. The AWG is also represented by a generic RWP and the RSOA operator is represented by the DEA operator described by (1). Exploiting the multiplicative properties of the Jones Matrix, this equivalent circuit reduces to the more compact form presented in Figure 4b.

The SOP evolution can be followed during its journey through the Poincaré sphere representation in Figure5. In the first circuit, Figure 5a, for high PDG RSOA, we can assume, without loss of generality, that the DEA produces a horizontal SOP, |H>, which is modified by the RWP in a generic SOP, |G>. The mirror transforms |G> into |G′ >, which retraces the circuit and enters the DEA with a generic SOP |G′ ′ >, which will not coincide with |H>, so that:

$$I = \langle H|G''\rangle \neq 1 \tag{2}$$

thus the round-trip gain may drop, inhibiting the laser action.

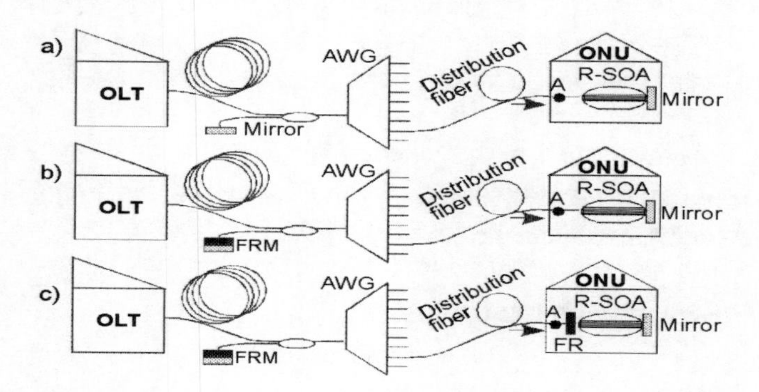

Figure 3. Retracing circuits.

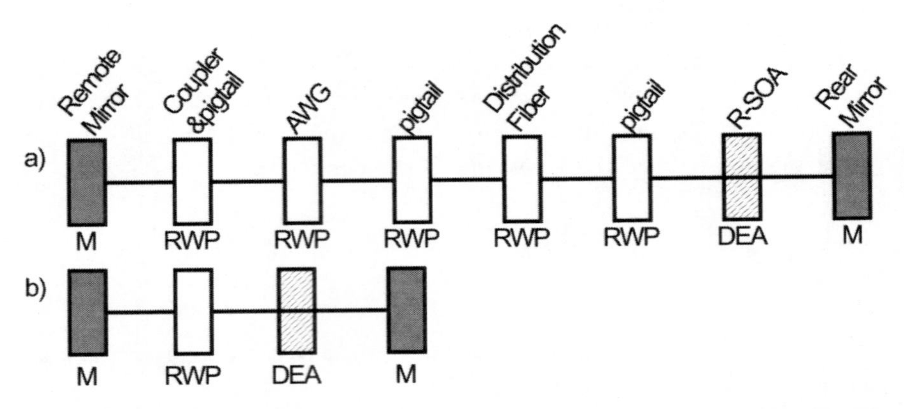

Figure 4. Equivalent circuit for the topology with remote standard mirror (a) and simplified circuit (b).

In Figure 3b (STCT circuit with remote Faraday Rotator Mirror) topology [2] a FRM substitutes the mirror. The equivalent circuit is shown in Figure 5b. Under the assumption of high PDG RSOA, again the DEA output SOP can be assumed |H⟩, which is transformed by the RWP in the generic |G⟩, which after the FRM is transformed into the orthogonal state |OG⟩ and reenters the DEA with state |V⟩, independently of the nature of the RWP. In this case:

$$I = \langle H | G'' \rangle = 0 \tag{3}$$

The signal SOP returning to the RSOA is stably re-traced, but orthogonally to the RSOA high gain transverse mode, causing the round-trip gain to drop well below the cavity losses and preventing lasing.

Finally the Figure3c-topology (STCT circuit with remote Faraday Rotator Mirror and Faraday Rotator at the RSOA) equivalent circuit is presented in Figure 5c, where a 45° Faraday Rotator is also inserted. Again for high PDG RSOA, the output SOP is assumed |H⟩ and is transformed by the FR into the state |R⟩, which is then transformed by RWP operator, $W(\Delta,\theta)$, into a generic |G⟩ SOP. After the FRM, |G⟩ is transformed into the orthogonal state |OG⟩ and returns to the FR with orthogonal SOP |L⟩, which is transformed by the FR into |H⟩, so that:

$$I = \langle H | G'' \rangle = 1 \tag{4}$$

Using the Jones matrix formalism the generic returning SOP $|G''\rangle$ can be expressed as the product of the launched SOP and the journey matrix *J*, which gives the retracing circuit action on the input SOP. *J* for Figure3c scheme is:

$$J = FR(-45°) \cdot W(\Delta,\theta) \cdot FRM \cdot W(\Delta,-\theta) \cdot FR(45°) \tag{5}$$

where $FR(45°)$ and FRM are the Faraday Rotator and Faraday Rotator mirror operator respectively. By using in (5) the properties of FRM and the definition of $FR(45°)$ [7], it is found that *J* is the identity matrix:

$$J = \begin{pmatrix} \sqrt{2}/2 & \sqrt{2}/2 \\ -\sqrt{2}/2 & \sqrt{2}/2 \end{pmatrix} \cdot i \begin{pmatrix} 0 & 1 \\ 1 & 0 \end{pmatrix} \cdot \begin{pmatrix} \sqrt{2}/2 & -\sqrt{2}/2 \\ \sqrt{2}/2 & \sqrt{2}/2 \end{pmatrix} = i \begin{pmatrix} 1 & 0 \\ 0 & 1 \end{pmatrix} \tag{6}$$

confirming (4) result.

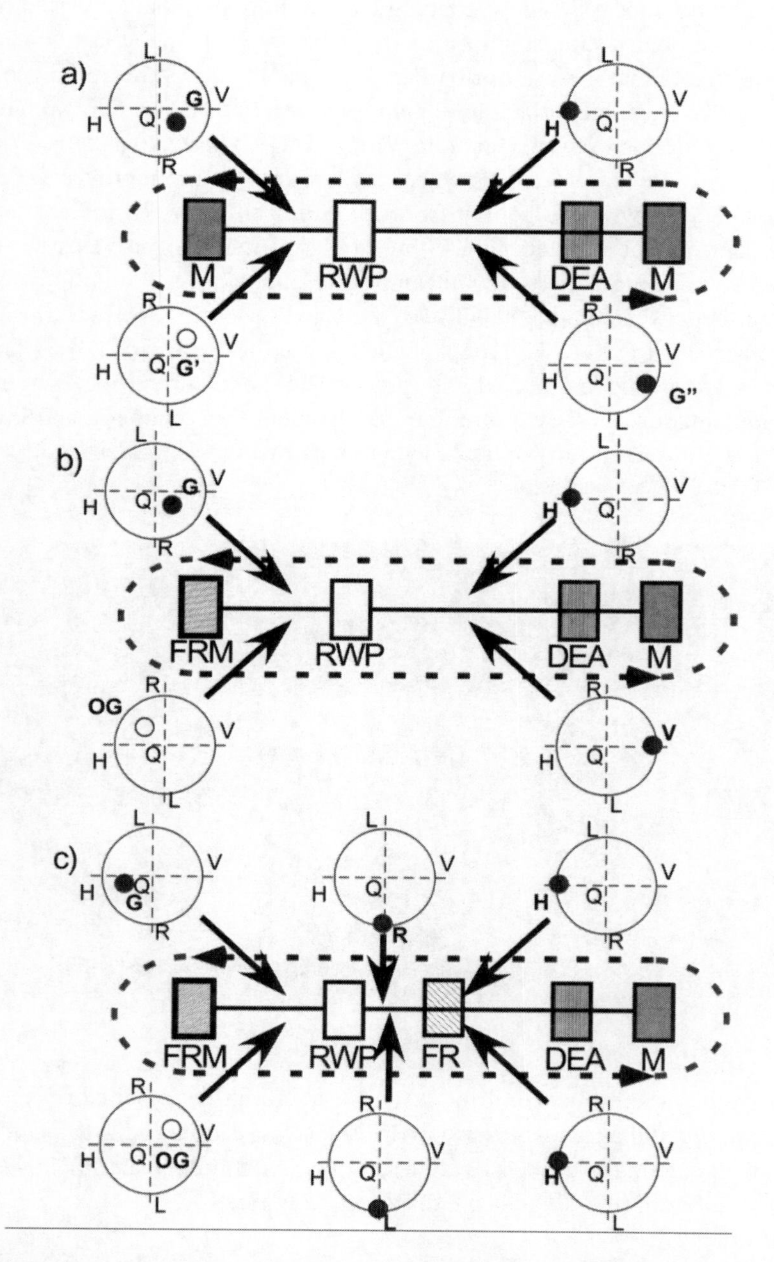

Figure 5. Equivalent circuits and SOP evolution on the Poincaré sphere representation for the retracing circuits presented in Figure1.

4. KEY DEVICE

Until recently, large gain RSOAs were made with bulk material to achieve polarization insensitivity of the gain.

However, polarization independence is no longer an issue to design the RSOA for self-seeding transmitters. Indeed, a new topology exploiting two Faraday rotators has been proposed to achieve polarization stabilization in the cavity [10], and at last OFC conference it has been demonstrated that the use of two Faraday rotators allows for a FP-based stable self-tuning cavity even with single-polarization gain sections [11]. As a result, we have decided to use compressively strained Multi Quantum Wells (MQW) instead of tensile strained bulk material. Similarly to the case of lasers where compressive strain improves the performances [12], we believe RSOA will also benefit from the large shift between hole bands, in order to reduce carrier density for a given gain. Single polarization also provides more degrees of freedom for designing the active section, aiming at either larger saturation power, smaller Noise Factor, or better temperature stability of the gain.

We have fabricated RSOA with the so-called Buried Ridge Stripe (BRS) process, where the active layers are etched and buried with p-doped InP, and current localization is obtained with proton implantation, as shown on Figure 6. Efficient fibre coupling and low reflectivity [9] is obtained by transferring the optical mode to a passive waveguide in order to increase its size.

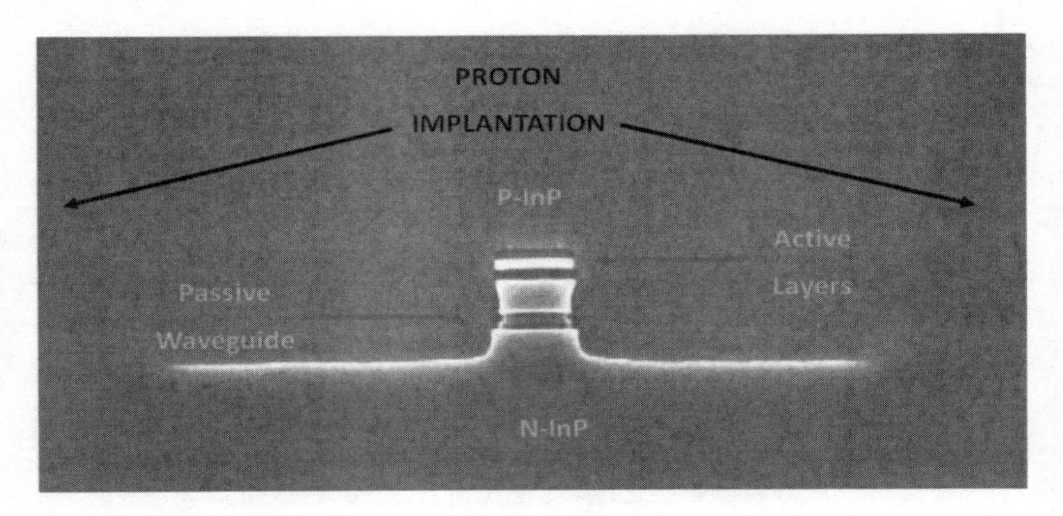

Figure 6. SEM image of the cross-section of a BRS RSOA.

In order to study the influence of the active structure on the modulation speed of RSOA, we have measured eye-diagrams of various RSOA having similar gain values (25 to 30 dB), made either with Quantum dashes, with MQW, or with bulk material having an optical confinement of 20% or 80 %. Results are displayed in Figure 7.

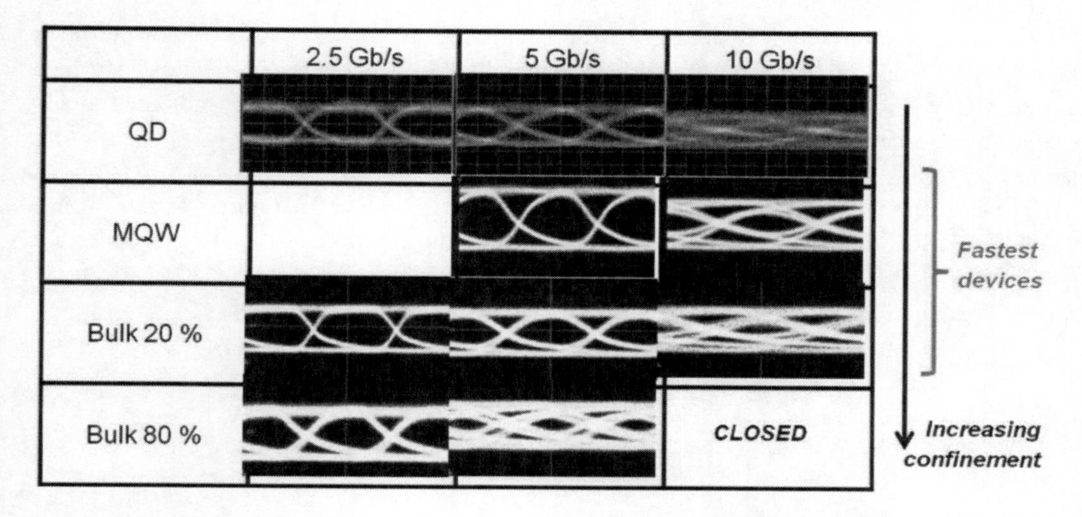

Figure 7. 2.5, 5 and 10 Gb/s eye diagrams (pure ASE modulation) of QD, MQW, or bulk (with an optical confinement of 20% or 80%)-RSOA.

The fastest devices are either made with MQW or with 20% bulk.

Another important feature of RSOA for the self-seeding cavity is the data cancellation, which is based on gain compression. To operate at large bit-rates and with low input powers, the gain should be as large as possible, and most importantly, the O/O modulation speed should be very large. Based on previous studies, we can deduce from these requirements that we should use very large optical confinements. To confirm this, we have studied the cancellation efficiency of various structures. Increasing the optical confinement drastically improves the efficiency and the speed of data cancellation. Figure 8 shows the static and dynamic transfer function of a bulk RSOA having an optical confinement of 80%. The input signal has an extinction ratio of 5 dB.

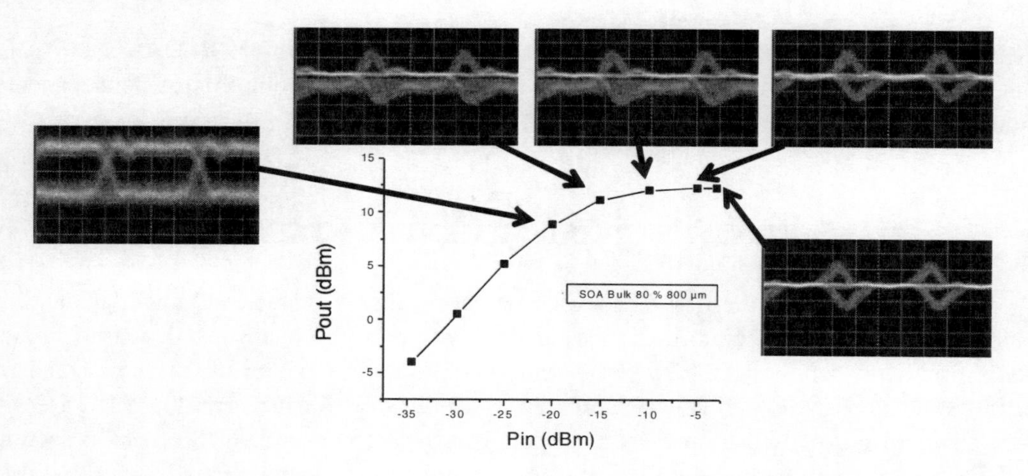

Figure 8. Transfer function and corresponding eye diagrams at 5 Gbit/s of a bulk RSOA with an optical confinement of 80%. Data cancellation is fast and efficient for an input power above -10 dBm.

Data cancellation with MQW RSOA is inherently less efficient because of a lower optical confinement. However, since large confinement bulk RSOA are slow modulators, we have chosen MQW RSOA with an optical confinement above 10 % as the ideal candidate for self-seeding transmitters. The lack of confinement is partially compensated by using 1 mm long devices.

The use of MQW also improves the performances at high temperature. We have put a MQW-based RSOA in a self-seeded cavity, and with an appropriate detuning between the gain peak and the AWG port, we have obtained uncooled operation from 20°C to 70 °C. Results are displayed in Figure 9.

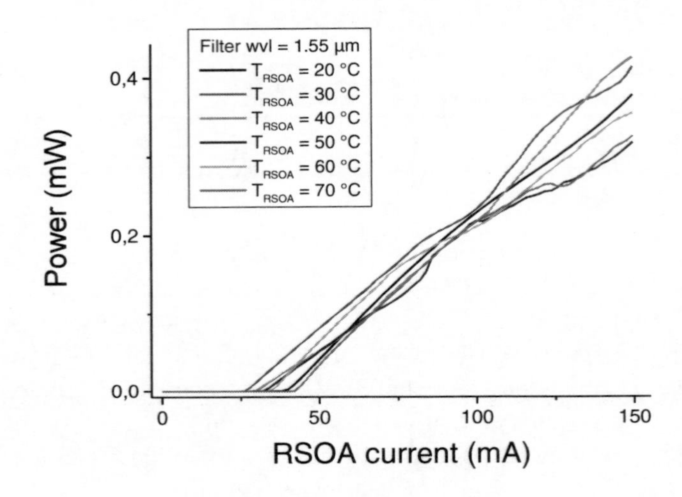

Figure 9. L(I) curves from 20 to 70 °C of a self-seeded cavity with a MQW-based RSOA, at 1.55 µm. The gain peak of the RSOA is 1520 nm at 20 °C.

Based on these studies, we have used specially designed MQW RSOA as the gain element for the self-seeded cavity, in order to obtain the best compromise between data modulation and data cancellation.

5. First Year Experimental Results

We will now present the first year experimental results evaluated in Orange labs and in Politecnico di Milano laboratories. First section will presents results achieved with low polarisation dependent gain (PDG) RSOA with a configuration exploiting a remote standard mirror (SM). The first experimental results up to 10 Gb/s obtained with a low-PDG RSOA and a standard mirror will be presented in detail in the first section. In the second section results achieved again with a low PDG RSOA exploiting also a remote Faraday mirror (FM) are presented for 32 channels at 1.25 Gb/s. Impact of very low PDG is also highlighted by the experimental results. Then we evaluated a first self-seeded WDM-PON prototype in order to compare the performances with the results obtained with RSOA chips specially developed for ERMES project. Moreover, this prototype shows the possibility to realize a symmetrical architecture using a self-seeded RSOA at each side of the network: OLT and ONU. This first prototype also permits to evaluate the availability of such a technology for the purpose of mobile front-haul backhauling. In the third section, we will present the evaluation results with the ERMES high-PDG RSOA chips.

5.1. Low-PDG RSOA with Standard Mirror

Figure 10 depicts the upstream transmission experimental set-up with a self-seeded ONU. The ONU contains only the RSOA which is directly modulated with a pulse pattern generator (PPG). The RSOA that we used has a low PDG of 2-3 dB. The laser cavity consists of the reflective facet of RSOA and an 80% reflective SM, which is connected to the AWG via a 30/70 splitter at the RN (see figure 10 represented in red). Here the AWG acts as wavelength multiplexer and demultiplexer and the self-tuning element. Thus, the ONU operating wavelength is chosen by the AWG channel. In these experiments two different 8-channels flat-top AWGs operating in the C band have been used for different modulation operation lower than 2.5 Gb/s and higher than 5 Gb/s. The first device is dense wavelength division multiplexing (DWDM) device with 100-GHz bandwidth at -3 dB for the 200-GHz grid with an insertion loss (IL) of 1.7 dB. The second one has a bandwidth of 200 GHz at -3 dB and a slightly lower IL of 1.3 dB. The total length of the cavity varies from few meters to 5 km. Also, 2 dB of extra losses are added inside the cavity to account for last distribution fiber connection losses. At the OLT, transmitted signals are received after several kilometres of feeder fiber by an avalanche photo-diode (APD) and a clock and data recovery (CDR). An error free transmission is considered for bit error ratio (BER) lower than 10-3 when a forward error code (FEC) is used.

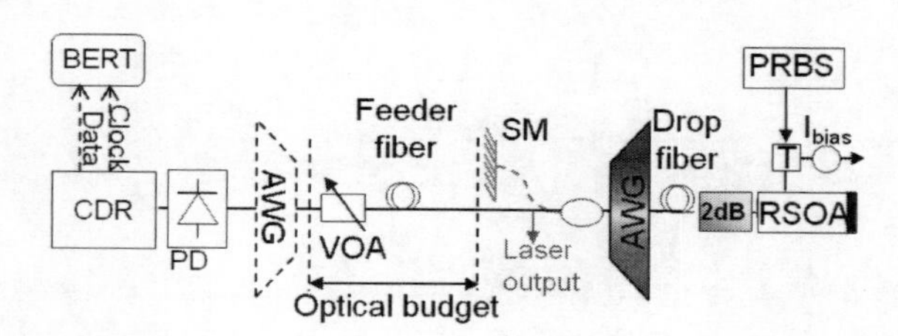

Figure 10. Experimental set up for BER measurement.

Several characteristics are evaluated in terms of spectrum with and without the mirror (SLED-MUX100GHz), relative intensity noise (RIN) and chirp parameter.

Figure 11 shows ASE slicing spectrum of RSOA and laser spectrum of RSOA-mirror. The laser emission wavelength is centered at 1554 nm. 30-dB gain is obtained in laser mode compared to slicing spectrum of RSOA (that is the SM is removed). The side mode suppression ratio (SMSR) of the laser is lower than 30 dB, which means that the laser obtained is multimode. In order to better understand the spectrum characteristics in particular the laser mode numbers, we measured laser spectrum with a high resolution (0.05 pm) optical spectrum analyzer with high dynamic range BOSA-C. The results show that laser is multimode and unstable, the mode numbers is very high as shown in the figure 13 and 14.

Figure 12 represents ASE power of RSOA versus bias current, laser power versus bias current, and power of sliced spectrum after the AWG filter.

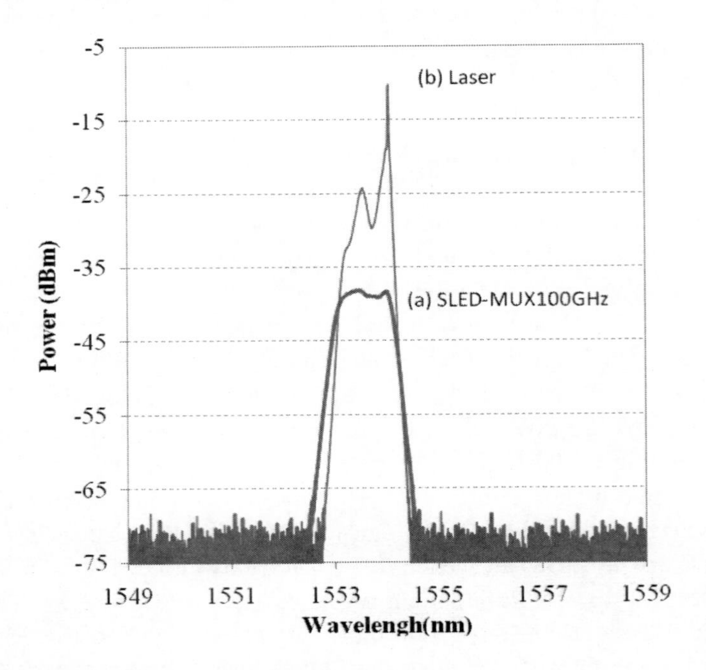

Figure 11. ASE spectrum of RSOA and ECL static response.

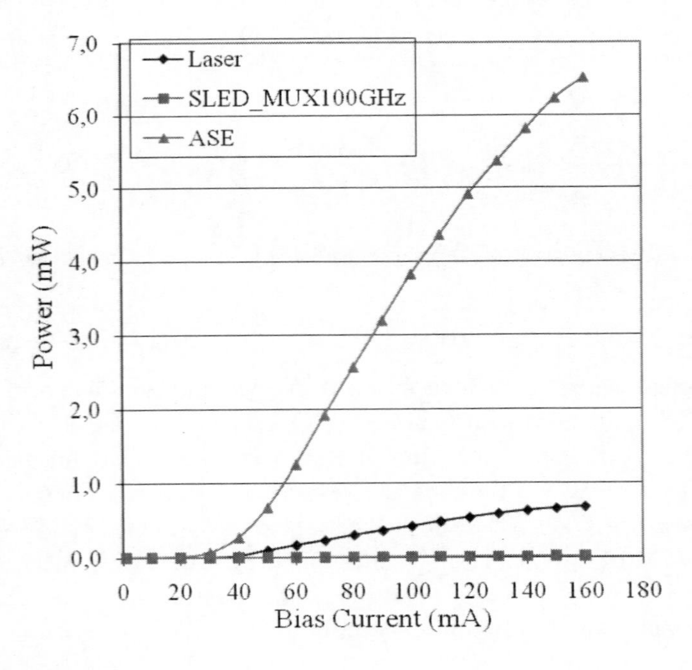

Figure 12. Laser power versus bias current.

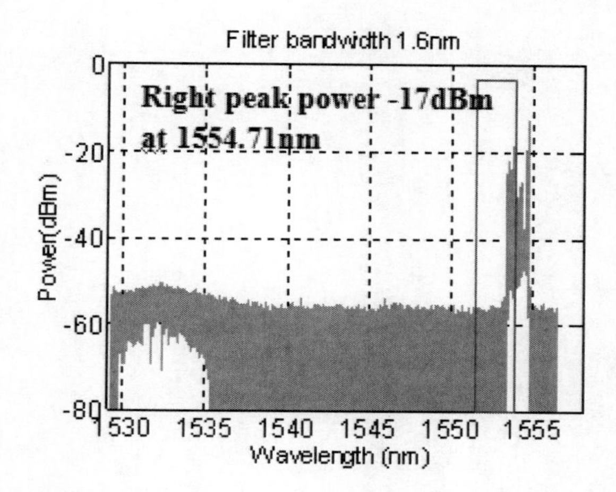

Figure 13. Laser spectrum with filter bandwidth of 1.6nm measured with BOSA.

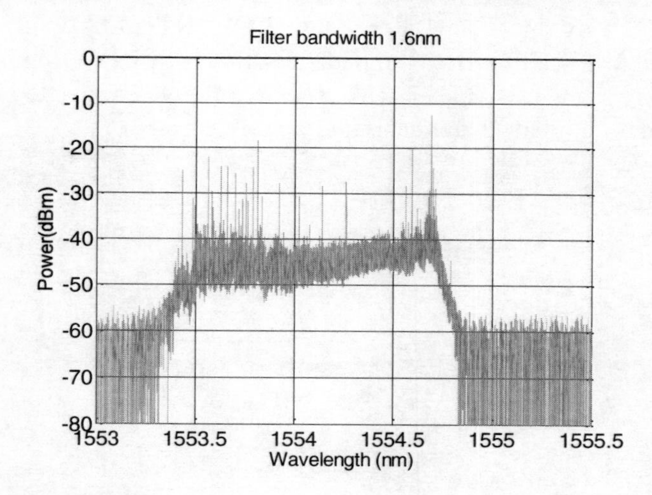

Figure 14. Zoom of laser spectrum.

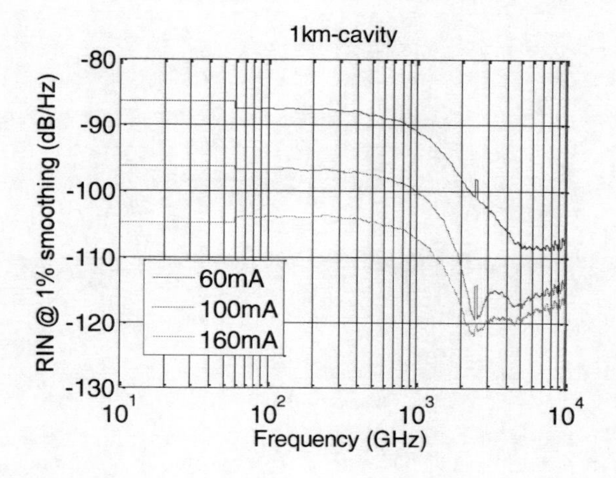

Figure 15. RIN versus frequency span for different biased currents.

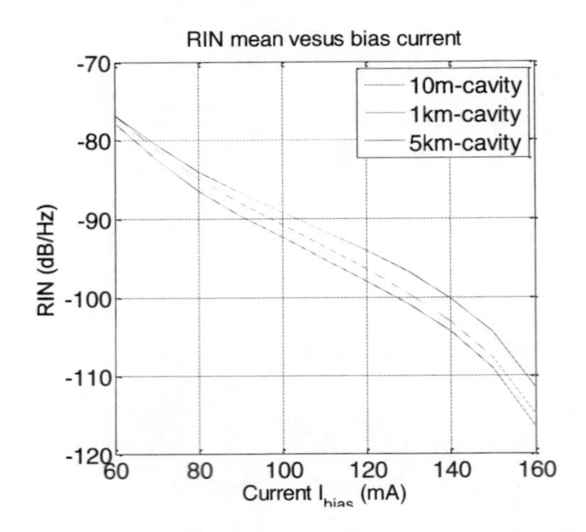

Figure 16. Mean RIN value versus biased current for different cavity length.

For a 1-km long cavity self-tuning laser, a high RIN level is measured at low frequency and can be optimized from -85dBc/Hz to around -115dBc/Hz by increasing the biased current of RSOA (see in Figure 15). A higher RIN level is observed when the cavity length is increased up to 5 km for the same biased current (see in Figure 16).

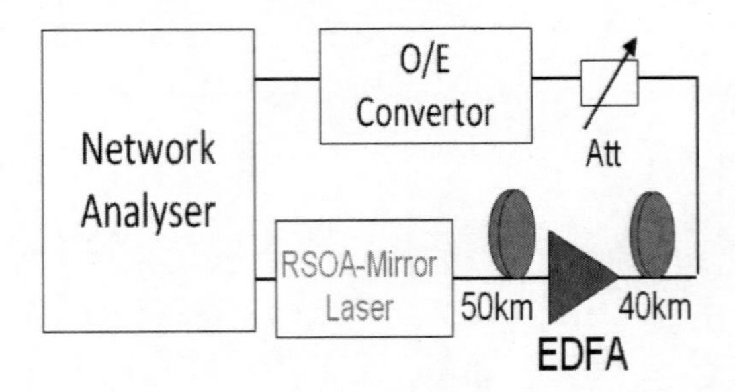

Figure 17. Experimental setup for measuring frequency response.

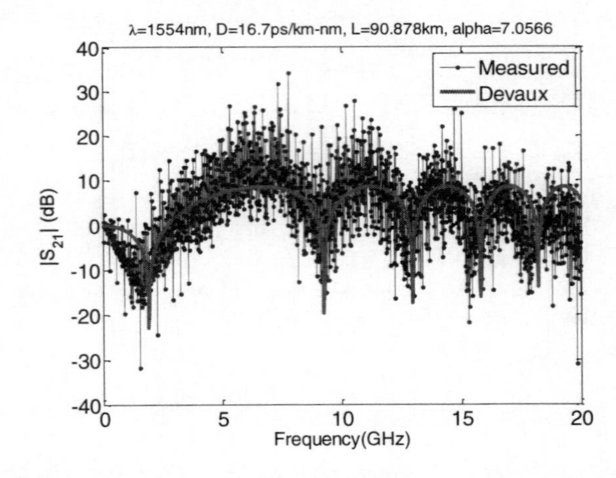

Figure 18. AM frequency response through 90km of SMF, measured in blue, model according to Devaux in red.

Figure 17 shows experimental setup for measuring frequency response. An optical amplifier of type EDFA is used to realize a long reach transmission in fiber up to 90 km. Output power of the EDFA is fixed at 10dBm.

The chirp parameter is then measured applying the fiber transfer function method originally proposed by F. Devaux in [10]. Figure 18 shows the measured AM frequency response through 90 km of SMF (in blue) and model according (in red). A first analysis shows two contributions of the phase and amplitude coupling: one coming from the gain section of the RSOA, and a second expresses itself through the large oscillation. We estimated to value of α=7 in case of 1-km cavity.

As the commercial RSOA has a reflection gain of about 12 dB, in order to determinate the capacity to overcome the overall cavity round trip losses, in this section we will at the first time evaluate the performance versus different cavity total round trip losses. Figure 19 shows the experimental setup where a variable optical attenuator (VOA) is inserted between RSOA and AWG to vary different added losses in the cavity. As the VOA has a symmetric IL of 2.2 dB, two large band MUX with IL of 0.5 dB and 1 dB are used to simulate a lower added losses in the cavity. Then we replace VOA by several lengths of standard monomode fiber (SMF) to compare the performance.

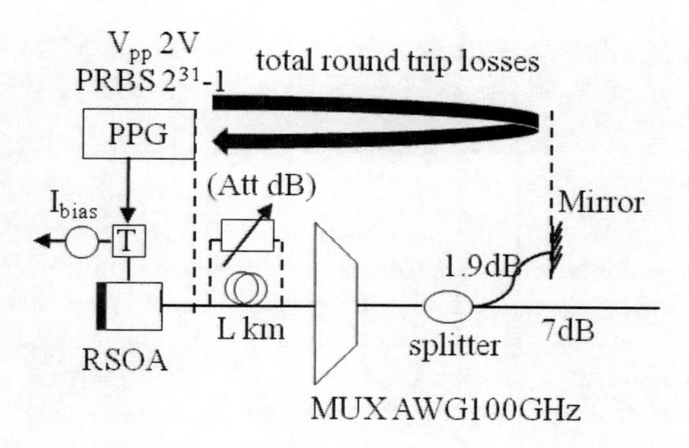

Figure 19. Experimental setup to evaluate BER performance versus cavity total round trip losses.

Table 1 shows the IL of different component in the cavity. For 1-km long cavity for example, the total round trip losses is twice of the sum of all the component IL in the cavity, which means 2x (0.8+1.7+1.9+0.9)=10.6dB.

Table 1. 1-km Cavity total round trip losses : 5.3*2 = 10.6dB

	Mirror	MUX-100GHz	Splitter 30/70	1km of SMF	5km of SMF	10km of SMF
Losses(dB)	0.8	1.7	1.9	0.9	1.4	3.2

We fix the bias current at 100 mA for all the BER measurement.

Table 2. Laser output power and ER of ED for different added losses

Added losses (dB)	0	0.5 (MUX1)	1 (MUX2)	1.8 (MUX1+2)	2.2 (VOA IL)	2.7	3.2	3.7	4.2	4.7
Pout (dBm)	-3.2	-3.3	-3.9	-4.2	-5.5	-5.9	- 6.5	-7.2	-7.8	-8.5
ER(dB) of ED	4.4	4.4	4.4	4.6	4.3	4.3	4.5	4.6	4.7	4.7

Figure 20 shows the laser output power versus different added losses in the cavity. As the bias current is fixed at 100mA, the RSOA has a fixed reflection gain. Laser output power decrease linearly versus added losses in the cavity.

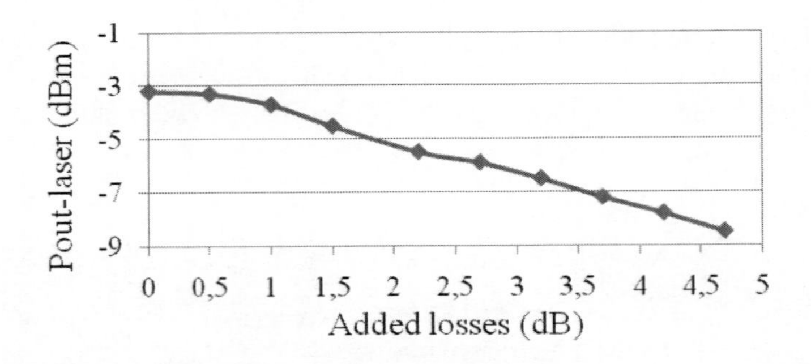

Figure 20. Laser output power versus different added losses.

Figure 21 depicts the BER measurement in case of BTB versus received power for different added losses from 0 dB to 4.7 dB in the cavity. Error floor start to appear from 1.5dB of added losses. We compared also BER performances between different added fiber length and added attenuation as shown in the figure 22. With 1km of added SMF having 1dB of attenuation, the BER curves are identical.

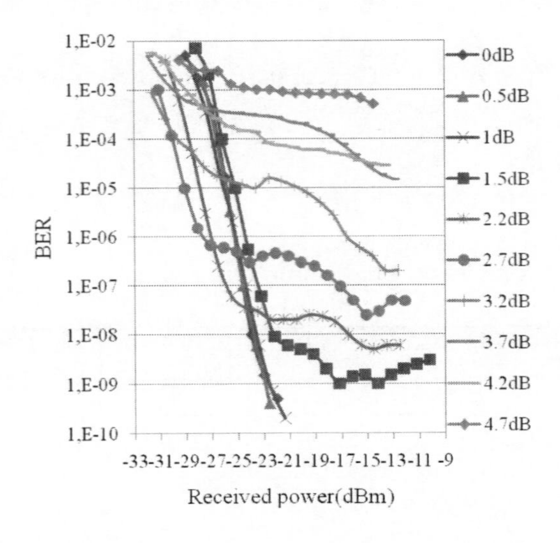

Figure 21. BER measurement versus received power for different added attenuation in the cavity.

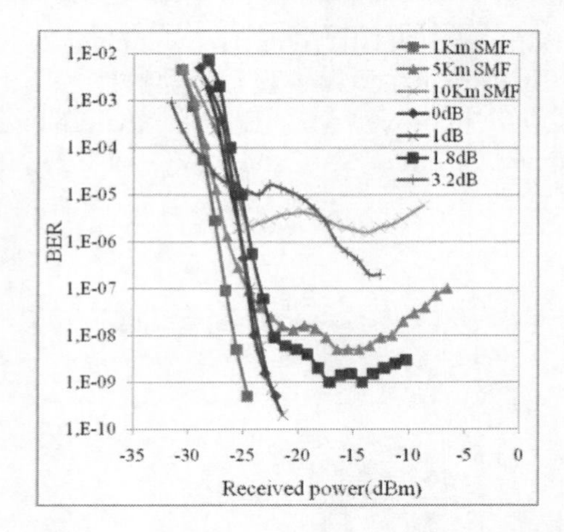

Figure 22. BER measurement versus received power for different distances and attenuation.

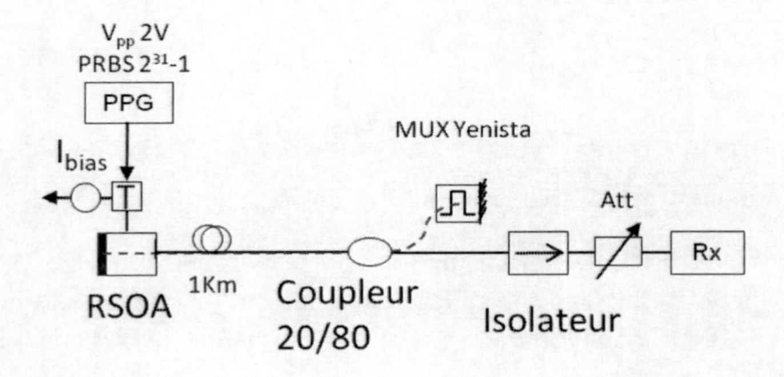

Figure 23. Experimental set-up to evaluate BER performance versus MUX bandwidth.

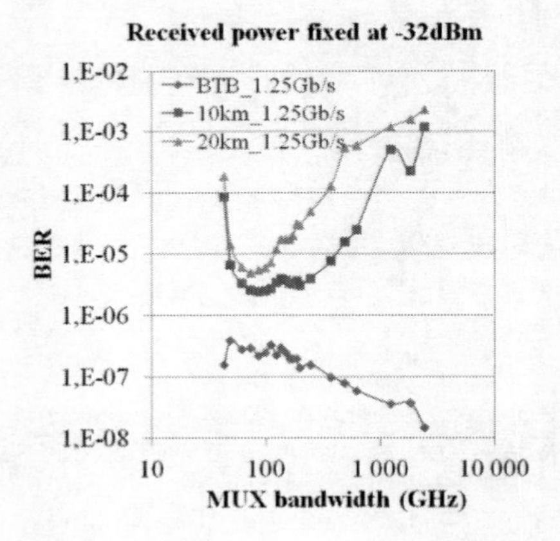

Figure 24. BER measurement versus filter bandwidth at 1.25Gb/s.

In Figure 23, the 100 GHz 8-channels AWG is replaced by a tunable filter with a bandwidth range from 40 GHz to 2.5 THz corresponding in wavelength from 0.35 nm to 20 nm. We measured BER performances versus filter bandwidth with a fixed received power of -32dBm at 1.25 Gb/s and -27.5 dBm at 2.5 Gb/s. In case of BTB at 1.25 Gb/s, the results show that BER performance is slightly improved when filter bandwidth increases. While the results in transmission after 10 km and 20 km of SMF show the best BER performance at about 100 GHz of filter bandwidth.

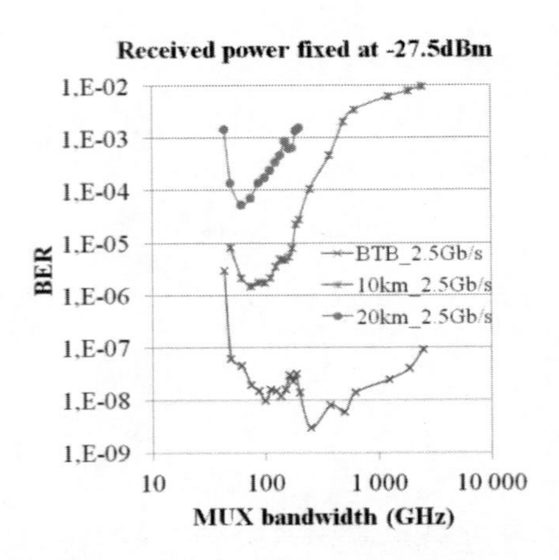

Figure 25. BER measurement versus filter bandwidth at 2.5Gb/s.

We firstly evaluate the lower speed modulation performance at 1.25 Gb/s and 2.5 Gb/s for different cavity length. The RSOA is biased at a current of 100mA and modulated by non return to zero (NRZ) PRBS data with a length of 2^{31}-1. Results are displayed in Figure 26.

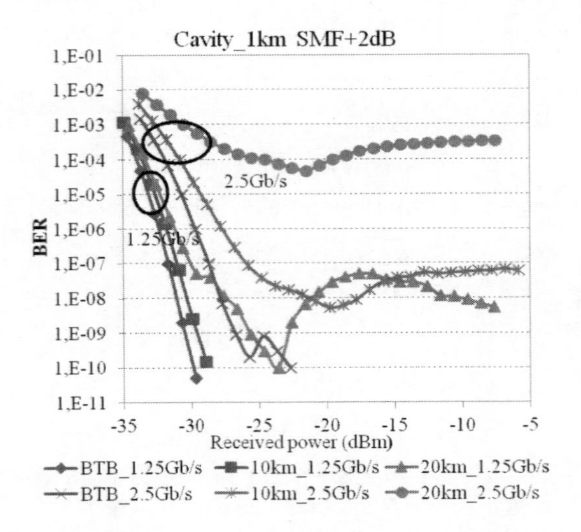

Figure 26. BER measurements for back-to back and after 10km/20km SMF link at 1.25Gb/s and 2.5Gb/s for 1-km long cavity.

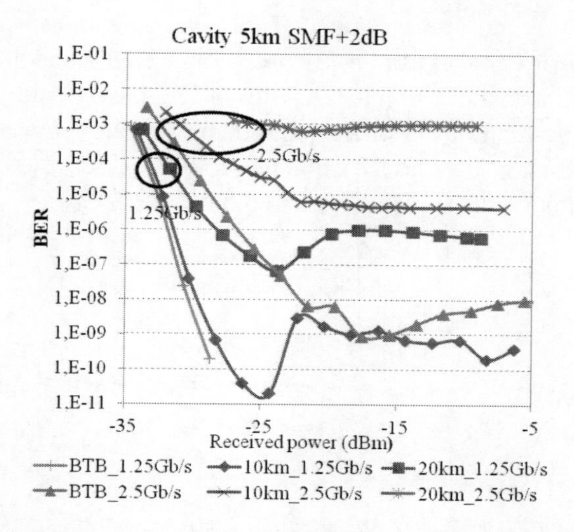

Figure 27. BER measurements for back-to back and after 10 km and 20 km of SSMF link at 1.25 Gb/s and 2.5 Gb/s for 5-km long cavity.

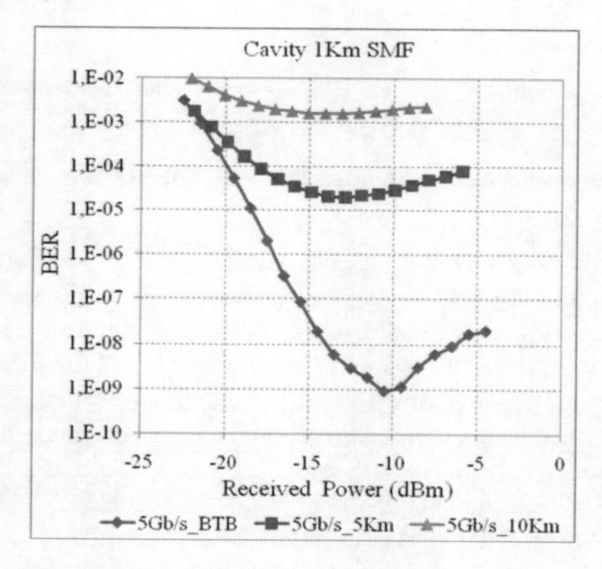

Figure 28. BER measurements for back-to back and after 5km/10km SMF links at 5Gb/s for 1km long cavity.

Figure 26 and Figure 27 represent BER measurements versus received power for 1-km and 5-km long laser cavities with an extra optical attenuation of 2 dB inside the cavity. In the former case, the laser output power is around -2 dBm with an extinction ratio (ER) around 4.5 dB (compare. Figures 14-15). The back-to-back (BTB) sensitivity at 1.25 Gbps and 2.5 Gbps for achieving a 10^{-3} BER is about -35 dBm and -33 dBm respectively, which means that an optical budget (OB) of 33 dB and 31 dB between OLT and laser can be tolerated. Both of two results prove the possibility of co-existence with standard class B+ GPON at 1.25 Gb/s and 2.5 Gb/s. The power penalties after 20 km of SMF (in the feeder) are less than 0.5dB at 1.25Gbps and 3dB at 2.5Gb/s.

In case of 5-km cavity, despite the higher RIN level and higher cavity total round trip losses, which lead an error floor appears in BTB case at 2.5GB/s and after fiber transmission case. We can still obtain an OB, which is 32 dB at 1.25 Gb/s and 29.5 dB at 2.5 Gb/s respectively with a laser output power of around -3dBm. Results prove equally the possibility of co-existence standard class B+ GPON with such a long cavity. The power penalties after 20 km of SMF are less than 1 dB at 1.25 Gb/s and 6.5 dB at 2.5 Gb/s.

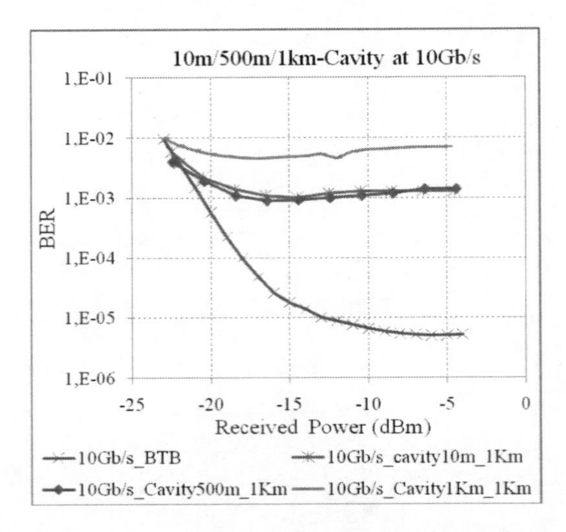

Figure 29. BER measurements for back-to back and after 1km SMF link at 10 Gb/s for 10m/1km long cavity.

As described previously, for higher bit rate modulation at 5Gb/s and 10Gb/s, an AWG with larger bandwidth and lower IL is required to reduce the cavity total round trip losses. The extra distribution losses of 2dB are equally removed to minimize cavity losses. Due to the optical bandwidth limitation of RSOA component and higher frequency loss at high bit rate transmission, we use for the first time a pre-emphasis technology to maintain eyes opening by correcting waveform amplitude to optimize transmission performance. As signal source, the PPG generate a NRZ data PRBS with a length of 2^{31}-1 and the clock is sent into a 4-tap emphasis, which generate an emphasis data to modulate our RSOA-based laser. Evaluation of 1km-cavity was done at a bit rate of 5Gb/s with the RSOA current biased at 150mA, which resulted in an output power of about 1.5dBm with 2.8 dB of ER. To achieve a 10^{-3} BER in B2B, the required power is about -22 dBm permitting an OB of 23.5dB. After 5km of SMF (see Figure 28), the power penalty is less than 0.5dB at a BER of 10^{-3}. When we increase the bit rate up to 10Gb/s, in case of 10m-cavity, we successfully obtained 1,6dB of ER and BTB curves with a sensitivity of -20.5dBm for a BER at 10^{-3} (see Figure 29), which means an OB of 22.3 dB can be tolerated. It is possible to realize a 1km-transmission in the feeder with a penalty of 5.5 dB at the considered BER. The cavity length can also be increased up to 500m to keep the same transmission performance. BER measurement for 1-km cavity is also performed at 10Gb/s. After propagation through 1 km of feeder fiber, the BER obtained was 4.10^{-3}.

5.2. Low PDG-RSOA with FM Mirror Optics Express

Figure 30(a) shows the experimented 32-channel WDM PON topology. Two-twin cyclic athermal NEL AWGs with 100-GHz spacing serve as multiplexer/demultiplexer at the OLT and at the RN respectively. DS transmitter is an externally intensity-modulated tunable L-band laser, coupled to the AWG by means of C-L WDM coupler (IL=0.8 dB), which also routes the incoming US signal to the US receiver. A feeder fiber of 24 km or 50 km of SSMF links the OLT to the RN. At the passive RN, the 80/20 output coupler and the mirror are the common portion of all the ONU network-embedded cavities. In particular two different mirror topologies have been exploited, the first, including an optical circulator and polarization controllers, allows for performance evaluation as a function of polarization variations inside the cavity; the second is a Faraday Rotator Mirror with 45° rotation at 1550 nm ±1.2° over the entire C-band, to realize a configuration recently demonstrated to give almost polarization-independent operation with low PDG RSOA [2]. The remaining passive section of the self-tuning cavity is constituted by the distribution fiber, ranging from 1 km to 5 km, and by the C L WDM coupler. Finally the gain and modulation medium is provided by two RSOAs by Alcatel-Thales III-V Lab [3], whose output gain spectra are shown in figure 30(b) and 30(c), compared with the channel allocation in C-band for the US (in black) and L-band for the DS (in red); the optical spectrum resolution is 0.5 nm. Both RSOAs are operated at room temperature by a thermo electric cooler and they have very low polarization dependent gain (PDG) in linear regime: lower than 1 dB for ONU1 and then 0.5 dB for ONU2. As can be seen RSOA2 gain spectrum does not adequately cover channels lower than 16. At each ONU the C-L WDM coupler also routes incoming DS signal to the DS receiver. All receivers include 1.25 Gb/s APD-TIA and clock and data recovery.

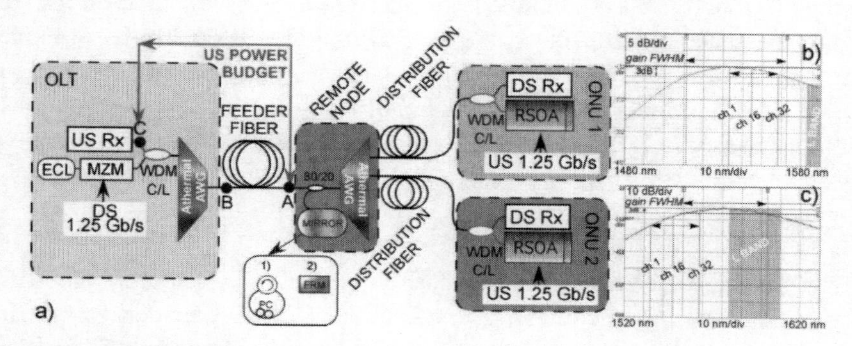

Figure 30. (a) Experimented WDM PON topology (b) III-V Lab RSOA1 output gain spectrum (c) III-V Lab RSOA2 output gain spectrum (0.5 nm resolution).

Fig 31(a) shows the power versus RSOA bias current curves for the three evaluated channels covering the C-bandwidth, for ONU1 and 1-km distribution fiber length. Optimized transmitter performance is around 160-mA bias current, corresponding to almost -4 dBm available output power. It should be reminded that output power is intended as measured at the RN, that is at point A of the Figure 1 setup: from the ONUs point of view, point A) can be called back to back. For all the considered channels the operating point is set in the nonlinear region of the power-versus-current curve, as the RSOA has to operate in saturation condition in order to bleach the incoming recirculating-modulated radiation. Figure 31(b)- (d) show the back to back eye diagrams at 1.25 Gb/s for the three channels at 160-mA bias current; CH32 shows a slightly noisier eye diagram due less effective residual modulation cancellation, being at the edge of RSOA1 3-dB bandwidth. Eyes are registered by the AC-coupled output of a linear PIN-TIA photodiode. Output ER is evaluated through the DC-coupled output of the same receiver: for all the considered channels it ranges from 6 to 6.5 dB. The output ERs are the results of the trade off between the desirable eye opening which increases performance and the capability of the RSOA to bleach the high ER of the recirculating modulation [4].

The first analysis of the embedded self-tuning cavities comprised BER measurements as a function of the received power, while simultaneously operating two ONUs respectively with 1.4 and 1 km distribution fiber, in the circulator-based mirror configuration. They have been performed for different channels in co presence of the corresponding DS signal. PRBS is 2^7-1, as this test pattern gives the closest match to the maximum run-length of a Gigabit Ethernet line code. Nevertheless a back to back evaluation of pattern dependence evidences for 2^{31}-1 pattern length no penalty at 10^{-4} BER and less than 1 dB at 10^{-11} BER with respect to the 2^7-1 pattern. Figure 31 shows BER curves for ONU1 channel 1 (CH1) at 1533.4 nm, which is close to RSOA1 gain peak. For the sake of clarity curves are displayed into two different charts. Figure 31(a) assesses the lack of crosstalk impairments both on US signal and on DS signal: no penalty is in fact observed when additional channels are ON, as can be seen comparing black full diamonds curve with open diamonds curve, for the DS signal, and the dark blue full circles curve with the light blue full circles curve for US signals. By means of the fiber polarization controllers (PC) inside the optical circulator mirror (see Figure 30) opposite SOP conditions within the cavity are analyzed: at 10^{-4} the difference between the two SOPs is around 1 dB, while at lower BER the error floor rises from 10^{-10} to $5 \cdot 10^{-7}$. Focusing on the best SOP condition, the comparison between results with 24 km and the 50 km feeder fiber, respectively light blue full circles and open circles curves, show little difference when CD load is almost doubled. In order to provide an insight on the sources of impairments in the scheme, the penalty due to OLT AWG and dispersive fiber have been measured alone. Results are shown in Figure 31(b). By comparing the back to back performance of the US signal (point A in Figure30) (full grey circles) with that at point B after 24-km transmission (full orange circles) and 50-km transmission (open orange circles), it can be seen that the propagation penalty is slightly higher than 1 dB for the longest distance, but no error floor appears. In these conditions the modulated output spectrum bandwidth is 11 GHz and thus it is not the limiting factor for typical PON feeder fiber lengths at this bit-rate. On the other hand, the comparison with CH1 filtered by the OLT AWG without propagation (open grey circles) shows that AWG filtering evidences the 10^{-11} BER error floor, as expected due to its action on the RIN associated to the US signal [5]. The combination of chromatic dispersion and OLT AWG filtering further increases this floor.

The same measurements have been performed both for ONU1, over the C band, and for ONU2, restricting to channels from CH16 to CH32. The results in figure 33 have been expressed in term of available US power budget for 24-km feeder fiber, measured as the difference between the output power of the self-tuning cavity (in A) and the received power necessary to achieve 10^{-4} BER (in C) for ONU1 (figure 33(a)) and ONU2 (figure 33b); 10^{-4} BER was chosen as it represents 239,255 Reed-Solomon pre-FEC limit. In these measurements PCs have been positioned to obtain best and worst polarization performance respectively for ONU1 (Figure 33(a)) and ONU2 (Figure33(b)).

In Figure 33, the red and blue areas represent the spread in presence of the second ONU and the DS transmission respectively in best and worst SOP conditions. The green area expresses the performance differences associated to polarization variations.

It can be seen that power budget for ONU2 CH16 and CH17 is 2dB lower than for ONU2 CH32, whose wavelength is close to RSOA2 gain peak experiencing nearly the same gain as ONU1 CH1, both allowing for a power budget of 27 dB. Similarly ONU1 CH32 behaves as ONU2 CH16, which sees a similar gain, as can be seen by comparing spontaneous emission spectra of Figure 30(b) and (c). The green area in the figure 33 represents the polarization spread, that is the difference between worst and best SOP performance. As already discussed for ONU1 CH1 even low PDG, can significantly impact on the transmitter performance as it is enhanced by multiple roundtrips during the cavity build-up [6] resulting in nearly 1 dB power budget difference at 10^{-4} and an increased error floor. The polarization spread is lower for ONU2, which relies on a lower PDG RSOA.

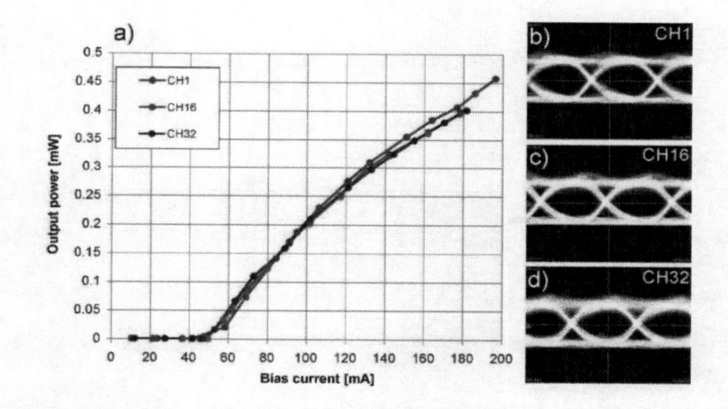

Figure 31. (a) Optical output power vs RSOA1 bias current, for US ch. 1, 16 and 32. Back to back 1.25-Gb/s eye diagrams with linear PD for (b) ch. 1, (c) ch. 16, (d) ch. 32.

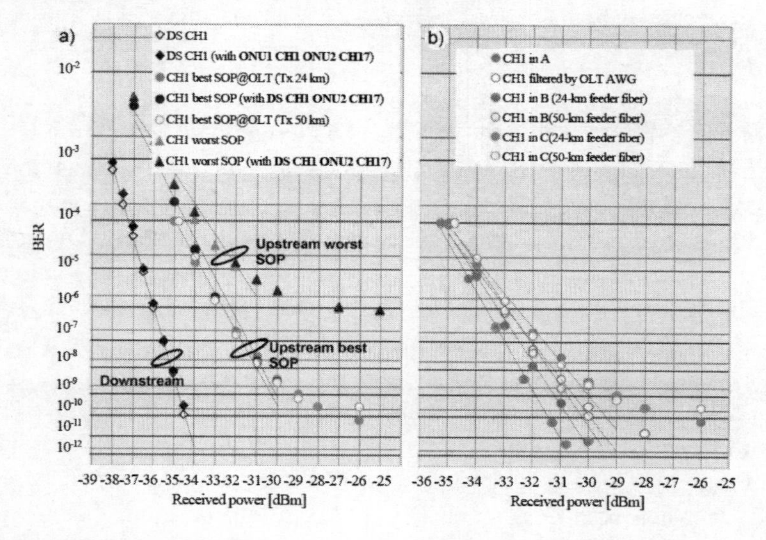

Figure 32. BER versus received power for channel 1 at ONU1: a) US and DS evaluation with simultaneous operation of ONU2 over 24 and 50 km b) analysis of channel 1 in different measurement point of the setup for 25 and 50 km with reference to Figure 30.

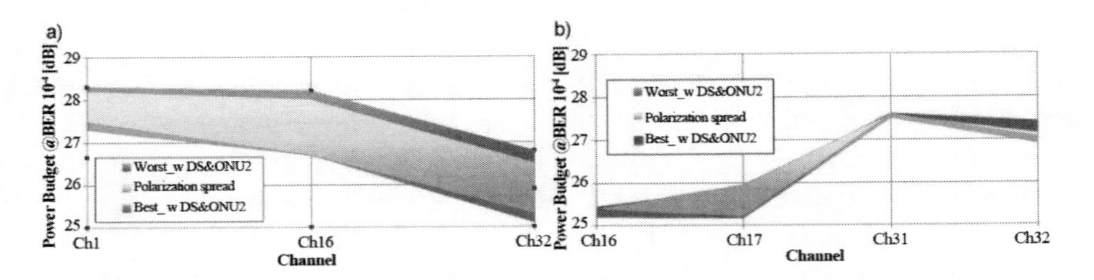

Figure 33. Power budget at 10^{-4} BER for different channels of ONU1 (a) and ONU2 (b).

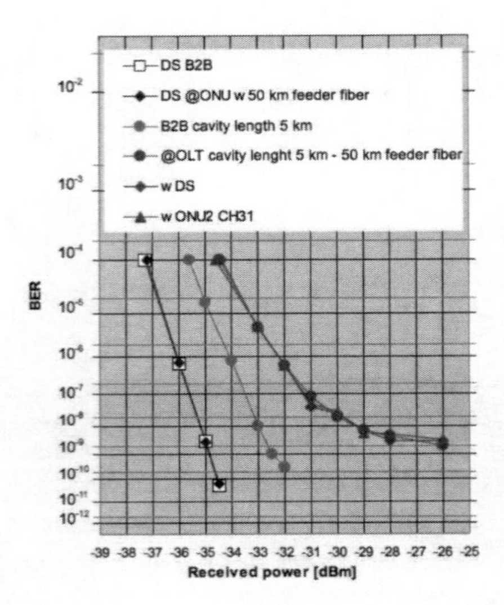

Figure 34. BER versus received power for channel 1 ONU1 with 5-km distribution fiber length for US in: back to back (green circles) and in point C (Figure1) after 50-km transmission (red circles), in point C after 50-km in presence of the second operating ONU (red triangles) and in presence of the DS (red diamonds). DS performance in back to back (black open squares) and after 50-km transmission (black full diamonds) is also displayed.

It has been demonstrated that birefringence and polarization issues can be coped with the insertion of a FRM at the RN [2] if the RSOA PDG is not significant. In this situation the cavity shows two alternating orthogonal polarization eigen states, thus the transmitter performance is stabilized despite polarization variations due to birefringence. This allows exploiting also longer distribution fiber for the network-embedded cavity. Due to fiber length, a 5-km cavity shows approximately 1.5-dB extra losses with respect to 1-km cavity, which are marginal with respect to experimented cavity losses due to a thermal AWG, output coupler ratio, C L WDM coupler, which average 16-17 dB. Cavity insertion losses referred to the gain balance define the performance of the self-tuning cavity.

Figure 34 presents the measurements while simultaneously operating ONU1 and ONU2 respectively with 5 and 1.4 km distribution fiber and exploiting a FRM as the common reflector. Results are quite similar to those commented in Figure 32. No penalty is found both for the DS and US when both ONUs and DS are taken into account. The US penalty with respect to DS is 2.5 dB in back to back (Figure30 point A), comparable with that of a shorter-length cavity. Moreover at 10-4 50-km transmission, at OLT receiver (Figure 30 point C), shows almost 1-dB penalty, while a 10-9 floor rises. Performance of the 5-km distribution fiber cavity transmitter after 50-km transmission is almost similar to best SOP performance for 1-km distribution fiber cavity, evidencing that exploitation of a FRM at the RN well overcomes the birefringence issue when exploiting low PDG RSOA.

5.3. High-PDG RSOA with FRM

In this section, we will show the first experimental results obtained with a high-PDG RSOA chips specially developed by III-V labs for ERMES project.

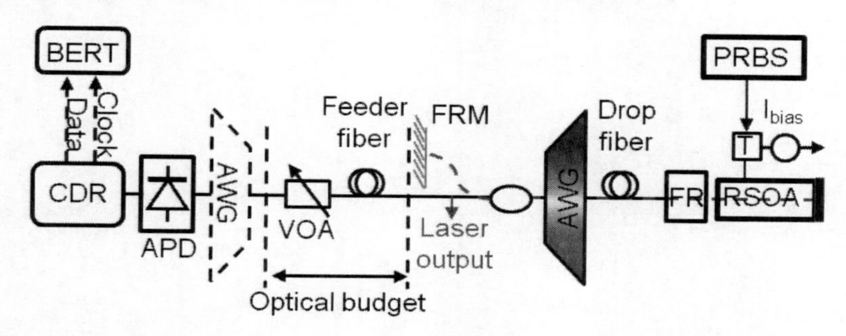

Figure 35. Experimental set up for BER measurement.

Figure 35 depicts the experimental setup of an upstream transmission with a self-seeded quantum-well (QW) C-band RSOA. The RSOA has a high signal TE mode reflection gain above 30 dB at 1540 nm (@ bias current=100mA). A 45° faraday rotator (FR) is set at the output of RSOA and a 90° FRM at the output of AWG via a splitter. Its splitting ratio will vary for the purpose of our experiments. This set up guarantees that the input polarization state at the RSOA is stable and aligned with the RSOA high gain, while the laser-cavity operation is unaffected by polarization effects within the light path. This assures a high round trip gain and a steady performance [10].

In our experiment, we use a single cyclic 16-channel gaussian AWG with a channel bandwidth of 100GHz and an IL of 2dB. The laser signal is send to a feeder fiber (up to 100km) and received by an APD coupled to CDR. The RSOA is directly modulated by a NRZ PRBS 2^{31}-1.

In order to investigate the OB, we emulate transmission losses using a VOA inside the cavity. We defined the optical budget between the laser output (output of the splitter at the RN) and the reception of the APD at the OLT.

For such an extra-long cavity source, the ratio between the gain and the round trip losses inside the cavity is an essential issue for system performances especially the optical budget. In the literature, more attentions are drawn to use a small coupling loss inside the cavity, consequently a large loss is introduced at the output of cavity (unsymmetrical splitter). In our experiment, four different splitters (50/50, 40/60, 20/80, 10/90 corresponding to coupling losses of 3.5/3.5, 3.7/4.2, 1.8/8.2, 1.6/10.7 dB) are inserted separately inside the cavity. The figure 30 represents the self-seeded source output power versus different coupling losses (laser output branch). We observed a linear evolution showing that the output power increases (from-7.9dBm to 0.3dBm) when the output splitting losses are reduced (from 10.7dB to 3.5dB), which means that by replacing a splitter of 10/90 by a 50/50, we can improve the output power from -7.9 dBm to 0.3 dBm. However, in the case of modulation at 1.25Gb/s, the extinction ratio decreases with the coupling losses (from 7.4dB to 6.1dB for a 10/90 replaced by a 50/50). Figure 37 shows the BTB BER performances of a10m cavity laser versus the optical budget and for several splitter ratios. For the splitter of 50/50, an OB of 35 dB is tolerated to achieve a BER of 10^{-3} (considering FEC operation), which represent 8 dB of improvement compared to the implementation with a 10/90 splitter.

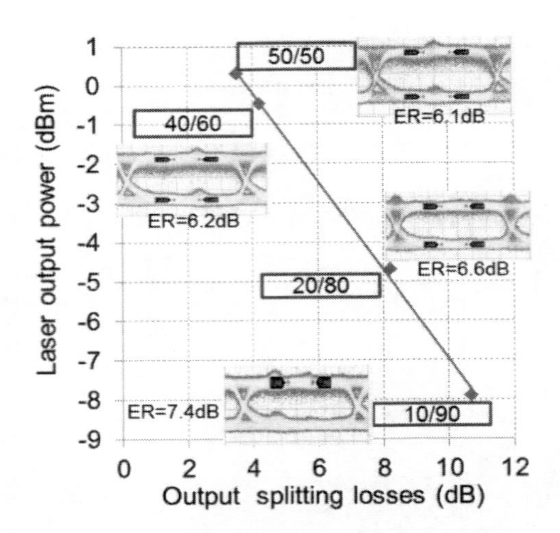

Figure 36. Laser output power versus different output splitting losses

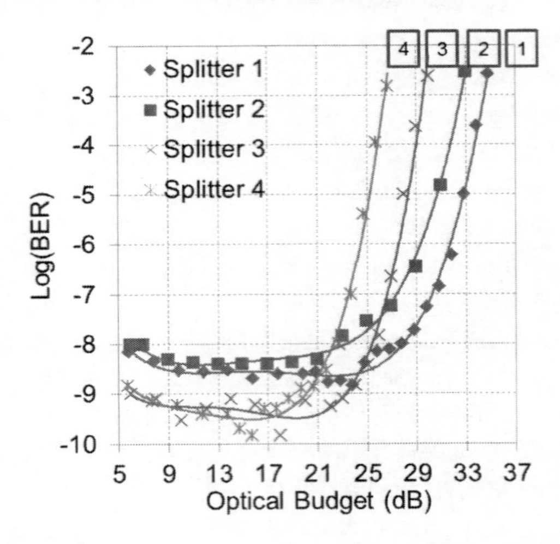

Figure 37. BTB BER performances versus optical budget for 4 splitters at 1.25Gb/s.

With the 50/50 splitter, we also evaluated the transmission performances at bit rates of 1.25Gb/s and 2.5Gb/s for different cavity lengths from 1km to 5km shown in figure 38. Table 4 represents the OB results for the considered BER of 10^{-3} at different bit rate and transmission reach. The green zone shows the OB value above 33dB which is the minimum required for co-existence with existing infrastructures demand (Class B+ GPON +5dB for MUX insertion). For a short cavity length, performances up to 10Gb/s was also achieved in BTB and for longer transmission reach thanks to pre-electronics signal treatments as in the section 5.1.

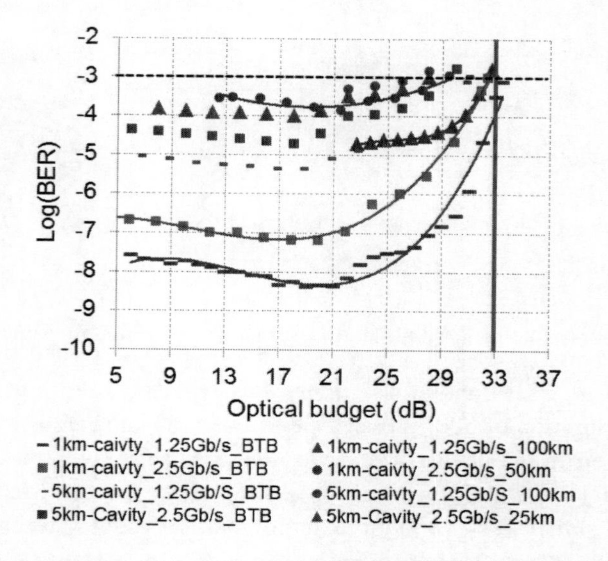

Figure 38. BER Transmission performances at 1.25 Gb/s and 2.5Gb/s for 1km& 5km-drop fiber

Table 4. OB measurements at 10^{-3} BER

	Cavity length	BTB	25km	50km	75km	100km
1.25 Gb/s	10m	35	35	34.8	34.5	34
	1km	33.3	33.1	33	32.9	32.8
	5km	30.8	30.2	30	29	27.8
2.5 Gb/s	10m	34	33.3	32.6	27.2	-
	1km	32.4	32	29.5	-	-
	5km	30	28	-	-	-

Table 5. Rise and fall time measurements

	Measured (µs)			
Cavity length	**10m**	**500m**	**1km**	**5km**
RT (10/90%)	7.6	7.7	7.8	25.8
RT (5/95%)	9.4	8.5	8.7	47
FT (10/90%))	2.8	2.3	2.4	2.9
FT (5/95%)	3.4	3.1	3.1	3.6

In order to reduce the energy expenditure of the WDM-PON system, the laser should be shut down following several sleep modes detailed in [11].

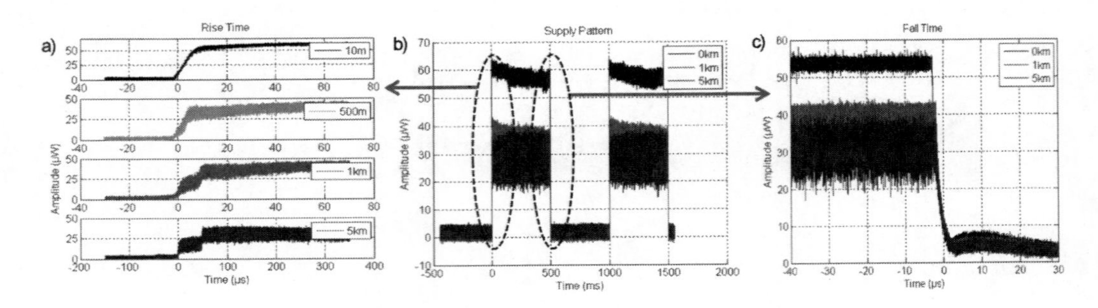

Figure 39. Measurement of rise time (a) and fall time (c) with applied supply pattern (b).

The external cavity laser continuous supply was replaced by a pulse pattern generator with a laser-on time duration of 500ms over a period of 1s (Figure 39-b). In our experiment a 10/90 splitter was used in the cavity. Then, we investigate the rise time (RT) (Figure 39-a) and fall time (FT) (Figure 39-c) of the laser for several cavity lengths. Measurement results of RT and FT at 5-95% and 10-90% of the maximum output level are presented in Table 2. We observed that the RT and the number of round trip needed to reach a stable laser operation depend on the cavity length. The longer the cavity is, the higher the rise time is. However, the fall time remains at a stable value around 3 µs. For sleep mode consideration, whatever the cavity length, these results show the possibility to awake and put back to sleep the laser within less than 8.19s as required in [11].

Such a laser could also be of interest for a burst mode implementation for hybrid time division multiplexing-WDM (TDM-WDM) approaches. However, for the shortest cavity (10m), the 10-90% rise time is measured at 7.6µs which is largely above the 16bits (12.8ns) limit for a 1.25Gbit/s burst mode transmission specified in ITU-T G984.2.

Self-seeded WDM-PON solution has been recently proposed for the mobile front-haul application based on common public radio interface (CPRI) [12].

One of the factors limiting the transmission performance is the jitter tolerance. The CPRI low voltage (LV) electrical specifications are guided by the 10G attachment unit interface (XAUI) electrical interface specified in Clause 47 of IEEE 802.3ae-2002. According to CPRI LV receiver AC timing specifications, at a BER of 10-12 the tolerated total jitter (TJ) is 0.65UI and the tolerated deterministic jitter (DJ) is 0.37UI. Back to back transmissions at 1.25Gb/s and 2.5Gb/s bit rates are realized with a 10/90 splitter in the cavity and for three different cavity lengths: 10m, 500m and 1km. The jitter measurement is performed with a digital oscilloscope at the Rx output of the APD with a 231-1 long NRZ PRBS. Figure 40 depicts the data eye diagram and the BER bathtub histogram at 1.25Gb/s for a 10m cavity. In several seconds, bathtub histogram measurements captured the most probable timing locations of data transitions. This histogram is also known as a jitter histogram. It is understood that the total jitter probability density function (PDF) is a convolution integral of bounded deterministic jitter (DJ) and unbounded Gaussian random jitter (RJ). BER is the integrated "tail" of the jitter distributions from each side of the data eye and is as a function of the sampling point.

Table 6. Jitter measurement results at 10-3 BER

Jitter (μs) (BER)		TJ (10-6)	TJ (10-12)	DJ (10-12)
1.25Gb/s	10m	0.29	0.41	0.05
	500m	0.3	0.41	0.05
	1km	0.26	0.36	-
2.5Gb/s	10m	0.52	0.71	0.11
	500m	0.6	0.85	0.12
	1km	0.62	0.85	0.12

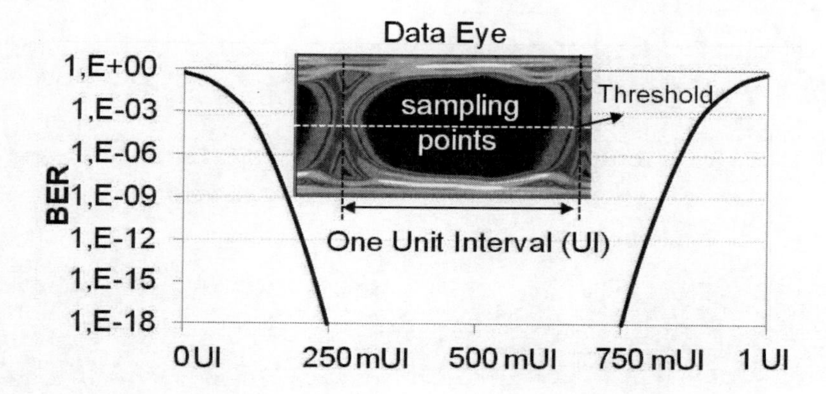

Figure 40. data eye and BER bathtub histogram at 1.25Gb/s.

Table 6 illustrates all the TJ values in unit interval (UI) measured at receiver. The TJ results at 1.25 Gb/s for a BER of 10^{-12} are compatible to the jitter specification values. However, at 2.5 Gb/s, jitter results are unfavorable. Considering a BER of 10^{-6} target, all jitter results remain in specified values.

5.4. Amplified Self-Seeded Architecture

In this section, we present for the first time a new WDM PON architecture based on amplified self-seeded RSOA to achieve a simple and standard WDM infrastructure (no mirror of specific device at RN). As shown in figure 46, the connection to one port of the AWG relocates the FRM from the RN to the OLT via an optical splitter. The extra-long self-tuning cavity source range from the OLT to the RRH, and has a length defined directly by the sum of feeder and drop fiber.

Figure 46 depicts the experimental setup for BER measurement with the same QW RSOA used in the section 1.3. A 45° FR is set at the output of RSOA and a 90° FRM at the output of AWG via an 80% splitter. An extra C-band bi-directional semiconductor optical amplifier (SOA) is inserted inside the cavity between FRM and the optical splitter in order to compensate extra-long cavity losses. The SOA offers a small signal gain >20dB and a low polarization dependent gain <1dB at 1550nm. In order to investigate the feeder optical budget, which is defined between the MUX at the OLT and the RN, we emulate transmission losses using a VOA inside the cavity. Then, the output power of the external cavity source as well as the feeder optical budget performances will depend on the cavity losses.

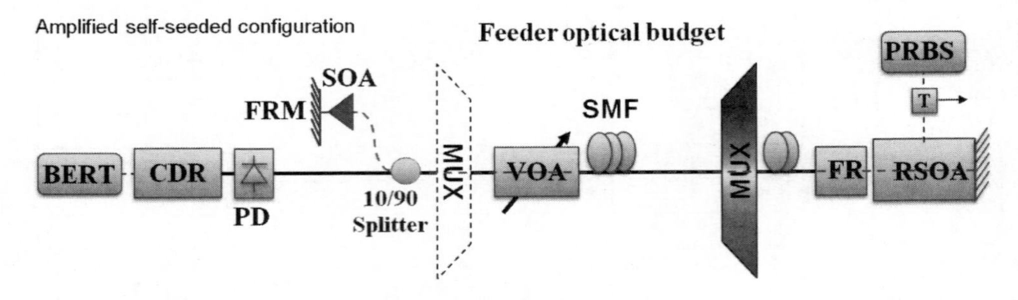

Figure 46. Amplified self-seeded RSOA experimental set up.

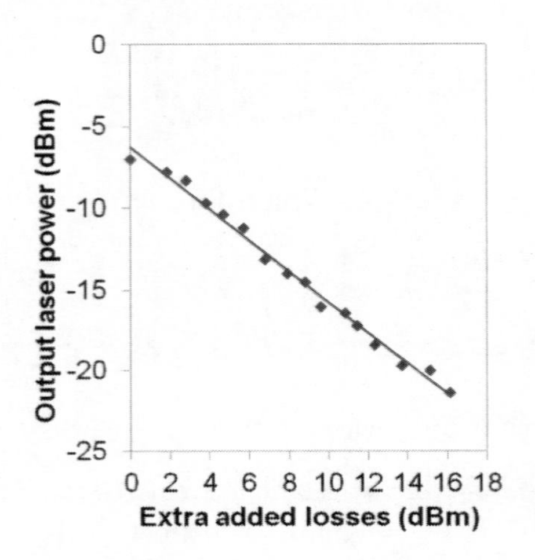

Figure 47. Output laser power vs. extra added losses at 1.25Gb/s.

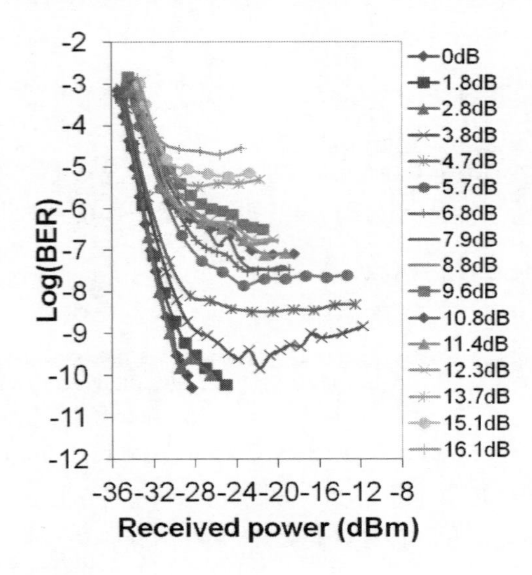

Figure 48. BER vs. extra added losses at 1.25Gb/s.

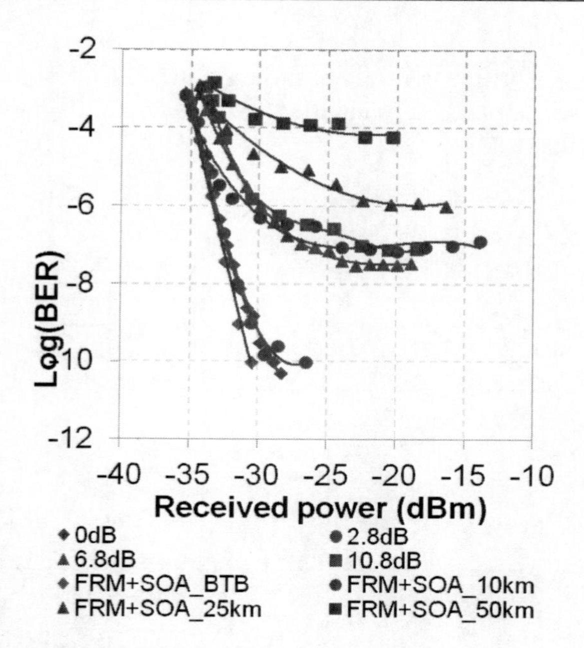

Figure 49. BER vs. added fiber and added losses at 1.25Gb/s.

At reception, the system consists in an APD, a CDR, and a BERT. This receiver is connected to the 20% output of the splitter.

The RSOA is directly modulated at 1.25Gb/s by a 2^{31}-1 long NRZ PRBS. In the first step, we evaluated the impact on the extra added losses emulated by the VOA inside de cavity for our proposed amplified self-seeded configuration showed in the figure 35. The RSOA is biased at 90mA and the SOA bias is 130mA.

In the figure 47 and figure 48, we observed the same tendency as described previously in the section 5.1.4. Laser output power decrease linearly versus added losses in the cavity. Figure 48 depicts the BER measurement in case of BTB versus received power for different added losses from 0dB to 16.1dB in the cavity. Error floor start to appear from 3.8 dB of the added losses. We compared also BER performances between different added fiber length and added attenuation as shown in the figure 49. With 10km of added SMF having 2.7 dB of attenuation, the BER curve degrades from 10^{-10} to 10^{-7} compared to that with 2.8 dB of added attenuation by the VOA.

We compare the performances with two configurations widely studied and presented in the literatures: spectrum sliced RSOA based configuration shown in the figure 50(a) and self-seeded RSOA based configuration shown in the figure 50(b). VOA1 and VOA2 are inserted in order to simulate the feeder optical budget and compare the performances in equivalent configurations. The RSOA is directly modulated at 2.5Gb/s by a 2^{31}-1 long NRZ PRBS.

The RSOA is biased at 70 mA for scheme (a) and 90mA for scheme (b) and (c). The main interest in comparing the RSOA spectrum-sliced configuration and our proposed amplified RSOA-based self-seeded one is that both of them stick to a WDM-PON regular architecture compatible with existing systems. A 12dB of optical improvement is obtained for amplified self-seeded source (c) compared to spectrum sliced source (a) shown in the figure 51. Figure 41 represents the BER measurement versus the feeder optical budget in the case of BTB (both feeder fiber and drop fiber are removed) at 2.5 Gb/s. For the amplified self-seeded configuration, an optical budget of 19dB can be tolerated in order to achieve a BER of 1.10^{-3}, which is the limit for an error free transmission when using FEC. Also, we observed a better ER and a better BER performance for amplified self-seeded configuration when the feeder optical budget is lower than 18dB.

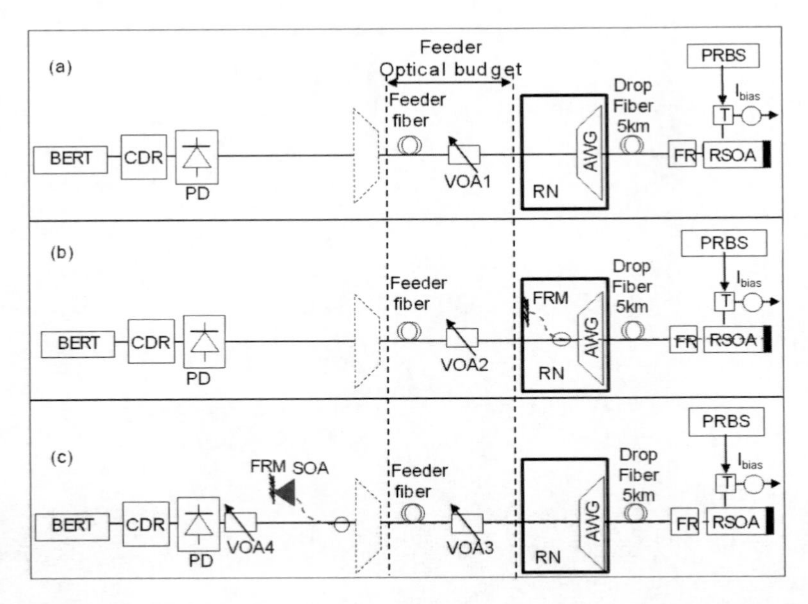

Figure 50. Experimental setup (a) RSOA spectrum-sliced configuration (b) RSOA-based self-seeded configuration (c) RSOA-based amplified self-seeded Configuration.

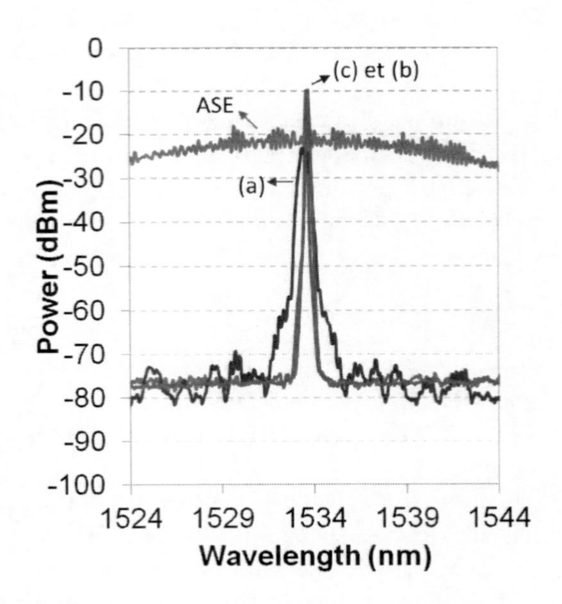

Figure 51. ASE and source spectra for three configurations.

Both configuration (b) and (c) create an extra-long external cavity formed between the FRM and the RSOA reflective facet. For RSOA-based self-seeded source seen in figure 50 (b), the cavity length is constituted of the drop fiber. While for the third configuration (c), the cavity includes both of feeder and drop fiber. Consequently, extra amplification (SOA) is required to overcome the total cavity round trip losses. Figure 53 and figure 54 depict the impact on the cavity length at 2.5 Gb/s for these two extra-long cavity sources. Varying from 10 m to several kilometers of drop fiber length, the BER performance of the self-seeded source degrades from 10^{-9} to 4.10^{-5}. By inserting a SOA inside the cavity at the output of the splitter, we observed a better ER and that the BER performances are improved for cavity lengths from 10 m to 5 km. Besides, for the first time, we can successfully reach up to 45 km of cavity length while achieving a 10^{-3} BER.

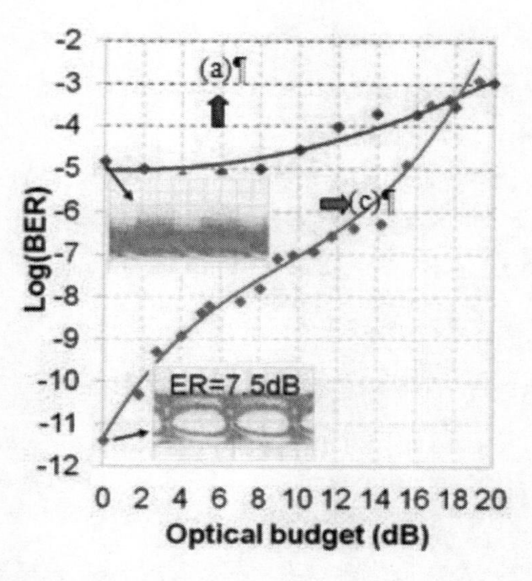

Figure 52. BER BTB performances at 2.5Gb/s for configuration (a) and (c).

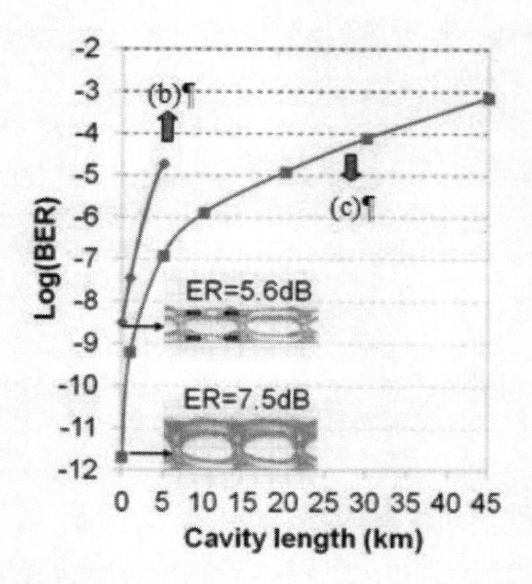

Figure 53. BER performance versus cavity length at 2.5Gb/s for configuration (b) and (c)

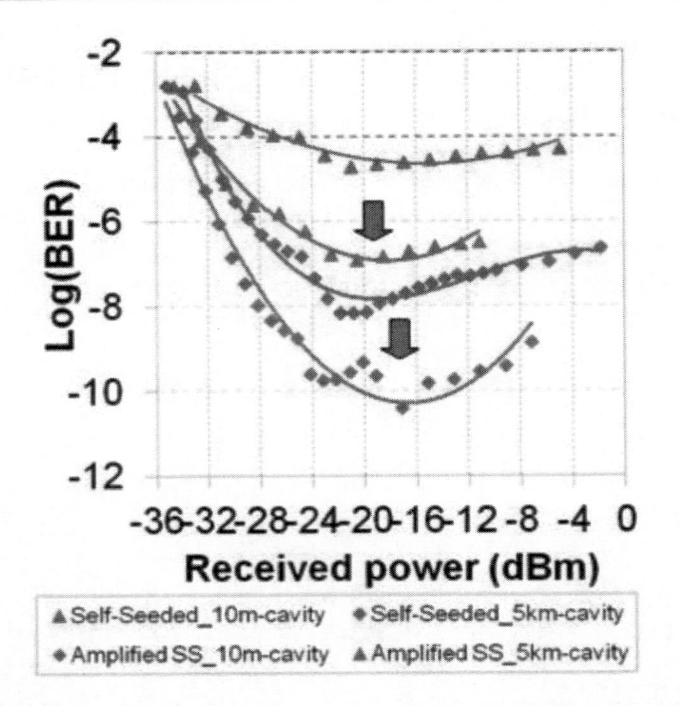

Figure 54. BER performance versus received power at 2.5Gb/s for configuration (b) and (c).

In order to compare the transmission performances of the three configurations, we choose a 5 km drop fiber. The BER measurements at 2.5 Gb/s for BTB and after transmission in feeder fiber of different lengths is presented versus received power in figure 55.

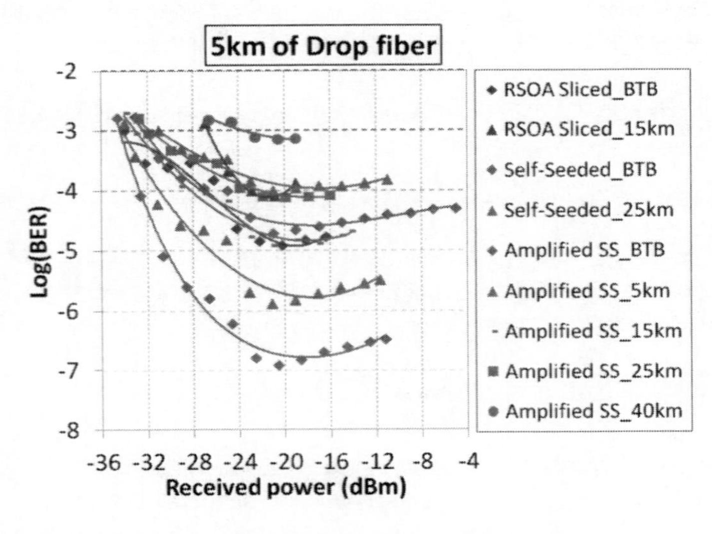

Figure 55. BER transmission performances versus received power at 2.5Gb/s.

Firstly for the RSOA spectrum sliced configuration (a), in the case of BTB, we obtain a -34 dBm sensitivity for a BER of 10^{-3}. Being spectrally sliced by the AWG, the ASE power is reduced to -15 dBm, which means that the optical budget obtained between the OLT and the

RN is equal to 19 dB. On the other hand, the transmission in the feeder fiber is limited to 15 km with a large 8 dB penalty due to chromatic dispersion effects. In comparison, the self-seeded source with an external cavity length up to 5km displays similar BER performances in the case of BTB with a -32 dBm sensitivity at a BER of 10^{-3}.

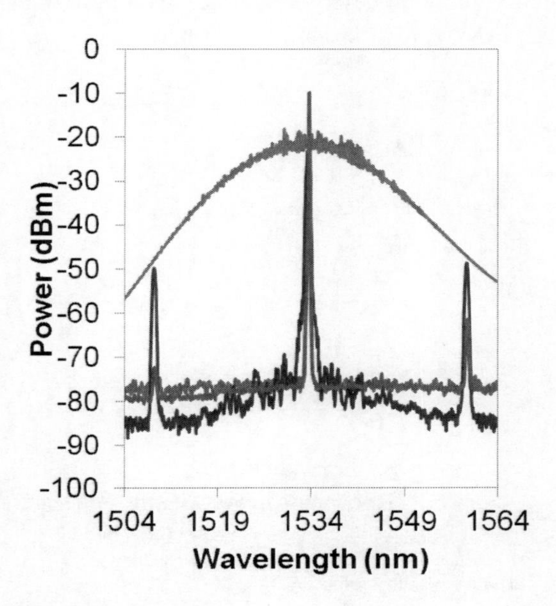

Figure 56. Spectra at 10Gb/s for 4 configurations.

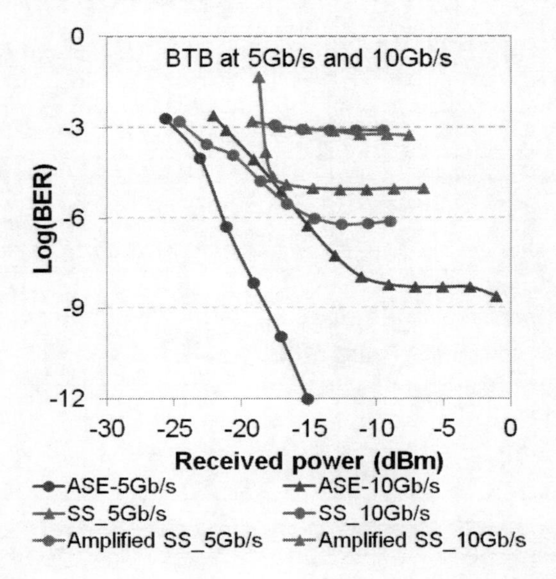

Figure 57. BER measurements versus received power at 5Gb/s and 10Gb/s for ASE, self-seeded and amplified self-seeded configurations.

The output power of the self-seeded source is about -4 dBm permitting a feeder optical budget of 28 dB. The transmission reach is increased to 25km of feeder fiber with a small penalty of 2 dB. Finally our proposed amplified self-seeded source with SOA displays the best sensitivity equal to -34.5 dBm in the case of BTB (5 km of drop fiber in the cavity). After 25 km of feeder fiber, the power penalty is below 2 dB for a BER of 10^{-3}. We also observe penalties due to chromatic dispersion effects by comparing BER at equivalent link losses with and without SMF. This can be explained by the higher level of RIN in the case of transmission in SMF. Finally, the total cavity length i.e. maximum transmission distance, can reach as much as 45 km of SMF with 40 km of feeder fiber.

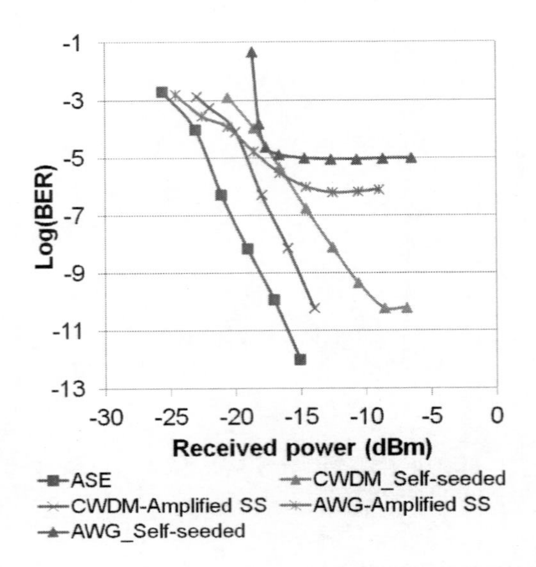

Figure 58. BER measurements versus received power at 5Gb/s with CWDM and AWG multiplexer.

Table 7. ER and ED at different bit rate for different configurations

	ASE	RSOA Sliced	Self-Seeded	Amplified SS
1.25Gb/s	ER=9dB		ER=6.5dB	ER=8dB
2.5Gb/s	ER=8.8dB		ER=6dB	ER=7.5dB
5Gb/s	ER=4.8dB		ER=4.9dB	ER=5dB
10Gb/s	ER=3.2dB		ER=2.8dB	ER=3.7dB

ERMES project aims to increase the bit rate up 10Gb/s. In the section 5.1, we obtained the experimental results up to 10 Gb/s with a low-PDG RSOA thanks to the emphasis technology. With the QW high-PDG RSOA, no improvement of BER performances is observed with the emphasis treatment.

We evaluated in this section only short cavity (10m) condition. The QW high-PDG RSOA can be ASE modulated up to 10 Gb/s with an ER of 3.2 dB showed in the table 7. No ER is measured above 5 Gb/s in the case of RSOA spectrally sliced configuration. We successfully obtained an ER of 5 dB at 5 Gb/s and 3 dB at 10 Gb/s for the self-seeded and amplified self-seeded configurations with no electronic treatment. Presented in figure 57, we obtained quasi-similar BER performance at 10 Gb/s in the case of these two configurations. No improvement in bit rate increase is observed by inserting a SOA inside the cavity.

5.5. Remotely-Pumped Self-Seeded Architecture

It has been experimentally demonstrated an unconventional hybrid TDM/WDM PON using self-seeded reflective semiconductor optical amplifiers (RSOA), providing a 512-split PON with 20-Gb/s aggregated upstream capacity [13] achieved by overlaying multiple WDM PONs on a legacy optical distribution network deployed for TDM PONs, as in figure 59(b). This unconventional scheme takes advantage of the properties of the RSOA-based self-seeded transmitter scheme, which is self-tuning, and thus colorless, and immune to the distributed backscatterings [7], which limit the ultimate bridgeable distance when a single feeder fibre is exploited in remotely-seeded solutions [14]. The unconventional hybrid topology is motivated by the fact that the self-seeded transmitter heavily suffer from cavity insertion losses [7], thus power splitting for TDM overlay is performed after wavelength demultiplexing (in upstream direction). We present in this section the experimental analysis of a colorless network-embedded self-tuning transmitter for a conventional hybrid stacked-WDM/TDM-PON [7]. To overcome the additional cavity losses due to power splitters the optical network unit (ONU) transmitter is assisted by a remotely-pumped optical amplifier, placed in the completely passive remote node (RN) and shared between all the users [15]. The pump is placed at the Central Office (CO), thus preserving the passive nature of the RN and in this sense differs greatly from solutions proposing Erbium doped fibre amplifiers (EDFA) at the remote node [14]. Though remote amplification has been already proposed for extending the reach of PON, the remotely pumped EDFA is here employed in a double pass configuration due to the properties of the self-seeded source. The discussed scheme sustains a 256-split PON with 80-Gb/s aggregate upstream (US) capacity.

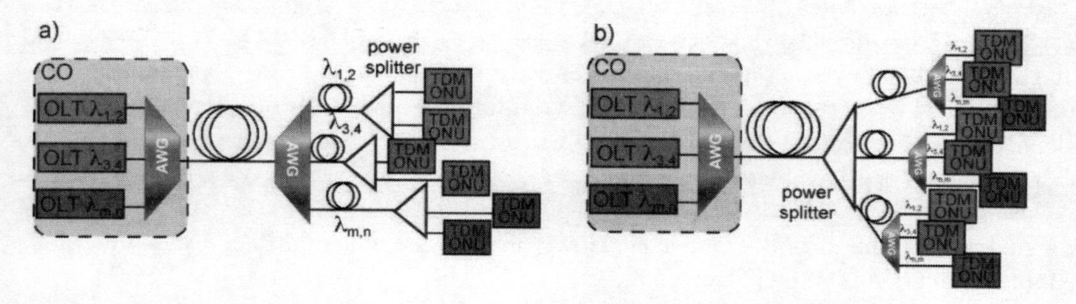

Figure 59. (a) Conventional stacked-WDM/TDM-PON. (b) unconventional hybrid TDM/WDM PON.

The self-seeding cavity performance depend on the cavity losses, the exploitation of a power splitter inside the ONU transmitter to implement Figure 59(a) conventional hybrid stacked-WDM/TDM-PON drastically raises the cavity losses of an amount equivalent to the doubled splitter rate. It is thus mandatory to overcome them with assisted amplification. In order to maintain the passive nature of the RN, a remotely pumped EDFA has been chosen. The topology [7] must be modified to integrate the remotely pumped EDFA in the RN, the final scheme is presented in Figure 60. The pump source, which is located at the central office (CO), after propagation in the feeder fibre, is coupled with the EDF by two 1550-1480 nm WDM couplers: the first one, WDM-1, also routes the DS signal to the RN cyclic AWG via the STC coupler output branch. The second one (WDM-2) common port is conversely connected to the EDF, while its 1550-nm port allows the US cavity signal to be coupled to the remotely pumped EDFA. Both the cavity signal and the pump are reflected by the RN cavity mirror, which can be for instance a Faraday Rotator Mirror (FRM) to overcome the polarization issue in a cavity exploiting a low polarization dependent gain (PDG) RSOA [2]. The realized amplifier is thus a double-pass EDFA.

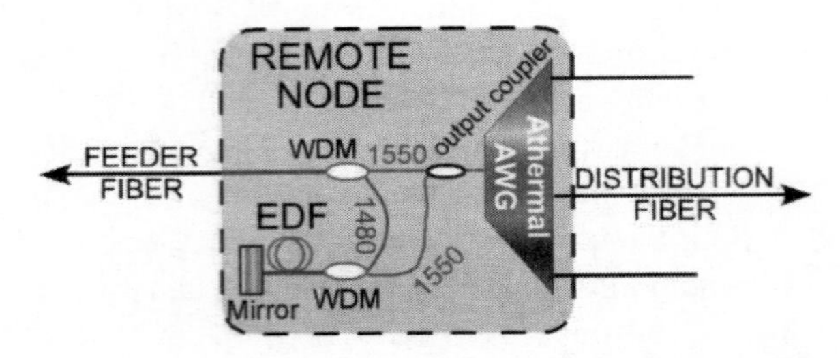

Figure 60. Self-tuning cavity topology for remotely pumped EDFA exploitation at the RN.

Figure 61 shows the experimented upstream transmitter based on the RSOA self-tuning cavity assisted by a remotely pumped EDF amplifier together with the whole experimental setup. The cavity is delimited by two mirrors: one belonging to the RSOA located at the ONU and one FRM located at the RN. The transmitter remaining passive part is embedded in the network itself: the WDM C/L coupler for US and downstream (DS) separation, the distribution fibre, here 420 m of SSMF, the 32-channel 100-GHz spacing cyclic athermal AWG and the 80/20 output coupler. A power splitter between the AWG and the ONU, here emulated by a variable optical attenuator (VOA), allows for hybrid TDM/WDM PON transmitter validation. At the RN the EDF length has been 20 m. The employed pump is a Raman-fibre laser at 1486 nm. The pump source is inserted into the feeder fibre through a 1550-1480 WDM coupler.

At the ONU the active element is constituted by a low PDG RSOA, with more than 2-GHz E/O bandwidth, which can be directly modulated at 2.5 Gb/s. The RSOA Electro/Optical (E/O) response has been measured for bias currents from 100 mA to 175 mA, which covers the exploited bias current range; the measurements are presented in Figure 62(a), showing for all the bias-current range, a 3-dB bandwidth larger than 2 GHz. The RSOA spectrum at the bias current of 100 mA is presented in Figure 62(b), as can be seen it is centered at 1580 nm, thus AWG channels lower than 20 are not within the RSOA full width half maximum (FWHM) gain bandwidth.

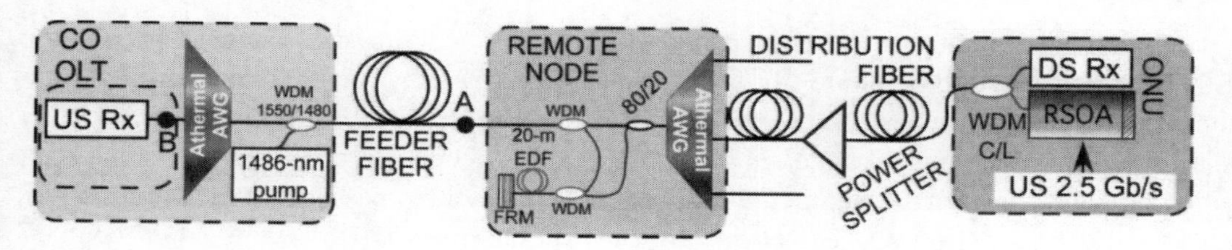

Figure 61. Experimental set up.

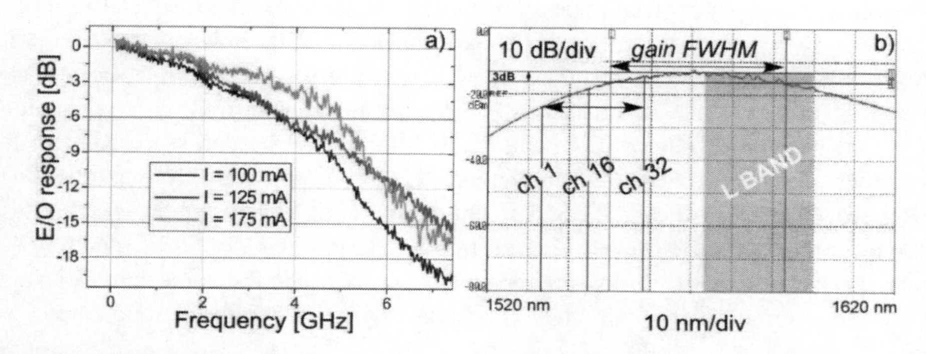

Figure 62. (a) RSOA E/O response for bias currents from 100 mA to 175 mA. (b) spectrum at 100-mA bias current.

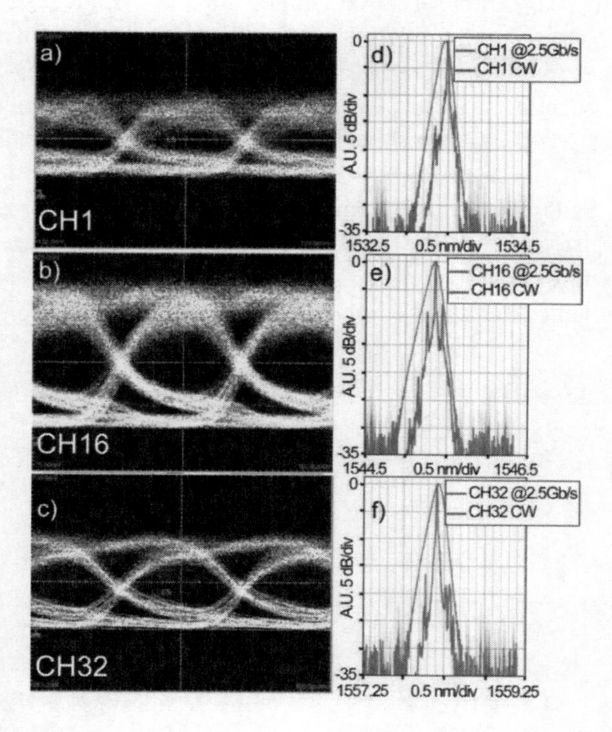

Figure 63. (a)-(c) Eye diagrams of channels 1-16-32 respectively. (d)-(f) respective CW and modulated spectra.

The exploited pump power is 25 dBm, which allows coupling an effective power in the EDF of nearly 16.6 dBm, due to fibre and coupler losses. As expected, pump propagation in the feeder fibre determines also Raman amplification in the L band, where the DS signal is spectrally located. In our set up, exploiting 25-km SSMF feeder fibre, the average net Raman gain in L band has been measured in 5 dB. The measured double-pass gain was 26 dB at 1543 nm at -10 dBm input power. The cavity losses, including the 1550-1480 nm WDM coupler, but excluding the doped fibre, with no VOA extra losses, have been estimated in 17.4 dB. Of course the emulation of 1:8, 1:16 or 1:32 splitters adds to the cavity losses of 18 dB, 24 dB and 30 dB respectively, which need to be recovered by the remotely pumped EDFA gain.

Figure 63(a)-(c) presents the measured back to back eye diagrams, i.e., taken at point A of Figure 61, for channels 1, 16 and 32. They correspond to different bias points, that is respectively 155 mA, 130 mA and 102 mA; the increasing bias currents for lower number channels are consistent with the RSOA gain bandwidth: to get the necessary gain for lower channels higher currents are needed. The applied data are 4 Vpp and the relative extinction ratios (ERs) are 4.7 dB, 6.7 dB and 8 dB. The operating points have been chosen to minimize the BER and represent a trade off between desirable high ER and cavity recirculating signal cancellation [4]. Figure 63(d)-(f) shows the output spectra for the same channels in CW regime (red narrower spectrum) and when applying 2.5-Gb/s modulation, modulated spectra (blue line).

Table 8 shows received powers to achieve the enhanced FEC limit of $4 \cdot 10^{-3}$ BER and the Reed-Solomon limit of 10^{-4} BER at 2.5 Gb/s. The measurements have been performed after 25-km SSMF propagation and after OLT AWG demodulation with a 2.5-GHz APD followed by a clock and data recovery module. With VOA set to emulate 1:8 split, i.e. 9 dB attenuation, all the 32 channels reach a BER lower than the enhanced FEC limit for received power lower than 29 dBm, as it is also demonstrated by Figure 64, where BER values for each of the 32 channels at -29 dBm received power are presented. This allows an aggregate upstream capacity of 80 Gb/s shared by 256 end-users. Table 8 first column also shows that, with reference to 10^{-4} BER, for channel numbers lower than 24 the necessary received power rapidly increases, consistently with the RSOA gain allocation. The same trend is confirmed by the performance with VOA emulating 1:16 and 1:32 split losses: only for the channel range 24-32 the performance reaches the Reed-Solomon or enhanced FEC limits [9].

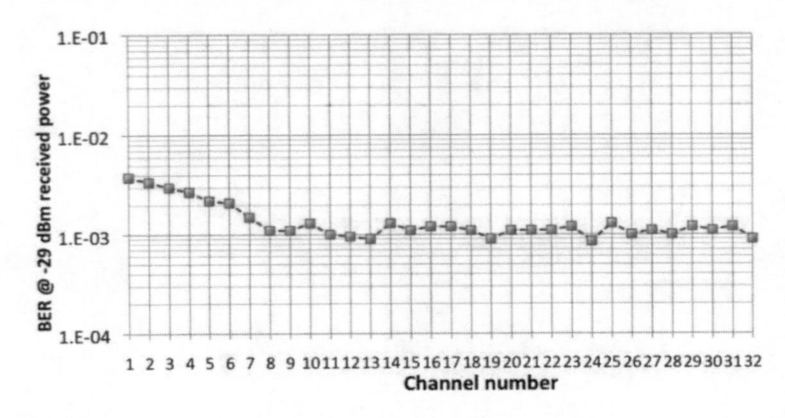

Figure 64. 2.5 Gb/s BER at -29 dBm received power for all the 32 channels.

Table 8. BER performance at 2.5 Gb/s

	Split 1:8		**Split 1:16**		**Split 1:32**	
	RX power	**BER**	**RX power**	**BER**	**RX power**	**BER**
CH32	-29.2 dBm	$<10^{-3}$	-28.2 dBm	$<10^{-3}$	-30.3 dBm	$<4\cdot10^{-3}$
	-24.6 dBm	$<10^{-4}$	-22.8 dBm	$<1.6\cdot10^{-4}$		
CH24	-29.5 dBm	$<10^{-3}$	-31.3 dBm	$<10^{-3}$	-30.5 dBm	$<4\cdot10^{-3}$
	-26.4 dBm	$<10^{-4}$	-24.5 dBm	$<10^{-4}$		
CH16	-30.3 dBm	$<1.5\cdot10^{-3}$				
	-20.4 dBm	$<10^{-4}$				
CH8	-30.2 dBm	$<1.5\cdot10^{-3}$				
	-18.4 dBm	$<10^{-4}$				
CH1	-29 dBm	$<4\cdot10^{-3}$				

With these splitting ratios, which imply heavier losses, the aggregate upstream capacity is thus limited to 20-Gb/s, differently from what has been demonstrated for the 256 users (1:8 split). The results seem to indicate that with the proper RSOA gain bandwidth allocation, up to 80 Gb/s capacity could be shared by up to 1024 users.

6. THE MODEL

The modelling task of the self-seeded transmitters was undertaken by the Karlsruhe Institute of Technology. Results of this task are presented in this section. Central to the self-seeded PON is the reflective SOA, and thus obtaining a good model for the RSOA is paramount to conducting the simulation task. The developed RSOA model is then incorporated into a larger model for the PON transmitter. As discussed in previous sections, the RSOA provides three important functions: amplification, data cancelation and direct modulation to imprint fresh data on the reflected wave. The RSOA is inherently an amplifier, albeit a nonlinear one. The nonlinear gain compression allows for the cancelation of the incoming modulation component while simultaneously allowing for fresh data to be modulated, via the injection current on the outgoing wave towards the central office.

The RSOA is a buried waveguide with various compositions of to form regions that confine both the injected carriers and the lateral profile of the propagating optical field [16]. The carriers are confined completely within the active region using double heterojunctions, the concentration of these carriers provides the optical gain the overlap of the optical lateral mode profile with the area of the active region is depicted in figure 65. The injected current into the RSOA provides the charged carriers for amplification via stimulated emission. The amount of amplification depends on the density of these carriers in the active region, amplification is provided by the injected charged carriers into the active region via the bias current. Simulations are performed for the RSOA operated within both the self and externally seeded configurations. In order to characterise the RSOA for operation within the self-seeded cavity, one must understand the gain saturation characteristic of RSOAs. The gain saturation behaviour can only be obtained by operating RSOAs within an externally-seeded configuration. The modelling requirements of RSOAs for the proposed self-seeded transmitter need to consider the following: spatially resolved counter propagating optical fields, spatially resolved carrier density dynamics, modulated input injection current

Homogeneous saturation of the RSOA due to the counter propagating optical fields in the RSOA internal cavity losses and RSOA-fiber coupling losses

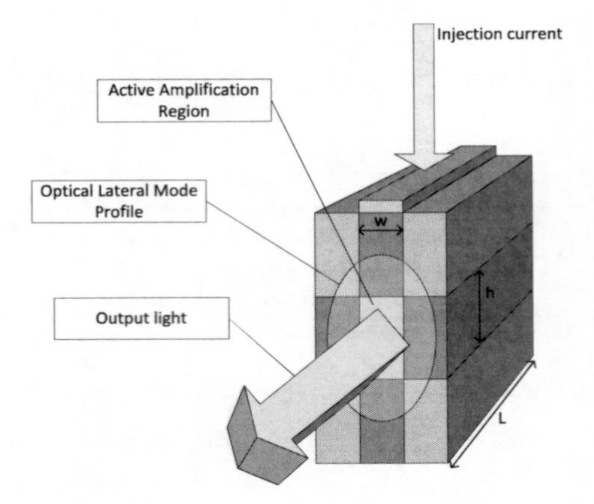

Figure 65. Schematic of RSOA showing the active region and lateral profile of the propagating optical field.

RSOA Model Description

The outline of the mathematical model of the RSOA is presented. The model is based on implementions of travelling wave amplifiers in [13,14]. A schematic of the amplification process is shown in figure 66. An initial field E^{+} is incident to the anti-reflecting (AR) facet of the RSOA. This wave gets amplified as it propagates along the RSOA; at the high-reflecting facet, a portion of the light is reflected back, this wave gets amplified and competes for carriers with the forwards travelling wave. When the wave reaches the AR, the wave leaves the amplifier and is collected into fiber. The wave leaving the RSOA is denoted E^{-}.

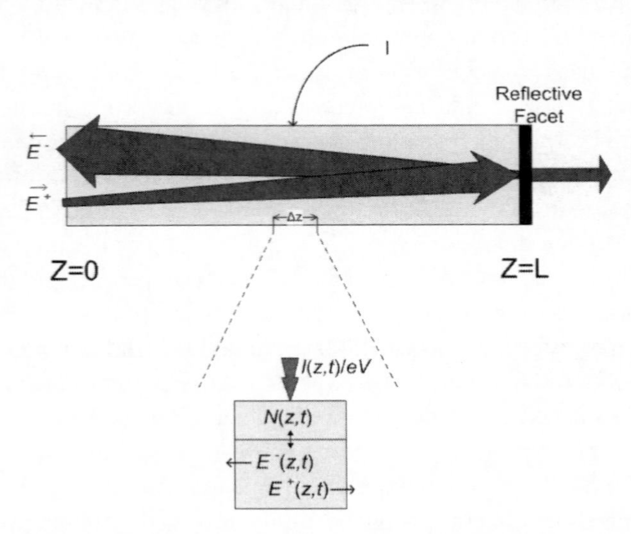

Figure 66. Modelling framework of RSOA showing the counter propagating fields highlighting interactions over a distance Δz.

The framework to model the RSOA is as follows: the RSOA is broken up into M separate sections such that the spatial granularity (distance step) defined in the calculations is $\Delta z = L/M$. The time step is then calculated as $\Delta t = \Delta z / v_g$, where v_g is the group velocity of the optical field in the RSOA. The carrier density rate equation is given hy:

$$\frac{\partial N(z,t)}{\partial t} = \frac{I_{RSOA}(z,t)}{eV} - R_{spon}(z,t) - R_{stim}(z,t) \tag{0.1}$$

where: z and t are the spatial and temporal variables respectively; N is the carrier density; I_{RSOA} is the injected current into the RSOA; e is the unit of elementary electronic charge; V is the volume of the active RSOA gain region; R_{spon} is the rate of spontaneous emission and R_{stim} is the rate of stimulated emission. The rate of stimulated emission is given by:

$$R_{stim}(z,t) = \frac{\Gamma a \left(N(z,t) - N_0\right)}{1 + \varepsilon_{nl} P_{tot}(t)} \frac{\left[\left|E^{+}(z,t)\right|^2 + \left|E^{-}(z,t)\right|^2 + \varepsilon^{+}(z,t) + \varepsilon^{-}(z,t)\right]}{wdh\upsilon} \tag{0.2}$$

where: Γ is the confinement factor of the light in the active region; a is the differential gain, N_0 is the carrier density at transparency; ε_{nl} is the nonlinear gain compression factor; wd is the area of the active region; $h\upsilon$ is the photon energy; $E^{\pm}$ represents the optical field of the counter-propagating optical signal with the directions defined in Figure 2(b); $\varepsilon^{\pm}$ represents the power of the propagating ASE in the RSOA; P_{tot} is the total power of the counter-propagating optical signal and ASE. The rate of spontaneous emission is given by:

$$R_{spon}(z,t) = AN(z,t) + BN^2(z,t) + CN^3(z,t) \tag{0.3}$$

where, A is the rate of nonradiative recombination; B is the rate of bimolecular recombination and C is the rate of Auger recombination.

The equations for the propagation of the counter-propagating fields are given by:

$$\frac{\partial E^{\pm}(z,t)}{\partial z} \mp \frac{1}{v_g} \frac{\partial E^{\pm}(z,t)}{\partial t} = \frac{1}{2} E^{\pm}(z,t) \left[-\gamma_{\text{int}} + (1 - j\alpha_{LE}) \frac{\Gamma a (N(z,t) - N_0)}{1 + \varepsilon_{nl} P_{tot}(t)} \right] \tag{0.4}$$

where γ_{int} is the internal scattering loss of the RSOA waveguide; α_{LE} is the linewidth enhancement factor. The RSOA boundary conditions (BC) are that the input lightwave to the RSOA is always known at all times, i.e. $E^{+}(0,t) = E_{in}(t)$. At the reflective facet the BC is given by $E^{-}(L,t) = \sqrt{R} E^{+}(L,t)$, with R being the reflectivity. For a given input, $E^{+}(0,t)$, the output lightwave $E^{+}(0,t)$ is calculated using Eq. (1) and Eq. (4) by employing theoretical SOA techniques [13, 14] and applying the RSOA boundary conditions. The RSOA input, $E^{+}(0,t)$, and output, $E^{-}(0,t)$, lightwaves can only be measured in fiber, therefore an additional fiber-RSOA coupling loss is used to determine the fiber-coupled input and output lightwaves at the RSOA input (AR) facet.

The RSOA is operated in the self-seeded configuration and a CW light is injected into the amplifier. The output power is measured in the coupled fiber. The static RSOA gain versus input power characteristic and the output power versus input power of an RSOA are shown in Figure 67 (a) and (b). The measured results (open circles) are plotted along with the simulation results (solid lines). The RSOA device parameters used are shown in Table 9. There is excellent between the calculated and measured results. The small signal gain is greater than 30 dB for low input powers and the gain saturates strongly with just a few dB gain at input powers of 0 dBm. The most important feature to observe from the plots is that the RSOA output power reaches a maximum; this is a feature of reflective amplifiers [18] in general due to the increased gain saturation from the reflective wave.

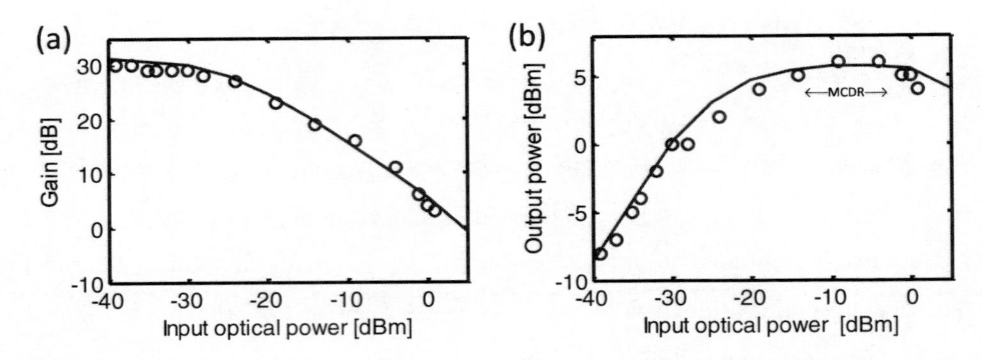

Figure 67. The (a) input power dependence of the fiber coupled RSOA gain (b) the RSOA input-output power transfer characteristic.

RSOA bias current is 160 mA. MCDR is the modulation cancelation dynamic range. Simulations are denoted by solid lines; open circles denote measured result. Measured results provided courtesy of PoliMi.

Table 9. RSOA parameters used in the simulations

RSOA Parameter	**Value**	**Parameter**	**Value**
RSOA length	800 μm	Confinement factor	20%
Area of Active region	0.18 μm2	Differential gain	4x10-20 m2
Reflectivity	10%	Carrier density at transparency	5x10-23 m-3
Rate of nonradiative carrier recombination	4x108 s-1	Bimolecular recombination coefficient	1x10-16 m3s-1
Alpha factor	4	Auger recombination coefficient	1x10-42 m6s-1
Nonlinear gain compression	2 W-1	Internal scattering loss	5,000 m-1
Fiber – RSOA coupling loss	6 dB		

The importance of the existence of a maximum output power is that the output power remains constant irrespective of the input power and this is precisely the condition to achieve modulation cancelation. For the employed RSOA, the input power at which modulation cancelation can be obtained is –6 dBm which makes RSOAs attractive devices to use for extended reach PONs. We define the range of input powers over which the difference in output power varies by less than 1 dB from the maximum output power to be the modulation cancelation dynamic range (MCDR). For any input signal whose intensity waveform is contained within the MCDR, the output signal will be compressed to a signal with minimal modulation component.

Dynamic RSOA Simulations: Modulation Cancelation

In the previous section we reported on the response of the SOA for CW optical input signal; however, the RSOA gain depends strongly on the carrier dynamics. Therefore in this section we show results of modulation cancelling experiments at 2.5 Gbit/s. A schematic outlining modulation cancelation is shown in figure 68.

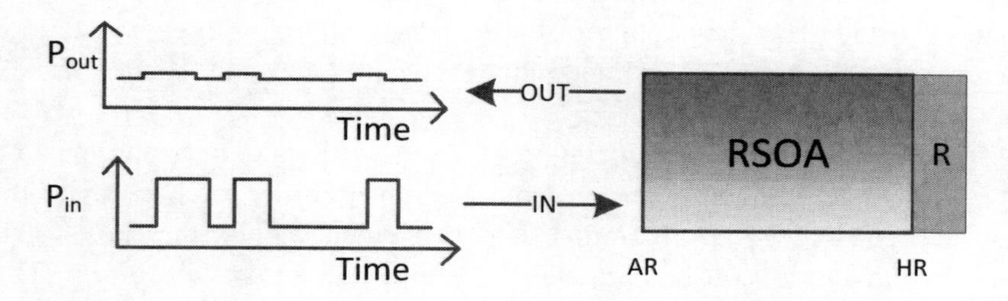

Figure 68. Depiction of the concept of modulation cancelation.

The RSOA is operated in the externally seeded configuration. The light source is externally modulated while the extinction ratio (ER) of the incoming signal is held constant at 6 dB. A 2.5 Gbit/s non-return-to-zero (NRZ) optical input signal with ER = 6 dB is launched into the RSOA. The average input power of the signal is controlled using a variable optical attenuator; this allows us to operate on different parts of the RSOA input-output transfer characteristic, Figure 3(b). The input eyediagram showing the 6 dB ER is shown in the top of Figure 5. The measured and calculated output eye diagrams for three different input optical powers are shown in figure 69 (a)-(c), the position on the input-output power transfer characteristic is shown in the insert of the simulated eye diagram. In Figure 69 (a) shows the output eye diagram when the average input power is -15 dBm, at this operating condition the gain is only moderately saturated and thus there is only a small closure of the signal eye diagram. When the average input power is increased to -6 dBm as is the case shown in figure 69 (b); the operating point being at the maximum output power of the input-output power transfer characteristic, that the modulation component of the outgoing wave has been completely cancelled apart from the transient spikes. The transient spikes occur because the RSOA gain cannot change instantaneously to changes in input power, thus spikes appear due to the transitions of the input signal power. When the input power is increased further to 0 dBm, there is a negative slope at this position on the power transfer characteristic, this the eye opens again however the output waveform is an inverted copy of the input signal and thus we term this operating region the signal self-inversion regime. There is excellent agreement between the measured and calculated results for the three operating regimes thus confirming the adequacy of the model to predict the static and dynamic behavior of RSOAs.

Modeling the Self-Seeded Transmitter

In this section we present results of the RSOA operating in a self-seeded cavity. The basic operation of the self-seeded transmitter is outlined in [3]. A laser field builds up due to recirculating ASE, and then the intensity of the outgoing wave is modulated via the RSOA injection current; the recirculating signal returning to the RSOA, that was modulated at a previous time (the delay τ being equal to the cavity transit time ~10 μs), is cancelled as it propagates through the RSOA. The reason is explained as follows: for typical semiconductor lasers with longitudinal dimensions of a few millimeters, the cavity transit time is much shorter than the photon lifetime due to cavity losses and the carrier lifetime; this condition leads to a carrier – photon resonance [16] under gain modulation because the gain equals losses condition is established over time scales of tens of picoseconds, much shorter than the gain modulation period. This decoupling of the gain modulation with gain saturation allows for the simultaneous cancelation of the incoming wave with the re-modulation of the intensity on the outgoing wave.

Central to obtaining satisfactory operation of a self-seeding source, one must obtain the cancelation condition as outlined in the previous section. In order to characterise the RSOA for operation in a self-seeded configuration, it is convenient to plot the RSOA gain versus output power for the currents at the extrema of the injection current waveform. The data to obtain these plots is to take the RSOA gain and output power graphs, such as those shown in Figure 68 (a) and (b), and plot RSOA gain versus output power with input power as parameter. Plotting the data in this fashion provides greater insight into the cavity. Plots of RSOA gain versus output power are shown in figure 71 for the RSOA current extrema at 80 and 180 mA.

The usefulness of plotting the RSOA gain versus output power is that the operating point of the self-seeded cavity can be read directly by noting that the RSOA gain equals the cavity losses. This is depicted in figure 71 by plotting the cavity loss line. The intersection of the cavity loss line with the gain-output power curve determines the circulating power in the self-seeded cavity. The losses are fixed, therefore the level of RSOA gain saturation adjusts until the gain equals the losses. In order to create a self-seeded transmitter that cancels the modulation component of the recirculating signal, the loss level must intersect the curves at the location of maximum output power. This is not possible to satisfy at two different values of injection current, though the loss value of 13 dB provides a good compromise. Such a high level of cavity loss would be expected given the insertion loss of the network components, i.e. the AWG and optical couplers.

The simulation platform is created by fixing a time delay of 400 ns (1,000 bits at 2.5 Gbit/s) between the time when the outgoing US signal leaves the RSOA and when the signal returns as a recirculating signal to the RSOA. The astute placement of Faraday rotators along the cavity leads to a retracing of the polarization state of the incoming lightwave at the RSOA input [10], therefore the scalar propagation equations can be used without modification. The laser field is built up from ASE, once a steady state field is established then the current is modulated. The standard laser output power versus current (L-I) curve is plotted in Figure 72.

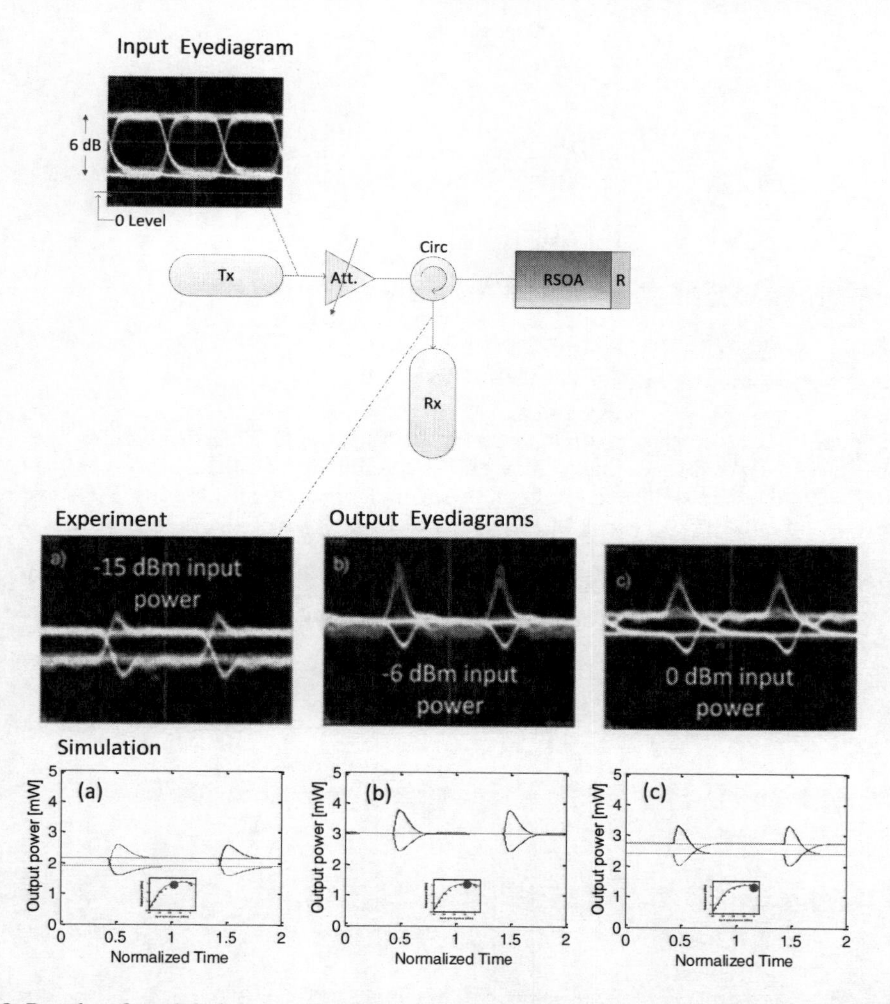

Figure 69. Results of modulation cancelation(a) signal squeezing (b) exact cancelation (c) self-inversion regime. Measured results provided courtesy of PoliMi. Acronyms: Tx transmitter, Rx receiver, Circ circulator.

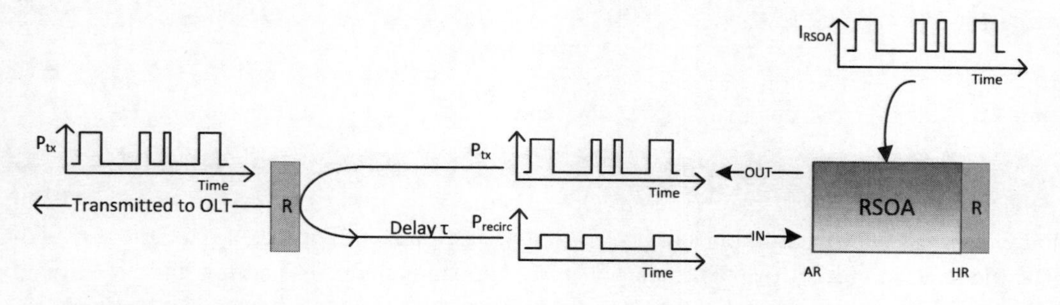

Figure 70. Schematic of self-seeded cavity, showing that the transmitted wave to the OLT is similar to that of the injected current waveform to the RSOA.

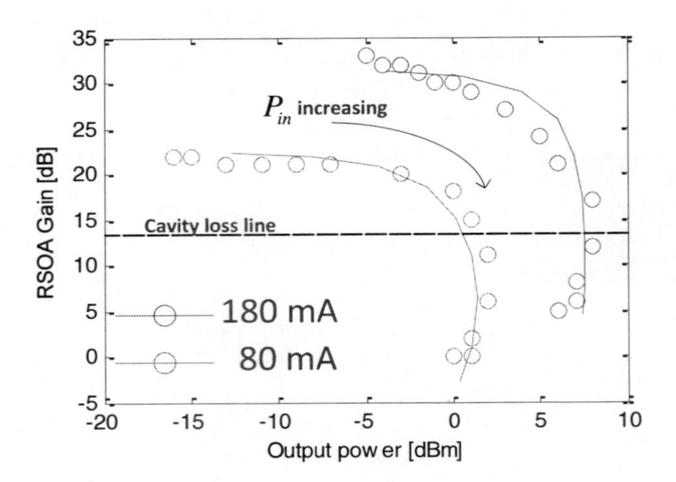

Figure 71. Plot of RSOA gain versus output power for two values of DC injection current. The intersection of the 13 dB cavity loss line with the curves determines the circulating power in the self-seeded cavity. Solid lines are simulated results and open circles are measured results. Measured results provided courtesy of PoliMi.

The L-I curve displays the typical threshold condition for lasers [4], for the self-seeded cavity shows a threshold current of about 45 mA, such a high threshold current is due to the large cavity and coupling losses.

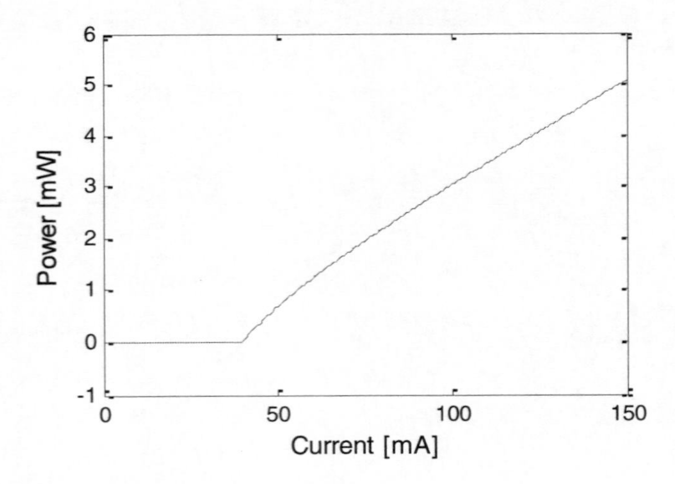

Figure 72. Static L-I curve of the self-seeded transmitter.

Now that the static behaviour of the cavity is established, the performance under modulated conditions will be presented. Operation as a transmitter is performed with direct modulation via the injection current at 2.5 Gbit/s. The bias current was set to 130 mA with NRZ modulation current of amplitude ±50 mA. The simulated eye diagram after 100 round trips with a unique modulation sequence applied after every roundtrip is shown in figure 73. The detected signal undergoes low pass filtering with a low pass bandwidth of 2.5 GHz, as was the case with the experiments. A clear open eye diagram with ER of 6 dB is clearly visible, highlighting the fact that the recirculating signal is completely cancelled and the only transmitted US signal pattern is the pattern applied via the injection current.

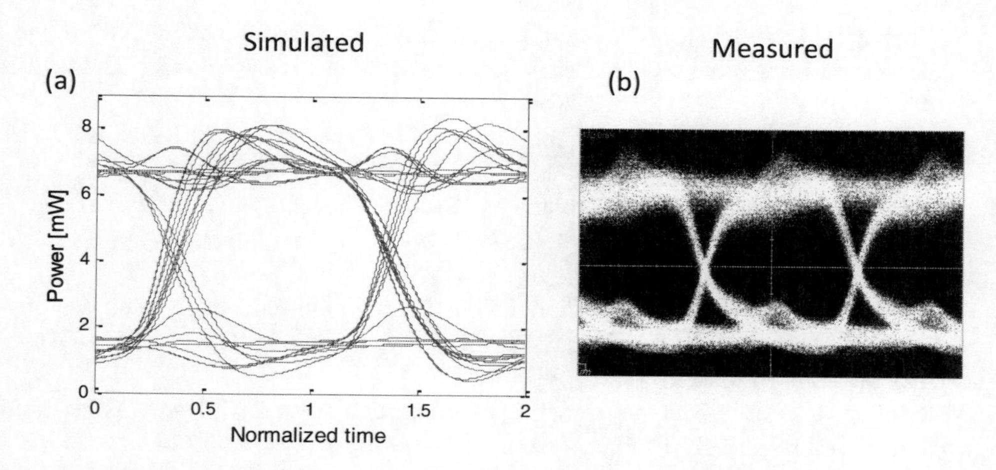

Figure 73. (a) Simulated eye diagram of transmitted signal from self-seeded cavity at 2.5 Gbit/s direct modulation with electronic low pass filter at the receiver. (b) Measured output eye diagram from self-seeded cavity. Measured result provided courtesy of PoliMi.

REFERENCES

[1] Q. Deniel, F. Saliou, L. Anet Neto, N. Genay, B. Charbonnier, D. Erasme, and P. Chanclou, "Up to 10 Gbit/s transmission in WDM-PON architecture using External Cavity Laser based on Self-Tuning ONU", *Proc. OFC/NFOEC,* Los Angeles, CA, 2011, Paper JTh2A.55.

[2] M. Presi, E. Ciaramella, "Stable Self-Seeding of Reflective-SOAs for WDM-PONs", in *Proc. OFC/NFOEC,* Los Angeles, CA, 2011, Paper OMP4.

[3] G. de Valicourt, D. Make, M. Lamponi, G. Duan, P. Chanclou, and R. Brenot, "High gain (30 dB) and high saturation power (11dBm) RSOA devices as colourless ONU sources in long reach hybrid WDM/TDM -PON architecture", *IEEE Photon. Technol. Lett.* 22, 191-193, Feb. 2010.

[4] K. Sato, and H. Toba, "Reduction of mode partition noise by using semiconductor optical amplifiers," *IEEE J. Sel. Topics Quantum Electron.* 7, 328–333, Mar./Apr. 2001.

[5] A. McCoy, P. Horak, B. Thomsen, M. Ibsen, and D. Richardson, "Noise suppression of incoherent light using a gain-saturated SOA: implications for spectrum-sliced WDM systems", *J. Lightwave Technol.* 23, 2399 – 2409, 2005.

[6] K.-Y. Liou, U. Koren, C. Chen, E. C. Burrows, K. Dreyer, and J. W. Sulhoff, "A 24-Channel wavelength-selectable Er-Fiber ring laser with intracavity waveguide-grating-router and semiconductor Fabry–Perot filter", *IEEE Photon. Technol. Lett.* 10, 1787-1789, Dec. 1998.

[7] L. Marazzi, P. Parolari, R. Brenot, G. de Valicourt, and M. Martinelli, “Network-embedded self-tuning cavity for WDM-PON transmitter”, *Opt. Expr.*, 20, 3781–3786 (2012).

[8] C. Michie, A. E. Kelly, J. McGeough, I. Armstrong, Ivan Andonovic and C. Tombling, “Polarization-Insensitive SOAs Using Strained Bulk Active Regions,” *J. Lightwave Technol.*, vol. 24, pp. 3920 - 3927, Nov. 2006.

[9] M. Martinelli, “Time Reversal for the Polarization State in Optical Systems” *J. Mod. Opt.*, vol. 39, pp. 451-455, Mar. 1992.

[10] M. Martinelli, L. Marazzi, P. Parolari, M. Brunero, and G. Gavioli, “Polarization in RetracingCircuits for WDM-PON”, *IEEE Photonics Technol. Lett.*, Vol. 24, N°. 14, July 15, 2012.

[11] M. Presi, A. Chiuchiarelli and E. Ciaramella, “Polarization Independent self-seeding of Fabry-Perot laser diodes for WDM-PONs”, Proceedings of OFC’12, Los Angeles (2012), paper OW1B.5.

[12] M.Krijn et al., “Improved performance of compressively as well as tensile strained Quantum well lasers”, *Appl. Phys. Lett.* 61, 1772 (1992), pp. 1772-1774.

[13] N. Cheng, Z. Xu, H. Lin, and D. Liu, “20Gb/s Hybrid TDM/WDM PONs with 512-Split Using Self-Seeded Reflective Semiconductor Optical Amplifiers,” in Optical Fiber Communications Conference (OFC), (2012), Anaheim, CA, Paper NTu2F.5.

[14] K. Y. Cho, U. H. Hong, Y. Takushima, A. Agata, T. Sano, M. Suzuki, and Yun C. Chung, “103-Gb/s long-reach WDM PON implemented by using directly modulated RSOAs,” *IEEE Photon. Technol. Lett.* 24, 209-211 (2012).

[15] J.-P Blondel, F. Misk, and P.M. Gabla, “Theoretical evaluation and record experimental demonstration of budget improvement with remotely pumped erbium-doped fibre amplification,” *IEEE Photon. Technol. Lett.* 5, 1430-1433 (1993).

[16] G. de Valicourt, G. H. Duan, C. Ware, M. Lamponi and R. Brenot, “Experimental and theoretical investigation of Mode Size Effects on Tilted Facet Reflectivity”, *IET optoelectronics* 5, pp. 175-180 (2011).

[17] F. Devaux, Y. Sorel, J.F. Kerdiles, “Simple measurement of fiber dispersion and of chirp parameter of intensity modulated light emitter,” *J. Lightwave Technol.*, vol. 11, pp.1937-1940, 1993.

[18] Jun-ichi Kani, “Power Saving Techniques and Mechanisms for Optical Access Networks Systems,” *IEEE Journal of Lightwave Technology*, Issue 99, October 3, 2012.

[19] F. Saliou, et al., “Up to 15km Cavity Self Seeded WDM-PON System with 90km Maximum Reach and up to 4.9Gbit/s CPRI Links”, *ECOC* 2012, We.1.B.6.

[20] L. Zhansheng, M. Sadeghi, G. de Valicourt, R. Brenot, and M. Violas, “Experimental validation of a reflective semiconductor optical amplifier model used as a modulator in radio over fiber systems,” *IEEE Photon. Techn. Lett.* 23, No. 9, 576 -578 (2011).

[21] J. Wang, A. Maitra, C. G. Poulton, W. Freude, and J. Leuthold, “Temporal dynamics of the alpha factor in semiconductor optical amplifiers,” *IEEE J. of Lightw. Tech.* 25, No. 3, 891-900 (2007).

[22] L. W. Casperson, and J. M. Casperson, “Power self-regulation in double-pass high-gain laser amplifiers”, *J. Appl. Phys.* 87, No. 5, 2079-2083 (2000).

[23] A. Siegman, “Lasers,” University Science Books, ISBN 978-0935702118.

In: Optical Fibers: New Developments
Editor: Marco Pisco

ISBN: 978-1-62808-425-2

Chapter 8

OPTICAL FIBER CONNECTION TECHNOLOGIES: CURRENT AND NOVEL FIELD INSTALLABLE CONNECTORS

Mitsuru Kihara*
Access Network Service Systems Laboratories,
Nippon Telegraph and Telephone Corporation, Japan

ABSTRACT

This chapter discusses the latest optical fiber connection technologies. In particular, current and novel field installable connectors are discussed in detail. First, after a brief introduction, the optical access network configuration and various optical fiber connections employed in Japan are explained. The section on connections includes a discussion of manufactured connectors. Then, current mechanical splices with refractive index matching material are introduced. These splices consist of a base with a V-groove guide, three coupling plates, and a clamp spring. A mechanical splice is suitable for joining optical fibers simply in the field. Current field installable connectors are also explained. These connectors have three main parts, a polished ferrule containing a short optical fiber (built-in optical fiber), a mechanical splice part, and a clamp. This connector holds an optical fiber drop cable or an indoor cable sheath. To install the connection, the optical fiber end must be stripped, cleaned, cleaved, and connected to the built-in optical fiber using a mechanical splice technique, and the cable sheath is then fixed in the clamp. The structure allows connection to another optical fiber connector in the field. Recently, two types of novel field installable connection techniques have been proposed. One is mechanical splicing, which is used to connect coated optical fibers without the need for stripping or cleaning procedures. The other is a field assembly connection technique, which employs a new type of field installable connector that makes it possible to realize a physical contact connection without a polishing procedure. The proposed connection techniques are also described. Mechanical splicing is achieved by precisely aligning and directly connecting coated fibers with a capillary. The assembled splice is installed with 1.3-μm single-mode fibers that have an 80-μm cladding and a 125-μm coating and they exhibit good optical performance with a low average insertion loss of 0.2 dB and a high

* Corresponding Author address Email: kihara.mitsuru@lab.ntt.co.jp.

return loss of over 46 dB. Connection is achieved with the developed field installable connector by using a chamfered fiber endface and the compression force of the buckled fiber. The assembled connectors achieve physical contact without the fiber endface being polished, which provides good optical performance with a low insertion loss of 0.11 dB and a high return loss of over 50 dB. These optical fiber connection technologies will be effective for connecting fibers in future optical network systems.

Keywords: Optical fiber splice, optical connector, coated optical fiber, physical contact connection, field installable connector

INTRODUCTION

There are now more than 34 million broadband service subscribers in Japan, and the number of subscribers to Fiber-to-the-Home (FTTH) services reached about 20 million in December 2011 [1]. Many single-mode optical fiber (SMF) connection techniques, such as fusion splicing, mechanical splicing, and the use of optical connectors, are currently employed in FTTH systems [2-3]. A fusion splice is fabricated using a fusion splice machine (splicer), which is a precision machine that provides fiber alignment, video monitoring, and arc discharge functions. A fusion splice provides the highest and most stable performance of all connections. Physical contact (PC) connectors are used in the central office and homes. These connectors require frequent reconnections. Manufactured connectors with a refractive index matching material are used in outside underground facilities to connect multifiber array ribbons and they have a small gap between two fiber ends that is filled with refractive index matching material to reduce Fresnel reflection. In contrast, mechanical splices and field installable connectors are used for fiber connections at aerial and residential sites. These connection techniques are the most suitable for wiring that corresponds to different aerial conditions and room arrangements.

Mechanical splices and field installable connectors, which also employ refractive index matching material between the fiber ends to reduce Fresnel reflection, are good optical components for fiber connection outdoors. A large number of mechanical splices and field installable connectors have been used in Japanese optical access networks. However, there are issues with these fiber connection components. Unexpected failures when installing these fiber connections might have a detrimental effect on performance. For instance, it has been reported that the insertion loss might increase greatly along with changes in temperature if there is a wide gap between the fiber ends as a result of the fibers being joined incorrectly. This is because silicone oil compound used as refractive index matching material moves into the wide gap and mixes with air [4]. This loss increase may not occur immediately after installation but intermittently over time. In the event of an unusual fault, it is difficult to find the defective connection point, and it takes a long time to effect a repair. To prevent such faults, it is important to conduct specified procedures correctly and use correct tools with field installable connectors. Therefore, we require easier procedures and a simpler structure than those of the current technique for field installable connections, which avoid the possibility of defective connections. Low-cost connectors are also required.

Recently, two types of field installable connection techniques have been proposed [5]. One involves a new mechanical splice that connects coated optical fibers without the need for stripping or cleaning procedures [6]. The other involves a new type of field installable connector that enables PC connection without a polishing procedure [7-8]. These connection techniques result in a simple component structure, ease of assembly, and excellent optical performance.

This chapter discusses the latest optical fiber connection technologies. In particular, current and new field installable connectors are discussed in detail. After a brief introduction, the optical access network configuration and various connection technologies currently used in Japan are described. This section introduces PC-type connectors and manufactured connectors with a refractive index matching material. Then, mechanical splices with refractive index matching material and current field installable connectors are explained. Finally, two types of novel field installable connection techniques are described. One is mechanical splicing, which is used to connect coated optical fibers without the need for stripping or cleaning procedures. The other is a field assembly connection technique, which utilizes a new type of field installable connector that enables physical contact connection without a polishing procedure. These optical fiber connection technologies will be effective for realizing fiber connections in optical network systems.

OPTICAL ACCESS NETWORK CONFIGURATION AND OPTICAL FIBER CONNECTIONS

Figure 1 shows the configuration of an optical access network in Japan, which is mainly composed of an optical line terminal (OLT) in a central office, underground and aerial optical fiber cables, and an optical network unit (ONU) on the customer's premises. The network requires fiber connections at a central office, underground, and at aerial and residential sites. PC connectors, such as miniature-unit coupling (MU) and single fiber coupling (SC) optical fiber connectors [9-10], are used in the central office and home. These connectors require more frequent reconnections than field installable connectors. Manufactured connectors with a refractive index matching material, such as mechanically transferable (MT) multifiber connectors [11], are used in the central office and underground facilities for connecting multifiber array ribbons and have a small gap between two fiber ends that is filled with refractive index matching material to reduce Fresnel reflection. In contrast, for fiber connections at aerial and residential sites, a field installable connection technique is required that best matches the wiring corresponding to different aerial conditions and room arrangements. Such a method should not require any adhesive, polishing, or electricity, and the connection procedure should be easy. Currently, there are two fiber connection techniques for field installation, mechanical splicing and field installable connection [12-14]. Field assembly (FA) termination connectors and field assembly small (FAS) connectors are types of field installable connectors, and many kinds of connectors have been developed and used in optical fiber networks. Each connector determines how or where it should be used. The typical connectors employed in optical access networks in Japan are described below.

First, the single fiber physical contact (PC) connectors are explained. Figure 2 shows the basic structure of a single fiber PC-type connector. With PC-type connectors, two ferrules are aligned in an alignment sleeve and connected using compressive force. Normally, the two fiber ends in the ferrules are connected without a gap and without an offset or tilt misalignment. The fiber coupling (FC) connector [15], SC and MU connectors are typical single fiber PC-type connectors. The SC connector consists of two plugs and an adaptor. The plug has a zirconia ferrule and a plastic-molded connector housing. The adapter excludes a split sleeve for alignment. The SC connector was developed based on an FC connector and uses a push-pull coupling mechanism. With single-mode optical fiber the connector has exhibited an average insertion loss of less than 0.1 dB and a return loss of more than 40 dB. The MU connector, which was based on the SC connector, has a 1.25-mm diameter ferrule and a miniature push-pull type plug. The MU connector is 4.4 mm wide and 5.6 mm high, and its cross-sectional area is more than 60 % smaller than that of an SC connector. The MU connector has also exhibited a low insertion loss and a high return loss with single-mode optical fiber.

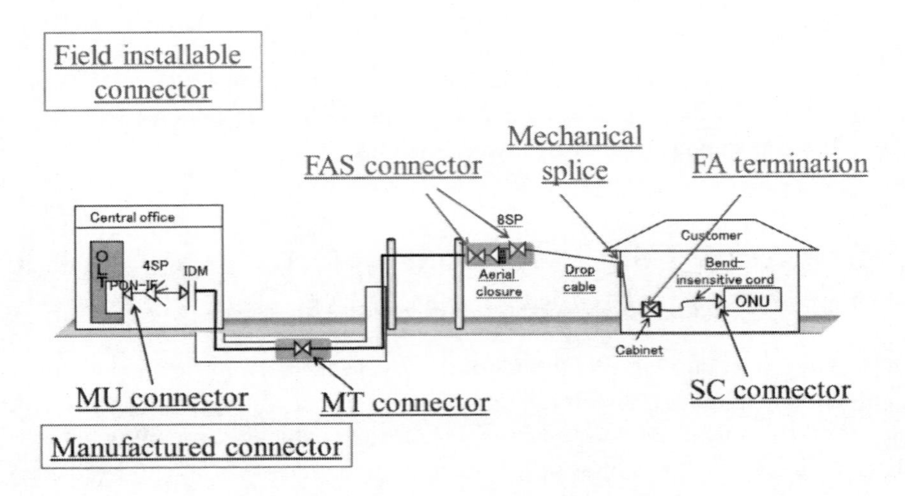

Figure 1. Optical fiber access network and connections.

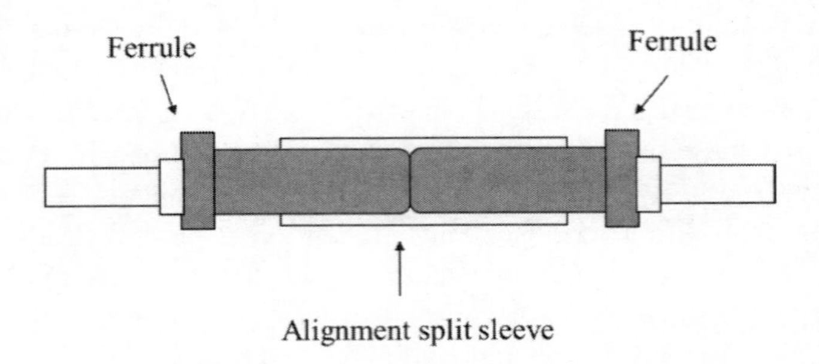

Figure 2. Basic structures of single fiber physical contact type connector.

Mechanically transferable (MT) connectors have been developed and used for connecting multifiber ribbon in such optical subscriber networks as FTTH systems. Figure 3 shows the MT connector structure. The MT connector consists of two plastic ferrules and two guide pins. The ferrules are aligned by the two guide pins and two guide holes and then held with a clamp spring. Figure 4 shows a cross-sectional diagram of an MT ferrule. The designed fiber positions between the two guide hole centers. If the fabricated fiber positions have large offset misalignments from their designed positions, this could result in insertion loss. It is important to minimize the offset misalignments. The fibers to be connected can be easily aligned by two guide holes and guide pins with refractive index matching material to achieve a low connection loss for multifiber array connection.

Figure 5 shows the multi-fiber push on (MPO) connector structure [16-17]. MPO connectors using the MT connector technique have been developed for use in a termination cabinet and in optical interconnections for very dense transmission systems. The MPO connector employs a push-on pull-off mechanism where the plug and adaptor are engaged by fitting a pair of elastic hooks into corresponding grooves, and the configuration of the ferrule end has been improved [18]. The ferrule endface is made oblique with an angle of 8 degrees to a plane perpendicular to the ferrule axis so that the reflected light is not transmitted in the reverse direction. In addition, the fiber ends are designed to protrude slightly at the ferule endface to allow direct fiber endface physical contact between multiple fibers [19]. With single-mode fiber ribbons the connector exhibited average insertion and return losses of less than 0.2 dB and more than 55 dB, respectively, without the use of refractive index matching material.

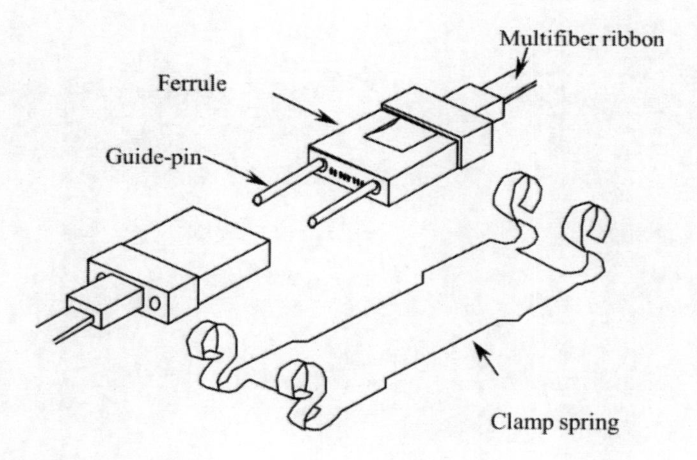

Figure 3. Structure of MT connector.

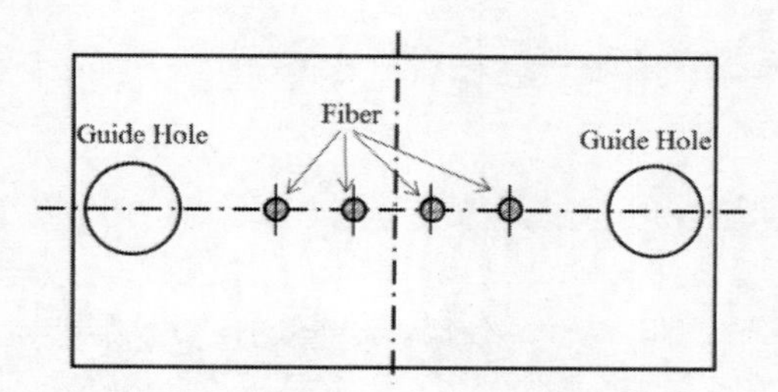

Figure 4. Cross-sectional diagram of MT ferrule.

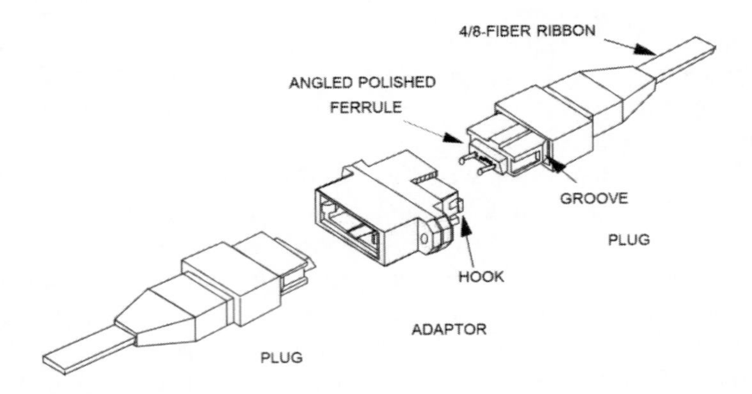

Figure 5. Structure of MPO connector.

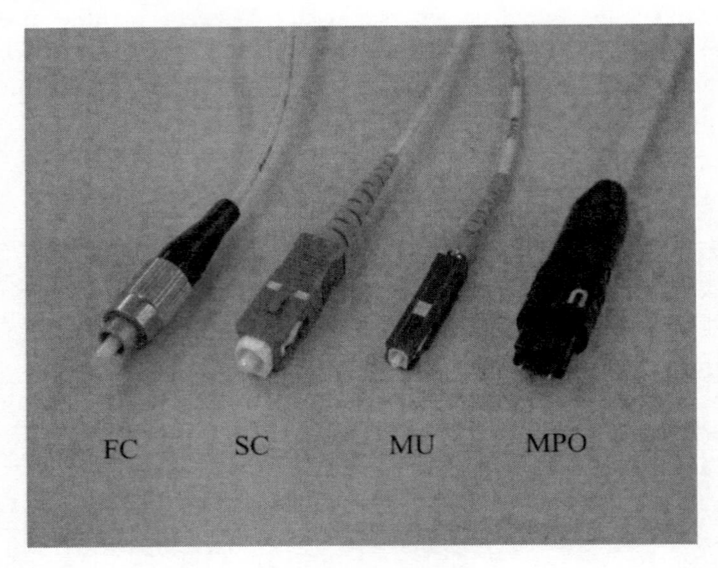

Figure 6. Photograph of manufactured PC connectors.

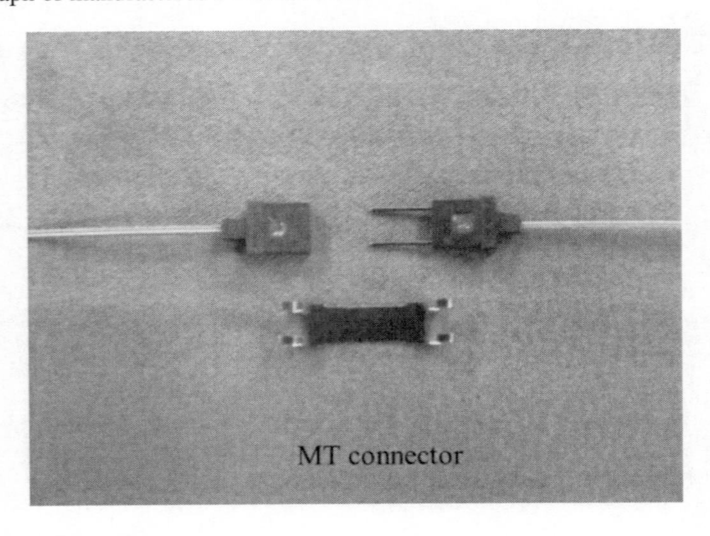

Figure 7. Photograph of MT connector.

Figures 6 and 7, respectively, show photographs of manufactured connectors that employ PC connection and refractive index matching material. Figure 6 shows FC, SC, and MU connectors designed for single fiber connection that use zirconia ferrules to enable PC connection. In contrast, the MPO connector is for multifiber ribbon connection and uses an oblique plastic ferrule endface to enable the PC connection of multiple fibers. Figure 7 shows an MT connector with refractive index matching material for the multifiber array connection of fewer than 12 fibers. These connectors are used in optical fiber networks.

CURRENT FIELD INSTALLABLE SPLICES AND CONNECTORS

Figure 8(a) and (b), respectively, shows the basic structure of a mechanical splice. A mechanical splice is suitable for joining optical fibers simply in the field. It consists of a base with a V-groove guide, three coupling plates, and a clamp spring [12]. When a wedge is inserted between the plates and the base, optical fibers can be inserted through the V-groove guide to connect them and fix them in position by releasing the wedge between the plates and base as shown in Figure 8(b). Refractive index matching material is used to reduce Fresnel reflection. This connection procedure requires no electricity.

Figure 9 shows the basic structure of an FA connector, which is used on a customer's premises. An FA connector is composed of three main parts, a polished ferrule containing a short optical fiber (built-in optical fiber), a mechanical splice, and a clamp. This connector holds the optical fiber drop cable or indoor cable sheath [20]. To assemble the connection, the optical fiber end is cleaved and connected to the built-in optical fiber using the mechanical splice, and the cable sheath is fixed in the clamp. The structure allows connection to other FA or SC connectors in the field because they all have the same connector interface. In addition, the FA connector is fabricated based on the above-mentioned mechanical splice technique; therefore, the connection can be assembled without the need for special tools or electricity.

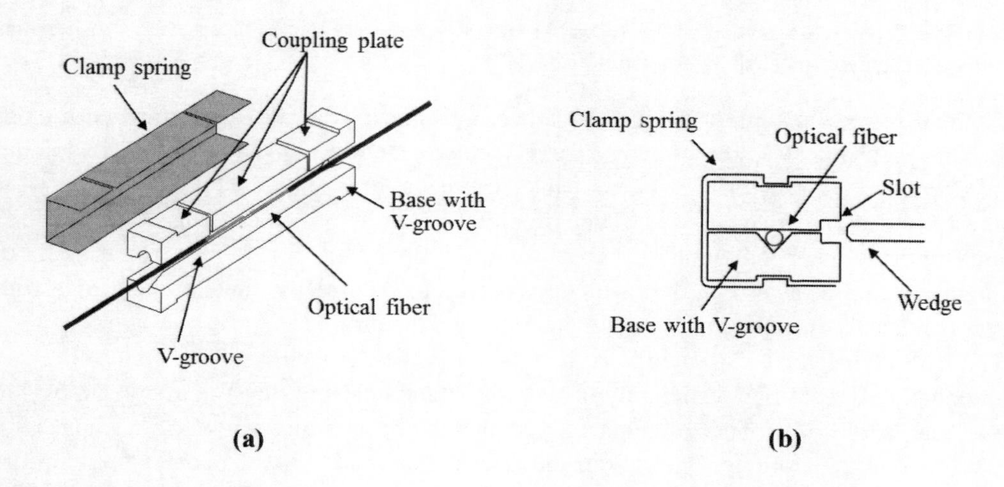

Figure 8. Structures of mechanical splice.

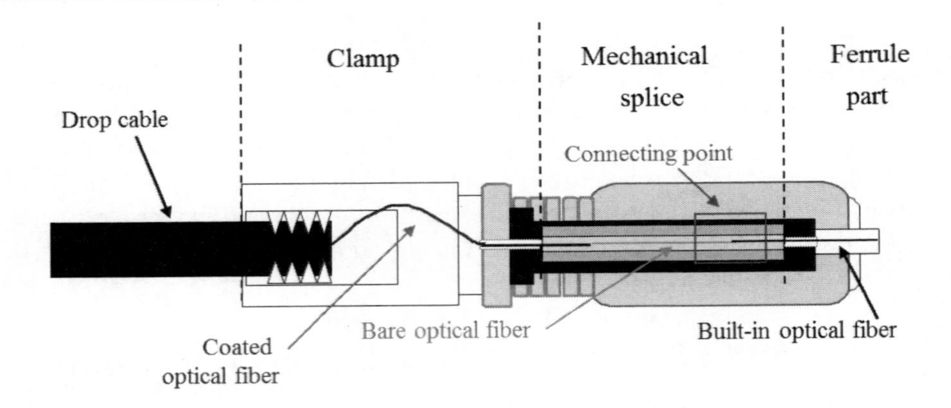

Figure 9. Structures of FA connector.

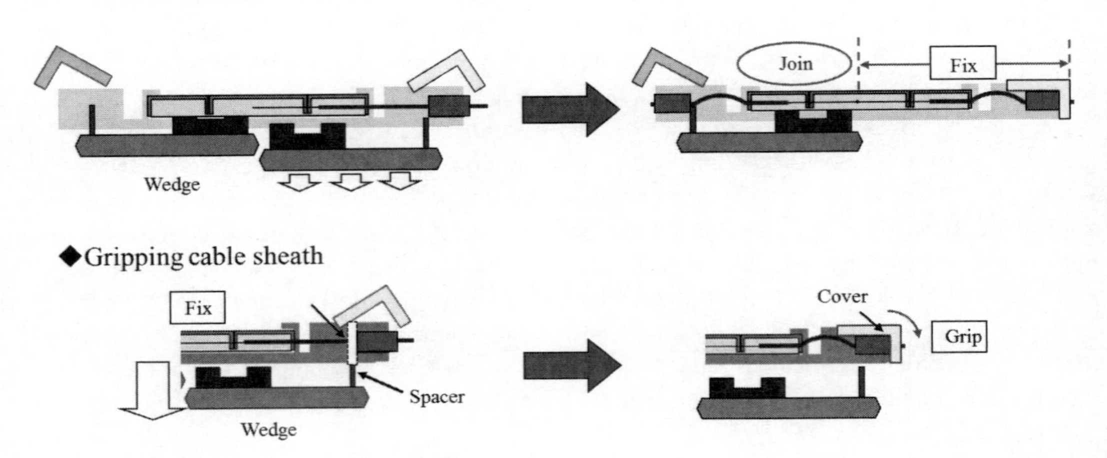

Figure 10. Latest mechanical splice for gripping cable.

Figure 10 shows the latest mechanical splice, which grips the sheath of a drop or indoor cable. This mechanical splice employs the FA connector technique in relation to joining optical fibers and gripping drop or indoor cable. There are three features. First, a conventional wedge is divided into two wedges. This structure makes it possible to insert and fix the prepared optical fibers individually. Next, joining a fiber to a fixed fiber results in fiber buckling, which informs us that the fibers have been joined correctly. Finally, the wedge with a spacer is released, the cover is placed in position, and the drop or indoor cable is gripped.

The FAS connector is designed to be assembled in the field and used for connecting and accommodating optical fiber efficiently in an aerial optical closure. Its structure is almost the same as that of the FA connector. Figure 11 shows the basic structures of FAS connectors. There are two kinds of plugs. One is a coated fiber type, which grips the coated optical fiber in a distribution cable, and the other is a drop cable type, which grips the drop optical fiber. The socket has an opening into which the plug is inserted. A polished ferrule containing a built-in optical fiber and a mechanical splice part is fixed in the connector. The plug is pushed into the opening of the socket to connect the two components, and the two ferrules are brought into contact. Then, the latch is engaged and the components are connected. To release the connection, a point on the latch is pushed. The latch is then disengaged and the plug is disconnected. Moreover, to prevent any accidental release, the mechanism only works when the pushing points on the latch on either side of the plug are pushed simultaneously.

Figure 12(a) and (b), respectively, show photographs of the mechanical splice for coated fiber connection and drop or indoor cable connection. Figure 13(a) and (b), respectively, show photographs of FA and FAS connectors. A large number of mechanical splices and FA and FAS connectors have been used in Japanese optical access networks.

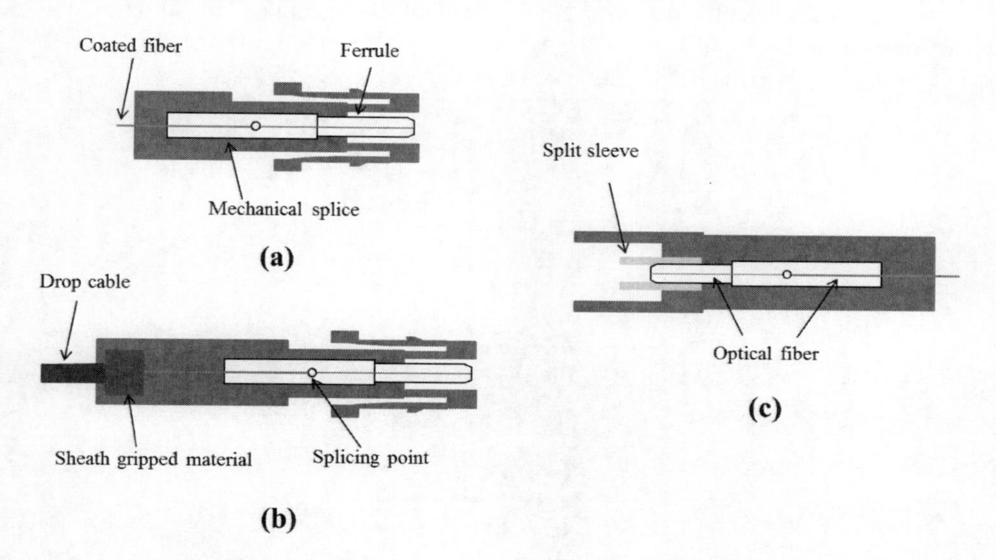

Figure 11. Structures of FAS connector; (a) coated fiber type plug, (b) drop cable type plug, and (c) coated fiber type socket.

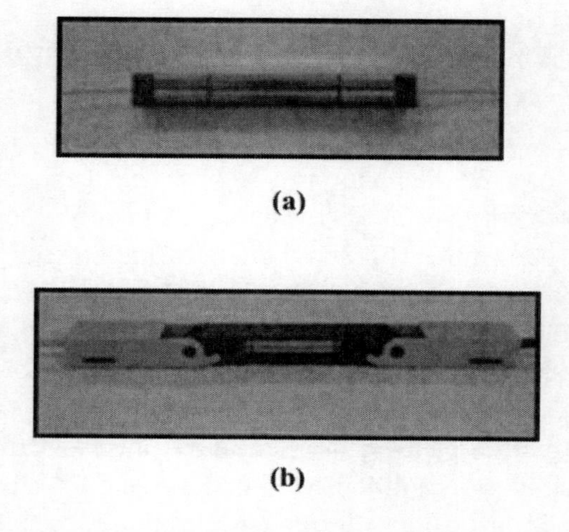

Figure 12. Photographs of current and latest mechanical splices.

The Technical Assistance and Support Center (TASC) of NTT East Corporation has investigated faults with mechanical splices and field installable connectors [4, 21-22]. The number of faults is not very large; however, certain faults might result in serious network problems. The TASC has been collecting defective mechanical splices and field installable connectors from various regions throughout Japan since 2001 and has investigated their outward appearance and optical characteristics by dismantling and testing them. The TASC has specified various causes for such faults and the most common are fiber breakdowns caused by accidentally scratched fibers or excessive stress, wide gaps between fiber ends caused by incorrect joins, and the incorrect cleaving of fibers. These faults result from fiber ends being incorrectly prepared and assembled or the use of inappropriate tools. Therefore, to reduce the number of defective connectors, we require easier procedures and a simpler structure than those currently used for field installable connections. In addition, low-cost connectors are needed.

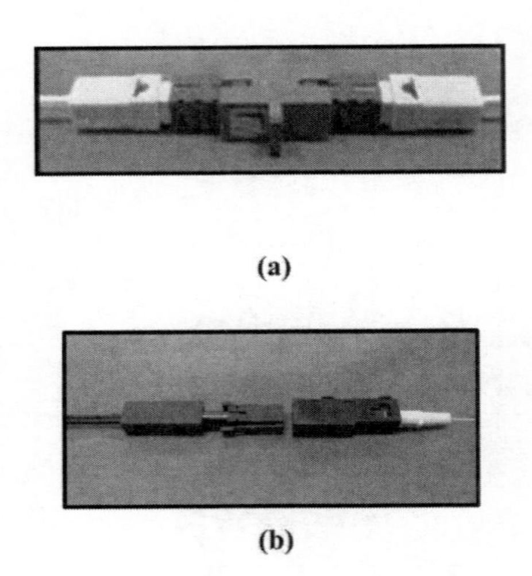

(a)

(b)

Figure 13. Photographs of FA and FAS connectors.

Development Concept of New Type of Field Installable Connection Technique

Two techniques have been studied and proposed for next generation mechanical splices and field installable connectors [5]. One is designed to greatly improve current fiber end preparation and assembly procedures [6], and the aim of the other is to greatly improve the structure of field installable connectors [7-8]. The two techniques are described below.

The same fiber end preparation procedures are employed for both mechanical splices and field installable connectors prior to fiber installation. Figure 14(a) shows the current fiber end preparation procedures. The fiber coating is stripped. Then the stripped fiber (bare fiber) is cleaned with alcohol, cut with a cleaver, inserted into a mechanical splice or a splice part inside a field installable connector, and joined to the opposite fiber or built-in fiber. Finally the inserted fibers are fixed in position. Stripping, cleaning, and cutting are important for successful fiber connection (to provide good performance) in the field. If any of these procedures are not conducted correctly, the fiber connection performance might deteriorate. All connection techniques involve handling and aligning bare glass fibers. Bare fibers are not easy to handle and may also be broken. Figure 14(b) shows the proposed technique for field installable connections. With this technique, coated fiber is directly cut without stripping or cleaning. The cleaved coated fiber is joined to the opposite fiber or built-in fiber. Finally the fiber is fixed in position. Consequently, the proposed technique results in an easier and safer connection procedure without the need to handle bare glass fibers.

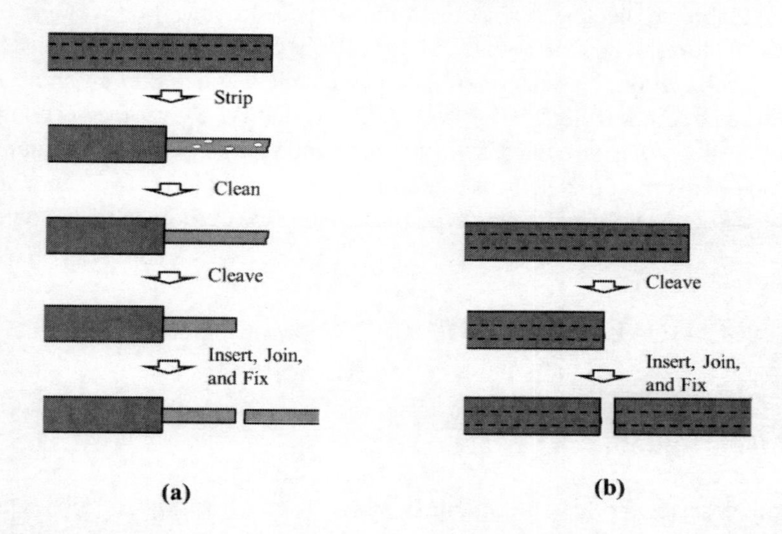

Figure 14. Optical fiber end preparation procedures; (a) current procedures and (b) new procedures.

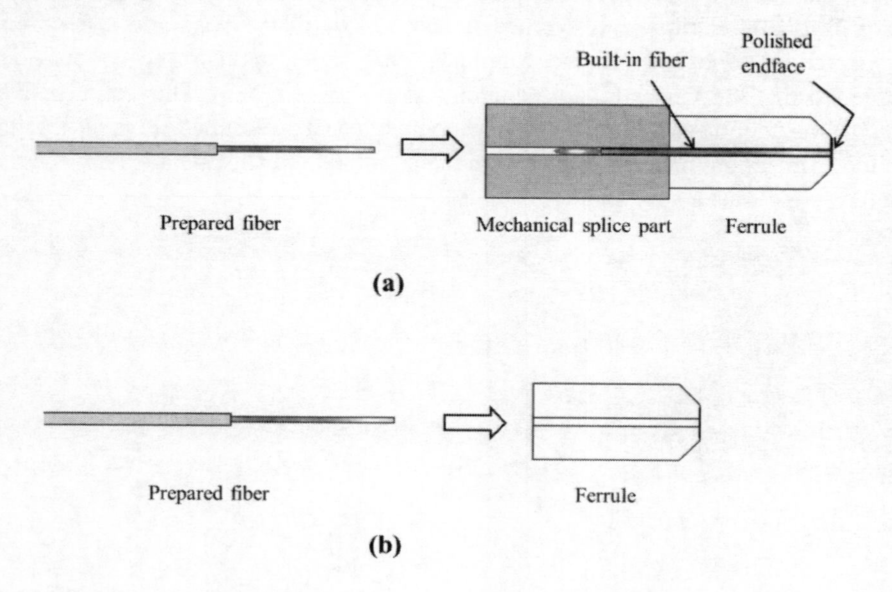

Figure 15. Basic structures of field installable connector; (a) current structure with built-in fiber with polished endface and mechanical splice and (b) novel structure without built-in fiber and mechanical splice.

Figure 15(a) shows the basic structure of a current field installable connector. The connector is composed of a polished ferrule containing a short optical fiber (built-in optical fiber) and a mechanical splice part. The polished ferrule enables PC connection. After the fiber-end preparation procedures (stripping, cleaning, and cleaving) are complete, the fiber is inserted, joined to the built-in optical fiber, and fixed in position in the mechanical splice part. This structure of the field installable connector makes connection possible with other PC connectors. However, the structure is not very simple. For a more economical connector, field installable connectors with a simpler structure are required. Figure 15(b) shows the structure of new field installable connectors for use with the proposed connection technique. The new structure has two features. The first is that the optical fiber inserted into the connector ferrule enables direct PC connection. Namely, there is no mechanical splice part and no built-in fiber or optical fiber fixed to the ferrule. The second feature is that the PC surface is compressed by the elasticity of the fiber. The structure is very simple and results in low-cost connectors. The following section details the proposed fiber end preparation techniques for mechanical splices and field installable connectors and a new type of structure for field installable connectors.

COATED OPTICAL FIBER CONNECTION TECHNIQUE WITHOUT STRIPPING OR CLEANING

Structure of New Splice

This section describes a new mechanical splice for coated optical fiber connections. Figure 16 shows its structure. The splice consists of a base, a capillary, three lids, and a clamp spring. In the capillary, coated fibers are precisely aligned and connected to each other. Fiber guide grooves for the capillary are formed on both sides of the base and a fiber buckling space is located on one side of the capillary. Fiber buckling is used to provide axial compressive force at the fiber ends to maintain a gap between them. This compressive force contributes to the long-term reliability of the connection. To eliminate Fresnel reflection, a refractive index matching material is placed in the capillary and fills the gap between the fiber ends. The new splice has a very simple structure.

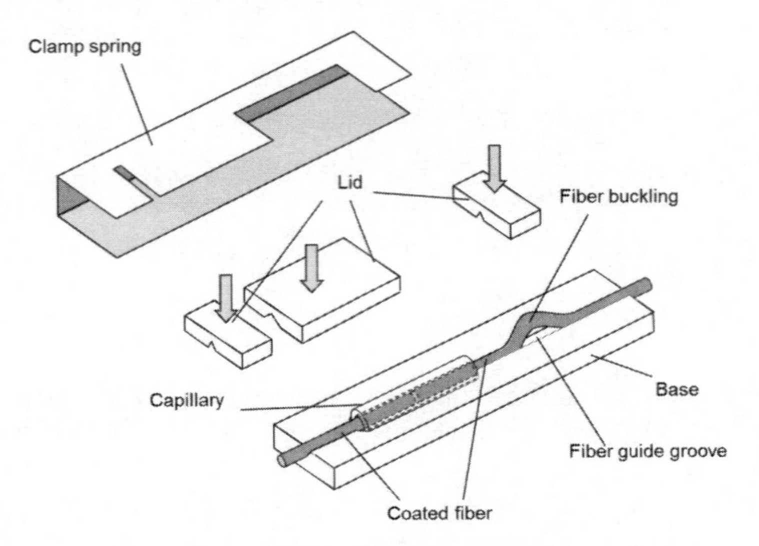

Figure 16. Structure of mechanical splice for coated fiber connection.

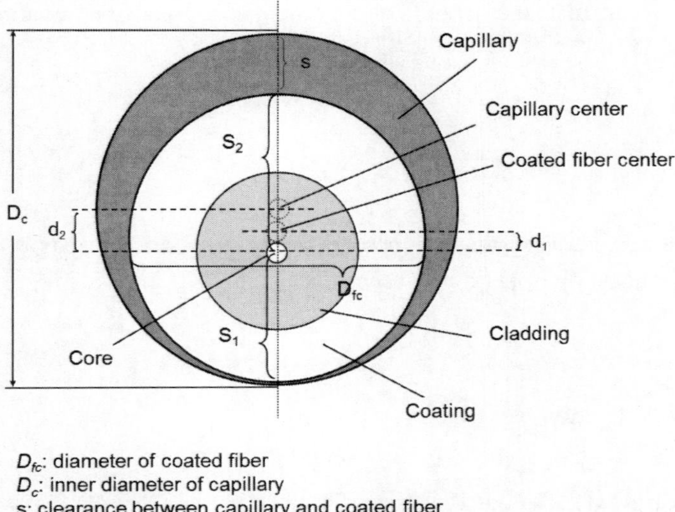

Figure 17. Cross-sectional diagram at capillary of mechanical splice for fiber core offset calculation.

To realize a connection between coated fibers with this splice, the fibers must be cleaved without removing their coatings. A conventional fiber cleaver blade is positioned so that it forms a correct initial crack on the fiber cladding surface through the coating. The cleaving method is described in the following section. A wedge is inserted between the base and the lids to form a space for fiber insertion above the fiber guide groove. The cleaved coated fibers are inserted into the capillary through the fiber guide groove. When the wedge is removed, the fibers are fixed and maintained mechanically with the base, the side lids, and the clamp spring. Consequently, the use of this structure and fiber installation procedure eliminates the need for the process normally used to strip the fiber coatings and clean and handle the stripped fibers. The tools needed for this procedure (stripper, alcohol, cleaning paper etc.) also become unnecessary. Moreover, this technique can increase the ease and speed of fiber installation and the reliability of long-term usage because no bare glass fibers are handled at any time during the procedures.

Performance Estimation and Prototype Design

The insertion losses of the mechanical splice were analyzed to confirm that the structure is suitable for connecting coated fibers. It is assumed that the insertion loss is caused solely by the offset of the fiber cores because the tilt and gap between the fiber ends are small enough for the losses they induce to be ignored. Figure 17 shows a cross-sectional diagram of our splice at the capillary. Here, D_c is the inner diameter of the capillary, D_{fc} is the diameter of the coated fiber, and S_1 and S_2 are the minimum and maximum distances from the core center in the coated fiber, respectively.

When aligning the coated fibers with the capillary, there is a fiber core offset, which is caused by two factors, namely the core-coating eccentricity C_{fc} of the coated fiber and the clearance between the capillary and the fiber coating. First, when C_{fc} is defined as S_1/S_2, the loss induced by C_{fc} is defined as follows. When two coated fibers spliced at the centers of their coatings meet, the distance from the core center to the fiber coating center d_1 is expressed as

$$d_1 = \frac{(1 - C_{fc})D_{fc}}{2(1 + C_{fc})}. \quad (1)$$

The longest distance between the core centers of the spliced fiber ends is $2d_1$. In contrast, the insertion loss α caused by offset d is estimated as [23]

$$\alpha = -10\log\left[\exp\left(-\frac{d^2}{w^2}\right)\right], \quad (2)$$

where w is the mode field radius of the fiber. The loss induced by C_{fc} can be calculated by using eqs. (1) and (2), and d=2 d_1. Figure 18 shows the estimated α value when the fibers are aligned using their coating centers. Calculations were performed for two fibers; a conventional 125/250-μm fiber with a cladding diameter of 125 μm and a coating diameter of 250 μm [24] and an 80/125-μm fiber [25]. To attain an insertion loss of less than 1.0 dB at 1.31 μm where w=4.3 μm, the C_{fc} value of the 125/250-μm fiber must exceed 0.984 and that of the 80/125-μm fiber must exceed 0.967. Based on this result, we used the 80/125-μm fiber for the first prototype splice with coated fiber.

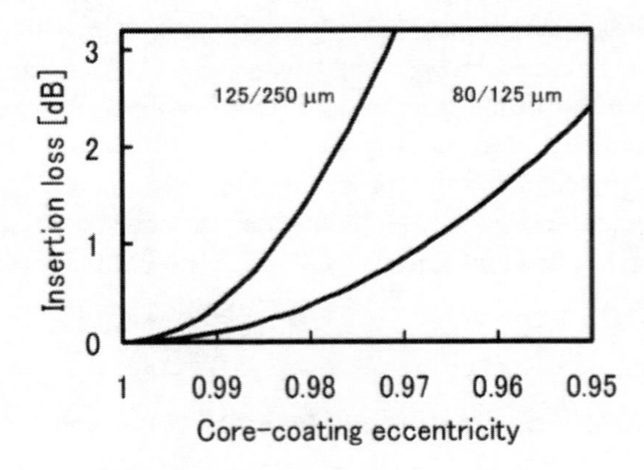

Figure 18. Insertion loss caused by core-coating eccentricity.

Next, the effect of the clearance between the capillary and the coated fiber on the insertion loss of the splice was analyzed. When the clearance is expressed as s, the offset between the capillary center and the coating center is s/2, and D_{fc} equals D_c - s. Therefore, the distance from the core center to the capillary center d_2 is expressed as

$$d_2 = \frac{(1 - C_{fc})(D_c - s)}{2(1 + C_{fc})} + \frac{s}{2}. \quad (3)$$

Figure 19 shows the calculated insertion loss of the splice for the 80/125-μm coated fiber taking s into consideration. The diameter of the capillary we used was 125.5 μm and C_{fc} was over 97%. From these results, a splice with an insertion loss of less than 1.0 dB must have a C_{fc} value of over 99%, and the s value between the capillary and the coated fiber should be in the 0-1.4 μm range (shown by the shaded area in Figure 19).

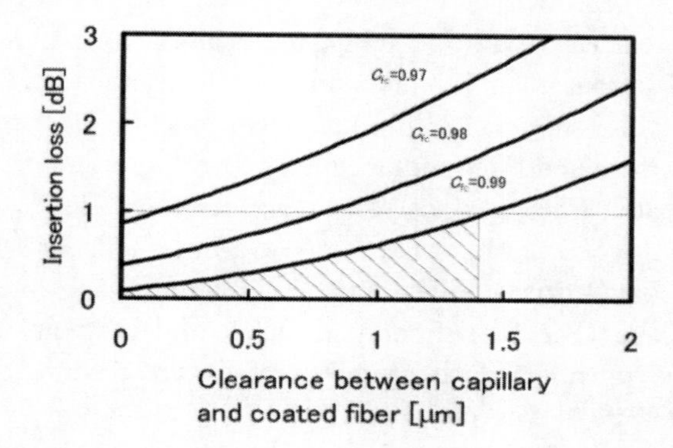

Figure 19. Insertion loss estimation for proposed mechanical splice.

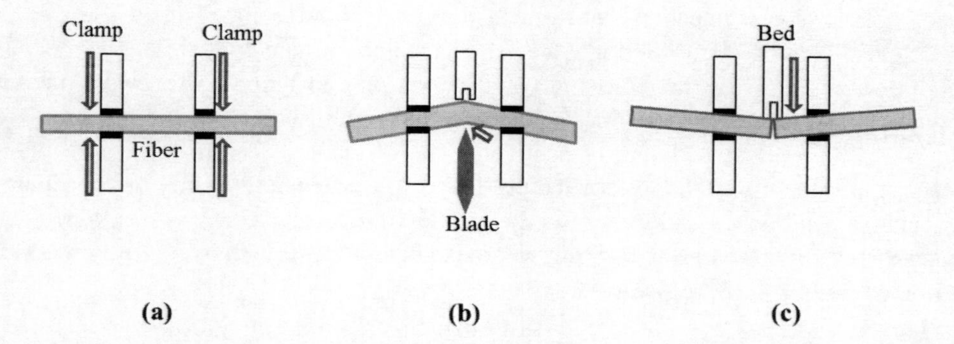

Figure 20. Procedures for cleaving optical fiber.

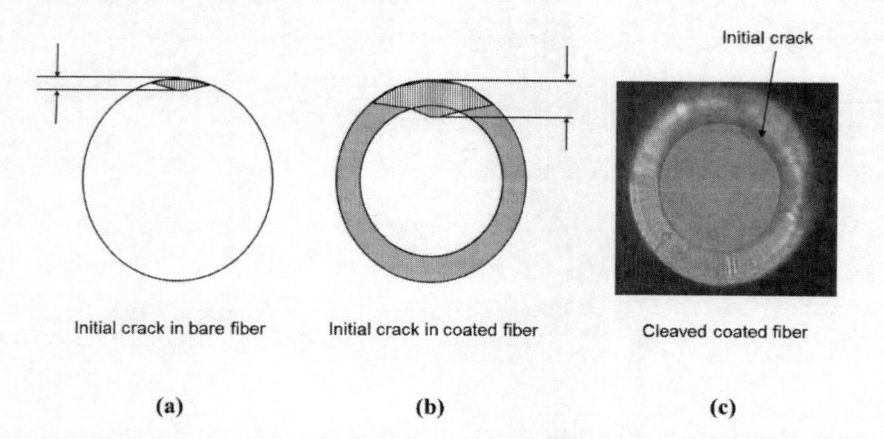

Figure 21. Cleaved fiber ends; (a) 125-μm bare fiber end, (b) 80/125-μm coated fiber end, and (c) photo of coated fiber end.

Performance of Assembled Splice

Figure 20 shows the procedure for cleaving an optical fiber [26-27]. A length of optical fiber is fixed with two clamps (step (a)). A blade is slid upwards and scratches the fiber (to initiate a fracture) (step (b)). The fiber is then bent at the scratch across a bed with an appropriate radius and pushed downward from the opposite side of the scratch until the fiber is cleaved (step (c)). To cut fiber with a coating, the position of the blade in a conventional fiber cleaver is arranged so that it can initiate a crack correctly through the coating. Figure 21 (a), (b), and (c), respectively, show a 125-μm bare fiber endface cleaved with a conventional fiber cleaver, a 80/125-μm coated fiber endface cleaved with the arranged fiber cleaver, and a photo of the cleaved coated fiber endface. The crack on the cleaved coated fiber covers a larger area than that on the cleaved bare fiber endface; however, the crack on the bare fiber part inside the cleaved coated fiber endface must reach almost the same depth as that on the cleaved bare fiber endface. The bare fiber part inside the correctly cleaved coated fiber endface forms the same mirror surface as that of the bare fiber endface cleaved with a conventional fiber cleaver, as shown in Figure 21(c).

Prototype splices for 80/125-μm coated 1.3-μm single-mode fiber were fabricated using the above structure and design. Figure 22 shows the outer components of the fabricated splice for coated optical fiber connections. The dimensions of the fabricated splice are 4.0 x 4.0 x 33 mm, which is as small as a conventional mechanical splice for bare optical fiber connections.

The insertion losses of the splices at wavelengths of 1.31 and 1.55 μm were measured, and Figure 23 shows the results. The average and maximum values of the insertion losses were 0.26 and 0.60 dB at 1.31 μm, respectively. The results at 1.31 μm agree well with the calculated result; therefore, the loss of the prototype splice depends mainly on the fiber core offset, which depend on C_{fc} and s between the capillary and coated fiber, as mentioned above. The measured insertion losses at 1.55 μm were also below 1 dB, with average and maximum values of 0.21 and 0.64 dB, respectively.

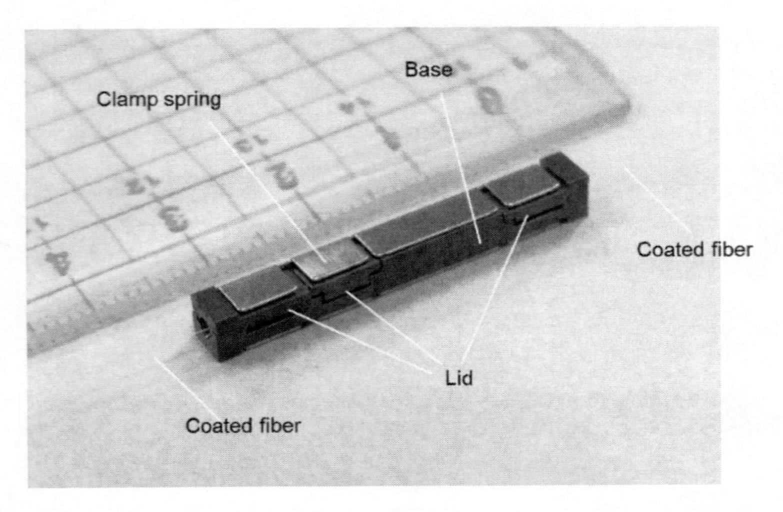

Figure 22. Photograph of assembled mechanical splice for direct coated fiber connection.

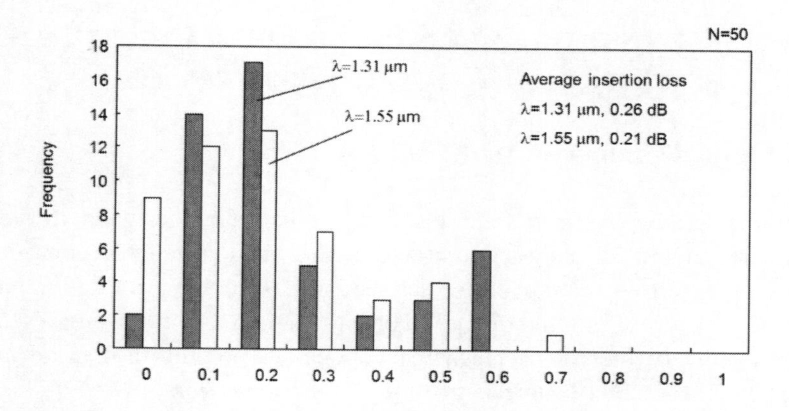

Figure 23. Measured insertion loss of proposed splice at wavelengths of 1.31 and 1.55 μm.

The return losses and changes in the insertion losses during a temperature cycle test based on Telcordia GR-326-Core were measured. The return losses were all over 46 dB at both 1.31 and 1.55 μm. The loss changes during the heat cycle test were less than 0.2 dB for all the samples. Therefore, the prototype splice exhibits good return loss and good environmental reliability. The time needed to cleave the coated fibers and install them in this prototype splice was also measured. The time needed to strip the coatings, clean and cleave the fibers, and install them in a conventional mechanical splice was also measured. It was found that the time needed for the prototype splices is 30% less than that needed for conventional mechanical splices. This splice can also be installed with multi-mode fibers for interconnection usage, and the measured results show that all the insertion losses were less than 0.1 dB. These experimental results reveal that our developed splice can connect coated single-mode fibers easily and quickly without the need for stripping and cleaning procedures. The optical performance and environmental reliability are good enough for the splice to be used in optical access networks and interconnections.

Recently, a field installable connector for 125/250-μm coated fiber connection has been studied [28-29]. The connector has a fiber coating removable part. Figure 24 shows the fiber coating removable part and coated fiber. The coating of the inserted coated fiber is removed by the fiber coating removable part inside the connector. The bare fiber is then aligned and joined to the built-in fiber.

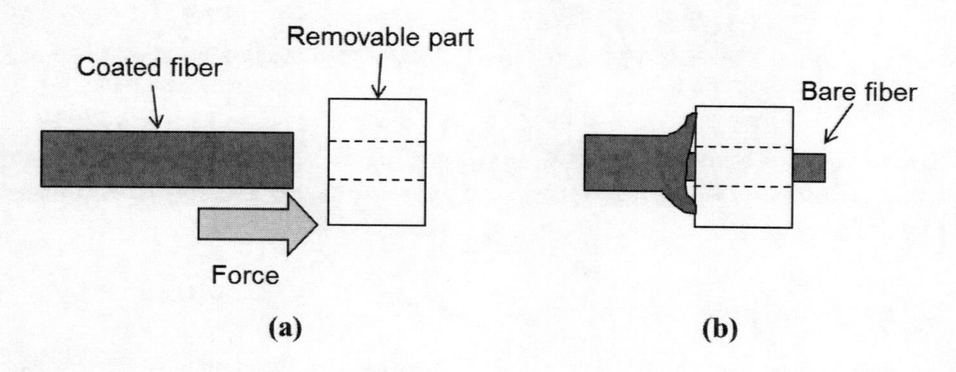

Figure 24. Novel fiber coating removable part and coated fiber.

NEW FIELD INSTALLABLE SAGGED FIBER CONNECTOR FOR PHYSICAL CONTACT CONNECTION

Structure of Optical Connector and Tool

This section describes our new field installable connectors designed to simplify the connector structure. Figure 25 shows the structure of a field installable connector, which mainly consists of a ferrule, clamp parts, and housing. The new configuration has two features. The first is that inserting the optical fiber into the connector housing results directly in PC connection. Namely, there is no mechanical splice and no built-in fiber or optical fiber fixed to the ferrule. The second feature is that the PC surface is compressed by the elasticity of the fiber. The connector assembly procedure is as follows. First, the fiber coating is removed, the fiber is cleaved, and the fiber endface is tapered with the chamfering tool. Then, the fiber is inserted into the connector ferrule, and the inserted fiber end protrudes from the ferrule. After that, the fiber is fixed mechanically to the clamp parts. When the two fibers are connected, the fibers buckle because the fiber end protrudes from the ferrule end. The contact surface is compressed by the elasticity of the fiber without the need for springs. The buckling force results in a PC connection between the two fiber endfaces. This connector uses the same fiber alignment structure as the conventional connector, namely the fibers are aligned using a ferrule and a split sleeve. Figure 26 shows a schematic diagram of the fiber endface chamfering tool, which consists mainly of a grinding wheel, a motor, and a fiber holder. The fiber is mechanically fixed to the fiber holder. The fiber end is pressed against the grinding wheel with slight bending. The optical fiber end but not the core comes into contact with the slanted surface of the grinding wheel. The slanted surface enables the fiber endface to be chamfered. The motor rotates the grinding wheel, which chamfers the edge of the fiber; thus, chamfering the edge of the fiber endface. There are two technical issues when designing the proposed configuration, as shown in Figure 25. The first is the design of the fiber end shape, which can realize a PC connection without the need to polish the fiber endface. The second is the design of the buckled fiber parameters, namely the buckling and protrusion lengths.

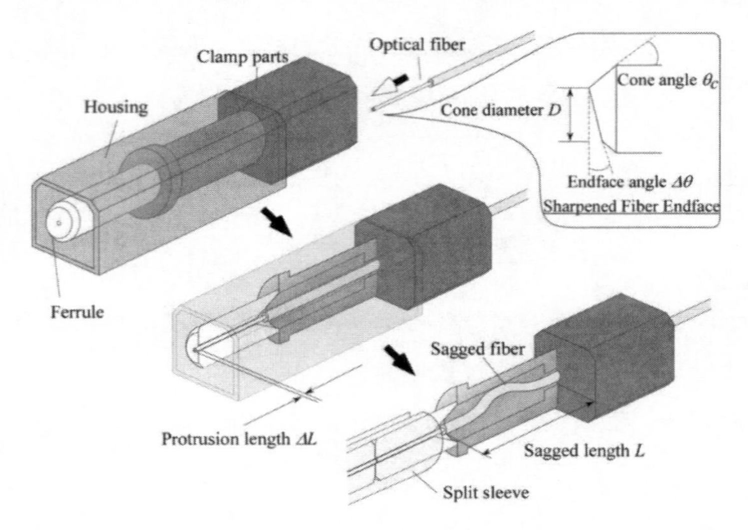

Figure 25. Structure of novel field installable connector.

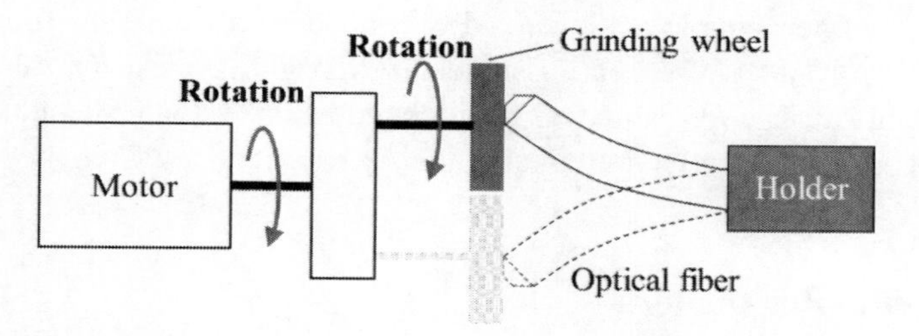

Figure 26. Structure of fiber endface chamfering tool.

Optical Fiber End Shape

The shape of the fiber endface was designed to achieve PC connection with chafering the fiber endface. A simple process was used to eliminate the polishing requirement. The process consists of stripping, cleaving, and chamfering. The stripping and cleaving processes are used with conventional fiber connection techniques such as fusion splicing and mechanical splicing. The cleaved fiber endface has a ripple at the edge of the cleaved surface around the point in contact with the cleaver blade. The ripple prevents the cleaved fiber endface from achieving PC connection. In the chamfering process, the tool chamfers the fiber endface and removes the ripple. It has been confirmed that chamfering the endface enables PC connection without the need to polish the fiber endface [30]. The chamfered fiber endface is conical, and the endface shape parameters are cone diameter D, cone angle θc, and endface angle $\Delta\theta$, as shown in Figure 25. D and θc are parameters related to the chamfering process conditions. $\Delta\theta$ is generated when the fiber is cleaved. This $\Delta\theta$ results in a gap between the two cores of the mated fibers. Axial compression force Fp is needed to overcome the gap that prevents PC connection. We examined the relationship between the endface shape and the Fp value needed to achieve PC connection.

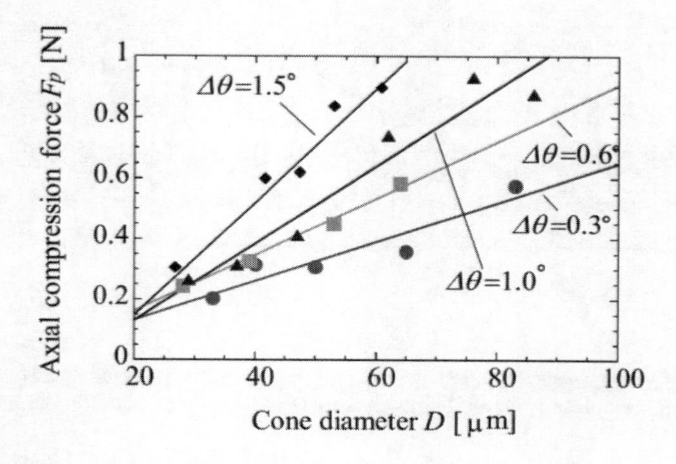

Figure 27. Fiber endface dependence on axial compression force.

Figure 27 shows the experimental results we obtained for Fp when we changed $\Delta\theta$ and D. Here, the axial forces that achieved a PC connection of a fiber sample were measured with a force gauge and a fiber holder that clamped the fiber sample. The reflectometer was used to monitor whether the fiber sample achieved a PC connection or not. We found that *Fp* was dependent on both $\Delta\theta$ and *D*. We can assume that the gap caused by $\Delta\theta$ increases as *D* increases at a certain $\Delta\theta$. The *Fp* value was less dependent on θc than $\Delta\theta$ and *D*. These results indicate that reducing *D* is effective in reducing the *Fp* required for PC connection.

Optical Connector Parameters

We designed the parameters of the connector. The buckling force *Pb* is expressed by the following equation with buckling length *Lb* [31]:

$$P_b = \frac{4\pi^2 EI}{L_b^{\ 2}}, \tag{4}$$

where *E* and *I* are Young's modulus and the cross-sectional moment of inertia, respectively. Figure 28 shows the calculated P_b values. We can control the *Fp* needed for PC connection by setting *Lb* according to Eq. (4). The connector requires the fiber to protrude from the ferrule end to generate the buckled fiber. The buckled amplitude increases as the protrusion length ΔL increases. Any excess increase in the buckled amplitude results in fiber bending loss inside the connector. *Lb* and ΔL were designed to realize a bending loss of less than 0.1 dB. The bending losses for different *Lb* and ΔL values were calculated. Figure 28 shows the ΔL value that causes a 0.1-dB bending loss for *Lb*. We can increase the ΔL limit by increasing *Lb*.

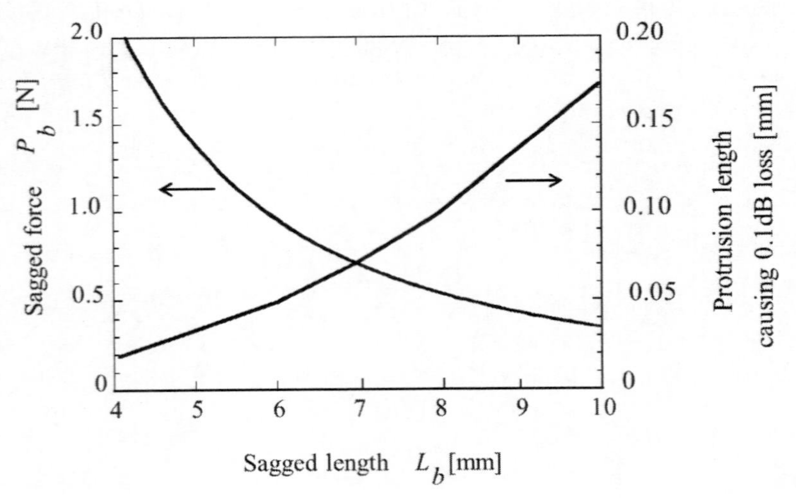

Figure 28. Buckling force and protrusion length causing 0.1-dB loss for buckling length.

Design for Field Installable Connector

As shown in Figure 27, it is important to clarify the $\Delta\theta$ range when the optical fiber is cleaved. The $\Delta\theta$ depends on the cleaver. We measured the $\Delta\theta$ of the fiber using commercial cleavers made by three different manufacturers. We measured $\Delta\theta$ by using the method for measuring the $\Delta\theta$ of an angle-polished ferrule in an optical connector. We measured the $\Delta\theta$ value of 300 cleaved fiber endfaces. The maximum $\Delta\theta$ value was 1.5 degrees with an average of 0.6 degrees. Therefore, we designed a fiber endface shape that enabled PC connection even if the maximum $\Delta\theta$ was 1.5 degrees. As a starting point, we confirmed that an endface diameter of over 30 μm does not affect the mode field diameter. Then Fp must exceed 0.5 N, as shown in Figure 27, because the PC connection of a fiber is guaranteed with a $\Delta\theta$ of 1.5 degrees. This compression force is obtained by setting an Lb of less than 8.3 mm, as shown in Figure 28. The second starting point is the minimum ΔL. It is assumed that the minimum target of ΔL is over 0.05 mm in terms of on-site assembly accuracy. Then, the requirement for Lb is over 6.2 mm for a bending loss of less than 0.1 dB, as shown in Figure 28. This Lb reveals that the maximum Pb that we can achieve is less than 0.9 N, as shown in Figure 28. The buckling force limit results in an endface diameter of less than 45 μm because the PC connection of a fiber with an $\Delta\theta$ of 1.5 degrees is guaranteed, as shown in Figure 27. When the target is a fiber PC connection with a $\Delta\theta$ of 1.5 degrees, an endface diameter of 30 to 45 μm and an Lb of 6.2 to 8.3 mm are specified.

Results

Field installable connectors and a chamfering tool were fabricated. Figure 29(a) shows a photograph of the fabricated connector. The connector consists of a modified SC-type connector and clamp parts. The dimensions are 6 × 6 × 20 mm. The cross-sectional size is the same as that of an SC-type connector plug. There is space for the buckled fiber inside the housing. The Lb is 7 mm based on the design for PC connection of the fiber with a $\Delta\theta$ of 1.5 degrees. Figure 29(b) is a photograph of the chamfering tool. The tool consists of a grinding wheel, a motor, and a fiber holder. The dimensions are 40 × 50 × 200 mm. The motor is powered by 3 AA batteries and rotates the grinding wheel. The grinding wheel chamfers the fiber endface at a rotation speed of 1500 min^{-1}. The installed SMF was cleaved with a conventional cleaver and tapered with a chamfering tool. The endface shape of the installed SMF had a θc of 30 to 45 μm based on the design for PC connection of the fiber with a 1.5-degree $\Delta\theta$. The connector was assembled with a ΔL of 0.075 mm. The optical characteristics were measured when the assembled connector was connected to an SC connector plug inside an SC-type adaptor at a wavelength of 1.31 μm.

(a)

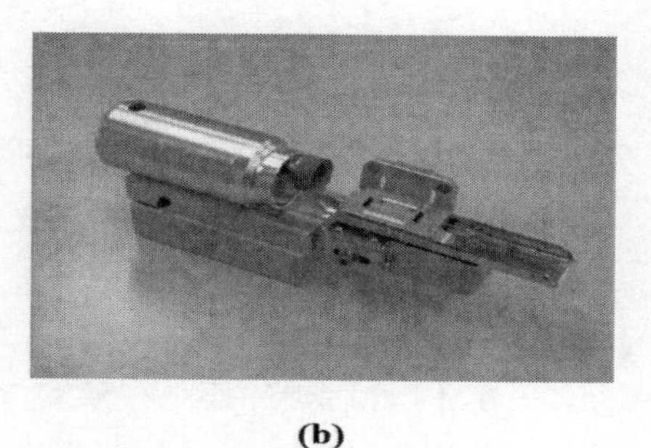

(b)

Figure 29. Photographs of (a) field installable connector (b) chamfering tool.

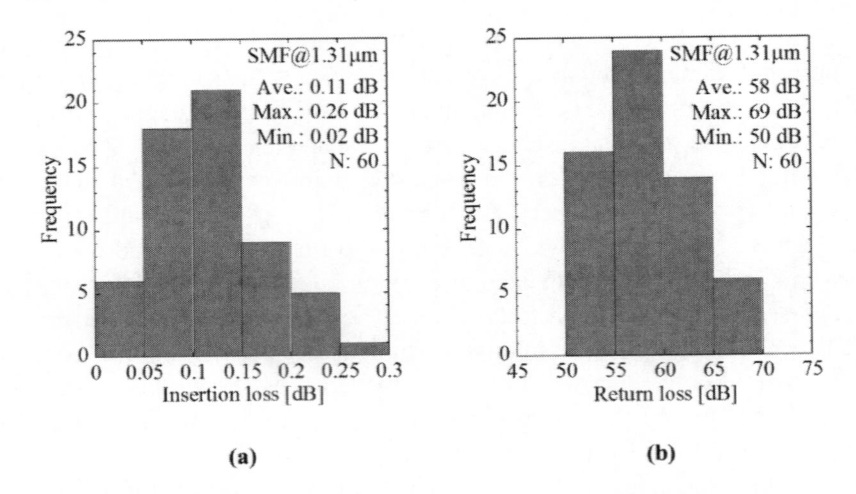

Figure 30. Optical characteristics of prototype field installable optical connector (a) insertion loss histogram (b) return loss histogram.

Figure 30(a) and (b), respectively, show histograms of the insertion loss and return loss for 60 pairs of assembled connectors. The insertion losses for 125/250-μm SMF connection were sufficiently small at less than 0.26 dB with an average value of 0.11 dB. The average return loss was 58 dB, which confirmed that all the connection points achieved PC connections. These results reveal that the proposed field installable connection technique achieves PC connection by using the chamfered fiber endface and the compression force of the buckled fiber.

CONCLUSION

This chapter reported the latest optical fiber connection technologies. In particular, it detailed current and novel field installable connectors. The current and latest mechanical splices with refractive index matching material were introduced. Current field installable connectors were also described. To install a connection, the optical fiber end must be stripped, cleaned, cleaved, and connected to an optical fiber using the mechanical splice technique, and the cable sheath is fixed in a clamp. The structure allows connection to another optical fiber connector in the field.

A recently proposed splicing technique for coated fiber connection was presented. The splice for this technique uses a structure in which coated fibers are aligned and connected precisely with a capillary. The splice can be used on site more easily and safely than conventional splices because it does not require the fiber coating to be stripped or the fiber to be cleaned. The fabricated splice for 80/125-μm coated SMF connection had an average insertion loss of 0.2 dB and a return loss of more than 46 dB. It is confirmed that this splice is effective for use in advanced optical networks and interconnections.

A new type of field installable 125/250-μm SMF connection technique was also reported. A connector was fabricated for this technique without using a polished fiber endface or a mechanical splice. The chamfered fiber endface and the buckling force of the fiber enabled PC connection without the need to polish the fiber endface. It was confirmed that the field installable connector achieved a PC connection with a low insertion loss of 0.11 dB and a high return loss of over 50 dB.

ACKNOWLEDGMENTS

This work was supported by NTT Laboratories, Japan. The author is deeply grateful to the following members of the NTT Laboratories; S. Tomita for his encouragement and suggestions; H. Son and R. Koyama for discussing the experiments related to the coated fiber connection technique, and Y. Abe, M. Kobayashi, S. Matsui, S. Asakawa and R. Nagase for discussions regarding the new field installable sagged fiber connector.

REFERENCES

[1] Ministry of Internal Affairs and Communications, 2011 White Paper Information and Communications in Japan [online], Available at: http://www.soumu.go.jp/johotsusintokei/whitepaper/eng/WP2011/2011-index.html, 2012.

[2] H. Shinohara, "Broadband access in Japan: Rapidly growing FTTH market," *IEEE Com. Mag*., vol. 43, 2005, pp. 72–78.

[3] D. B. Keck, A. J. Morrow, D. A. Nolan, and D. A. Thompson, "Passive components in the subscriber loop," *J. Lightwave Technol*., vol. 7, Nov. 1989, pp. 1623-1633.

[4] M. Kihara, R. Nagano, H. Izumita, and M. Toyonaga, "Unusual fault detection and loss analysis in optical fiber connections with refractive index matching material," *Optical Fiber Technol.*, vol. 18, 2012, pp. 167–175.

[5] M. Kihara, Y. Abe, R. Koyama, K. Saito, and S. Tomita, "Study on easily field-assembled optical fiber connector technology," *Proc. IEICE Gen. Conf.* 2008, B-13-13, March 2008 (in Japanese).

[6] H. Son, M. Kihara, and S. Tomita, "New easily assembled mechanical splicer for direct coated optical fiber connection without stripping and cleaning," Proc. ECOC2007, Berlin, Germany, 2007, vol. 5, pp. 43-44.

[7] Y. Abe, M. Kihara, M. Kobayashi, S. Matsui, R. Nagase, and S. Tomita, "Novel field installable sagged fiber connector realizing physical contact connection without polishing fiber endface," OFC/NFOEC, NThC2, 2008.

[8] Y. Abe, M. Kihara, M. Kobayashi, S. Matsui, S. Asakawa, R. Nagase, and S. Tomita, "Design and performance of field installable optical connector realizing physical contact connection without fiber endface polishing," *IEICE Trans. Electron*., vol. E93-C, no. 9, pp. 1411-1415, 2010.

[9] E. Sugita, R. Nagase, K. Kanayama, and T. Shintaku, "SC-type single-mode optical fiber connectors," *IEEE/OSA J. Lightw. Technol*., vol. 7, 1989, pp. 1689–1696.

[10] R. Nagase, E. Sugita, S. Iwano, K. Kanayama, and Y. Ando, "Miniature optical connector with small zirconia ferrule," IEEE Photon. Technol. Lett., vol. 3, No. 11, 1991, pp. 1045–1047.

[11] T. Satake, S. Nagasawa, and R. Arioka, "A new type of a demountable plastic molded single mode multifiber connector," *IEEE J. Lightwave Technol.*, vol. LT-4, 1986, pp. 1232-1236.

[12] For example, "Optical fiber mechanical splice", US patent No.5963699.

[13] T. Nakajima, K. Terakawa, M. Toyonaga, and M. Kama, Development of optical connector to achieve large-scale optical network construction," in Proceedings of the 55th IWCS/Focus, 2006, pp. 439-443.

[14] K. Hogari, R. Nagase, and K. Takamizawa, "Optical connector technologies for optical access networks," *IEICE Trans. Electron.*, Vol. E93-C, No. 7, 2010, pp. 1172–1179.

[15] N. Suzuki, Y. Iwahara, M. Saruwatari, and K. Nawata, "Ceramic capillary connector for 1.3 single-mode fibers," *Electron. Lett.*, vol. 15, no. 25, pp. 809-811, 1979.

[16] S. Nagasawa, T. Tanifuji, M. Matsumoto, and M. Kawase, "Single-mode multifiber connectors for future large scale subscriber networks," ECOC1993, MoP1.5, 1993.

[17] T. Satake, S. Nagasawa, M. Hughes, and S. Lutz, "MPO-type single-mode multi-fiber connector: low-loss and high-return-loss intermateability of APC-MPO connectors," *Optical Fiber Technol.*, vol. 17, 2011, pp. 17-30.

[18] M. Takaya, M. Kihara, and S. Nagasawa, "Design and performance of a multifiber backpanel type connector," *IEEE Photon. Technol.* Lett., vol. 8, no. 5, 1996, pp. 655-657.

[19] M. Kihara, S. Nagasawa, and T. Tadatoshi, Design and performance of an angled physical contact type multifiber connector," *IEEE/OSA J. Lightw. Technol.*, vol. 14, no. 4, 1996, pp. 542-548.

[20] T. Shimizu, M. Ida, A. Daido, Y. Aoyagi, and K. Takamizawa, "Optical wiring technology using single-mode hole-assisted fiber," *Proc. IEICE Gen. Conf.* 2012, BI-3-2, March 2012 (in Japanese).

[21] TASC of NTT East Corporation, "Fault cases and countermeasures for field assembly connectors in optical access facilities," *NTT Technical Review*, vol. 9, No. 7, 2011.

[22] M. Kihara, M. Okada, M. Hosoda, T. Iwata, Y. Yajima, and M. Toyonaga, "Tool for inspecting faults from incorrectly cleaved fiber ends and contaminated optical fiber connector end surfaces," *Optical Fiber Technol.*, vol. 18, 2012, pp. 470–479.

[23] D. Marcuse, "Loss analysis of single-mode fiber splices", *Bell Syst. Tech. J.*, Vol. 56, No. 5, pp. 703-718, 1977.

[24] IEC 60793-2-50, Product specifications – Sectional specification for class B single-mode fibres.

[25] T. Murase, N. Nakamura, and Y. Morishita, "Small-diameter optical fibers with coating diameter of 125 μm", in *Proc. International Wire & Cable Sympo. (53rd IWCS)*, pp. 145-149, 2004.

[26] D. Gloge, P. W. Smith, D. L. Bisbee, and E. L. Chinnock, "Optical fiber end preparation for low-loss splices," *The Bell Sys. Tech. J.*, vol. 52, No. 9, 1973, pp. 1579–1588.

[27] T. Haibara, M. Matsumoto, and M. Miyauchi, "Design and development of an automatic cutting tool for optical fibers," IEEE/OSA J. *Lightwave Technol.*, vol. 4, No. 9, 1986, pp. 1434–1439.

[28] R. Koyama, K. Nakajima, T. Sekiguchi, and T. Kurashima, "A study on splice method for coated optical fibers," *Proc. IEICE Gen. Conf.* 2011, BCS-1-5, Sept. 2011 (in Japanese).

[29] M. Ida, T. Sasaki, A. Daido, K. Takamizawa, and T. Numata, "Development of new connector for coated optical fiber using coating removal technology," Proc. The 60th IWCS Conf., Nov. 7-9, Charlotte, North Carolina, USA, 2011, pp. 88-92.

[30] M. Kobayashi, S. Iwano, R. Nagase, S. Asakawa, and S. Mitachi, "Multiple fiber PC connector with 0.2 dB insertion and 60 dB return losses," Proc. OFC'97, WL11, 1997.

[31] S. Timoshenko, "Strength of Materials, Part II, Advanced Theory and Problems, Third Edition," New York: D. Van Nostrand Company, Inc., pp. 145-152, 1956.

In: Optical Fibers: New Developments
Editor: Marco Pisco

ISBN: 978-1-62808-425-2

Chapter 9

NEW ADVANCES IN TAPERED FIBER SENSORS

Ricardo M. André, Manuel B. Marques and Orlando Frazão*
INESC Porto and Faculty of Science, University of Porto, Porto, Portugal

ABSTRACT

The emergence of optical fiber tapers allowed for the development of new sensing heads to measure physical and external parameters. In the last five years, investigation became more intense with the rapid development of new types of fiber such as microstructured fiber. This chapter aims to present the latest work based on tapers: single tapers, interferometers based on tapers, tapers combined with other structures such as Bragg gratings or surface plasmon resonance. Finally, it also describes new taper geometries and tapers in microstructured fiber.

Keywords: Optical fiber sensors, tapers, interferometers and gratings

INTRODUCTION

Tapers fabricated in single-mode fibers are optical devices that have become important throughout the years. Many applications in optical communications, fiber lasers [1], biosensors [2] and interferometry [3] make use of optical fiber tapers. An optical fiber taper presents, as its fundamental characteristic, the interaction of the evanescent field of the guided mode with the surrounding medium. This property enables the use of these structures as an alternative to core-exposed fibers, commonly used in sensing. These devices are also very important in the conception and optimization of directional couplers [4, 5], in the expansion of laser beams [6] and naturally, the measurement of physical parameters such as curvature [7, 8].

* Corresponding Author address Email: ofrazao@inescporto.pt.

In 1981, Kawasaki *et al.* [4] demonstrated for the first time the importance of tapers in the fabrication of single-mode fiber couplers. In 1984, Kumar *et al.* [9] presented a taper in multimode fiber as an optical intensity refractometer. In 1986, Boucouvalas *et al.* [10] studied the response of taper-based couplers to the refractive index of the external medium. In 1993, Brophy *et al.* [11] built a Mach-Zehnder interferometer using two tapers in series. In 1995, Xu *et al.* [12] developed a sensor capable of measuring strain whilst insensitive to temperature; the structure consisted of a fiber Bragg grating written in the biconical section of the taper. The combination of fiber Bragg gratings (FBG) or long period gratings (LPG) with tapers has also been explored to increase the sensitivity of these structures to certain physical parameters [13, 14]. Also worth mentioning are the interferometric configurations based on tapers that, in 2006, allowed new sensing possibilities: a Michelson interferometer in a twin-core fiber where the coupling between cores is assisted by a taper [15]; a modal Mach-Zehnder interferometer based on a taper in series with a long period grating [16]; a modal interferometer based on a nonadiabatic taper with dimensions around 10 mm [17]. Continuing this line of investigation, in 2007, a modal Michelson interferometer that uses solely a taper in a single-mode fiber and a mirror at the fiber end was presented [18]. Two refractometers, adequate for biochemical sensing, based on Michelson and Mach-Zehnder interferometric configurations were demonstrated in 2008 [19, 20]. The configuration based on the Mach-Zehnder was also applied as a strain sensor in 2009 [21].

In the last 30 years, tapers have had a very important role in fiber sensing. With the development of new fabrication techniques and new types of fiber, taper technology has seen a resurgence and the number of publications has grown in the last few years. This chapter intends to show this growth mainly in the last two years.

STATE OF THE ART

In this section, a brief review of the state-of-the-art in tapers as sensing elements in the last five years is presented. The section is divided into six topics, namely single-taper devices, tapers in series creating interferometers, tapers combined with fiber Bragg gratings, tapers manufactured in microstructured fibers, new taper geometries, and finally their application in biosensors.

Single-Tapers

Many types of sensors have been based on single-tapers as they are one of the simplest structures in optical fiber. In the last couple of years the main focus has been on the enhancement of sensitivities and optimization of configurations (see Figure 1).

In 2012 several refractive index sensors were proposed based on single-tapers. An abrupt taper on a single-mode thin-core fiber section was proposed [25]. The section of tapered thin-core fiber was spliced between single-mode fibers resulting in an intermodal interferometer as cladding modes are excited at the first interface.

Year	Sensing Configurations
2013	Taper + Core offset MZI for refractive index sensing [22]; Uptaper + downtaper MZI for refractive index sensing [23]; LPG + Taper + LPG as a MZI for refractive index sensing [24].
2012	Tapered thin-core fiber for refractive index sensing [25]; Tapered modal interferometer for refractive index sensing [26]; Modal single-taper MZI for refractive index sensing [27]; Single-taper acoustic vibration sensor [28]; Single-taper force sensor [29]; Single-taper temperature-independent micro-displacement sensor [30]; Microfiber loop micro-displacement sensor based on resonance condition variation [31]; Single taper in multimode fiber for simultaneous strain and temperature sensing [32]; MZI with bi-taper on thinned cladding as a refractometer [33]; Thinned cladding sandwiched between two abrupt tapers MZI for refractive index sensing [34]; Micro-MZI with two abrupt tapers for picoliter refractive index sensing [35]; Refractive index sensor based on a MZI with an uptaper – downtaper – uptaper [36]; MZI with uptaper and core offset for bending-vector sensing [37]; Bi-taper MZI for lateral stress sensing [38]; Microfiber Fabry-Pérot interferometer with two FBGs for refractive index sensing [39]; Humidity sensor based on a taper coated with polyvinyl alcohol and a FBG [40]; Tapered PCF for refractive index sensing [41, 42]; Temperature-independent curvature sensing based on tapered PCF [43]; Temperature-independent MZI strain sensor based on two waist-enlarged SMF concatenated to a PCF section [44]; Tapered suspended-core fiber for strain sensing [45]; Modal interferometer with C-shaped taper for refractive index sensing [46]; S-taper as refractive index and axial strain sensor [47]; Micro-taper with gold coating for refractive index sensing through surface plasmon resonance [48].
2011	Single-taper acoustic vibration sensor [49]; 3-taper Mach-Zehnder refractive index sensor [50]; Bi-taper on thinned fiber MZI for refractive index sensing [51]; MZI with two waist-enlarged tapers for temperature sensing [52]; Bi-taper MZI for curvature sensing [53]; Inclinometer based on a taper and cleaved end Michelson interferometer [54]; Taper + FBG high force sensor [55]; Study of the fabrication order of FBG and taper [56]; Chemically tapered PCF by acid microdroplets for refractive index sensing [57]; MZI bi-taper on birefringent boron-doped PCF for strain and temperature sensing [58]; S-tapered MZI for refractive index and axial strain sensing [59]; Single-taper label-free detection of biomolecules [60]; Single-taper for detection of single nanoparticles [61]; Multiple fluorescence sensing with tapered polymer fiber [62]; Tapered plastic fiber for refractive index sensing [63]; Multi-tapered fiber with gold coating for surface plasmon resonance sensing [64].
2010	FBG in microfiber for refractive index sensing [65]; Bi-taper + gold-coated cleave double-pass MZI for refractive index and temperature sensing [66].
2009	Bi-taper MZI for strain sensing [21]; Tapered MZI for the simultaneous measurement of refractive index and temperature using two different sources [67]; Accelerometer based on a taper, a tilted FBG and a cleaved end [68].

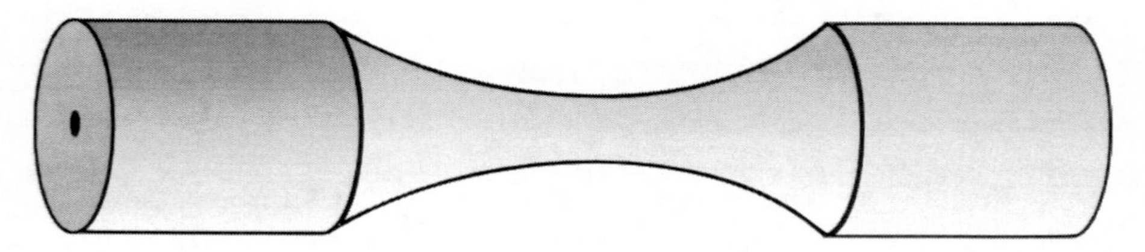

Figure 1. Schematic representation of a single-taper in optical fiber.

A Mach-Zehnder based on a non-adiabatic taper is also demonstrated [27]. The interference occurs between the first order cladding mode, excited in the downtaper, and the fundamental cladding mode. A modal interferometer based on a single 10 μm-diameter taper was reported [26]. This refractive index sensor is temperature independent and the index is taken by measuring the period of the measured spectra.

Two interesting micro-displacement sensor configurations were proposed in 2012. One is based on a locally bent microfiber taper with a diameter of 1.92 μm [30]. This is a bi-modal interferometer that relies on the specific bending configuration for transforming a micro-displacement into a spectral shift. The other forms a loop with a microfiber where the loop diameter is changed as a function of the micro-displacement [31]. The resonance wavelengths consequently shift as a function of the loop diameter resulting in a simple yet cunning configuration.

Acoustic vibration sensors based on single-tapers have also been studied in recent years. In 2011, a sensor based on a Mach-Zehnder with a separate reference arm was proposed [49], while in 2012, one using non-adiabatic tapered fibers simply analyzed in transmission is demonstrated [28]. Another single-taper sensor developed in the last year is a high-sensitivity force sensor that relies on a wavelength shift [29].

Tapers in single-mode – multimode – single-mode fiber structures were proposed for strain and temperature discrimination with increased sensitivity [32]. Tapers from the standard 125 μm-diameter down to 15 μm were fabricated on coreless multimode fiber (see Figure 2).

A tenfold increase in strain sensitivity was achieved when tapering the SMS structure from 125 μm (-2.08 pm/με) to 15 μm (-23.69 pm/με) (see Figure 3).

When analyzing in temperature, a slight decrease in sensitivity is found when decreasing taper diameter. The fact that the sensitivity to both parameters has different responses allows for a simultaneous measurement scheme using two tapered structures with different diameters. Considering the 87 μm-taper and the 25 μm-taper, resolutions of 1.6°C and 5.6 με were achieved. Combining two different tapered SMS structures with different strain and temperature sensitivities it is possible to measure strain and temperature simultaneously.

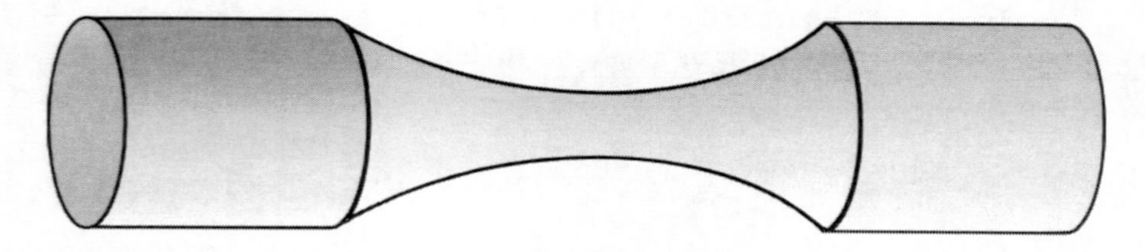

Figure 2. Schematic representation of a taper in a coreless multimode fiber.

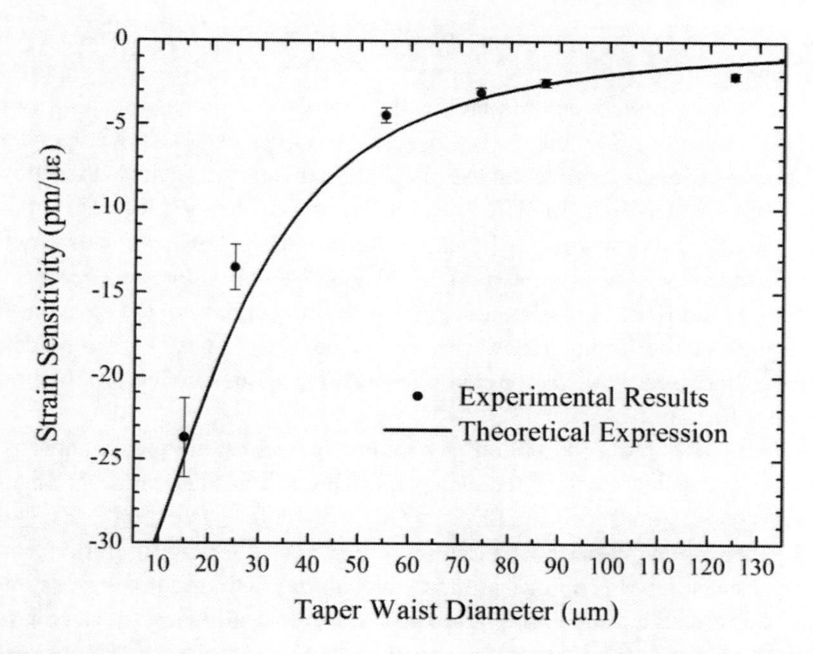

Figure 3. Relationship between strain sensitivity and taper waist diameter. For the theoretical approximation a value of κ_0 = -1.28 pm/µε was used [32].

Mach-Zehnder Interferometers

Mach-Zehnder interferometers (MZI) are very popular configurations. The most basic configuration is the fabrication of two tapers in series along a fiber [11] (see Figure 4). In the last two years, most of the papers published on optical fiber sensors based on tapers are Mach-Zehnder configurations.

Many Mach-Zehnder interferometers as refractometer configurations were proposed in the last years. In 2011, a bi-taper created on an already taper-thinned fiber section was reported [51]. This MZI configuration allows for a maximum sensitivity of 2210.84 nm/RIU at an external refractive index of 1.40.

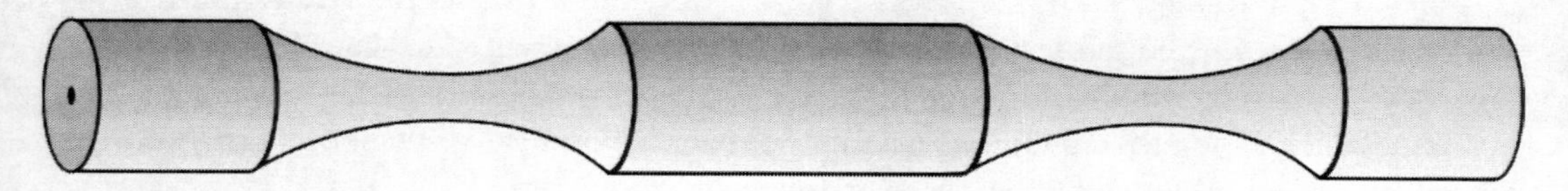

Figure 4. Schematic representation of a bi-taper in optical fiber, very common configuration for a Mach-Zehnder interferometer.

A similar bi-taper configuration where the fiber is thinned by chemical etching instead of tapering allows for a sensitivity of approximately 580 nm/RIU [33]. A MZI with two ultra-abrupt tapers (1:1), fabricated by concatenating instead of tapering, and where the cladding between tapers is thinned by chemical etching was published in 2012 [34]. This configuration reaches a sensitivity of 665 nm/RIU. A MZI formed by three cascaded single-mode fiber tapers where the central weak taper is sandwiched between two stronger tapers to enhance sensitivity to refractive index is proposed in 2011 [50]. A micro MZI formed by two abrupt tapers with a separation of 179.5 μm was proposed in 2012 [35]. This very small MZI can be used to measure the refractive index of picoliter-volume drops and has sensitivities of 600 and 4000 nm/RIU at around 1.34 and 1.61 μm respectively. Besides downtapers, interesting configurations can be made with uptapers. In 2012, an uptaper pair with a downtaper in between was implemented as a MZI for refractive index sensing [36]. The uptapers were produced by fusion and the downtaper by stretching. In 2013, a pair uptaper-downtaper also combined to form an MZI was shown to have a high-quality interference spectra and a sensitivity of 82.8 nm/RIU in the range 1.333-1.3869, similar to the previous one [23]. Combining a single-mode abrupt taper with an off-core section, a MZI was demonstrated [22]. At the taper, the core mode is coupled into cladding modes which are then recoupled at the offset back to the core.

MZI based on tapers are also applied as bending sensors. A simple example is a bi-taper that when kept straight has no interference pattern but when bent presents an interference pattern in the transmission spectrum [53]. In 2012, a more interesting MZI configuration allows for the measurement of the bending vector [37]. This configuration consists of a lateral-offset and an uptaper. The lateral offset breaks the cylindrical symmetry and defines two directions along which the bending response is different allowing for vector sensing. An inclinometer was also proposed where the sensing head is composed of a single-taper and a cleaved fiber end [54]. This works as a Michelson interferometer since the light passes twice through the taper. Local lateral stress can also be measured using a bi-taper MZI [38].

A MZI consisting of two uptapers fabricated by fusion was implemented as a temperature sensor with a sensitivity of 70 pm/K [52]. A different approach to a bi-taper MZI was taken by Lu *et al.* using two different sources (S-band and C/L-band light sources) to discriminate refractive index and temperature simultaneously [67]. A double-pass bi-taper MZI was characterized as a refractive index and temperature sensor [66]. Here, a tip with a bi-taper and a gold-coated mirror end allows light to traverse the bi-taper twice, effectively doubling the sensitivities of the sensor.

Tapers and Fiber Bragg Gratings

Fiber Bragg gratings (FBG) have been successfully employed as fiber sensors for many years and are one of the most successful fiber sensors implemented to date. In recent years, new fiber Bragg grating configurations have become possible due to new fabrication techniques such as femtosecond laser inscription. Fabricating FBGs in tapers and microfibers can enhance sensitivities and allow for new sensors (see Figure 5).

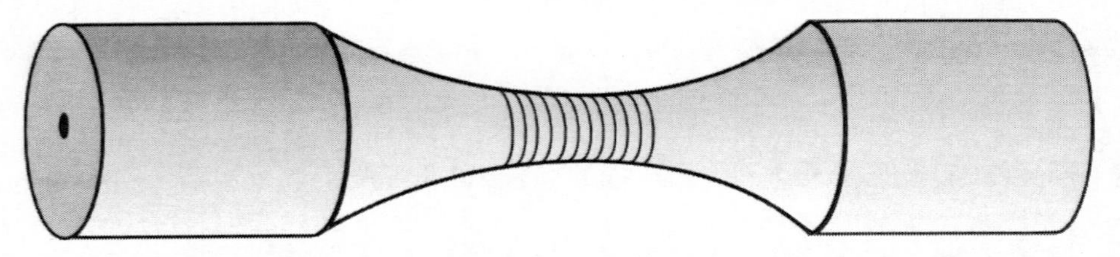

Figure 5. Schematic representation of a fiber Bragg grating written on the tapered region of a fiber.

In 2010, a FBG fabricated by femtosecond laser in a microfiber with 2 μm-diameter was proposed for refractive index sensing [65]. Due to the microfiber diameter and the fabrication method, the FBG is exposed to the surrounding medium without any etching process common in other configurations. In 2012, also for refractive index sensing, a Fabry-Pérot cavity is produced by fabricating a taper in the center of a FBG [39]. Using a flame-brushing technique, a microfiber is created in the middle of the FBG resulting effectively in two FBGs separated by a taper, and thus forming a Fabry-Pérot interferometer.

Combining FBGs and tapers, other sensors have been demonstrated. In 2009, an accelerometer based on an abrupt taper and a tilted FBG was reported [68]. A force sensor where an FBG is fabricated in a uniform waist taper of a photosensitive fiber was reported in 2011 [55]. By writing a FBG on a taper, the force sensitivity can be increased in several orders of magnitude when compared with FBGs in conventionally sized fibers. In 2012, a relative humidity sensor was built combining a multimode fiber taper coated with polyvinyl alcohol and a fiber Bragg grating [40]. Humidity changes the refractive index of the coating and thus the transmission losses. In 2011, the chronological order of tapering and FBG inscribing was studied [56]. It was ascertained that tapering the fiber Bragg grating region results in a Fabry-Pérot cavity while inscribing the FBG in the tapered region results in a phase-shifted Bragg grating.

In 2013, a MZI consisting of taper sandwiched between two long-period gratings is proposed for refractive index sensing [24]. This sensor is applied for real-time fuel conformity analysis.

Tapers in Microstructured Optical Fibers

Tapering microstructured optical fibers (MOF) is not as linear as tapering standard single-mode or multimode fiber. Hole collapse and cross-section deformation can change the guiding conditions which can be undesirable. Sometimes, this deformation is the objective of tapering (see Figure 6).

Again, refractive index sensing is the main sensing application of tapers in MOFs. A tapered standard Photonic Crystal Fiber (PCF), chemically etched with acid microdroplets was demonstrated to have a sensitivity of 750 nm/RIU in the range 1.3577 – 1.3739 [57].

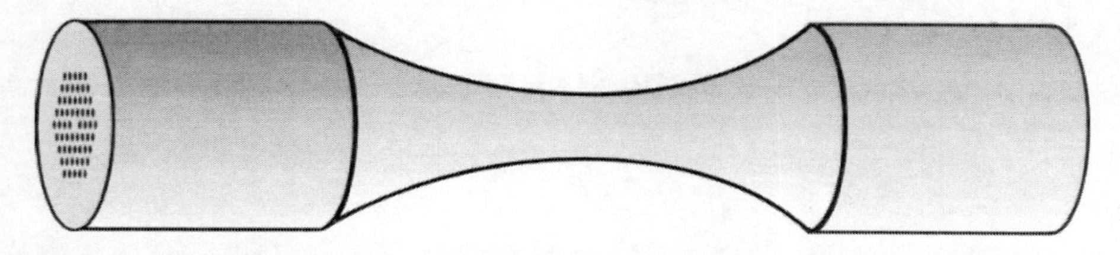

Figure 6. Schematic representation of a tapered photonic crystal fiber.

A slightly tapered PCF by flame-brushing was studied and an optimal sensitivity of 1600 nm/RIU is predicted [41]. A temperature independent refractometer based on a tapered PCF Mach-Zehnder interferometer was proposed in 2012 [42]. A very high sensitivity of 1529 nm/RIU was obtained in the range 1.3355 – 1.413.

Other sensors have been developed such as a temperature-independent curvature sensor based on a tapered PCF Mach-Zehnder interferometer not unlike the previous one [43]. A different configuration where two waist-enlarged tapers are spliced to a PCF section to form a MZI for strain sensing [44]. A simultaneous strain and temperature sensor was proposed using a bi-taper on a birefringent boron-doped fiber [58]. This configuration can discriminate between strain and temperature taking use of the visibility and fringe displacement parameters, as they are both sensitive to both physical parameters.

Tapers in suspended-core fiber with different cross-section diameters were produced by filament heating (see Figure 7) [45]. Tapering the suspended-core fiber leads to the reduction of the number of guided modes in the silica-air core. This is visible in the change from a typical multimode interference spectrum at low diameter reduction to an interference between few modes at higher diameter reduction. Analyzing the strain sensitivity of the tapered structures, an increase in sensitivity is readily detectable even for a small diameter reduction. A 73% increase in sensitivity from the 120 μm-taper to the 70 μm-taper is detected (see Figure 8). The sensing heads are also insensitive to temperature.

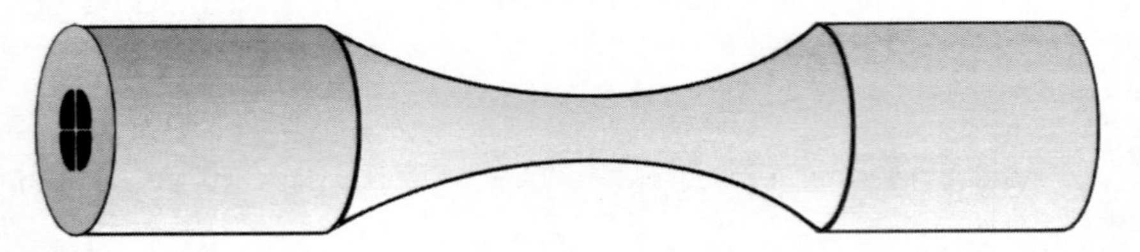

Figure 7. Schematic representation of a tapered suspended-core fiber.

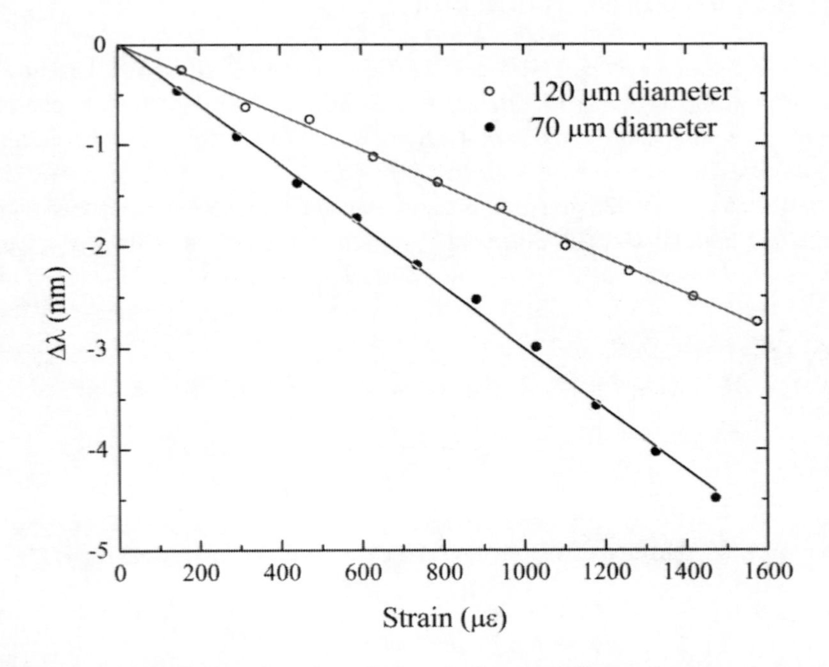

Figure 8. Wavelength shift as a function of applied strain for two different diameter tapers in suspended-core fiber [45].

These tapered suspended-core fiber structures have advantages when compared to normal fiber tapers as sensing elements. Besides having enhanced strain sensitivity, they also have very low temperature sensitivity and are protected from external medium perturbations due to the strong confinement of light in an air-silica waveguide.

New Taper Geometries

In recent years, tapers with unconventional shapes, whether along the length of the taper or as a cross-section, have started to appear in literature. A new line of investigation may be starting. In 2011, an S-shaped fiber taper Mach-Zehnder interferometer was proposed and fabricated by applying non-axial strain while pulling the fiber [59] (see Figure 9).

This configuration was applied as a refractometer and a strain sensor. Enhanced sensitivities were achieved when compared to traditional two-taper-based MZI sensors.

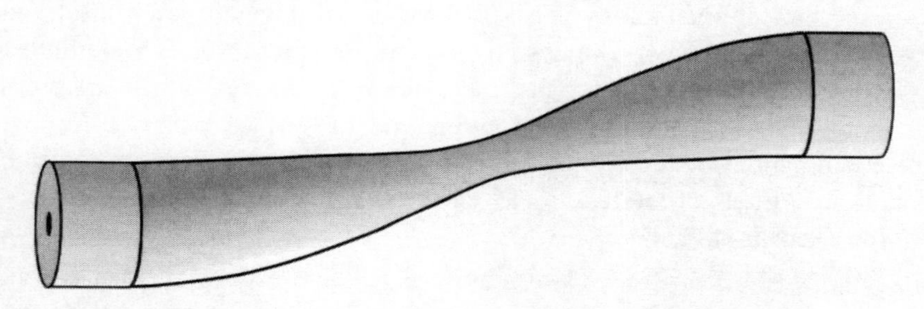

Figure 9. Schematic representation of an S-shaped taper in optical fiber.

In 2012, perfected fabrication of these structures led to an increase in both sensitivities taking the refractive index sensitivity from 1590 nm/RIU to 2066 nm/RIU in the refractive index range of 1.407-1.421, and the strain sensitivity from 60 pm/με to 183.4 pm/με [47]. These structures were produced in a commercial fusion splicer. A similar approach, but using a flame-brushing technique, lead to a C-shaped fiber taper modal interferometer for refractive index sensing with sensitivities much higher than two-taper-based sensors [46].

Also in 2012, fiber tapers etched using a microdroplet of hydrofluoric acid were produced [69]. These tapers have a cross section the shape of a teardrop which leads to a high group birefringence of approximately 0.017.

Tapers as Biosensors

Tapers in optical fiber also have biological applications. Some examples are the label-free detection of biomolecules using a taper in single-mode fiber [60]. The biosensor was evaluated with an Immune globulin G antibody-antigen pair. A subwavelength taper was used for the detection of single nanoparticles [61]. Here, with each nanoparticle that binds to the taper surface, the transmitted light power shows an abrupt jump. Fluorescence measurements were demonstrated using a tapered polymer fiber [62]. Two cases were studied: one where the fluorescent dyes were deposited on the taper and one where the dye is diffused into the fiber taper by heating. A tapered plastic fiber was studied as a biosensor for the measurement of bacteria concentration through the measurement of refractive index [63].

Surface plasmon resonance (SPR) based sensors have become increasingly popular to study biochemical reactions and identify biological entities. In 2011, a multi-tapered fiber coated with a gold film was analyzed as a refractive index sensor based on SPR [64]. Here, the influence of multiple tapers was studied in depth. In 2012, a similar refractive index sensor based on SPR with a gold coated micro-taper was developed [48].

CONCLUSION

This chapter shows that taper-based technology can still solve many problems in fields like engineering and biomedical sciences. The combination with other sensors can solve the cross-sensitivity to temperature for the simultaneous measurement of several physical parameters. Using current technology it is still possible to develop new sensors namely by reducing even further the taper dimensions down to nanosensors. Nevertheless, many applications are still to be explored. The last section of the state-of-the-art presents new applications in biosensing including the combination of surface plasmon resonance with tapers. This combination of a physical effect with a taper can not only lead to new sensitivities and resolutions of interest to the scientific community, but also to new products in the optical fiber sensor market.

REFERENCES

[1] G.J. Pendock, H.S. MacKenzie, and F.P. Payne, *Applied Optics* 32, 5236 (1993).

[2] A.J.C. Tubb, F.P. Payne, R.B. Millington, and C.R. Lowe, *Sensors and Actuators B: Chemical* 41, 71 (1997).

[3] F. Gonthier, S. Lacroix, X. Daxhelet, R.J. Black, and J. Bures, *Applied Physics Letters* 54, 1290 (1989).

[4] B.S. Kawasaki, K.O. Hill, and R.G. Lamont, *Optics Letters* 6, 327 (1981).

[5] T.A. Birks, *Applied Optics* 28, 4226 (1989).

[6] K.P. Jedrzejewski, F. Martinez, J.D. Minelly, C.D. Hussey, and F.P. Payne, *Electronics Letters* 22, 105 (1986).

[7] C. Caspar and E.-J. Bachus, *Electronics Letters* 25, 1506 (1989).

[8] L.C. Bobb, P.M. Shankar, and H.D. Krumboltz, *Journal of Lightwave Technology* 8, 1084 (1990).

[9] A. Kumar, T.V.B. Subrahmanyam, A.D. Sharma, K. Thyagarajan, B.P. Pal, and I.C. Goyal, *Electronics Letters* 20, 534 (1984).

[10] A.C. Boucouvalas and G. Georgiou, *Optics Letters* 11, 257 (1986).

[11] T.J. Brophy, L.C. Bobb, and P.M. Shankar, *Electronics Letters* 29, 1276 (1993).

[12] M.G. Xu, L. Dong, L. Reekie, J.A. Tucknott, and J.L. Cruz, *Electronics Letters* 31, 823 (1995).

[13] W. Du, X. Tao, and H.-Y. Tam, *IEEE Photonics Technology Letters* 11, 596 (1999).

[14] J.-F. Ding, A.P. Zhang, L.-Y. Shao, J.-H. Yan, and S. He, *IEEE Photonics Technology Letters* 17, 1247 (2005).
[15] L. Yuan, J. Yang, Z. Liu, and J. Sun, *Optics Letters* 31, 2692 (2006).
[16] O. Frazão, R. Falate, J.L. Fabris, J.L. Santos, L.A. Ferreira, and F.M. Araújo, *Optics Letters* 31, 2960 (2006).
[17] K.Q. Kieu and M. Mansuripur, *IEEE Photonics Technology Letters* 18, 2239 (2006).
[18] O. Frazão, P. Caldas, F.M. Araújo, L.A. Ferreira, and J.L. Santos, *Optics Letters* 32, 1974 (2007).
[19] Z. Tian, S.S.-H. Yam, and H.-P. Loock, *Optics Letters* 33, 1105 (2008).
[20] Z. Tian, S.S.-H. Yam, J. Barnes, W. Bock, P. Greig, J.M. Fraser, H.-P. Loock, and R.D. Oleschuk, *IEEE Photonics Technology Letters* 20, 626 (2008).
[21] Z. Tian and S.S.-H. Yam, *IEEE Photonics Technology Letters* 21, 161 (2009).
[22] G. Yin, S. Lou, and H. Zou, *Optics & Laser Technology* 45, 294 (2013).
[23] S. Zhang, W. Zhang, P. Geng, and S. Gao, *Optics Communications* 288, 47 (2013).
[24] J.H. Osório, L. Mosquera, C.J. Gouveia, C.R. Biazoli, J.G. Hayashi, P.A.S. Jorge, and C.M.B. Cordeiro, *Measurement Science and Technology* 24, 015102 (2013).
[25] J. Shi, S. Xiao, L. Yi, and M. Bi, *Sensors* (Basel, Switzerland) 12, 4697 (2012).
[26] G. Salceda-Delgado, D. Monzon-Hernandez, A. Martinez-Rios, G.A. Cardenas-Sevilla, and J. Villatoro, *Optics Letters* 37, 1974 (2012).
[27] L. Xu, Y. Li, and B. Li, *Applied Physics Letters* 101, 153510 (2012).
[28] B. Xu, Y. Li, M. Sun, Z.-W. Zhang, X.-Y. Dong, Z.-X. Zhang, and S.-Z. Jin, *Optics Letters* 37, 4768 (2012).
[29] X. Wang, W. Li, L. Chen, and X. Bao, *Optics Express* 20, 14779 (2012).
[30] H. Luo, X.X. Li, W. Zou, Z. Hong, and J. Chen, *IEEE Photonics Journal* 4, 772 (2012).
[31] A. Martinez-Rios, D. Monzon-Hernandez, I. Torres-Gomez, and G. Salceda-Delgado, *Sensors* (Basel, Switzerland) 12, 415 (2012).
[32] R.M. André, C.R. Biazoli, S.O. Silva, M.B. Marques, C.M.B. Cordeiro, and O. Frazao, *IEEE Photonics Technology Letters* PP, 1 (2012).
[33] B. Li, L. Jiang, S. Wang, Q.C. Mengmeng Wang, and J. Yang, *Optics and Lasers in Engineering* 50, 829 (2012).
[34] B. Li, L. Jiang, S. Wang, J. Yang, M. Wang, and Q. Chen, *Optics & Laser Technology* 44, 640 (2012).
[35] N.-K. Chen, T.-H. Yang, Z.-Z. Feng, Y.-N. Chen, and C. Lin, *IEEE Photonics Technology Letters* 24, (2012).
[36] S. Gao, W. Zhang, P. Geng, X. Xue, H. Zhang, and Z. Bai, *IEEE Photonics Technology Letters* 24, 1878 (2012).
[37] J. Zhang, Q. Sun, R. Liang, J. Wo, D. Liu, and P. Shum, *Optics Letters* 37, 2925 (2012).
[38] P. Lu, G. Lin, X. Wang, L. Chen, and X. Bao, *IEEE Photonics Technology Letters* 24, 2038 (2012).
[39] S. Zhang, W. Zhang, S. Gao, P. Geng, and X. Xue, *Optics Letters* 37, 4480 (2012).
[40] T. Li, X. Dong, C.C. Chan, C.-L. Zhao, and P. Zu, *IEEE Sensors Journal* 12, 2205 (2012).
[41] C. Li, S.-J. Qiu, Y. Chen, F. Xu, and Y.-Q. Lu, *IEEE Photonics Technology Letters* 24, 1771 (2012).
[42] K. Ni, C.C. Chan, X. Dong, C.L. Poh, and T. Li, *Optics Communications* null, (2012).

[43] K. Ni, T. Li, L. Hu, W. Qian, Q. Zhang, and S. Jin, *Optics Communications* 285, 5148 (2012).

[44] F. Xu, C. Li, D. Ren, L. Lu, W. Lv, F. Feng, and B. Yu, *Chinese Optics Letters* 10, 70603 (2012).

[45] R.M. André, S.O. Silva, M. Becker, K. Schuster, M. Rothardt, H. Bartelt, M.B. Marques, and O. Frazão, *Photonic Sensors* 1 (2012).

[46] H. Luo, X. Li, W. Zou, W. Jiang, and J. Chen, *Applied Physics Express* 5, 012502 (2012).

[47] R. Yang, Y.-S. Yu, C. Chen, Y. Xue, X.-L. Zhang, J.-C. Guo, C. Wang, F. Zhu, B.-L. Zhang, Q.-D. Chen, and H.-B. Sun, *Journal of Lightwave Technology* 30, 3126 (2012).

[48] T. Wieduwilt, K. Kirsch, J. Dellith, R. Willsch, and H. Bartelt, Plasmonics 1 (2012).

[49] Y. Li, X. Wang, and X. Bao, *Applied Optics* 50, 1873 (2011).

[50] D. Wu, T. Zhu, M. Deng, D.-W. Duan, L.-L. Shi, J. Yao, and Y.-J. Rao, *Applied Optics* 50, 1548 (2011).

[51] J. Yang, L. Jiang, S. Wang, B. Li, M. Wang, H. Xiao, Y. Lu, and H. Tsai, *Applied Optics* 50, 5503 (2011).

[52] Y. Geng, X. Li, X. Tan, Y. Deng, and Y. Yu, *IEEE Sensors Journal* 11, 2891 (2011).

[53] D. Monzon-Hernandez, A. Martinez-Rios, I. Torres-Gomez, and G. Salceda-Delgado, *Optics Letters* 36, 4380 (2011).

[54] L.M.N. Amaral, O. Frazao, J.L. Santos, and A.B. Lobo Ribeiro, *IEEE Sensors Journal* 11, 1811 (2011).

[55] T. Wieduwilt, S. Brückner, and H. Bartelt, *Measurement Science and Technology* 22, 075201 (2011).

[56] S.F.O. Silva, L.A. Ferreira, F.M. Araújo, J.L. Santos, and O. Frazão, *Fiber and Integrated Optics* 30, 9 (2011).

[57] S. Qiu, Y. Chen, J. Kou, F. Xu, and Y. Lu, *Applied Optics* 50, 4328 (2011).

[58] G. Statkiewicz-Barabach, J.P. Carvalho, O. Frazão, J. Olszewski, P. Mergo, J.L. Santos, and W. Urbanczyk, *Applied Optics* 50, 3742 (2011).

[59] R. Yang, Y.-S. Yu, Y. Xue, C. Chen, Q.-D. Chen, and H.-B. *Sun, Optics Letters* 36, 4482 (2011).

[60] Y. Tian, W. Wang, N. Wu, X. Zou, and X. Wang, *Sensors* (Basel, Switzerland) 11, 3780 (2011).

[61] J. Zhu, Ş.K. Ozdemir, and L. Yang, *IEEE Photonics Technology Letters* 23, 1346 (2011).

[62] C. Pulido and Ó. Esteban, *Sensors and Actuators B: Chemical* 157, 560 (2011).

[63] C. Beres, F.V.B. de Nazaré, N.C.C. de Souza, M.A.L. Miguel, and M.M. Werneck, *Biosensors & Bioelectronics* 30, 328 (2011).

[64] S.K. Srivastava and B.D. Gupta, *IEEE Photonics Technology Letters* 23, 923 (2011).

[65] X. Fang, C.R. Liao, and D.N. Wang, *Optics Letters* 35, 1007 (2010).

[66] Y. Li, L. Chen, E. Harris, and X. Bao, *IEEE Photonics Technology Letters* 22, 1750 (2010).

[67] P. Lu, L. Men, K. Sooley, and Q. Chen, Applied Physics Letters 94, 131110 (2009).

[68] T. Guo, L. Shao, H.-Y. Tam, P.A. Krug, and J. Albert, *Optics Express* 17, 20651 (2009).

[69] J.C. Mikkelsen and J.K.S. Poon, *Optics Letters* 37, 2601 (2012).

In: Optical Fibers: New Developments
Editor: Marco Pisco
ISBN: 978-1-62808-425-2

Chapter 10

SMART CFRP SYSTEMS WITH EMBEDDED FBG FOR STRUCTURAL MONITORING AND RETROFITTING

Maria-Barbara Schaller[1]*, Stefan Kaeseberg[2] and Torsten Thiel[3]
[1]GGB mbH, Espenhain, Germany
[2]HTWK Leipzig, Germany
[3]AOS GmbH, Dresden, Germany

ABSTRACT

The fields of activity in civil engineering are subjected to a constant change. Thereby maintenance, strengthening and monitoring of existing buildings have become more and more important. During the last ten years an increasing amount of Carbon Fiber Reinforced Polymer (CFRP) applications to rehabilitate damaged concrete or steel elements was observed. Thereby some important disadvantages of the brittle materials must be considered, for example the low ductility of the bond between CFRP and concrete and brittle failure of FRP.

With embedded sensor systems it is possible to measure crack propagation and strains. In this paper a sensor based CFRP system will be presented, that can be used for strengthening and measuring. The used optical fibers with Fiber Bragg Gratings (FBG) have a large number of advantages in opposite to electrical measuring methods. Examples are small dimensions, low weight as well as high static and dynamic resolution of measured values. A Bragg Grating consists of a periodic sequence of artificial and equidistant refraction switches in the core of an optical fiber. It can be produced over emblazing of an interference pattern of ultraviolet light. The core is surrounded by cladding. The main problem during the investigations was the fixing of the glass fiber and the small FBG at the designated position.

In this paper the possibility of setting the glass fiber with embroidery at the reinforcing fiber material will be presented. The direct embroider of the optical fiber (and the FBGS) clearly simplifies the fixing. An embroidery machine, using computerized

* Company: GGB mbH Address: Leipziger Straße 14, D-04579 Espenhain Germany Telephone: +49 / 342066460 Fax: +49 / 3420664678 Email: mbschaller@ggb.de.

support, is able to fix the fiber optical system accurately fitting at the carbon fiber material.

By using computer-controlled machines it is possible to achieve a very high degree of prefabrication as well as a high productiveness. The economic industrial fabrication of smart structures can be realized. Another possibility is the direct converting at the building site by handmade lamination with an epoxy resin. On the basis of four point bending tests on beams (dimensions of 700 x 150 x 150 mm) and confined columns the potential of the Smart CFRP system is introduced.

Keywords: SHM applications (aerospace, marine, railway, automotive, pipelines, civil engineering, energy generation and distribution, production, etc.)

Introduction - Monitoring and Smart Composite Structures Monitoring in Civil Engineering

The fields of activity in civil engineering are subjected to a constant change. Thereby maintenance, strengthening and monitoring of existing buildings have become more and more important. This tends to result in smaller investment for new buildings and significant increase for cost for maintenance and observation. These arrangements should start as early as possible and must be carefully maintained. In many cases this convention was not hold in the past. A lot of cases of damage at buildings at the age of 30 till 50 years underline this fact. The reasons for damages are manifold and reach from faulty construction till unpredictable natural phenomena [1]. Among other things natural facts have caused a stronger focus on building monitoring by publicity. One contribution for a safe structural monitoring can be given by modern measurement techniques. They allow a blanket assessment of the actual situation additionally to visual controls. Thereby a clear distinction between temporary and permanent measurement should be made. For permanent measurement rugged measurement systems are needed. Electrical systems like strain gauges are not the best alternative. Hence optical measurement systems move over to the foreground.

Structural Situation in Europe – Building Situation

European-wide examinations prove the high importance of older existing buildings. Studies in big European countries like France or Great Britain show that the fraction of apartments in older buildings (more than 20 years old) takes a portion of 75 percent of the whole floor space. It is very important to protect these fundamental economic resources. One possibility of an effective and sustainable maintenance can be smart composite structures, which contain Fiber Reinforced Polymer and integrated fiber optical measurement systems.

From FRP to Smart Composite Structures

Fiber Reinforced Polymers (FRP) have got more and more important during the last decade. The fields of application are widespread and not only focused on civil engineering. In civil engineering the main usage of FRP is the repair of concrete structures. But also for other building materials, for example wood constructions, FRP can be used. In the majority of cases carbon fibers are used as reinforcing material. Reasons are the superior technical properties of Carbon fibers compared to other high-strength fibers like glass fibers or aramid fibers [1]. Table 1 shows the properties of different fiber types. The high modulus of elasticity and the great tensile strength of carbon can be seen clearly. The Carbon Fiber Reinforced Polymer (CFRP) arises when circa 70 vol.-% high-strength carbon fibers are embedded into an epoxy matrix. The number of products of CFRP for reinforced concrete (RC) constructions is huge. Examples are at the surface bonded CFRP laminates or sheets as well as laminates which are placed in slots at the concrete surface. Furthermore it is possible to prestress CFRP laminates. Another favorable way is to combine these different types.

Table 1. Properties of different fiber types

Fiber type	Density	Axial tensile strength	Axial modulus of elasticity	Breaking strain
[-]	[g/cm³]	[kN/mm²]	[kN/mm²]	[%]
Glass	2.57	2.60	75	3.50
Aramid	1.45	3.00	110 - 125	2.40 - 2.70
Carbon	1.80	3.53 - 4.90	230	1.50 - 2.10

For example the combination between prestressed and in slot bonded laminates can be realized. Main fields of application are strengthening of RC beams under flexural tension and shear as well as the retrofitting of columns with wrapped CFRP sheets.

Failure types of FRP materials can differ. Main types of failure can be the cracking of the reinforcing fiber or matrix and the bond failure or delamination between fiber and matrix. The stiffness of the fiber material, form, amount and orientation of fiber, the bond between fiber and epoxy matrix as well as the matrix properties affect the stiffness and resistance of the FRP material. Besides the types of failures of the composite material, the bond behavior between FRP and concrete surface is very important. In particular this contact zone is critical for the design of CFRP strengthened concrete structures. Above all in structures under bending moment the small tensile strength of concrete is the most important parameter for the bond bearing strength. Figure 1 shows the delamination of a CFRP sheet during a displacement controlled four point bending test. The delamination started in this case at the last bending crack. In figure 1 also the failure mode of a wrapped concrete column can be seen. In this case the cracking of the reinforcing fiber (after reach of ultimate strain) was the failure type.

The described failure types happen very fast and often without any previous notice. A ductile behavior of the strengthened structural element can not be achieved under this term. But assurance is essential in civil engineering. Furthermore it is not comprehensively clarified if the constancy of the bonding between the different partners will be assured over a long time or under dynamic load.

With adequate measurement systems it might be possible to realize a safe monitoring to control the powerful but brittle CFRP strengthening systems. With optical measurement systems, which base on glass fibers, one can integrate the sensor system in the FRP material

[2]. The name of such materials is smart composite. These structures have the ability to measure their own mechanical behaviors and to give a solid feedback. The most important representatives under the fiber optic measurement systems are the Fiber Bragg Grating Sensors (FBGS).

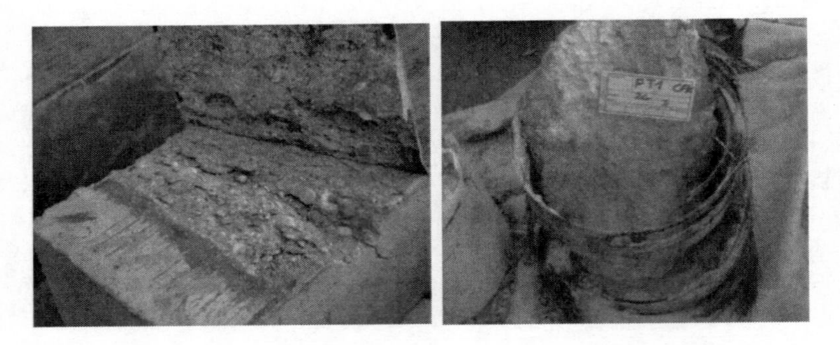

Figure 1. Destroyed concrete beam and column after failure of CFRP material.

Properties of Fiber Bragg Grating Configuration and Assembly

A Bragg Grating consists of a periodic sequence of artificial and equidistant refraction switches in the core of an optical fiber [3]. It can be produced over emblazing of an interference pattern of ultraviolet light. The core is surrounded by cladding. The refraction index of both is different and this results in total reflection of inducted light λ.

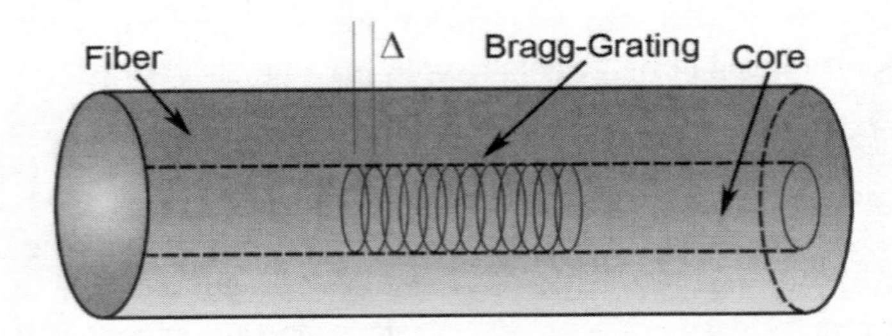

Figure 2. Glass Fiber with Bragg Grating.

For mechanical protection the glass fiber gets an additional coat of synthetic material. This coating can consist of polyimide and prevents the infiltration of water and hydrogen. The maintenance of the safety function is very important, in order to guarantee an error free and durable FBG unit. For the production of FBG in optical fibers it is necessary to have powerful ultra violet laser with wavelength of circa 240 ... 250 nm. These will be apportioned in two bales. This generates a pattern with a period Δ. This period is addicted by the angle of the laser. Δ describes thereby the distance between two interference maxima in the pattern. At every maximum a change of the refraction index will happen, whereby the actual pattern develops. The coating must be removed before producing the FBG. This means an additional stress for the optical fiber. Particularly a decrease of strength in these areas might be possible.

It is also possible to write the FBG into the optical fiber during the production of the optical fiber.

THEORETICAL BACKGROUND

Because of the emblazed interference pattern a reflection of inducted appointed light wavelength is possible. Light with the Bragg wavelength λB will be reflected [3]. This means that light of inducted spectrum will be reflected according to equation (1).

$$\lambda_B = 2 \cdot n_{eff} \cdot \Delta \qquad (1)$$

where λ_B = Bragg wavelength, n_{eff} = effective refraction index and Δ = period of diffraction grating.

This term of the light spectrum will be missing in the penetrated array. With equation (1) it is possible to clarify the measuring principle of FBG. A change of the period of diffraction grating results in an adjustment of the Bragg wavelength. Now other spectra of light will be reflected. These modifications can be activated by strain or temperature and then changes can be measured. The change of strain in the optical fiber can be explained with equation (2).

$$\Delta\lambda_B = \lambda_B \cdot (1 - p_e) \cdot \Delta\varepsilon \qquad (2)$$

where $\Delta\lambda_B$ = change of Bragg wavelength, p_e = photo elastic component $\approx$ 0.22 and $\Delta\varepsilon$ = change of strain.

Besides the monitoring of strain in structural elements the temperature measurement will be of interest. Also in this area FBGS can be used. With equation (3) the change of temperature can be considered.

$$\Delta\lambda_B = \lambda_B \cdot (\alpha + \xi) \cdot \Delta T \qquad (3)$$

where α = thermo-elastic coefficient, ξ = thermo-optic coefficient and ΔT = change of temperature.

Equation (3) shows that the simultaneous change of strain and temperature results in a modification of wavelength. With equation (2) and equation (3) the following combination is feasible in which strain and temperature are considered.

$$\Delta\varepsilon = \frac{1}{1 - p_e} \cdot \left(\frac{\Delta\lambda_B}{\lambda_B} - (\alpha + \xi) \cdot \Delta T \right) \qquad (4)$$

For explicit strain measurements it will be necessary to perform a compensation of the temperature influence. One possibility is an additional FBG only for temperature measurement in mechanically decoupled areas. These FBG can be emblazed at the same optical fiber.

MULTIPLEXING OF SEVERAL FIBER BRAGG GRATING

One important reason for continuing propagation of Fiber Bragg Gratings should be seen in the ability of Multiplexing [4]. One FBG can measure strain or temperature changes only at one point. But there is the possibility to distribute several Bragg Gratings at one optical fiber. The geometrical arrangement along the length of the fiber can be varied. In figure 3 such a disposition is shown.

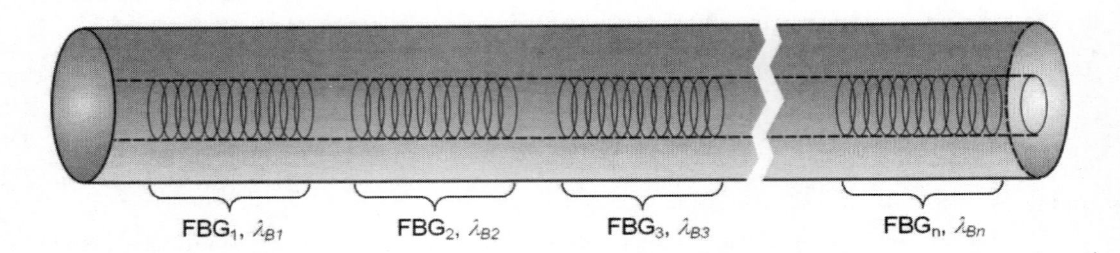

Figure 3. Multiplexing.

At every single Bragg Grating light spectrums will be reflected, which obey equation (1). A determination of the different Bragg Gratings is possible because of the variation of the period of diffraction grating between the various FBG. In such a way a distributed and wavelength coded sensor is developed.

The variation of the period of diffraction means that a preferably wide-band light spectrum must be inducted into the optical fiber. This works with strong light emitting diodes and with laser technique.

The analysis of the reflected light spectrums can be done with different methods. Thereby a differentiation in active and passive optical filters as well as interferometric and spectral measurement systems is possible.

The spectral analysis is possible with a spectrometer and allows the direct measurement of the Bragg wavelength.

USE IN FRP FOR CONCRETE STRUCTURES POSSIBILITIES FOR CONCRETE CONSTRUCTIONS

The advantages of fiber-optic measurement systems compared to classical electric measurement procedures are great.

Examples are the small dimensions and the low weight as well as the high static and dynamic resolution of measurement values. Other advantages are the insensibility towards electromagnetic radiance and the most chemicals. By the use of glass fibers, in which the FBG are emblazed, it is possible to integrate the sensor system directly into building materials like concrete or FRP.

In figure 4 the relative dimensions of a strain gauge and a FBGS plus reinforcement steel are shown.

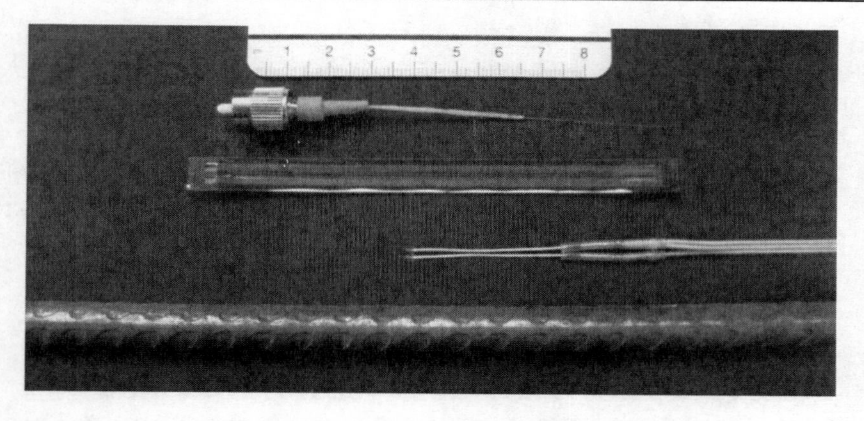

Figure 4. Relative dimensions of reinforcement steel, glass fiber and strain gauge.

Because of the favorable properties of FBGS, different applications for reinforced concrete constructions become possible [5, 6]. So it is possible to integrate the FBG sensor system in reinforcing bars. This can be done by making a groove in the reinforcing bar in which the optical glass fiber can be placed and fixed with epoxy resin. The distribution of strain along the length of the steel bar can be measured by multiplexing of several FBG as introduced in chapter 2.3. The results can be used to explore non-linear effects like tension stiffening. But there is a problem, because the bound between reinforcing bar and optical fiber/FBG must be realized by the epoxy resin. So careful arrangement of the FGB's is very important for realistic results.

The direct use in the concrete matrix is still harder because of the small long term durability of the coating around the glass fiber in alkaline medium. Moreover the optical glass fibers have to be especially protected because of their small proportions. The fitting and adjustment of the optical fibers/FBGS in concrete must be done with care.

USE IN FRP

The problems discussed in chapter 3.1 can be avoided with the use of Fiber Bragg Gratings in Fiber Reinforced Polymers [2]. The direct embedding in the epoxy resin allows exact strain measurement in the material. So mistakes are minimized during the monitoring. The epoxy resin is thereby an effective protection for the optical fiber.

CFRP systems for retrofitting of concrete structures with optical sensors have already been discussed in several publications. It could be shown that the reinforcing function of the CFRP can be ideally combined with the measurement and monitoring functions of the optical sensors like FBGS. Lu and Xie [7] accomplished strain measurements in smart CFRP sheets with FBGS. The sheets reinforced the tension zone of concrete beams, which were also arranged with FBGS sensors in the compressive zone and the reinforcing bars. For control extra electrical measurement with strain gauges was carried out. The results showed a very good agreement and it was possible to monitor the whole strain distribution of the profile. So it is possible to get better understanding of the non-linear behavior of CFRP reinforced concrete beams. It is a chance for more safety and sustainability.

It also was possible to get first results with fiber optical measurement systems at real constructions. Bastianini et al. [8] were able to localize and monitor failures between concrete surface and the used CFRP system at inaccessible sites. FBGS can measure lateral stress and cracks in cross direction in FRP.

There is no information about damage of the CFRP laminate in publications. Consequently it is possible to develop intelligent smart structures, which are able to reinforce concrete structures because of their potentials to measure a multitude of mechanical states over a long time.

SETTING THE GLASS FIBER WITH EMBROIDERY SENSOR-BASED TEXTILE CLUTCH

For an effective production of smart structures, it is very important to fix the optical fiber sufficiently during the production and the lamination of the FRP material. Especially the placing of the fiber in a particular design is complex and must be done carefully.

One possibility to realize any designs of sensor arrangements can be seen in embroidering the optical fiber directly on a carrier material. In this case the carrier materials are the reinforcing fibers which are often arranged as webs or clutches (for example carbon fiber clutch, like in figure 6). If only a uniaxial state of stress has to be measured, the fixing of the fiber can be easily done. The fixing can be realized with epoxy resin, because of the linear direction of the fiber. But if it is necessary to measure biaxial stress conditions (or if temperature compensation is needed) more difficult fiber courses will be required. Some minimum radiuses must be applied if the direction of the fiber changes. So the stress, caused by breaks, for the fragile fibers can be minimized. The adaptation at the designated courses of the optical fiber by hand is now very difficult. The selective fixing (epoxy resin) causes a lot of problems, and often it does not bring the optimal results. The direct embroider of the optical fiber (and the FBGS) clearly simplifies the fixing. An embroidery machine, using computerized support, is able to fix the fiber optical system accurately fitting at the carbon fiber material.

By using computer-controlled machines, like shown in figure 5, it is possible to achieve a very high degree of prefabrication as well as a high productiveness. The economic industrial fabrication of smart structures can be realized.

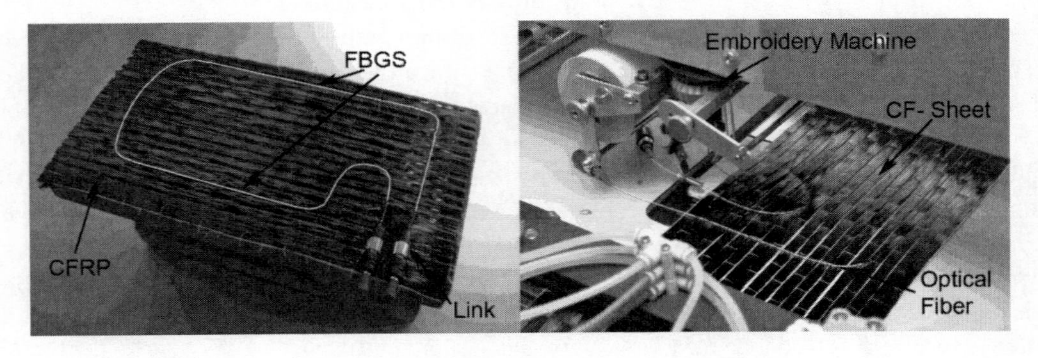

Figure 5. Courses of optical glass fibers with FBG on smart CF-composites.

Figure 6. Production of sensor based CFRP (First: embroider of optical fiber; Second: lamination by hand at building site).

In figure 5 the direction of an optical fiber with FBG for strain and temperature measurement is demonstrated. With this method it is possible to fix the Fiber Bragg Gratings close to these locations where strain monitoring is wanted. Through the embroidery method the direct mechanical bond between optical fiber and carbon fiber clutch is possible. Now the sensor based carbon fiber textile can be easily industrially laminated. Another possibility is the direct converting at the building site by hand made lamination (cf. figure 6).

Broadband light spectrum can be inducted and the reflected light can be analyzed with connections shown in figure 5 (link). The transport of the light spectrum over long distance is possible by further using glass fibers.

PROCESSING

The most important question during the embroider tests was, if the optical fibers or the reinforcing fibers would be damaged or influenced through the embroider procedure. With tension tests at CFRP sheets, without FBGS and with FBGS, it was possible to detect that there were only marginal losses of bearing strength and stiffness. Damages at the glass fiber and the FBG will be less if computer-controlled machines are used. With directly following spectral measurements at the fiber after the embroider procedure an effective quality check is possible. In fact the laminating processes with epoxy resin as well as the application of the sensor-based sheets at concrete structures (figure 6) were possible without problems.

EXPERIMENTS WITH CFRP REINFORCED BEAMS AND COLUMNS

In the context of the „regional Wachstumkern highStick“ technical textiles like carbon fiber sheets (CF clutches) are to be embroidered with optical fibers, to test the effectiveness of the strain measurement of the developed smart composites. For this purpose it was necessary to create reinforced concrete beams as well as short concrete columns. In 4 point bending tests (beams) and compression tests (columns) the significant changes of strength and strain were researched.

FOUR POINT BENDING INVESTIGATIONS WITH RC BEAMS (TEST PROGRAM 1)

The length of the used specimens conducted 70 cm (dimensions of 700 x 150 x 150 mm) and they consisted of high strength concrete as well as two reinforcing bars (diameter 6 mm) as bending reinforcement. All in all 3 beams were produced. At first the test beams were loaded in a deflection controlled four point bending test up to a crack width of 0.40 mm. Thereby it was possible to simulate a realistic measure of redevelopment and to bring the beams into the cracked status. Thereafter the application of the sensor based CF sheets (tensile strength: 3 900 N/mm², elastic modulus: 230 000 N/mm², thickness: 0.11 mm) followed with an epoxy resin at the surface of the tensile zone of the concrete beams. For the deflection controlled tests displacement transducer were used as well as load cells to get load deflection curves and load strain curves of the bending tests. Figure 7 explains the whole test set-up and the used equipment.

The CFRP sheets were glued (like shown in figure 7) at the underside in the tensile zone of the concrete beams. The CFRP material was appointed with an optical fiber with one FBG by use of the embroider technique. The orientation of the optical fibers carried out in longitudinal direction in the middle of the sheet. Furthermore the optical fiber was arranged parallel by use of a kink as turning point (beam 1). For checking of the optical strain measurement strain gauges were used additionally. They can be seen in figure 7 at the right hand side.

Every strain gauge was ranged alongside the particular FBG. Furthermore the nitriding optical fibers and the arrangement of the used FBG can be seen. It is also possible to recognize the good integration in the epoxy matrix. Through the arrangement of the FBG in the middle of the beams it was possible to compare the strain measurements of the FBG. Furthermore it was possible to compare the electrical measurement system (strain gauge) directly with the results of the FBG. The results of the strain measurement with the bragg gratings, because of bending stress, can be seen in figure 8.

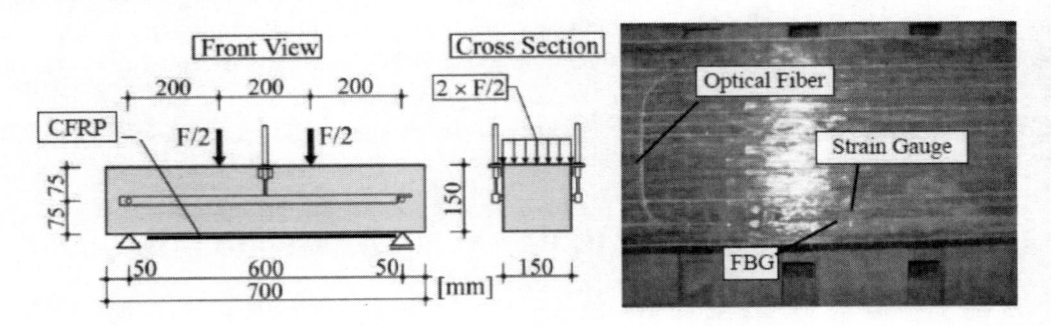

Figure 7. Test set-up of the four point bending test and arrangement of the optical fiber measurement system (Test program 1).

The agreement between the FBG strain measurements of the three beams was good by trend. The beams 1 and 2, which were loaded up to the failure, presented a good agreement, especially before and after the failure. The beams 2 and 3, in these cases the arrangements of the optical fibers were the same, showed in the strain area in range from 1 000 to 4 000 μm/m a very good agreement. The comparison of the results of the FBG strain measurement with the results of the strain gauges (figure 8) show, that an efficient monitoring of the strain development inside of the CFRP Sheets was possible. The curves present a good convergence between electrical and optical measurement methods at the three test beams (B1, B2, B3). The achieved results clarify the capability of the nitrogenized FBG sensor system. A damage or detraction of the optical glass fiber because of embroider or of the laminating did not appear.

FOUR POINT BENDING INVESTIGATIONS (TEST PROGRAM 2 AND 3)

After the satisfying tests with optical fibers with only one FBG, new experiments with three FBG at one fiber should be investigated at three concrete beams with the same setting as in the chapter before. For this purpose the optical fiber was again arranged parallel by use of a kink as turning point. The control of the optical strain measurement was once more realized with a strain gauge, which was arranged in the middle of the CFRP sheet. FBG 1 and 2 were used for strain measurement in the middle of the beam at each side. The FBG 3 was arranged in the turning point to realize temperature compensation.

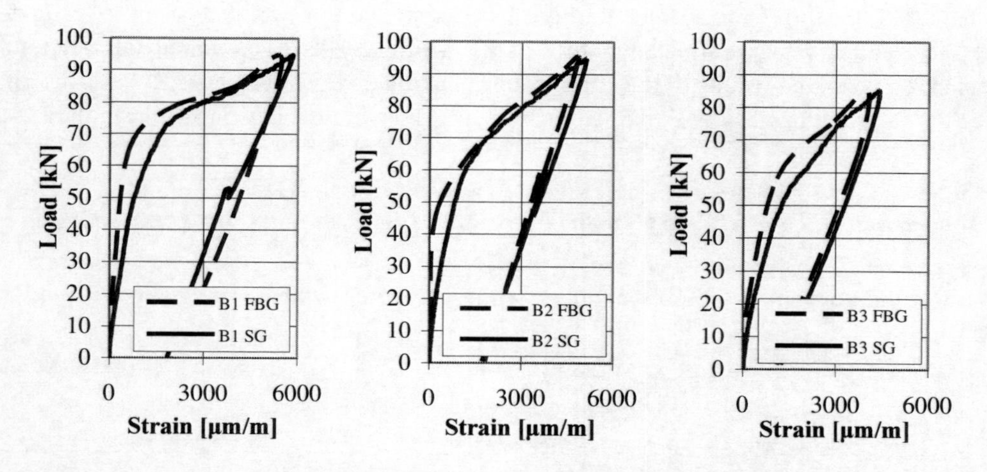

Figure 8. Load strain curves of strain gauge (SG) und Fiber Bragg Grating (FBG) (Test program 1).

Thereby it was to analyze if the arrangement in the breaking point assures an adequate mechanical decoupling to realize an independent temperature measurement. The load strain curves explained in figure 9 (left hand side) show a good agreement between the results of the strain gauges and the average values (FBG av) of the two in lengthwise placed FBG 1 and 2. However between the FBG 1 and 2 arranged at beam 1 irregularities between the measurement results were assessed. Reasons for these results can be bending cracks near the FBG position, which create different strain states at the FBG measurement points because of the non-linear behaviour of the reinforced concrete beams (concrete, reinforcing bars, CFRP material). The results attest that it is possible to nitrogenize several FBG, grouped at one optical fiber, at a CF clutch free of failure. The analysis of all FBG was possible. Thereby it worked to create temperature compensation by the use of FBG 3 in the breaking point.

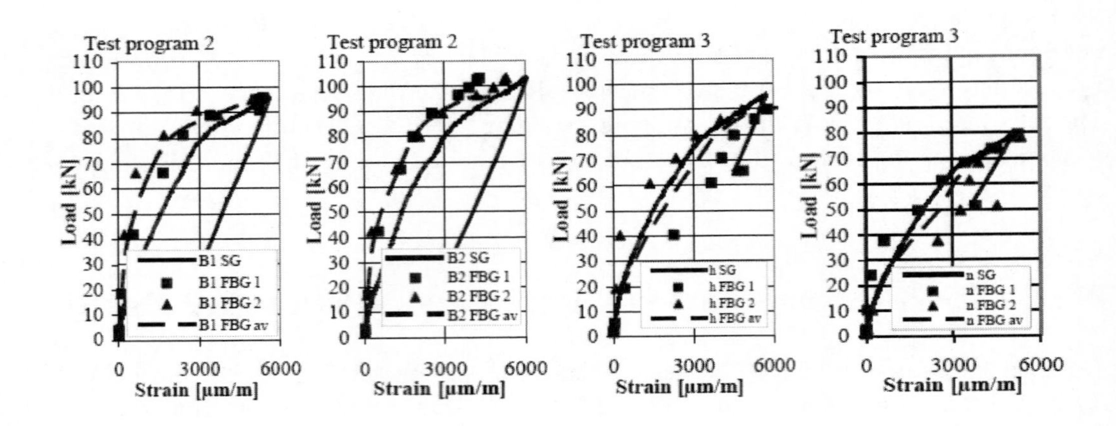

Figure 9. Load strain curves of strain gauge and FBG by use of concrete C60/75 (h) (Test program 2 and 3) and C30/37 (n) (Test program 3).

In test series 3 the effects of a concrete with normal strength (C30/37) (n) were monitored. Furthermore a new sensor arrangement at the CF sheet was sampled (figure 9 right hand side). More over all, bending cracks of concrete tension zone were investigated. These facts and a very careful laminating of the CFRP System guarantied the very good agreement between the results of the strain gauges and the average values (FBG av) shown in figure 9 on the right hand side. Bending cracks near FBG were the reasons for irregularities.

EXPERIMENTS WITH CONFINED SHORT CONCRETE COLUMNS

To test the new smart CFRP system, different concrete members were strengthened and tested.

Table 2. Experimental results

	f_{cc}	f_{cc}/f_{co}	ε_{cc}	$\varepsilon_{cc}/\varepsilon_{co}$	$\varepsilon_{l,FBG}$	$\varepsilon_{l,SG,max}$	Short Concrete Column strength category: C30/37; FBG 2; FBG 1; Pressure; CFRP-Sheets: Tensile strength: 3900 N/mm; Elastic modulus: 230.000 N/mm; thickness: 0,11 mm; Layer: 2; width: 150 mm; Strain Gauge; Integrated Optical Fiber with 2 FBG; 7.5; 15; 7.5; 30; 15; [cm]
	[N/mm²]	[-]	[‰]	[-]	[‰]	[‰]	
1	72.7	1.74	15.69	7.57	14.87	9.26	
2	66.4	1.59	12.40	5.99	14.10	7.72	
3	68.2	1.63	14.00	6.76	16.54	7.83	
	69.10	1.65	14.03	6.77	15.17	8.27	

So for example short concrete columns were confined and the significant changes of strength and strain (of the used concrete) researched. In table 2 the test set-up and the main properties of the columns can be seen.

In figure 10 the stress strain curves of the first specimen are shown. It can be seen, that an efficient monitoring of the strain development inside of the CFRP sheet was possible. The curves present a very good convergence between electrical and optical measurement methods.

Thereby only the FBG system guarantied a save measurement till the ultimate strain of the carbon fibers at nearly 1.5 % (see also table 2 $\varepsilon_{l,FBG}$). All strain gauges (SG) showed failure at approximate 0.8 % (table 2: $\varepsilon_{l,SG,max}$). Furthermore it also works to find out the area of failure inside the CFRP material, like shown in figure 10 (right hand side). In this case the ultimate strain inside the CFRP material was reached near FBG 2. In table 2 the ultimate stress (f_{cc}) and strain (ε_{cc}) of the three specimens in opposite to the results of plain concrete columns (f_{co}, ε_{co}) are shown. It clearly can be seen, that a high raise of strength and strain is possible with the used sensor based CFRP sheets.

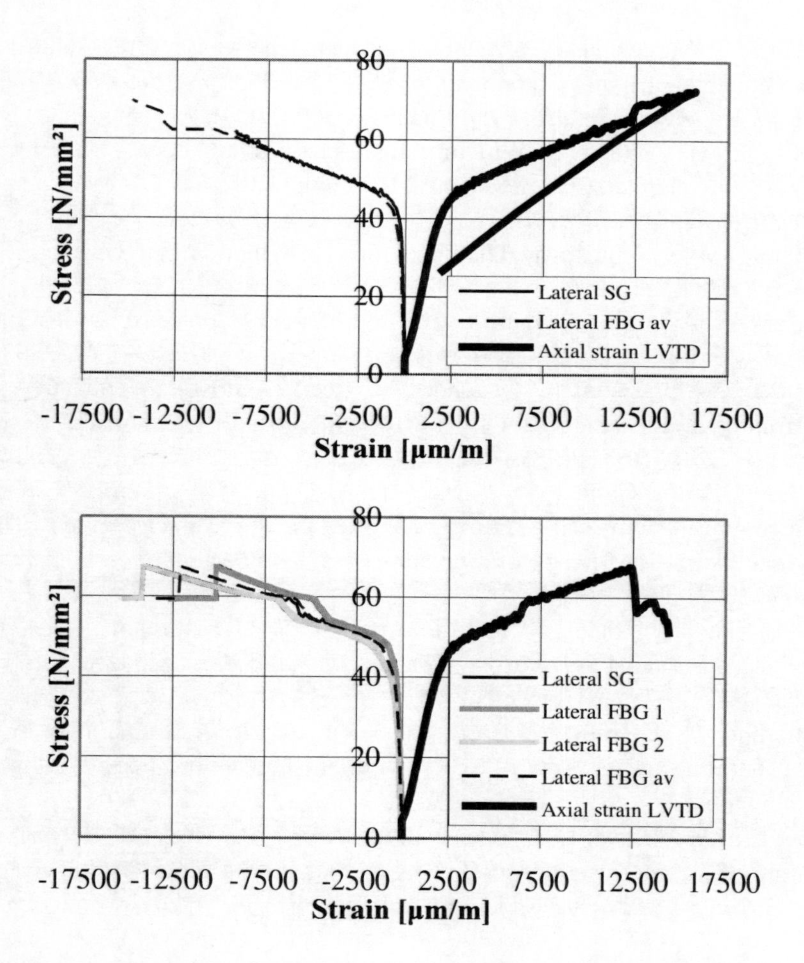

Figure 10. Stress strain curves of strain gauge (SG) and FBG average (FBG av) of column 1 and 2.

CONCLUSION

Structural monitoring will become more important because of the increasing age of the existing buildings. The precious economical resource building-asset must be monitored and repaired carefully. One possibility to monitor and reinforce existing concrete structures is given by the embedding of optical fibers with FBG in Carbon Fiber Reinforced Polymer (smart composite). The small proportions of optical glass fibers allow very flexible strain and temperature measurements inside of the CFRP material. Advantages like the high tensile strength of the CFRP material will not be influenced by the FBG sensor system. But the arrangement and fixing of the optical fibers and the FBG at the right position is problematic. The presented embroider method enables an accurate and reproducible fixing method.

REFERENCES

[1] S. Käseberg; T. Müller; H. Kieslich; K. Holschemacher; Faser-Bragg-Gitter-Sensoren: Einsatzmöglichkeiten im Betonbau, *pp 233 - 256 in Betonbau im Wandel, Holschemacher, Editor, Bauwerk Publishers, Berlin* (2009).

[2] K.-T. Lau; L.-M. Zhou; J.-S. Wu; Investigation on strengthening and strain sensing techniques for concrete structures using FRP composites and FBG sensors, *Materials and Structures 34,* 42 – 50 (2001).

[3] S. Kurtaran; M.-S. Kiliçkaya; The modelling of Fiber Bragg Grating, *Opt Quant Electron 39, 643 – 650 (2007)*

[4] E. Mehrani; A. Ayoub; A. Ayoub; Evaluation of fiber optic sensors for remote health monitoring of bridge structures, *Materials and Structures 42,* 183 – 199 (2009)

[5] V. Saouma; D. Anderson; K. Ostrander; B. Lee; V. Slowik; Application of a fiber Bragg grating in local and remote infrastructure health monitoring, *Materials and Structures 31,* 259 – 266 (1998).

[6] C.L. Wong; P. A. Childs; R. Berndt; T. Macken; G.-D. Peng; N. Gowripalan; Simultaneous measurement of shrinkage and temperature of reactive powder concrete at early-age using fibre Bragg grating sensors, *Cement and Concrete Composites 29,* 490 – 497 (2007)

[7] S. Lu; H. Xie; Strengthen and real-time monitoring of RC beam using "intelligent" CFRP with embedded FBG sensors, *Construction and Building Materials 21*, 1839 – 1845 (2007).

[8] F. Bastianini; M. Corradi; A. Borri; A. Tommaso; Retrofit and monitoring of an historical building using "Smart" CFRP with embedded fibre optic Brillouin sensors, *Construction and Building Materials 19,* 525 – 535 (2005).

[9] W. Moerman; W. Waele; C. Coppens; L. Taerwe; J. Degrieck; R. Baets; M. Callens; Monitoring of a Prestressed Concrete Girder Bridge with Fiber Optical Bragg Grating Sensors, *Strain 37,* 151 – 153 (2001).

In: Optical Fibers: New Developments
Editor: Marco Pisco

ISBN: 978-1-62808-425-2

Chapter 11

OPTICAL FIBER SENSORS FOR VITAL SIGNS MONITORING

Zhihao Chen[1], Doreen Lau[2] and Pin Lin Kei[2]

[1]Neural and Biomedical Technology Department, Institute for Infocomm Research, Agency for Science, Technology and Research (A*STAR), Singapore
[2]Department of Radiology, Singapore General Hospital, Singapore

ABSTRACT

Every rhythm, rate and force in the form of heart beat, respiratory rate, blood pressure, temperature, etc. are vital signs of life required for monitoring, as they offer clues to a medical condition, and are harbingers of critical health situations or death once on the decline. Today's escalating healthcare costs and increasingly rigorous regulatory requirements for medical device quality, reliability and safety prompt an emerging market demand for better sensor technologies that can facilitate the early diagnosis and prevention of diseases. Since their first successful implementation for endoscopic imaging in the 1960s, optical fibers have once again revolutionized medicine in the area of vital signs monitoring.

Over the past few decades, the high sensitivity of these hair-thin strands of fibers to external perturbations has been exploited extensively for applications in the physical, biological, chemical and imaging aspects of healthcare sensing. These fibers are ideal material for development and integration into the new era of minimally-invasive, miniaturized and cost-effective vital signs monitors due to their intrinsic biocompatibility, chemical inertness, small-sized dielectric nature and low cost. In particular, their immunity from electromagnetic interference and radiofrequency allows for their safe use without any electrical interference with other hospital electrical equipment. These are all important aspects for vital signs monitoring in various hospital and home-based settings.

This chapter illustrates the latest technological innovation in the development of optical fiber sensors for vital signs monitoring by both the commercial and research sectors in the medical device industry. There will be a particular highlight on discussing the emerging experimental breakthroughs in developing non-invasive optical fiber sensors for the monitoring of vital signs in chronic diseases such as cardiovascular and pulmonary illnesses. Our work on micro-bend optical fiber sensors for non-invasive monitoring of respiratory rate, heart rate and ballistocardiography (BCG) will be described. This optical fiber sensor mat technology has been successfully commercialized for home use. Challenges met in the development and market penetrance of optical fiber sensors for vital signs monitoring will be examined, and the future prospects of utilizing optical fibers in healthcare sensing will be discussed.

I. INTRODUCTION

Vital signs, or signs of life are important physiological parameters of the body's most basic functions. A single skip of the heartbeat or a momentary drop in blood oxygen level detected during an intraoperative procedure are vital parameters that can offer indications of the patient's health status and progress. The detected vital parameters also alert the medical professionals of any signs of delayed recovery or adverse health outcomes of the monitored patient (Figure 1). In this era of unprecedented growth in the world's ageing population and increased burden of chronic diseases, there is an emerging demand for more advanced biomedical instrumentation to enable more efficient diagnosis, monitoring and treatment of patients. In recent years, vital signs monitors have become essential applications in your familiar hospital settings such as the operating theatres, temergency rooms, intensive care units, diagnostic imaging suites, etc. It has also enabled better physiologic monitoring of elderly and palliative in nursing homes and hospices, as well as for outpatient monitoring in the home environment and occupational monitoring in the army, aeronautic or other industries that require strenuous physical work.

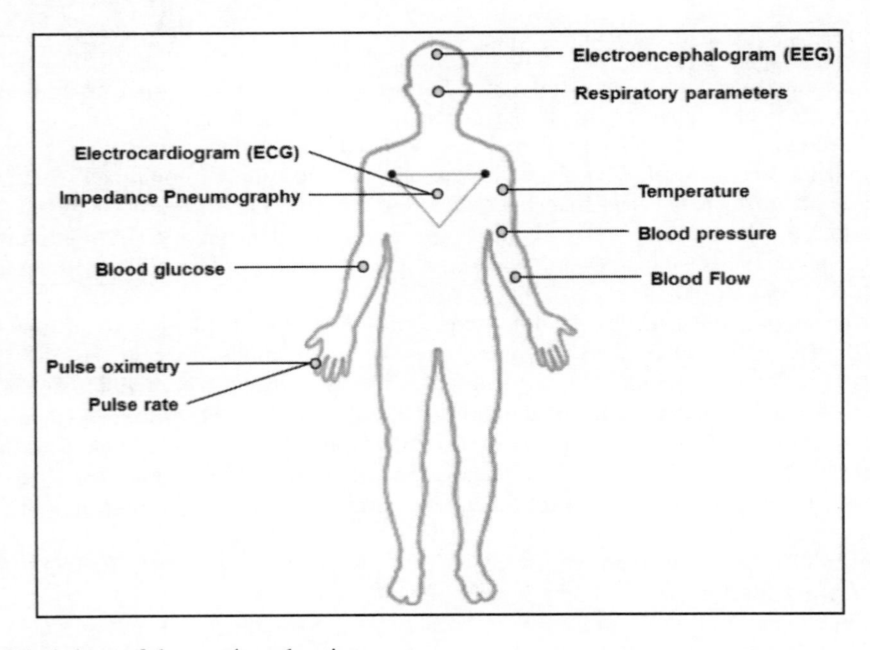

Figure 1. Vital signs of the monitored patient.

Recent advances in fiber optic technology have significantly changed the medical device industry. Ever since their first successful implementation for endoscopic imaging and safe laser delivery during surgical therapy in the 1960s [1], optical fibers have once again revolutionized medicine in the area of vital signs monitoring. The ability to carry gigabits of information at the speed of light has increased the potential for further research in optical fibers for medical applications. At the same time, significant improvements and cost reductions in optoelectronic components have combined fiber optic telecommunications with optoelectronic devices to create optical fiber sensors that can be used for the measurement of vital parameters [2].

Over the past few decades, numerous research has been conducted to evaluate the high sensitivity of these hair-thin strands of fibers to external perturbations in the physical, biological, chemical and imaging aspects of healthcare sensing [3]. The use of optical fiber sensors for vital signs monitoring is based on the unique features of optical fibers: small size, mechanical flexibility, chemical inertness, biocompatibility, and immunity to electromagnetic and radiofrequency interference. The small size of the optical fibers enables the construction of miniaturized sensors for *ex vivo* applications, whereby only a small amount of the sample material (for example the blood) is needed for analyte measurement (for example, for the detection of blood glucose). The small size and dielectric nature of optical fibers, coupled with their intrinsic mechanical flexibility also allow for their easy installation into miniaturized devices for vital signs sensing during intra-operative procedures. The chemical inertness and biocompatibility of optical fibers also ensure their safe use on patients, especially with regard to invasive procedures or skin contact during the monitoring sessions. Where electronic monitoring instruments are involved, such as in intensive care, or where high voltage (defibrillation), heating (radiofrequency ablation) and high magnetic fields (magnetic resonance imaging) are applied, the immunity of optical fibers to electromagnetic and radiofrequency interference enable their use in various vital signs monitoring and medical treatment sessions. In addition, the relative inexpensiveness and ecological compatibility of optical fibers allow for their disposal after use and maintenance of hygiene in medical settings.

The general structure of an optical fiber sensor system is shown in Figure 2. The system consists of an optical source, a sensing or modulating element (that transduces the measurand i.e. the vital sign being measured into an optical signal), an optical detector and an electronic processor to extract sensing information. Depending on the sensing mechanism (phase, intensity, wavelength and polarization state) used to modulate the optical property of guided light when incident light is carried to and fro the site of measurement, optical fiber sensors can be classified as intrinsic, extrinsic and spectral.

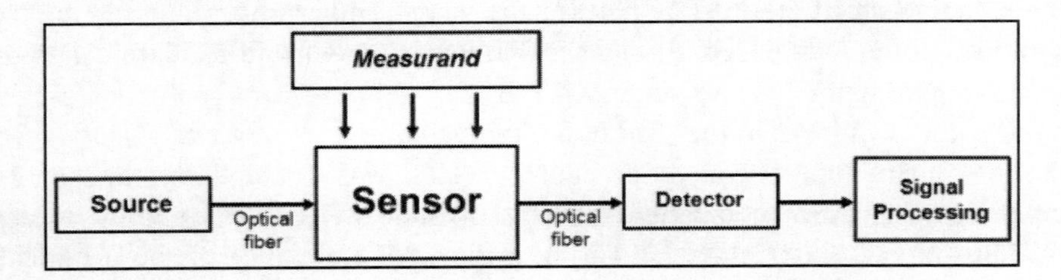

Figure 2. General structure of an optical fiber sensor system.

Applications of optical fiber sensors in vital signs monitoring can be broadly classified into two main categories: 1) Physical sensors for temperature, heart rate, respiratory rate, blood pressure, blood flow and velocity measurement, etc.; and 2) Chemical sensors that use spectroscopic, fluorescence and indicator techniques to detect and measure the level of specific chemical compounds and metabolites such as the blood pH, pO_2, pCO_2, blood glucose, etc. The measurement and continuous monitoring of changes in these physiologic parameters are tell-tale signs of a person's heath status and organs' function, especially important in the existence of chronic diseases such as cardiovascular illnesses and diabetes.

Recent breakthroughs in optical fiber research have encouraged the commercialization of better sensor technologies so as to satisfy the consumers' needs and solve medical problems. In this chapter, we will illustrate the latest technological innovation in the development of optical fiber sensors for vital signs monitoring by both the commercial and research sectors in the medical device industry. A particular highlight will focus on the emerging experimental breakthroughs at improving the sensitivity of these optical fiber sensors to the vital parameters, as well as the latest technological trends in devising more user-friendly and commercializable optical fiber based vital signs sensors. Our work on non-invasive, contactless optical fiber sensors for respiratory rate, heart rate, and ballistocardiography based on intensity modulated micro-bending effect will be described. This optical fiber sensor mat technology has been successfully commercialized for home use. Challenges met in the development and market penetrance of optical fiber sensors for vital signs monitoring will be examined, and the future prospects of utilizing optical fibers in healthcare sensing will be discussed.

II. COMMERCIAL OPTICAL FIBER SENSORS FOR VITAL SIGNS MONITORING

Over the past decades, the market has witnessed a tremendous growth in the number of optical fiber sensors invented to serve the need for vital signs monitoring (Table 1). A majority of these commercial discrete sensors based on Fabry-Perot interferometry and fiber Bragg gratings, as well as distributed sensors based on optical scattering techniques such as Raman and Brillouin scattering have been rapidly introduced into the market, together with their compatible interrogation instruments. The use of optical fiber sensors in healthcare is fueled by a constant need for advancements in vital signs monitors that can better enabled to "sense" and "monitor" the physiologic status of patients in a wide range of diagnostic and therapeutic applications. A recent global market report by Global Industry Analysts has projected a lucrative profit of US$1 billion for the medical fiber optics industry by the year 2017. According to the market report, there is an increasing demand for surgical equipment which can aid in minimally invasive surgery [4]. By far, the most common optical fiber based vital signs monitors available in the market are temperature and pressure sensors developed for monitoring during intraoperative procedures. This was enabled by the small-sized, mechanically flexible and biocompatible features of optical fiber, which allow the sensors integration into miniaturized catheters. The intrinsic EMI immunity of these optical fiber sensors has also allowed for their use in more advanced surgical procedures such as MRI image-guided therapy.

One of the earliest pioneering sensors on the market for vital signs sensing during surgery is the Camino® OLM intracranial pressure (ICP) sensor (Camino Laboratories, San Diego, CA; now acquired by Integra LifeSciences, Plainsboro, NJ, USA). The small dimension of the Camino® ICP sensor allows its easy insertion into a Bolt catheter for the monitoring of changes in the pressure of cerebrospinal fluid in patients suffering from post-traumatic brain injury. Similar to many physical optical fiber sensors in the market, the working principle of the Camino® ICP sensor (Figure 3) is simple. It is based on the concept of intensity modulation. Changes in bellows (diaphragm) position due to pressure shift modulates the intensity of the reflected light and the output light signal variation is related to the pressure causing the displacement of the bellows [5]. In the brain-injured patient, ICP monitoring allows for the early detection of intracranial hypertension and facilitates timely therapeutic interventions to minimize ischemic injury. Since its introduction in 1984, the Camino® ICP sensor has become one of the most commonly used ICP monitoring systems in the world, with a sales volume of 60,000 units per year. In recent years, a newer version of the Camino® sensor was introduced with even better sensor performance and safety features that enable simultaneous and continuous measurement of intracranial pressure, cerebral perfusion and brain temperature [6], in response to an increasing market demand for sensors that can perform multi-parametric measurements, and strong competition from the newer generation of companies such as Opsens, Inc. (Québec, Canada), Neoptix (Québec, Canada) and Samba Sensors (Västra Frölunda, Sweden). Of special mention is the OPP-M25 pressure sensor (Opsens, Inc., Québec, Canada), which has been marketed as the smallest microelectromechanical system (MEMS) based optical fiber pressure sensor at 0.25 mm OD capable of tissue pressure monitoring during the performance of even less invasive procedures. The system has been reported with good robustness in a wide variety of applications including cardiovascular and intraocular pressure sensing, and is unaffected by temperature shift and moisture induced drift [7].

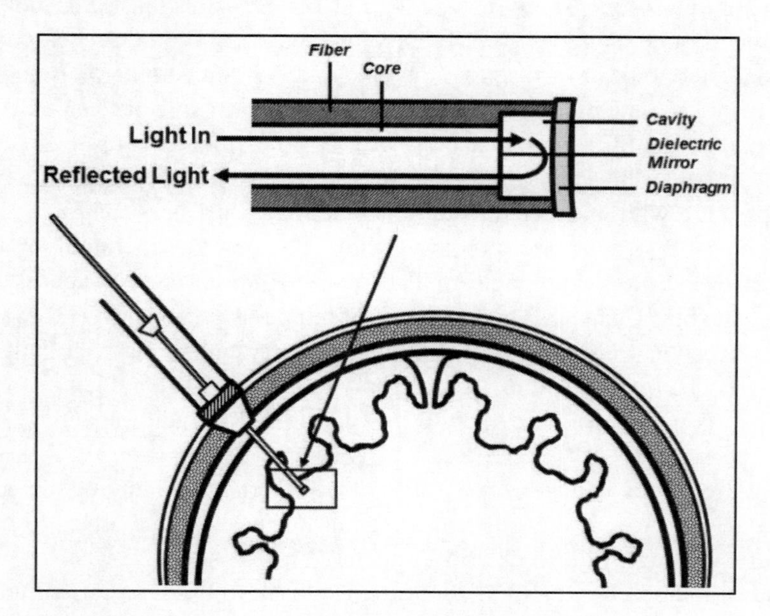

Figure 3. Schematic representation of an interferometry-based optical fiber sensor that has been used for the monitoring of physical parameters (e.g. tissue/blood pressure and temperature) during surgical procedures.

Table 1. Examples of some optical fiber sensors for vital signs monitoring available in the market

Vital Parameter	Sensor Name	Company
Pressure	Camino® OLM ICP	Integra Life Sciences (Plainsboro, NJ, USA)
	FOP-M	FISO Technologies, Inc. (Québec, Canada)
	OPP-M40, OPP-M25	Opsens, Inc. (Québec, Canada)
	Samba® Preclin	Samba Sensors (Västra Frölunda, Sweden)
Temperature	FOT-M	FISO Technologies (Québec, Canada)
	FOT-STB	LumaSense Technologies (Santa Clara, CA)
	Neoptix T1™	Neoptix (Québec, Canada)
	OPT -M	Opsens, Inc. (Québec, Canada)
Pulse Oximeter	7500FO Fiber Optic Tabletop Pulse Oximeter	Nonin Medical, Inc. (Plymoth, MN, USA)
Blood flow	Blood Flowmeter	ADInstruments Ltd. (Sydney, Australia)
	moorVMS-LDF	Moor Instrument (Devon, UK)
	PreSep oximetry catheter	Edward Life Sciences LCC (Irvine, USA)
Force	TactiCath force-sensing catheter	EndoSense, Inc. (Geneva, Switzerland)
ECG/EEG	Phototrode ™	Srico, Inc. (Columbus, OH)

Another common range of optical fiber based sensor used during surgical monitoring is the blood temperature sensors. These include the FOT-STB (LumaSense Technologies, Santa Clara, CA), which is one of the earliest temperature sensor to be introduced commercially, and has been marketed to be able to sense in a wide range of temperature between 0 to 120°C within a quick response time of 0.25 seconds. The working principle of FOT-STB is based on fluoroptic thermometry [8], which makes use of the inherent optical properties of phosphorescent materials to emit luminescence of certain characteristics according to the change in temperature with time. Temperature is determined by measuring the decay time of the emitted light in fluoroptic thermometry, while the newer generation of temperature sensors function based on the principle of Fabry-Perot Interferometry such as the OTP-M sensor (Opsens, Québec, Canada). This new type of temperature sensor (OTP-M) is based on the temperature-dependent birefringence of a pure monocrystalline material built into the tip of the fiber optic probe for temperature sensing and transduction [9] and has been designed to provide accurate tissue temperature monitoring during MRI, radiofrequency ablation and hyperthermia processes.

During radiofrequency catheterization, it is also important to monitor the contact force exerted on the tissue to prevent the overheating of tissues and char formation. The TactiCath® force-sensing irrigated ablation catheter (EndoSense, Inc., Geneva, Switzerland) improved catheter ablation by providing real-time feedback to the surgeons on the amount of force they have exerted on the heart tissue during radiofrequency ablation. This sensor has already obtained approval from the European regulatory authority and its function is based on 3 fiber Bragg grating fibers (FBGs) mounted on the tip of an intra-aortic catheter that serves as a laser-ablation delivery probe for the treatment of atrial fibriliation. The FBGs detect the force exerted against the heart wall by the stress induced on them. Measurements of the contact force are calculated based on the changes in the wavelengths of light between the FBGs when pressure is being applied on the tissue [10]. Force control is essential for delivering appropriate laser ablation pulses needed to produce lesions that are induced in the heart walls to reduce abnormal electric activity. Optical fiber sensors are inherently useful materials for use in force-sensing catheters during radiofrequency ablation as they are immune from RF and EMI, and can be disposed after use due to their low cost.

Another type of optical fiber sensor that has been introduced commercially for cardiovascular monitoring is the blood flow sensors such as the PreSep oximetry catheter (Edward Life Sciences LCC, Irvine, USA), the moorVMS-LDF (Moor Instrument, Devon, UK) and the Blood Flow Meter (ADInstruments Ltd., Sydney, Australia). These blood flow sensors allow for the measurement of blood cell perfusion in the microvasculature of tissues and organs based on the concept of Laser Doppler blood flowmetry [11]. A "Doppler" shift occurs when light is scattered by moving red blood cells. By illuminating a vascularised tissue sample with a monochromatic single frequency of light, and processing the frequency distribution of the backscattered light, an estimate of the blood perfusion can be achieved (Figure 4).

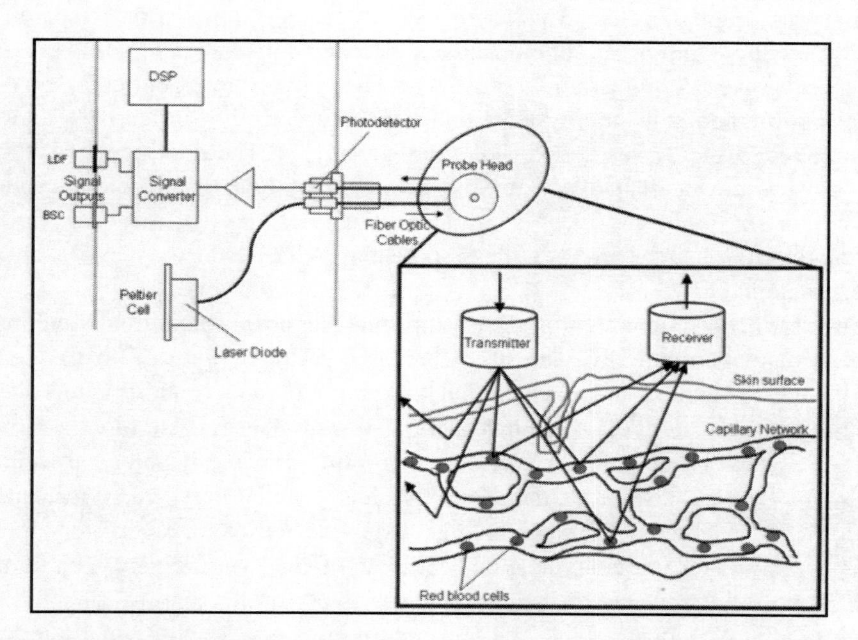

Figure 4. Laser Doppler blood flowmetry. The transmitting optics illuminate laser light onto the tissue. In return, the receiving optics receive the reflected/scattered laser light and transmit to a photodetector for further signal processing.

Among the latest optical fiber based vital signs monitor newly introduced into the market is the 7500FO fiber optic tabletop pulse oximeter (Nonin Medical, Inc., Plymoth, MN, USA). This optical fiber based pulse oximeter was specially developed as a compact and portable system for use in MR facilities to monitor neonates, pediatric and adult patients in the pulse range of 18-321 bpm.

The working principle of this sensor is based on reflectance pulse oximetry that allows for the monitoring of blood oxygen saturation (SpO_2) [12]. Organs and tissues must be sufficiently perfused with oxygenated blood in order to carry out their normal cellular functions for survival. In cases of severe hypoperfusion or extreme hypoxia, dysfunction or failure of organs may ensue. By shining two different wavelengths of light, red and infrared light, through the vascular tissues, the backscattered light received by the photodetector is measured for the intensity. The variations in current due to light received by the photodetector are assumed to be related to changes in blood volume underneath the probe. These variations in current are electronically amplified and recorded as photoplethysmograph (PPG). As the absorbance of light by oxygenated hemoglobin and deoxygenated hemoglobin are different at these two wavelengths of light, the amplitude of the red and infrared PPG signals are sensitive to changes in the oxygen saturation. SpO_2 can then be estimated from the ratios of these amplitudes.

Similar to many optical fiber companies which originally developed optic fiber instruments for physical and chemical monitoring purposes such as strain analysis and toxicity measurement in the construction and mining industry, an increasing market demand for more sensitive biomedical instrumentation have prompted the companies to devolve in further research and development effort to create optical fiber sensors for medical use. Most intriguingly, the Photrode™ (Srico Inc., Columbus, OH), an optical fiber based voltage sensor which was originally created under the Small Business Innovation Research (SBIR) venture between the Ohio-based photonics engineering company and the U.S. National Aeronautics and Space Administration (NASA)'s Glenn Research Centre for non-medical related fiber-optic communication systems and lightning detection in avionics and mining, has been exploited further for use in physiologic sensing [13]. The Photrode™ is an extrinsic fiber-optic sensing device that allow for the detection of both the electroencephalogram (EEG) signals of the brain and electrocardiogram (ECG) signals of the heart through the capacitive pickup of voltage changes in the biopotentials (EEG and ECG) and Mach-Zehnder intensity modulation.

ECG monitoring is important for measuring and diagnosing abnormal rhythms of the heart, in particular abnormal rhythms that occur as a result of damage to the conductive tissues of the heart such as in a myocardial infarction or abnormal rthythms caused by electrolyte imbalances. In fact, the Photrode™ has now been used by the U.S. Army Aeromedical Research Laboratory for assessing their army pilots' physiological wellness and flight-readiness. Researchers under the Neuropsychiatry and Surgery Departments at the Walter Reed Army Institute of Research are also using the sensor for ambulatory monitoring of alertness as well as for triage applications. EEG and ECG measurements can be made by placing the biopotential contacts of the sensor respectively on the skin of the skull or chest without the use of any conductive gel. The sensor comprises of a Mach-Zehner interferometer that has been fabricated in a lithium niobate electro-optic crystal substrate attached to the optical input and output fibers to concentrate an ambient biopotential signal at the waveguides (Figure 5).

Thin-film gold electrodes deposited on the crystal operate in a push-pull configuration to modulate the light entering the sensor from a continuous wave laser source connection via optoelectric effect as the two sections of the interferometer are subjected to opposite voltages from the biopotentials. The modulated light output is then further processed for the biopotential signal measurements [14].

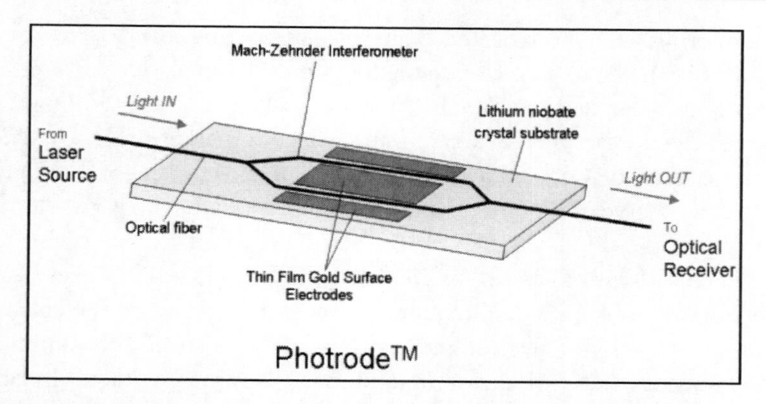

Figure 5. Schematic representation of Phototrode^{TM} (Srico Inc., Colmbus, OH).

III. Optical Fiber Based Vital Signs Monitors in Research and Development

A Medline analysis on publications and patents reported over the past few decades showed a significant increase in the number of research work on medical fiber optics (Figure 6). In fact, many of the recent research work on medical fiber optics are actually catered to the development of optical fiber sensors for diabetes, cardiovascular and respiratory monitoring of vital signs that can enable a better quality in management of chronic illnesses such as diabetes, cardiovascular and pulmonary diseases that are becoming more prevalent in our affluent and sedentary society [15]. A trending fabrication and design of these newer sensors is to develop non-invasive, less bulky and low cost sensor system that can be portable or even worn on the human body for continuous monitoring in various setting outside the hospital, such as in the home of the patient or for the physiologic monitoring of militants or workers handling heavy operational workload.

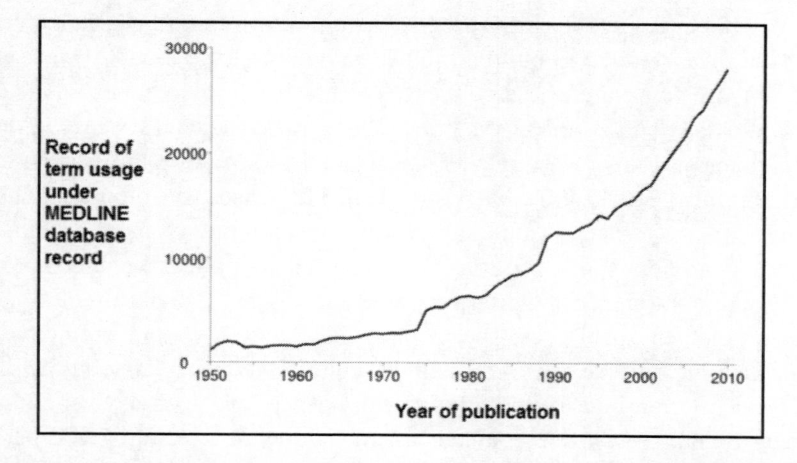

Figure 6. Trend graph showing a remarkable increase in the number of research and development work on optical fiber sensor for medical applications. Source Data: 2012 Medline Baseline Distribution, released December 14, 2011 [16].

Glucose monitoring is important for diabetes care as improved glycemic control in diabetic patients has been shown to decrease the long-term complications of type 1 and type 2 diabetes. Maintaining near normal levels of blood glucose is also important for the management of patients in the intensive care unit. A recent market report by BCC Research showed a projected profit of US$356.5 million in the global market for continuous glucose monitoring devices by the year 2016, with a five-year compound annual growth rate (CAGR) of 14.2% from the year 2011 to 2016 [17]. In fact, there is an expected rise to more 330 million people diagnosed with diabetes by 2030 [18]. However, a tight level of control over blood glucose is difficult to achieve and requires frequent glucose measurements.

As of current, a majority of the commercially available sensors for continuous glucose monitoring e.g. Glucoday® S (Medicon Ireland Ltd., Newry, Northern Ireland) and the implantable sensor from MiniMed Paradigm (Medtronic, Inc., MN, USA) are based on electrochemical principles that usually rely on the enzyme glucose oxidase for molecular recognition of the glucose molecules. Hence, there may be limited electrochemical selectivity due to the electrooxidation of other species such as ascorbate and uric acid or interference by drugs taken by the patients [19]. The current methods for glucose determination also require patients to obtain blood by pricking their fingers, which can be a pain and burden on diabetics who require daily monitoring of their blood glucose level and might be problematic for diabetics wound-healing difficulties.

Of latest development efforts are attempts to fabricate optical fiber sensors that are portable and can allow for the non-invasive measurement of blood glucose with better sensitivity based on fluorescence intensity and lifetime monitoring using near-infrared spectroscopy. Near infra-red spectroscopy is a technique that is dependent on photo-induced electron transfer in which the binding of glucose to fluorescent conjugates in the form of enzymes such as glucose oxidase and hexokinase or the lectin concanavalin A (Con A) will quench and affect the emission of light (Figure 7) [20].

Based on the concept of Fourier transform infrared spectroscopy of attenuated total reflection (FT-IR-ATR), Tamura et al. (2004) [21] developed an optical fiber based glucose sensing system for the non-invasive measurement of blood glucose at the middle finger and back of the ear lobe by ATR spectroscopy using mid-infrared. Partial Least Square Regression (PLSR) was applied to build a prediction model of the blood glucose level using in-vitro blood glucose (measured by a conventional hand-held glucometer) as the reference values. Cross-validation and further data processing showed adequate correlation in blood glucose measurement in the wave number range from 1500 to 950 cm^{-1} (which contain the absorption peak of glucose i.e. 1400 cm^{-1}, 1100 cm^{-1}). The sensor was reported to have a good accuracy in measurement that is superior to the conventional method of blood glucose measurement. In another study, bearing in mind that optical methods for glucose measurement can be confounded by basal properties of tissues, Yamakoshi et al. (2006) [22] addressed this problem by using fast spectrophotometric analysis and predictive modeling with PLSR to create the glucose sensor termed the Pulse Glucometry, that has been proven to measure blood glucose non-invasively with clinically acceptable accuracy using the finger as the measurement point. The system enabled fast transmittance spectra of 100 per second to resolve the optical spectra (900 to 1700 nm) of blood volume pulsations throughout the cardiac cycle, which allowed for the calculation of the difference spectra from the pulsatile signals to eliminate the optical effects from other tissues.

It was hopeful that further development of this technology could eventually produce a noninvasive and easily portable home glucose monitor that would improve the quality of life for many diabetic patients. Nevertheless, there is a need to conduct further human trials and predictive modeling for the respective fiber-based glucose sensing methods to account for the influence of individual differences as the current trials were predominantly based on measurements done on small number of healthy subjects.

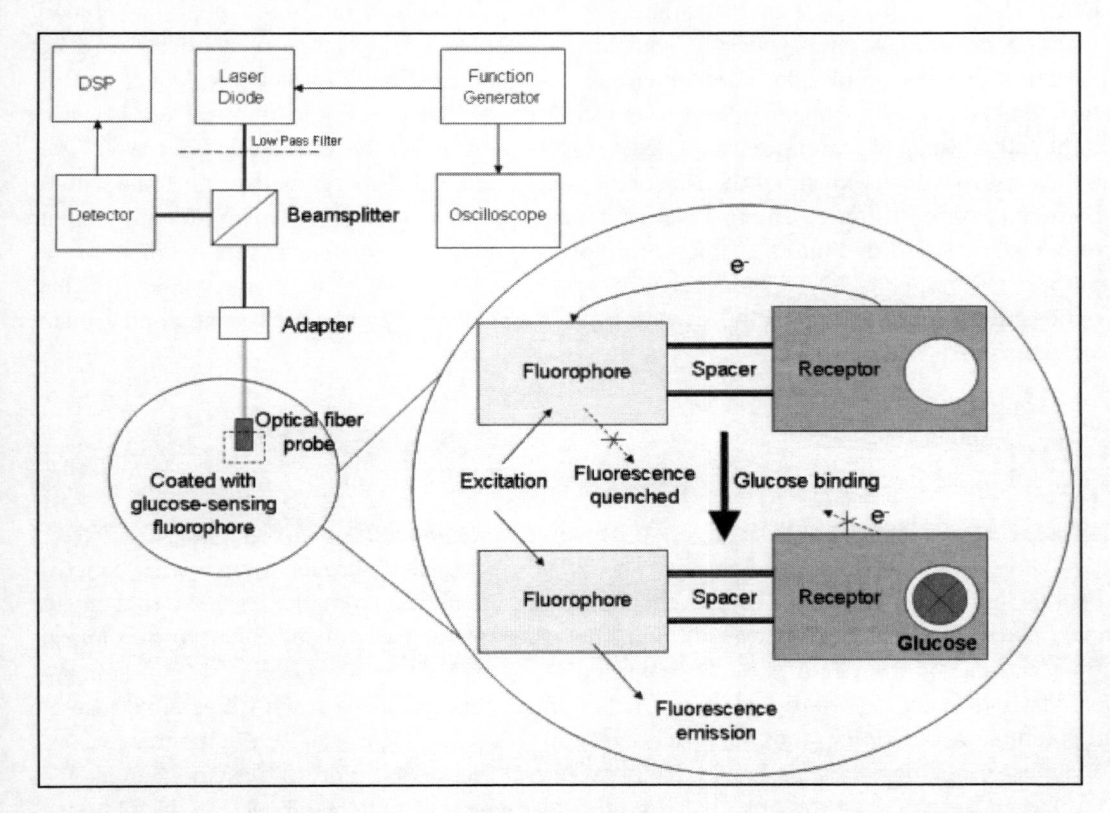

Figure 7. Blood glucose monitoring based on the principle of photo-induced electron transfer in near infra-red spectroscopy based on de Silva et al. (2001). In a complex consisting of glucose receptor linked to a fluorophore via a spacer, laser excitation of the fluorophore induces electron transfer from the unoccupied receptor to the fluorophore, quenching the fluorescence signal. On blinding of a glucose ligand, the receptor undergo changes in its redox/ionization state, preventing the transfer of electron to the fluorophore. As a result, fluorescence emission occurs.

The monitoring of cardiac and respiratory activities are important aspects of vital signs monitoring. Many of the optical fiber based cardiac and respiratory activity sensors published in literature based on the concepts of macro-bending, fiber Bragg grating and optical time-domain reflectometer (OTDR) [23-24], fiber optic statistical mode (STM) and high order mode excitation (HOME) [25] and photo-plethysmography (PPG) [26] were not feasible for industrial acceptance and clinical use due to their bulkiness, design complexity and high fabrication costs. In contrast, the optical fiber sensor system developed in our laboratory based on the working principle of micro-bending effect [27] is of low fabrication cost and has a simple design but robust performance, readily applicable for various clinical and home use.

The performance of our sensor in cardiac and respiratory monitoring has also been studied extensively in clinical trials conducted in hospitals for in-bed monitoring, and patient monitoring during MRI. We were also fortunate to have the opportunity to commercialize our sensor system as home-based devices for respiratory sensing in elderly and infants in Singapore and other countries.

Our sensor system allows for the non-invasive and continuous monitoring of vital signs that include breathing rate and breathing movement, heartbeat, ballistocardiography, blood pressure in the monitored patient (Figure 8) [28-32] for a wide range of applications. The system utilized a simple, but comprehensive assembly consisting of an optical fiber sensor mat (embedded with optical fibers), a photo-electronic transceiver to transmit and receive light and a digital signal processing (DSP) algorithm to detect the vital signs (Figure 9). The optical fiber used to construct the sensor is a standard graded-index multimode fiber with a core diameter of 100 μm. The mat can be constructed to a suitable dimension for placement on a bed, or embedded into a cushion or pillow, or mounted into a smart vest for monitoring purposes. Maximum micro-bend sensitivity was achieved by proper construction of the optical fiber such that the spatial frequency Λ of the micro-benders satisfy the approximate relationship:

$$\Lambda = \frac{2\pi a}{\sqrt{2\Delta}}$$

where a is the fiber core radius and Δ is the relative refractive index difference.

The sensor design is based on micro-bending effect created through a "sandwich" micro-bender structure (Figure 10). Under mechanical perturbation that include periodic movement such as breathing and heartbeat vibration, deformer plates (transducer) squeeze the optical fiber and induce a series of micro-bending points along the axis of the fiber. Micro-bending causes light coupling from guided into radiation modes, resulting in the irreversible loss of light and a modulation in light intensity detected by the light detecting unit in the transceiver. By measuring the modulated light intensity over a period of time, information about the monitored person's vital signs such as the breathing rate and heartbeat can be obtained through the DSP algorithm.

In our system, the interrogation unit for the sensor is in the form of a transceiver that comprises of a LED light source operating at 1310 nm, a light detector that can detect light in the range of 1100-1650 nm, a micro-processing unit and other circuits, and a USB interface or wireless interface for connection to a computer or hand held device running the DSP algorithm or display final results.

For example, the electrical output of the transceiver connected to the computer is sampled at a rate of 10 Hz for breathing rate detection and 50Hz for heart rate measurement. Peak detection is performed in the time domain for calculation of the vital signs (Figure 11). For instance, breathing rate has a distinct typical frequency range of about 0.05 -0.5 Hz that can be extracted and bandpass filtered for calculation (Figure 13).

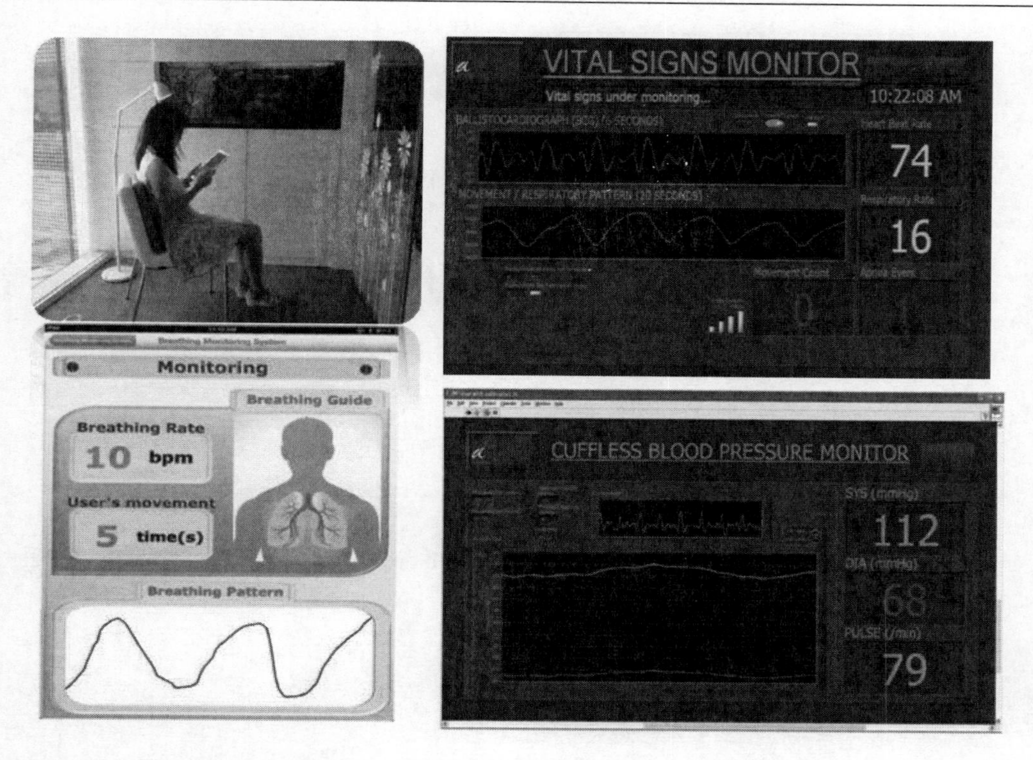

Figure 8. Various applications can be explored using our microbend optic fiber sensor system.

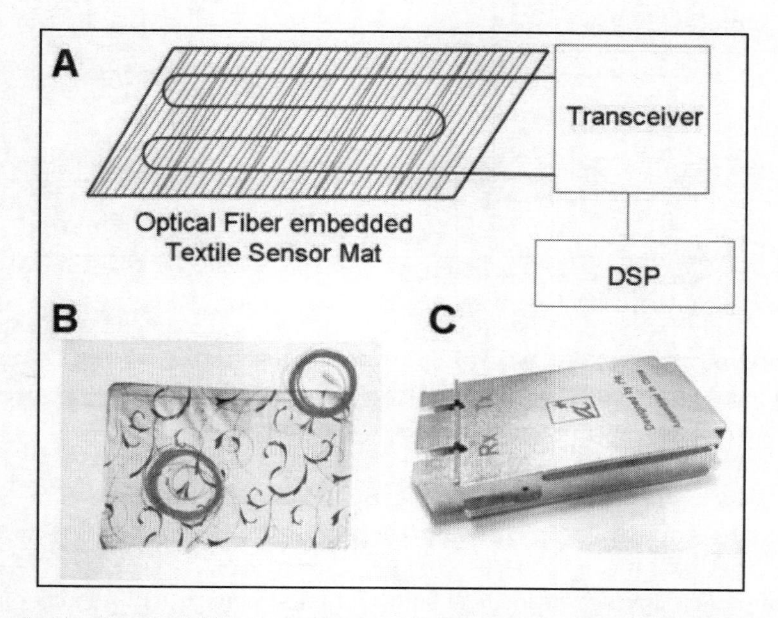

Figure 9. Our microbend optical fiber sensor system (A) for vital signs monitoring comprises of a 2-mm thin optical fiber embedded sensor mat (B), a photo-electronic transceiver (C) and a data processing algorithm.

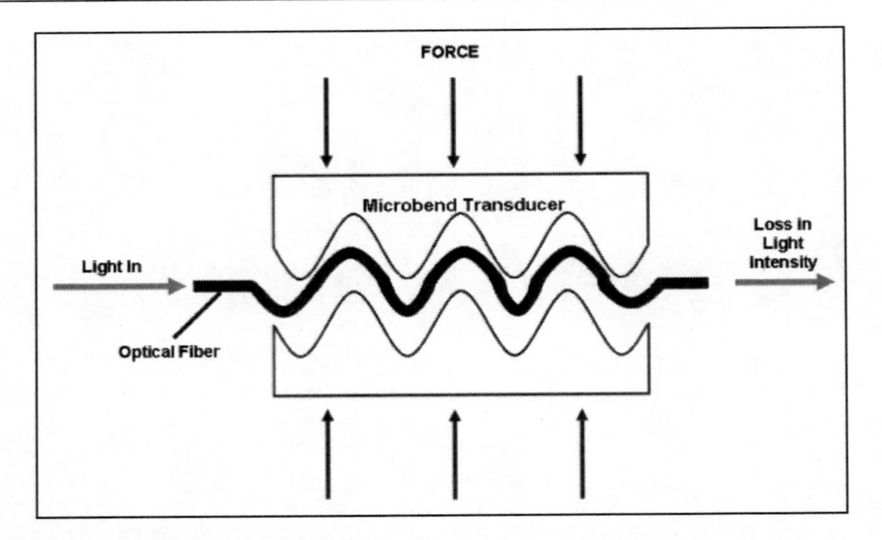

Figure 10. Optical fiber based vital signs sensors developed in our laboratory are based on the working principle of intensity modulation due to micro-bending effect.

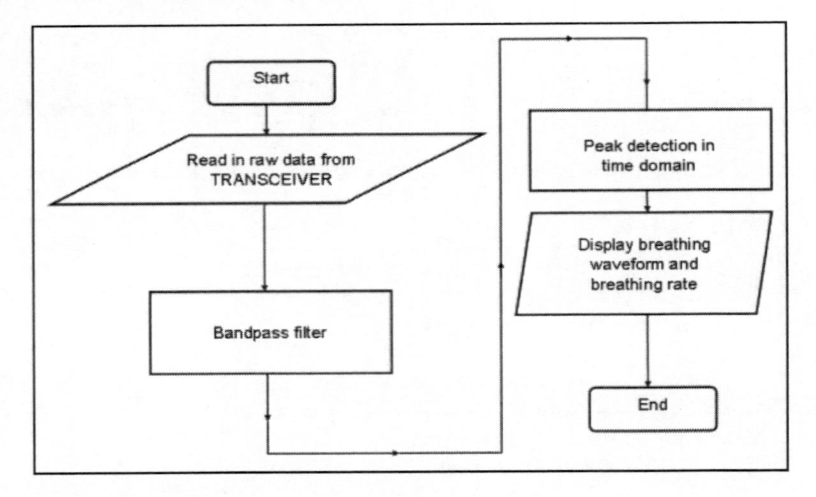

Figure 11. An example of workflow for processing a modulated signal

In the preceding paragraphs, we will provide an illustration of various optical fiber sensors (including our micro-bend optical fiber sensors) that have been used in some emerging application areas that require physiologic monitoring.

"Smart" Textiles for Physiologic Monitoring

With a continuing shortage of medical staff and a looming patient population, there is a need to develop non-intrusive physiologic monitors that can track the patient's condition and alarm the medical staff of any adverse medical situation. The fibrous nature and biocompatibility of optical fibers have allow for their embodiment into fabric or medical textile materials that can be in the form of a bed, a pillow or even a vest worn on the human body for physiologic sensing. We and others [28-35] have utilized this special property of optical fibers to create "smart" textiles that can be used to sense important vital parameters. Spillman et al. (2004) [25] developed a "smart bed" that enabled the nonintrusive monitoring of patients' breathing rate, heart rate and body movement using a statistical mode (STM) sensor and a high order mode excitation (HOME) sensor. Whilst fiber Bragg grating (Figure 13) has been suggested for patient respiratory monitoring as a mattress [33] and body temperature monitoring as a bedsheet [34]. The OFSETH has also suggested the integration of Bragg grating fibers into wearable textile for the monitoring of various vital signs [35].

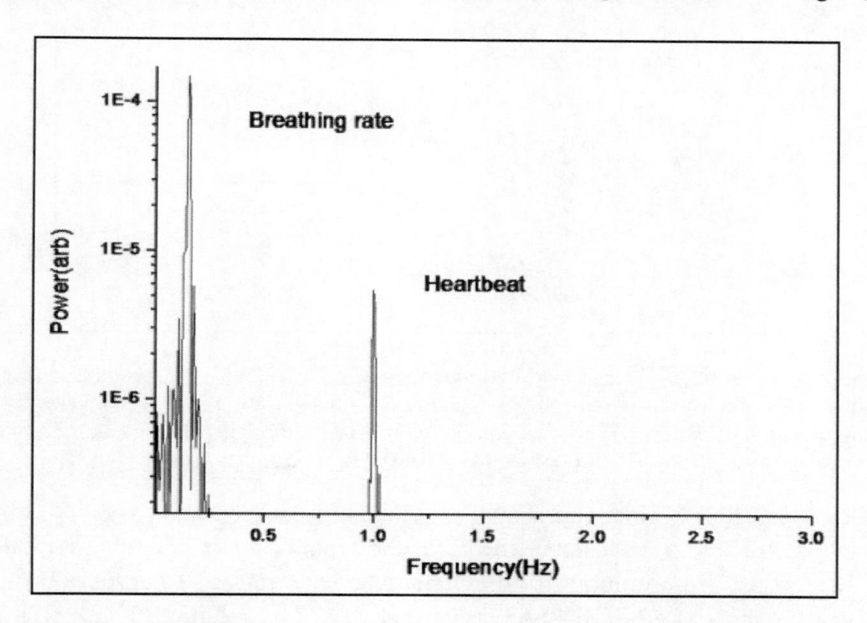

Figure 12. Frequency spectrum graph showing a measured breathing signal and a heartbeat signal. Bandpass filtering is used to distinguish between the two physiologic signals.

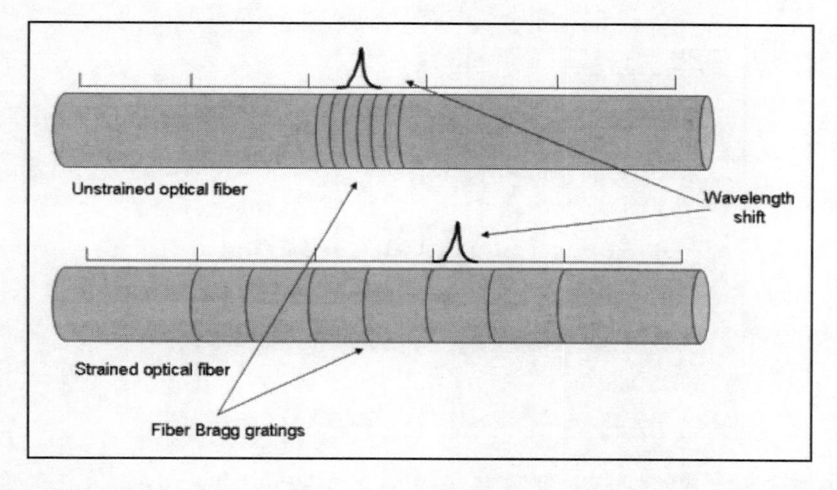

Figure 13. Shear force from cardiac and respiratory motion can be detected using fiber Bragg gratings (FBG). A shift in Bragg wavelength occurs when there is a change in the modal index or grating pitch of the fiber.

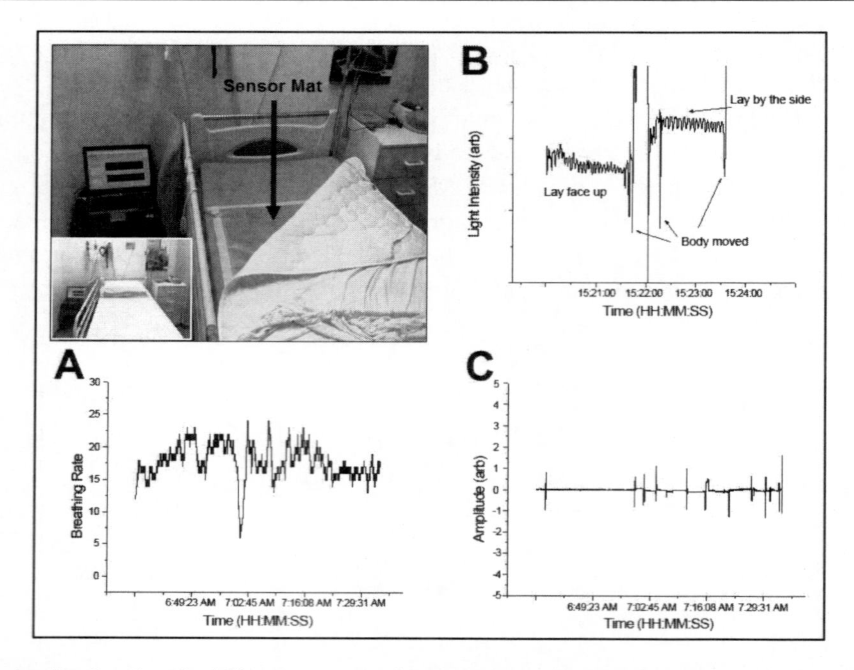

Figure 14. A microbend optical fiber sensor mat integrated as a "smart bed" in a hospital setting. (B) Our microbend optic fiber sensor is sensitive enough to detect the variation in light intensity according to the sleeping position of patients (lying face up, body movement and lying by the side). The system is able to detect the breathing rate (A) and body movement (C) of the same person during sleep.

The micro-bend optical fiber sensor mat developed in our laboratory was also integrated into bedding structures such as a "smart bed" for the monitoring of breathing rate and body movement [28, 32] , a "smart pillow" [29] to detect the heart rate, a "smart cushion" [30] for heart rate measurement based on ballistocardiography (BCG), and a cuffless blood pressure monitor for blood pressure measurement [31], as shown in Figure 14-17.

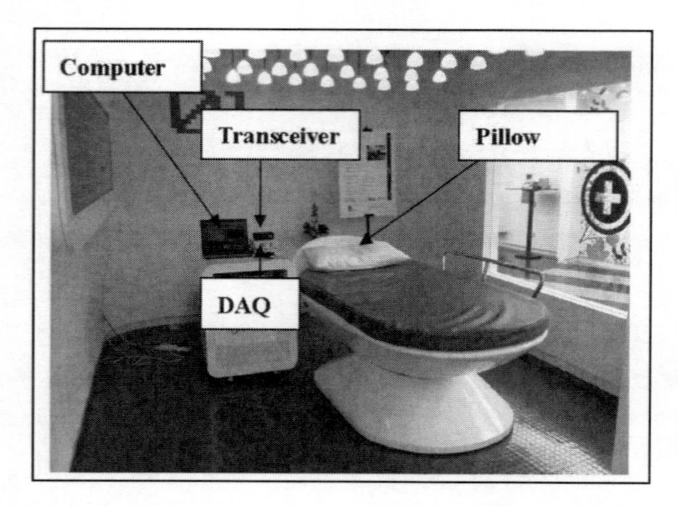

Figure 15. "Smart pillow" for heart rate monitoring based on micro-bending effect.

Neonatal Monitoring

Neonatal monitoring is especially crucial in the neonatal intensive care units (NICU) where the infants were normally borne premature or are critically ill [36]. The global fetal and neonatal care market has been estimated to grow at a CAGR of 5.9% and is projected to achieve a profit of $797 million in the year 2016 [37]. Our microbend fiber sensor mat technology for vital signs monitoring has been transferred to an industry partner to manufacture the baby monitor (Figure 16), a non-invasive and portable home-based device that monitors the breathing rate and movement of infants. The accuracy of the in-reading breathing rate of infants has been validated in clinical trials conducted in Singapore's largest paediatric hospital prior to commercialization. This technology is also licensed to other industry partner for other applications. Optical fibers are ideal as sensing elements to be used on infants and paediatrics due to the absence of electrode or use of electricity. In addition, the use of wireless remote technology i.e. no cord required to connect the sensor mat to the parent control unit minimizes the occurrence of safety hazards for the monitored infants and paediatrics.

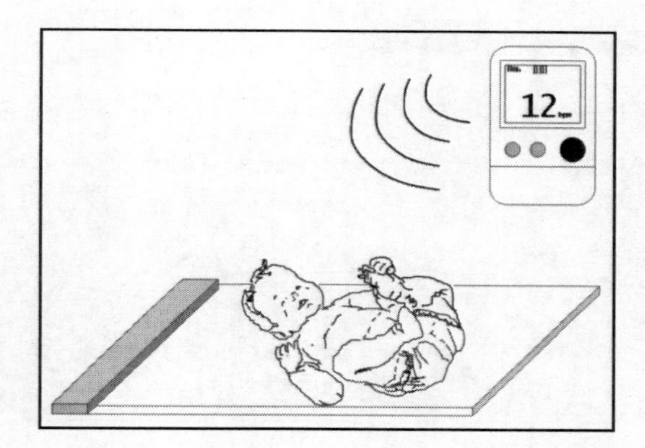

Figure 16. The Baby Monitor allows for the remote monitoring of infants' breathing rate and breathing movement.

Patient Monitoring during MRI

The remote sensing for vital signs of patients inside the MRI gantry is especially crucial for the monitoring of unresponsive patients such as the sedated, comatose and critically ill. The use of electronic-based physiologic sensors for patient condition monitoring is potentially hazardous as the presence of ferromagnetic components in these sensors cause them to be prone to EMI and may also risk radiofrequency (RF) burns on the patients' skin due to electrical currents induced in the metal components by the changing magnetic field gradients during MRI [38, 39]. In addition, physiologic sensors in recent development are mostly electronic-based e.g. piezo-electronics and capacitive coupled electrodes, and may not be suitable for use with future MRI scanners operating at higher magnetic field strength. Thus, the increased market demand for better patient safety during MRI physiologic monitoring has prompted various research groups to make use of the inherently EMI and RF immune nature of optical fibers in the development of MRI-safe vital signs monitors.

Our group has investigated the use of optical fiber sensor system in breathing rate and body movement monitoring during MRI (Figure 17) [32]. In a trial conducted on sixty healthy subjects between the age range of 21-70 years old and variations in breathing rate in the range 8-22 breaths per minute, our system was able to detect a good and comparable breathing rate to the predicate electronic respiratory detecting device used under current hospital settings. The breathing movement detected by the sensor was also useful as a respiratory movement trace during MRI for liver imaging at the end-expiratory phase of the breathing cycle. Our sensor system also showed good performance in heart rate, BCG and blood pressure detection that can potentially be used in multi-parametric monitoring of patients during MRI. In addition, the embedment of sensing fibers into a thin sensor mat for placement on the MRI bed, rather than a sensing Velcro strapped around the chest of the monitored patient allows for extra patient comfort and ease of operation.

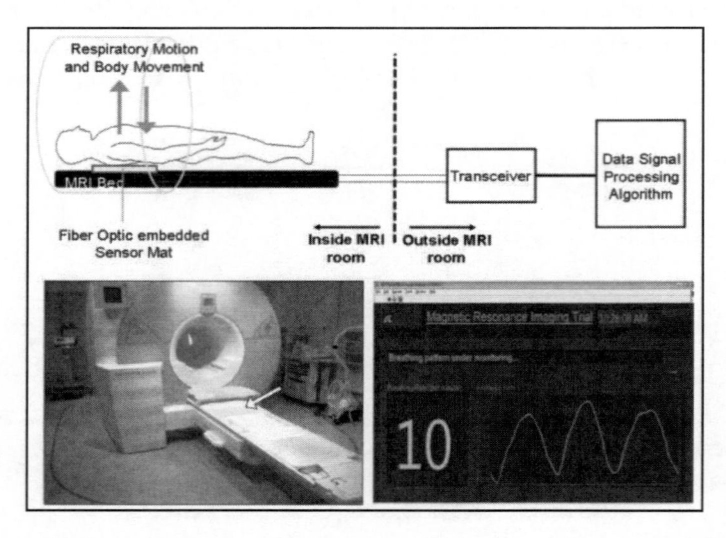

Figure 17. An experimental setup for the monitoring breathing rate and body movement in sixty healthy subjects in the MRI suite.

IV. Future Prospects and Challenges

In conclusion, future applications of optical fiber sensor for vital signs monitoring will reply heavily on cost reduction and the development of effective and suitable packaging to meet the needs of each sensing application. Discrete physical sensors for the physiological monitoring during surgical procedures are expected to continue to evolve and grow to acquire greater market shares as consumers continue to demand for better sensing devices that can aid in minimally invasive surgery. Whilst emerging sensor types such as non-invasive sensors for cardiac and respiratory monitoring and for the monitoring of blood glucose level as well as blood pressure monitoring without cuffs and with minimum contact, and innovative sensor design and packing into smart textiles and portable monitors are expected to carve a niche in the global market in the next decade as consumers continue to demand for better sensing technologies at lower cost and with better end-user characteristics that can aid in the management of chronic illness and provision of more quality care.

Regardless of the sensor type, further strides and engineering need to ensure reliability and standards in packaging of the monitoring devices, in which significant technical hurdles and market barrier (as listed below) need to be overcome.

- Unfamiliarity with the technology
- Conservative attitude of some industries and customers
- Provision of a proven field record and clinical testing
- Cost involved in sensor development
- Availability of trained personnel
- Development of a complete sensing solution
- Lack of standards
- Quality, performance, packaging and reliability
- Obtaining of regulatory approval and device certification

By overcoming these technical hurdles and market barriers, optical fiber sensors will be expected to gain more commercial momentum and achieve faster market growth in the future.

ACKNOWLEDGMENTS

The authors wish to thank Soon Huat Ng, Ju Teng Teo and Xiufeng Yang for their help in this chapter. This work was supported under A*STAR Biomedical Engineering Programme Grant 102 148 0011.

REFERENCES

[1] Kapany N. S. (1967). *Fiber Optics. Principles and applications*. Academic Press, NY, USA, 88-99.

[2] Udd E., Spillman, Jr. W. B. (2011). *Fiber Optic Sensors: An Introduction for Engineers and Scientists*, 2nd edition. John Wiley & Sons, Inc., Hoboken, NJ, USA.

[3] Mignani A. G. & Baldini F. (1997). Fiber-optic sensors in health care. *Phys. Med. Biol.*, vol. *42*, 967-979.

[4] Global Industry Analysts. (2012). M*edical Fiber Optics: A Global Strategies Business Report*. Published 1 May 2012. Retrieved from http://www.companiesandmarkets.com/Market/Healthcare-and-Medical/Market-Research/Medical-Fiber-Optics-A-Global-Strategic-Business-Report/RPT926176. Accessed on 3rd December 2012.

[5] Brett Trimble. (1993). *Fifty thousand pressure sensors per year: a successful fiber sensor for medical applications*. 9th Optical Fiber Sensors Conference, 457-462.

[6] Integra LifeSciences. (2012). IntegraTM Camino® Intracranial Pressure Monitoring Kit. Retrieved from http://www.integralife.com/Neurosurgeon/Neurosurgeon-Product-Detail.aspx?Product=45&ProductName=Integra%3Csup%20class=%22prodSup%22%3E%E2%84%A2%3C/sup%3E%20Camino%3Csup%20class=%22prodSup%22%3E%C2%AE%3C/sup%3E%20Intracranial%20Pressure%20Monitoring%20Kit&ProductLineName=ICP%20Monitoring&ProductLineID=11. Accessed on 3rd December 2012.

[7] OpSens, Inc. (2012). *OPP-M Fiber optic miniature physiological pressure sensor*. Retrieved from http://www.opsens.com/en/industries/products/pressure/opp-m/. Accessed on 3rd December 2012.

[8] Zhang Z. Y. & Grattan L. S. (1994). *Fiber Optic Fluorescence Thermometry*. Springer, New York, USA.

[9] OpSens, Inc. (2012). *Fiber optic temperature sensor OTP-M*. Datasheet. Retrieved from http://www.opsens.com/pdf/products/OTP-M%20REV%201.5.pdf. Accessed on 3rd Dec 2012.

[10] EndoSense, Inc. (2012). *The TactiCath®: A new standard in ablation catheters*. Retrieved from http://www.endosense.com/uploads/tacticath/Endosense_Folder_Insert.pdf. Accessed on 3rd December 2012.

[11] Shepherd A. P., Öberg (1990). *Laser Blood Flowmetry*. Springer, New York, USA.

[12] Nonin Medical, Inc. (2012). *Nonin 7500 Series Tabletop/Portable Pulse Oximeter*. Retrieved from http://www.nonin.com/documents/7500_brochure.pdf. Accessed on 3rd December 2012.

[13] NASA Spinoff (2012). Superior Sensor Making Sense in Military, Medicine Retrieved from http://spinoff.nasa.gov/Spinoff2004/p_6.html. Assessed on 01 November 2012.

[14] Kingsley S. A., Sriram S., Pollick A. & Marsh J. (2004). Photrodes for physiological sensing. Proceedings of SPIE, In: Optical Fibers and Sensors for Medical Applications IV, I. *Gannot ed.*, vol. *5317*, 158-417.

[15] World Health Organization (WHO). (2011). *Global atlas on cardiovascular disease prevention and control. Policies, strategies and interventions*. WHO Press, World Heart Federation and World Stroke Organization.

[16] Palidwor et al. (2010). *J. Biomed. Discov. Collab.*, vol. *5*, 1-6.

[17] Blonde L. & Karter A. J. (2005). Current evidence regarding the value of self-monitored blood glucose test. *Am. J. Med.*, vol. *118*, no. 9A, 20S-26S.

[18] BCC Research (2012). *Global Market for Continuous Glucose Monitoring Devices to Grow to $656.5 Million by 2016*. Report Code: HLC102A. Published June 2012. Retrieved from http://www.bccresearch.com/report/continuous-glucose-monitoring-cgm- technology-markets-hlc102a.html. Accessed on 3rd December 2012.

[19] Pickup J. C., Hussain F. & Evans N. D., et al. (2005). Fluorescence-based glucose sensors. *Biosensors and Bioelectronics*, vol. *20*, 2555-2565.

[20] Pasic A., Koehler H., Schaupp L. & Pieber T. R. (2006). Fiber-optic flow-through sensor for online monitoring of glucose. *Anal. Bioanal. Chem.*, vol. *386*, 1293-1302.

[21] Tamura K. (2004). Noninvasive measurement of blood glucose based on optical sensing. Proceedings of the 21st *Instrumentation and Measurement Technology Conference* 2004, vol. *3*, 1970-1974.

[22] Yamakoshi K. & Yamakoshi Y. (2006). Pulse glucometry: a new approach for noninvasive blood glucose measurement using instantaneous differential near-infrared spectrometry. *J. Biomed. Opt.*, vol. *11*, no. 5, 054028-1 – 054028-9.

[23] Grillet A., Kinet D., Witt J., Schukar M., Krebber K., Pirotte F., Depre A. (2008). Optical fibre sensors embedded into medical textiles for healthcare monitoring. *IEEE Sensors J.*, vol. *8*, no. 7, 1215-1222.

[24] Dzuida L., Skibniewski F. W., Krej M. & Lewandowski J. (2012). Monitoring respiration and cardiac activity using Fiber Bragg Grating-based sensor. *IEEE Trans. Biomed. Eng.*, vol. *59*, no. 7, 1934-1942.

[25] Spillman W. B., Mayer M., Bennett J., Gong J., Meissner K. E., Davis B., Claus R. O., Muelenaer A. A. & Xu, X. A smart bed for non-intrusive monitoring of patient physiological factors. *Meas. Sci. Technol.*, vol. *15*, 1614-1620.

[26] Lindberg L. Ugnell G., H. & Öberg P. A. (1992). Monitoring of respiratory and heart rates using a fibre-optic sensor. *Med. Biol. Eng. Comput.*, vol. *30*, 533-537.

[27] Berthold J. W. (1995). Historical review of microbend fibre-optic sensors. *J. Lightwave Technol.*, vol. *13*, 1193.

[28] Chen Z. H., Teo J. T. & Yang X. F. (2009). In-bed fibre optic breathing and movement sensor for non-intrusive monitoing. In: Optical fibers and sensors for medical diagnostics and treatment applications IX, edited by Israel Gannot, *Proc. Of SPIE*, vol. *7173*, 7173P.

[29] Chen Z. H., Teo J. T., Ng S. H. & Yim H. Q. (2011). Smart pillow for heart rate monitoring using a fiber optic sensor. Optical fibers and sensors for medical diagnostics and treatment applications IX, edited by Israel Gannot, *Proc. Of SPIE*, vol. *7894*, 789402.

[30] Chen Z. H., Teo J. T., Ng S. H. & Yang X. F. (2012). Portable fiber optic ballistocardiogram sensor for home use. *Proc. Of SPIE*, vol. *8218*, 82180x.

[31] Chen Z., Yang X. F., Teo J. T. & Ng S. H. (2013). Noninvasive Monitoring of Blood Pressure Using Optical Ballistocardiography and Photoplethysmograph Approaches, *Proc of 35th Annual International IEEE EMBC 2013*. THD16.3

[32] Lau D., Chen Z., Teo J. T., Ng S. H., Rumpel H., Lian Y., Yang H., Kei P. L. (2013). Intensity-modulated microbend fiber optic sensor for respiratory monitoring and gating during MRI. *IEEE Trans. Biomed. Eng.* (Epub ahead of print).

[33] Foo V. F. S., Leong K. P., Hao E. J. Z., Jayachandran M., et al. (2008). Non-intrusive respiratory monitoring system using fiber Bragg grating sensor. *10th IEEE International Conference on e-Health Networking, Applications and Service (HEALTHCOM 2008)*, 160-164.

[34] Matin M., Hussain N. & Shoureshi R. (2005). Fiber Bragg sensor for smart bed sheet. In: Photonic Devices and Algorithms for Computing VII, edited by Iftkharuddin K. M and Awwal A. A. *S. Proc. SPIE*, vol. *5907*, 590706.

[35] De jonckheere J., Jeanne M., Grillet A., Weber S., et al. (2007). OFSETH: Optical fibre embedded into technical textile for healthcare, an efficient way to monitor patient under magnetic resonance imaging. *Proc. of the 29th Annual International Conference of the IEEE EMBS*, 3950-3953.

[36] Murković I., Steinberg M. D., Murković B. (2003). Sensors in neonatal monitoring: current practice and future trends. *Technol. Health Care*, vol. *11*, no. 6, 399-412.

[37] Markets and Markets. (2011). Fetal (Labor & Delivery) and Neonatal Care Equipment Market – Nasal CPAP, Oxygen hoods, ECMO, NO delivery units, ventilators, Fetal monitors, Capnographs, Pulse Oximeter, Vital Signs Monitors, Resuscitators – Global Trends, Competitor Analysis & Forecasts up till 2016. Report Code: MD 1629. Published on September 2011. Retrieved from http://www.marketsandmarkets.com/Market-Reports/fetal-neonatal-care-equipment-market-412.html. Accessed on 3rd December 2012.

[38] Shellock F. G. (2001). Monitoring during MRI. An evaluation of the effect of high-field MRI on various patient monitors. *Med. Electron.*, vol. *17*, no. 4, 93-97.

[39] Holshouser B. A., Hinshaw D. B. & Shellock F. G. (1993). Sedation, anesthesia and physiologic monitoring during MR imaging: evaluation of procedures and equipment. *J. Magn, Reson. Imaging*, vol. *3*, 553-558.

INDEX

#

A

B

C

D

E

F

G

H

I

J

L

M

N

O

P

Q

R

S

T

U

V

W

Y

Z